T0215725

Fundamentals of Partial Differential Equations

Atul Kumar Razdan · V. Ravichandran

Fundamentals of Partial Differential Equations

Springer

Atul Kumar Razdan
Department of Applied Sciences
Engineering
MIET
Meerut, Uttar Pradesh, India

V. Ravichandran
Department of Mathematics
National Institute of Technology
Tiruchirappalli
Tiruchirappalli, Tamil Nadu, India

ISBN 978-981-16-9867-5   ISBN 978-981-16-9865-1 (eBook)
https://doi.org/10.1007/978-981-16-9865-1

This Springer imprint is published by the registered company Springer Nature Singapore Pte Ltd.
The registered company address is: 152 Beach Road, #21-01/04 Gateway East, Singapore 189721,
Singapore

To my wife Anju and children, Anuja and Aman

—Atul Kumar Razdan;

To Prof. Dato' Indera Rosihan M. Ali and my wife Kalaiselvi

—V. Ravichandran

Preface

Differential equations provide a unifying theme to study physical systems. A single differential equation involves the derivatives of a scalar function of one or more independent variables that, in practical terms, is an approximation of a law governing natural processes of the system. We obtain a (coupled) system of differential equations while dealing with a phenomenon defined in terms of a vector function of one or more independent variables. In general, differential equations are fundamental to model many different types of phenomena. Together with some deeper concepts in analysis and geometry, they help solve important practical problems in science and engineering [1–3].

A study of *univariate phenomena* such as related to planetary motion and gravitational forces among large bodies conducted in the seventeenth century led Newton to apply ordinary differential equations as an important tool to analyse related practical problems. In 1747, d'Alembert was first to formulate differential equation model for a *multivariate phenomenon* such as *vibrating string*, which is known as the wave equation in dimension one. Subsequently, several eminent mathematicians developed differential equation models for various other types of multivariate phenomena such as concerning *fluid flow* (Euler, 1755), *minimal surfaces* (Lagrange, 1760), *hydrodynamics* (Daniel Bernoulli, 1762), *potential theory* (Laplace, 1780), *heat conduction* (Fourier, 1805), *three-body problems* (Poincáre, 1892), *continuous symmetry* (Lie, 1888), and many more.

Applications of differential equations in diverse scientific investigations, and certain related developments, have proved to be as effective to the physical sciences as it has been to the evolution of modern mathematics [4]. The discovery of some mathematical theories ensued in the twentieth century facilitated wide spectrum applications of differential equations to dynamical systems in almost all disciplines of science and engineering [5]. The present book is mainly about basic aspects of some important partial differential equations that find applications to problems in physics and engineering such as concerning mechanical vibrations, wave motion, heat conduction, fluid flow, and electrodynamics. We shall be dealing with differential equations involving two or three independent variables. Throughout the book, we have attempted to maintain a proper balance between the mathematical concepts introduced and their applications to the physical world. The reader interested in abstract theory of partial differential equations may like to read excellent texts such as [6] or [7] or [8].

The work on this book started sometime ago with class notes of the second author on the topic real analysis and partial differential equations that he taught to undergraduate engineering students at the NIT Tiruchirappalli. The first part of the book develops mathematical tools related to *classical vector analysis and ordinary differential equations*. In Chap. 1, we review some relevant concepts of multivariable analysis, basic theory of surfaces and curves, and also classical vector calculus. In Chap. 2, we discuss the theory of ordinary differential equations inasmuch as needed in the subsequent chapters. The former is used in Chap. 3 to derive fundamental differential equation models of some multivariate phenomena concerning transport of physical quantities such as the mass, momentum, energy, and the charge. The treatment of most topics covered in the first part of the book is mainly *illustrative*, with an occasional digression to related theory.

In Chap. 4, we introduce basic concepts such as *order* and *linearity type* of a general partial differential equation. The focus of the chapter is to discuss mathematical issues related to a *classical solution* of initial-boundary value problems. In Chap. 5, we discuss Lagrange–Charpit method of finding the *complete integral* of general first-order partial differential equations. In Chap. 6, we apply the *method of characteristics* to solve Cauchy problems for some hyperbolic type differential equations.

The main part of the book discusses mathematical concepts relevant to some standard analytical *solution methods* such as separation of variables, Fourier series, eigenfunctions expansion, and also transform techniques due to Fourier and Laplace. We apply aforementioned methods to solve some interesting initial-boundary value problems for prototypical second order linear partial differential equations such as wave, heat, and Laplace equations. The book is largely self-contained excepting the details related to some important theorems of analysis that are used in various chapters. The reader may refer companion volume "*Fundamentals of Analysis with Applications*" [9], or any other standard text such as [10], for prerequisites related to real analysis.

The main audience for the present book are students pursuing an advanced level undergraduate course in physics and engineering. However, an appropriate selection of topics may also be useful to the beginner-level graduate students in mathematics. For example, a suitable amplification of topics presented in the first four chapters may be offered as an introductory course on the subject to the graduate students interested in certain specific applications of differential equations.

We are thankful to two reviewers for their comments. We are thankful to Dr. Shamim Ahmad, Publishing Editor, Mathematics, Springer India, for his continued support; he has insisted us to write two separate books instead of our original plan of a single book on Real Analysis and PDE! The authors are thankful to the team at Springer and Mr. T. Parimel Azhagan and Mr. Naresh Kumar Mani for their continuous support during the production process. We will be happy to receive comments for improvements including errors/misprints.

Meerut, India   Atul Kumar Razdan
Tiruchirappalli, India   V. Ravichandran
December 2021

References

1. Keshet, L. E., Mathematical Modeling in Biology, Classics in Applied Mathematics, Vol.46, SIAM, 2005.
2. Murray, J.D., Mathematical Biology – An Introduction, Springer, NY, 2002.
3. Perthame, B., Transport Equations in Biology, Birkhäuser, 2007.
4. Kline, M., Mathematical Thought from Ancient to Modern Times, Vol. 1, 2, 3, Oxford University Press, London, 1972.
5. Brezis, H. and Browder, F., Partial Differential Equations in the 20th Century, Adv. in Mathematics, 135, 76-144, 1998.
6. Arnold, V. I., Lecturers on Partial Differential Equations, Springer, 2004.
7. Evans, L. C., Partial Differential Equations, AMS GTS, Vol.19, 2010.
8. Jost, J., Partial Differential Equations, GTM-214, Springer, New York, 2002.
9. Razdan, A. and Ravichandran, V., Fundamentals of Analysis with Applications, Springer, to appear.
10. Rudin, W., Principles of Mathematical Analysis, third edition, McGraw-Hill Book Co., New York, 1976.

Contents

List of Figures

Chapter 1
Introduction

If one looks at the different problems of the integral calculus which arise naturally when one wishes to go deep into the different parts of physics, it is impossible not to struck by the analogies existing. Whether it be electrostatics or electrodynamics, the propagation of heat, optics, elasticity, or hydrodynamics, we are led always to differential equations of the same family.

Henri Poincáre (1854–1912)

In this book, we discuss some applications of linear partial differential equations to problems in physics and engineering. The ideas presented are used to classify equations and discuss certain important solution methods for the related mathematical problems. Let an open set $\Omega \subseteq \mathbb{R}^n$ represent some physical volume. An n-tuples $(x_1, \ldots, x_n)$ is used to denote the *spatial position* of a point $x \in \Omega$, and $t \geq 0$ is a time variable. By a *model* of an n-dimensional evolutionary physical system we mean an equation that involves derivatives of a *sufficiently smooth*[1] function $u = u(x, t)$ defined over a domain $\Omega_1 = \Omega \times \mathbb{R}^+$. To avoid geometric complexities, we may take Ω to be *simply connected* set. Further, the open set Ω is assumed to have a *piecewise smooth boundary* written as $\Gamma = \partial\Omega$. Whenever necessary, the open set Ω is assumed bounded. The first and second derivatives of a C^2-function $u = u(x, t)$ are written as

---

[1] Recall that a function is said to be sufficiently smooth if enough partial derivatives exists over a domain so that any statement involving u, and some of its derivatives, remain valid over the domain.

© The Author(s), under exclusive license to Springer Nature Singapore Pte Ltd. 2022
A. K. Razdan and V. Ravichandran, *Fundamentals of Partial Differential Equations*,
https://doi.org/10.1007/978-981-16-9865-1_1

$$\frac{\partial u}{\partial x_i} = \partial_{x_i} u = u_{x_i} = p_i; \qquad\qquad \frac{\partial u}{\partial t} = \partial_t u = u_t = q;$$

$$\frac{\partial^2 u}{\partial x_i \partial x_j} = \frac{\partial^2 u}{\partial x_j \partial x_i} = \partial^2_{x_i x_j} u = u_{x_i x_j} = r_{ij}. \tag{1.0.1}$$

For the *del operator* ∇ in n variables $x_1, \ldots, x_n$ given by

$$\nabla := \left(\frac{\partial}{\partial x_1}, \ldots, \frac{\partial}{\partial x_n} \right), \tag{1.0.2}$$

we may write the (scalar) *Laplacian* $\Delta := \nabla \cdot \nabla$ as

$$\Delta := \nabla^2 = \frac{\partial^2}{\partial x_1^2} + \cdots + \frac{\partial^2}{\partial x_n^2}. \tag{1.0.3}$$

Recall that the *gradient* of a C^1-function $\varphi : \Omega \to \mathbb{R}$ at point $a \in \Omega$, denoted by grad $\varphi(a)$, is a vector function given by

$$\operatorname{grad} \varphi(a) := \left(\frac{\partial \varphi}{\partial x_1}\Big|_a, \ldots, \frac{\partial \varphi}{\partial x_n}\Big|_a \right). \tag{1.0.4}$$

The beginning of the subject is attributed to *Jean d'Alembert*'s work related to vibrations of a string such as in a violin. In 1747, d'Alembert formulated the one dimensional *wave equation*[2]

$$\frac{\partial^2 y(x, t)}{\partial t^2} = \frac{\partial^2 y(x, t)}{\partial x^2}, \qquad x \in [0, \ell] \text{ and } t \geq 0, \tag{1.0.5}$$

to study the vibrations of a distorted string of length ℓ that is taken to be stretched between the points $x = 0$ and $x = \ell$ along the x-axis, where $y(x, t)$ is the *transverse displacement* of the string at point $x \in [0, \ell]$ and at a time $t > 0$. We may take a *displacement function* $d(x) = y(x, 0)$ to represent the initial deflection at time $t = 0$. For simplicity of the involved details, it may be assumed that the string is sufficiently thin and perfectly elastic. As the ends are fixed at $x = 0$ and $x = \ell$, the function $y = y(x, t)$ must satisfies the *boundary conditions* given by

$$y(0, t) = y(\ell, t) = 0, \qquad \text{for all } t \geq 0. \tag{1.0.6}$$

As for d'Alembert himself, the *initial velocity* $v(x)$ is taken to be zero. That is,

$$v(x) = y_t(x, 0) = 0. \tag{1.0.7}$$

---

[2] As the equation was derived by using Taylor series argument, d'Alembert took $y = y(x, t)$ to be a C^2-function (*loi de continuité*). As such, the possibility of displacement function $d(x)$ to have a *kink* (the case of plucked string) has been avoided tacitly.

It follows easily that a unique solution of Eq. (1.0.5) is given by

$$y(x, t) = F(t + x) + G(t - x),   \qquad (1.0.8)$$

where F and G are arbitrary C^2 functions (Theorem 6.1). The condition $y(0, t) = 0$ implies $G \equiv -F$ so that we can write (1.0.8) as

$$y(x, t) = F(x + t) + F(x - t),$$

subject to that F is an *odd function*. Also, by condition $y(\ell, t) = 0$, it follows that F is a *periodic function* of period 2ℓ. Finally, by using the *initial condition* $y(x, 0) = d(x)$, we conclude that the function $y(x, t)$ is given by

$$y(x, t) = \frac{1}{2}\Big[d(x + t) + d(x - t)\Big], \qquad x \in [0, \ell] \quad \text{and} \quad t \geq 0.$$

Therefore, we have that the function $y(x, t)$ is given by

$$y(x, t) = \frac{1}{2}[\hat{d}(x + t) + \hat{d}(x - t)],$$

where $\hat{d}$ denote the odd periodic extension of the function $d : [0, \ell] \to \mathbb{R}$ to the line $\mathbb{R}$, having the period 2ℓ (see Sect. 8.2 for details). Soon after the publication of *travelling wave solution* by d'Alembert, an unclear connection between a physical problem and related mathematical description led to the *vibrating string controversy*[3] that engaged eminent scientist such as Daniel Bernoulli, Euler, and Lagrange. In 1759, the main issue was resolved partially by Lagrange.

A general partial differential equation for a sufficiently smooth function $u : \Omega_1 \to \mathbb{R}$ is an implicit equation of the form

$$F(x; \; t; \; u; \; p_i; \; q; \; r_{ij}; \cdots) = 0, \qquad (1.0.9)$$

where F is some nice function of variables x, t, u, p_i, q, r_{ij}, ... such that at least one derivative F_{p_i}, F_q, $F_{r_{ij}}$, ..., is nonzero over an appropriate domain. More generally, a *system* of differential equations for a vector function $u = u(x, t) : \Omega_1 \to \mathbb{R}^p$ is a collection of p differential equations of the form (1.0.9), one each for the p coordinate functions $u_i : \Omega_1 \to \mathbb{R}$ of the vector function u, for $i = 1, \ldots, p$.

In practical terms, a function $u = u(x, t)$ represents a scalar quantity such as the mass or charge density, temperature, linear displacement, etc., at a point $x \in \Omega$ and time t. And, for $n = 3$, a vector function $u = u(x, y, z, t) \in \mathbb{R}^4$ may represent a vector field such as the *flow velocity* of a viscous fluid or the *force* of a gravitational/electromagnetic field. Therefore, a large number of differential equations arise as a mathematical formulation of a universal law governing some phenomenon. The

---

[3] Nonetheless, the related developments helped the future mathematician to have a better understanding about many other boundary value problems for partial differential equations.

significance of differential equations to practical problems in science and engineering is mainly due to the fact that such types of equations provide efficient mathematical tools to analyse many different types of physical phenomena such as concerning physical, chemical, mechanical, electrical, social, and even biological processes.

The part of function F in Eq. (1.0.9) that contains terms of highest order derivatives only is called the *principle part* of the differential equation. In majority of tractable situations, the terms containing functions p_i, q, r_{ij}, ... have functions $a_1, a_2, \ldots \in C^0(\Omega_1 \times \mathbb{R})$ as the *coefficients*, which in practical terms represent adjustable parameters such as *speed, decay rate, specific heat, wave number, viscosity, elasticity coefficients*, etc. We also come across *nonhomogeneous* differential equation of the form

$$F\left(x; \ t; \ u; \ p_i; \ q; \ r_{ij}; \cdots \right) = g\left(x, t\right), \tag{1.0.10}$$

where $g(x, t)$ is called a *source function* (or a *sink function*). When $g(x, t)$ is identically a zero function, we say (1.0.10) is a *homogeneous* differential equation.

The *order* of a differential equation is the order of the highest derivative appearing in the equation. Clearly, for a differential equation of order $k \geq 1$, we must have $u \in C^k(\Omega_1)$. In addition to the order of a differential equation, linearity of function F with respect to variables u, p_i, q, r_{ij}, ... provide a sort of *algebraic classification* of differential equations. We say (1.0.9) is a

1. *semilinear equation* if F is linear with respect to variables u, p_i, q, $r_{ij}, \ldots$, whichever is present in the equation, and the coefficients appearing in the principle part of the equation are functions of the independent variables $x_1, \ldots, x_n$, t only;
2. *quasilinear equation* if the principle part is semilinear, and at least one coefficient in lower order terms is a function of independent variables $x_1, \ldots, x_n$, t and also of some lower order derivative of the function u;
3. *fully nonlinear equation* if it is not a quasilinear equation.

The first order differential equations for a function $u = u(x, t) \in C^1(\Omega_1)$ are largely well understood. In particular, for a subinterval $I \subset \mathbb{R}$, a general first order differential equation for a function $u = u(x, t) \in C^1(I \times \mathbb{R})$ is given by

$$F\left(x, t; \ u; \ p, q\right) = 0, \quad \text{where} \quad p = u_x \text{ and } q = u_t. \tag{1.0.11}$$

Notice that, in particular, (1.0.11) is a *semilinear equation* if we have

$$F\left(x, t; \ u; \ p, q\right) = a\left(x, t\right)p + b\left(x, t\right)p + c\left(x, t; \ u\right). \tag{1.0.12}$$

We say (1.0.12) is a *linear equation* if the function c is given by

$$c\left(x, t; \ u\right) = c_1\left(x, t\right)u + d\left(x, t\right). \tag{1.0.13}$$

Further, (1.0.11) is a *quasilinear equation* if we have

$$F(x, t; u; p, q) = a(x, t; u)p + b(x, t; u)p + c(x, t; u). \qquad (1.0.14)$$

In general, we assume $F \in C^2(I \times \mathbb{R}^3)$ is such that

$$\left(\frac{\partial F}{\partial p}\right)^2 + \left(\frac{\partial F}{\partial q}\right)^2 \neq 0, \qquad \text{over } I \times \mathbb{R}^3. \qquad (1.0.15)$$

Some important first order differential equations for a C^1-function $u = u(x, t)$ listed below find applications in phenomena such as mentioned alongside.

1. Suppose $u = u(x, t) : \mathbb{R}^+ \times \mathbb{R}^+ \to \mathbb{R}$ represents the concentration of a chemical in a fluid flowing at a constant speed c through a thin tube of uniform cross-sectional area, and let the decrease in the concentration level of the chemical due to some reaction mechanism is proportional to the instantaneous concentration. It can be shown that $u = u(x, t)$ is given by the linear differential equation

$$u_t + c u_x = -\lambda u, \qquad \text{for } x \in \mathbb{R}^+, \text{ and } \lambda > 0, \qquad (1.0.16)$$

   where λ denotes the *decay rate*. The above differential equation is called the *advection–decay equation* (Example 6.2).

2. Suppose $u = u(x, t) : \mathbb{R} \times \mathbb{R}^+ \to \mathbb{R}$ represents the speed of a nonviscous fluid flowing through a thin pipe of uniform dimension. It can be shown that the function u is given by the quasilinear differential equation

$$u_t + u u_x = 0, \qquad \text{for } -\infty < x < \infty, \text{ and } t \geq 0, \qquad (1.0.17)$$

   known as the *inviscid Burgers' equation*, which can also be viewed as the *logistic equation* with variable speed given by $u = u(x, t)$. This *differential equation model* was first introduced in 1915 by English mathematician *Harry Bateman* (1882–1946).

3. Suppose $u = u(x, t) : \mathbb{R} \times \mathbb{R}^+ \to \mathbb{R}$ represents the vehicle density along a busy road at a point $x \in \mathbb{R}$ and time $t \geq 0$. It can be shown that the function u is given by the quasilinear differential equation

$$u_t + (1 - 2u) u_x = g(x, t), \qquad \text{for } -\infty < x < \infty, \text{ and } t \geq 0, \qquad (1.0.18)$$

   where the function $g = g(x, t)$ gives the *flux density* at an entry point (source) or at an exit point (sink) (Example 5.3).

The second order differential equation is more frequent in mathematics because a large number of practical problems are modelled by such types of equations. In general, a second order differential equation for a function $u = u(x, t) \in C^2(I \times \mathbb{R})$ is given by

$$F(x, t; u; p, q; u_{xx}, u_{xt}, u_{tt}) = 0, \qquad \text{where } p = u_x \text{ and } q = u_t, \qquad (1.0.19)$$

and F is some nice function satisfying the *nondegeneracy* condition (1.0.15) over a domain in $I \times \mathbb{R}^7$. As discussed later in Chap. 4, some of the most fundamental mathematical problems in physics and engineering are related to the second order partial differential equations such as given below.

1. As discussed in Sect. 4.3, many important phenomena such as the heat conduction, Brownian motion, fluid dynamics, population genetics, neurology, etc., are best described by the fundamental *diffusion equation* given by

$$u_t + \nabla \cdot \mathbf{Q} = 0, \qquad \text{for} \quad u = u(x, t) \in C^2(\Omega_1),$$

which represents the mass density of the diffusing material at position $x \in \Omega_1$ and time $t \geq 0$, and $\mathbf{Q}$ is the associated *flux vector*. Notice that, when $\mathbf{Q} = \mathbf{Q}(x)$ at a point $x \in \Gamma = \partial\Omega_n$, we obtain the usual *conservation law*. For example, by *Fick's law*, we have $\mathbf{Q} = -k \nabla u(x, t)$, where the diffusion coefficient k may be assumed constant. Therefore, the second order differential equation given by $u_t = k\Delta u$ describes the concentration fluctuations of the diffusing material through the region Ω_1. In the case of Fourier's law, the constant k is replaced by $-c$, where c represents the *thermal conductivity* of the medium. More complex models are involved when the diffusion coefficient k is allowed to be a function of u, x or t. For example, with $k = k(u)$, the general equation above yields the *fully nonlinear* diffusion equation given by

$$u_t = \nabla k(u) \cdot \nabla u + k(u)\nabla^2 u.$$

2. The *Ronald Fisher* equation for function $u = u(x, t) \in C^2(I \times \mathbb{R})$ given by

$$u_t - u_{xx} = r\, u(1 - u),$$

is a reaction–diffusion equation that was initially formulated to study the spatial distribution of an advantageous genes and explore its travelling wave solutions. The same equation also models natural processes related to wave propagation, chemical kinetics, nuclear reactors, combustion, etc. Notice its resemblance to the first order *logistic equation*. Unlike a first order transport equation, the determination of speed of the *travelling waves* for the Fisher equation is actually part of the problem.

3. In 1948, *Johannes Burger* studied a more general Burger equation given by

$$u_t + uu_x = \nu\, u_{xx}, \qquad \text{with } \nu \text{ being the kinematic viscosity},$$

which is also used to model natural processes related to turbulence, nonlinear acoustic, traffic flow, etc. It is a perfect prototype conservation equation to study *shock waves*. It can also be viewed as a heat equation, with the nonlinear term uu_x representing the *convection*.

The present book is mainly about the first and second order differential equations that involve scalar functions associated with two fundamental processes, namely, the *flow* of a material and the associated *flux*. In Chaps. 3 and 4, we discuss derivation of *differential equation models* for some univariate and multivariate phenomena. Our present-day understanding about some of the most challenging mysteries of the nature has been made possible due to the contributions from the fundamental (system of) partial differential equations such as given below.

1. A quantum analogue of Newton's second law gives the quantum state $\psi(x, t)$ of a particle of mass m at time t moving in a force field defined by a real potential function $\varphi(x)$, $x \in \mathbb{R}^n$, such that

$$i\hbar\psi_t + \frac{\hbar^2}{2m}\nabla\psi = \varphi(x)\psi,$$

where the complex-valued wave function $\psi = \psi(x, t)$ represents the *probability distribution*, and $\hbar$ is the Planck's constant. This is called the *n*-dimensional Schrödinger equation. It is a typical example of a dispersive equation[4] such as the following quasilinear *Korteweg–De Vries equation* (KdV):

$$u_t + u_{xxx} + uu_x = 0,$$

where $u : X \times \mathbb{R} \to \mathbb{R}$, with $X \in \{\mathbb{R}, \mathbb{T}\}$. For such types of partial differential equations, there is usually a competition between dispersion that over time smooths out the initial data (in terms of extra regularity and/or in terms of extra integrability) and the nonlinearity that can cause concentration, blow up, or even let the problem to be *ill-posed* in Hadamard sense.

2. At point $x \in \Omega \subset \mathbb{R}^3$ and time $t \geq 0$, $\mathbf{E}$ be the electric field; $\mathbf{D}$ the field of electric flux density; $\mathbf{H}$ the magnetic field; $\mathbf{B}$ the field of magnetic flux density; and, $\mathbf{J}$ the field of current density of current flow through a medium. We also take $\rho = \rho(x, t)$ to be the scalar field of electric charge density of the medium. Then the system of partial differential equations as given below are known as the *Maxwell Equations*:

$$\nabla \times \mathbf{H} = \mathbf{J} + \frac{\partial \mathbf{D}}{\partial t};$$
$$\nabla \times \mathbf{E} = -\frac{\partial \mathbf{B}}{\partial t};$$
$$\nabla \cdot \mathbf{D} = \rho;$$
$$\nabla \cdot \mathbf{B} = 0.$$

The significance of above types of equations in science and engineering is mainly due to the fact they apply to all areas of scientific investigation concerning elec-

---

[4] Informally speaking, a partial differential equation is called *dispersive* if, when no boundary condition is imposed, its solutions spread out in space as they evolve over time.

tromagnetic waves, which includes optics. A brief review of Maxwell equations, and also some other related equations, is provided as Appendix A.2.

3. The first three *Navier–Stokes equations* model conservation of momentum that describes *viscous flow* of Newtonian fluids, with a dynamic viscosity $\mu > 0$, and the fourth expresses the *incompressibility* of the fluid:

$$u_t + (u \cdot \nabla)u = -\nabla\varphi + \nu\,\nabla^2 u + f;$$
$$\operatorname{div} u = 0$$

where $u = (u_1, u_2, u_3)$ is the fluid velocity vector, $\nu = \mu/\rho_0$ is called the kinematic viscosity, with ρ_0 being the uniform density. Notice that, in terms of *material derivative*

$$\frac{D}{Dt} \equiv \partial_t + u \cdot \nabla,$$

the left side Du of the first equation (in vector form) represents the per unit volume inertia. The scalar function φ, as internal source, represents the per unit mass thermodynamic work. The term $\nu\,\nabla^2 u$ represents the diffusion viscosity. And, the vector field f, as the body acceleration, is the external source. These equations are very significant mainly due to applications in many areas of science and engineering dealing with viscous fluid flow.

4. *Black–Scholes equation* for an unknown price $v = v(s, t)$ of an option at stock price s and at time $t > 0$,

$$v_t + \frac{1}{2}\sigma^2 s^2 v_{ss} = rv - rsv_s,$$

where r is the risk-free interest and σ refers to the volatility of the stock. Technically, time decay term v_t is called the *theta*, the second term on the left giving the convexity of the derivative value is called *gamma*, and the right side represents the risk-free return from a long position in the derivative and a short position consisting of v_s shares of the underlying.

For a differential equation (1.0.9), a *classical solution* is a sufficiently smooth scalar function $\varphi = \varphi(x, t)$ that satisfies the equation identically over a subdomain of the set Ω_1. Unlike an ordinary differential equation, the general solution of a partial differential equation involves functions as parameters, and the *solution space* in this case is an infinite-dimensional function space defined on some domain. It is in general important to find ways to establish the *existence* of a solution of lower order partial differential equations of simpler linearity type. Further, to analyse a physical problem in terms of a solution of the underlying differential equation model, it is equally important to determine conditions that ensure uniqueness of a solution.

In most situations, *uniqueness* is subject to that a solution φ needs to satisfy certain assigned *auxiliary conditions* specified for the involved unknown function. In practical terms, there are mainly two types of auxiliary conditions: An *initial condition* is given by a functional specification that φ, and some of its derivatives, must satisfy at an initial time, say $t = 0$, over a subdomain of Ω_1. On the other hand, a *boundary condition* is given by functional specifications for φ, and also some of its derivatives, on the boundary $\Gamma = \partial\Omega$. For example, suppose $y = y(x, t)$ represents the *transverse displacement* of a vibrating string of length ℓ with ends fastened at points $x = 0$ and $x = \ell$ along the x-axis. It can be shown the function y is given by the differential equation

$$y_{tt} = c^2 y_{xx}, \quad \text{for all } 0 \leq x \leq \ell, \text{ and } t \geq 0,$$

which is d'Alembert's wave equation. Now, suppose the string is *plucked* at time $t = 0$ so that the *initial deflection* $y(x, 0)$ of the string is given by a function $f(x)$, for $x \in (0, \ell)$. We may also assume that, when the string in equilibrium state is *struck*, the *initial velocity* $y_t(x, 0)$ is given by a function $g(x)$, for $x \in (0, \ell)$. In this case, we say $y = y(x, t)$ satisfies the *nonhomogeneous* initial conditions given by

$$y(x, 0) = f(x), \quad \text{and} \quad y_t(x, 0) = g(x), \quad \text{for } 0 < x < \ell;$$

and also the *homogeneous* boundary conditions given by

$$y(0, t) = 0, \quad \text{and} \quad y(\ell, t) = 0, \quad \text{for } t > 0.$$

The above boundary conditions hold because of the fact that the two ends are fixed. In general, the main concern in most applications is to find an explicit solution of a differential equation that is unique subject to assigned *auxiliary conditions*. When no analytical method is available to serve the purpose, a numerical technique is applied to obtain a sequence of *approximate solutions* that converges in some appropriate function space.

An *initial value problem* (or simply an IVP) is about finding a solution of a differential equation that also satisfies specified initial conditions. A *Cauchy problem* is a special type of an initial value problem wherein it is given that $u(x, 0)$ is some specified function defined for $x \in \Omega$. In Chap. 7, we apply the method of characteristics to solve Cauchy problems for first order partial differential equations in two variables Also, a *boundary value problem* (or simply an BVP) is about finding a solution of a differential equation that also satisfies specified functional relations over the boundary Γ. More generally, an *initial-boundary value problem* (or simply an IBVP) is about finding a solution of a differential equation that also satisfies the specified initial and boundary conditions. According to Jacques Hadamard [1], an IBVP is said to be *well posed* if the three conditions given below hold:

1. (**Existence**) There exists *at least one solution* of the IBVP.
2. (**Uniqueness**) There exists *at most one solution* of the IBVP.
3. (**Stability**) The solution is *stable* in the sense that it depends continuously on the assigned data. That is, for any perturbation in initial-boundary conditions, and also in the values of involved *parameters*, change in the solution is very small.

In some cases, we also need to verify the *regularity properties* of the solution. Finally, we do analysis of the solution to express its physical interpretations in not-so-mathematical terms. An initial-boundary value problem is said to be *ill-posed* if any one of the above three conditions fails to hold. In certain situations such as control theory, ill-posed problems arise naturally, which happens if the model equations are not homogeneous or when part of the initial data renders IBVP ill-posed. In this book, we are mainly dealing with IBVPs for the three fundamentals differential equations related to wave motion, heat conduction, and potential theory. We study the effect of various types of initial and boundary conditions on a solution of a differential equation. In last four chapters, we will solve some IBVPs for second order linear differential equations, by using *separation of variables, Fourier theory, eigenfunctions expansion*, and *transform techniques* due to Fourier and Laplace.

In Chap. 2, we review some relevant concepts and related theorems of multivariable analysis, geometry of surfaces and curves, and vector calculus. To be more specific, we describe three fundamental concepts of multivariable analysis inasmuch as we need in the sequel, and give statements of related theorems. The second part of the chapter is mainly about the geometry of 2-parameter family of curves represented by level surfaces over a domain in $\mathbb{R}^3$. The focus in the last part s to discuss theorems of Gauss, Stokes and Helmholtz concerning the *del operator* ∇ as given by (1.0.2). Notations introduced in Chap. 2 are used throughout the book. The reader may refer the companion volume "*Fundamentals of Real Analysis with Applications*" [2], or any other standard text such as [3], for details.

The above mentioned *solution methods* for IBVP use some parts of the theory of ordinary differential equations. The method of characteristics uses technique of solving first order system of initial value problems for ordinary differential equations. The method of separation of variables leads to boundary value problems for second order linear ordinary differential equations, with Sturm–Liouville types of boundary conditions. In Chap. 3, we review some relevant *solution methods* applicable to ordinary differential equations inasmuch as needed in the sequel. In addition, we also discuss some prototypical *differential equation models* for univariate phenomena such as concerning the motion of a body in a *spring-mass-dashpot system*, and also the one used to study *growth pattern* of a specie in a competitive environment. As has been elaborated nicely by Morris Kline in his three-volume book [4], most paradigms of knowledge have been benefited from the applications of ordinary differential equations to certain intricate practical problems. Our main source for the content developed in this chapter have been the standard texts on the topic [5–10].

In Chap. 4, we derive *continuity equation* governing the mass transport, *equation of motion* governing the momentum transport, and *diffusion equation* governing the energy flow. The main focus of the chapter is to help reader appreciate the physical aspects of some prototypical second order linear partial differential equations modelling vibrations of a string and a membrane, heat conduction in a bar and cube, fluid flow in a thin pipe and cylindrical tube, and certain natural processes concerning electromagnetism.

In Chap. 5, we introduce some concepts such as *order* and *linearity* for a general partial differential equation, and fix notions for use in the sequel. We also describe *classification* of linear partial differential equations in two variables. The chapter also includes a brief description about issues related to the *existence* and *uniqueness* of a solution of certain simple *initial-boundary value problems*. Unlike ordinary differential equations, there is no unified general theory for linear partial differential equations of order ≥ 2 [11], and the main reason what makes the situation difficult is attributed to the involved geometric complexities [12]. In this chapter, we surely have a little more to say about the related issues.

In Chap. 6, we discuss three methods of finding a general solution of a first order partial differential equation. For a semilinear equation, we use *transformation of coordinates*. For a quasilinear equation, we use *Lagrange's method*. And, we use *Charpit's method* to deal with the general case. In Chap. 7, we discuss first order partial differential equations in two variables. The *method of characteristics* is applied to solve some interesting Cauchy problems for such types of equations. In Chaps. 8 and 9, we use *separation of variables, Fourier series*, and *eigenfunctions expansion* to solve some important IBVPs for second order linear partial differential equations. In most of these four chapters, the supplied physical reasoning and mathematical argument suffice to accept the fact that a solution so obtained is indeed appropriate.

The *transform techniques* developed by Fourier and Laplace offer powerful tools to solve many different types of initial value problems, and also some boundary value problems, defined over an *unbounded domain*. Both use the linearity of the underlying integral transform. The Laplace transform converts a linear differential equation with constant coefficients into an algebraic equation, whereas the Fourier transform converts a linear partial differential equation into an ordinary differential equation. In some cases, we find it much easier to solve the resulting algebraic/differential equation(s) by simple *manipulation* or by applying some known *solution method*. Finally, we obtain a solution of the original problem by applying the respective *inverse transform*.

A real turning point in the history of mathematics is attributed to Fourier's work on heat conduction in solids, which motivated Bernhard Riemann to introduce fundamental concepts such as integration, surfaces, and manifolds. The subsequent developments in modern mathematics led to the introduction of important subjects such as *variational calculus, functional analysis, differential and algebraic topology*, and also *geometry of smooth manifolds*, which motivated further applications of differential equations to complex and diverse forms of practical problems in science and engineering.

In Chap. 10, we introduce Fourier integral of integrable aperiodic functions f : $\mathbb{R} \to \mathbb{R}$ as a natural extension of the Fourier series of the periodic extension of the truncated function, satisfying Dirichlet's conditions. We compute the Fourier integrals of some interesting functions, considering our needs with regard to solving a class of IBVPs over *unbounded domains*. We discuss Gaussian heat kernel and fundamental solution of Laplace equation in terms of Green's function. For additional motivation, we give physical interpretation of abstract facts in the context of *signal analysis*.

In Chap. 11, we discuss Laplace transform, and their properties, to facilitate our needs to use the related techniques as a potent method of solving some IBVPs over semi-infinite domain $[0, \infty)$. For additional motivation, we give physical interpretation of abstract ideas in the context of linear time-invariant system as applicable in the control theory.

Throughout the book, the term *differential equation* refers to an ordinary or a partial differential equation, and also to a *system* of such types of equations. We can always figure out which type of equation is being referred to in any particular situation. For the main part of the book, we shall be dealing with differential equations in two or three independent variables.

References

1. Hadamard, J. (1952). *Lectures on Cauchy's problem in linear partial differential equations.* Yale University Press.
2. Razdan, A., & Ravichandran, V. (in press). *Fundamentals of analysis with applications.* Springer.
3. Rudin, W. (1976). *Principles of mathematical analysis* (3rd ed.). McGraw-Hill Book Co.
4. Kline, M. (1972). *Mathematical thought from ancient to modern times* (Vols. 1, 2, 3). Oxford University Press.
5. Arnold, V. I. (1992). *Ordinary differential equations* (3rd ed.). Springer.
6. Birkhoff, G., & Rota, G.-C. (1969). *Ordinary differential equations* (2nd ed.). Wiley.
7. Braun, M. (1993). *Differential equations and their applications* (4th ed.). Springer.
8. Coddington, E. A., & Levinson, N. (1987). *Theory of ordinary differential equations.* Tata McGraw-Hill Publishing Co.Ltd.
9. Hartman, P. (1964). *Ordinary differential equations.* Wiley.
10. Myint-U, T. (1978). *Ordinary differential equations.* Elsevier.
11. Klainerman, S. (2000). *PDE as a unified subject* (Special Volume). GAFA: Geometric And Functional Analysis.
12. Arnold, V. I. (2004). *Lecturers on partial differential equations.* Springer.

Chapter 2
Classical Vector Analysis

A mathematician may say anything he pleases, but a physicist must be at least partially sane.

Josiah Willard Gibbs (1839–1903)

In this chapter, we review relevant concepts of multivariable analysis, basic geometric properties of curves and surfaces, and also some important theorems of classical vector calculus. The concepts reviewed in the first two sections are fundamental to every aspect of topics discussed in the subsequent chapters. We also need some variants of the fundamental divergence theorem to formulate *differential equation models* of certain important phenomena. In Sect. 2.1, we recall theorems related to three important analytical properties of a scalar or vector function defined on an open set in $\mathbb{R}^n$ and fix notations for use in the sequel. In Sect. 2.2, we recall some important geometric notions related to classical theory of curves and surfaces. In Sect. 2.3, we discuss three fundamental theorems related to the *del operator*

$$\nabla := \left(\frac{\partial}{\partial x_1}, \cdots, \frac{\partial}{\partial x_n} \right)$$

for the case when $n = 3$. Throughout this book, a symbol in boldface font represents a vector quantity, which we may view as a column vector, whenever necessary.

2.1 Multivariable Calculus

The set $\mathbb{R}^n$ of n-tuples of real numbers is a linear space over the field $\mathbb{R}$, where the *addition* and *scalar multiplication* are defined, respectively, as

© The Author(s), under exclusive license to Springer Nature Singapore Pte Ltd. 2022   13
A. K. Razdan and V. Ravichandran, *Fundamentals of Partial Differential Equations*,
https://doi.org/10.1007/978-981-16-9865-1_2

$$x + y = (x_1 + y_1, \ldots, x_n + y_n);$$
$$a \cdot x = (a\,x_1, \ldots, a\,x_n),$$

for $x = (x_1, \ldots, x_n)$, $y = (y_1, \ldots, y_n) \in \mathbb{R}^n$ and $a \in \mathbb{R}$. The usual dot product of vectors x and y defines an *inner product* on $\mathbb{R}^n$, which is written as $\langle x, y \rangle$. That is,

$$\langle x, y \rangle := x \cdot y = x_1 y_1 + \cdots + x_n y_n. \tag{2.1.1}$$

It can be shown that the function $\langle\,,\,\rangle : \mathbb{R}^n \to \mathbb{R}$ given by (2.1.1) is a *positive definite, symmetric, bilinear functional* (Exercise 2.1). Therefore, $\mathbb{R}^n$ is an inner product space. In general, a linear space X over the field $\mathbb{R}$ is called an *inner product space* if there exists a positive definite, symmetric, bilinear functional $b : X \times X \to \mathbb{R}$.

Further, a linear space X over the field $\mathbb{R}$ is called a *normed space* if there exists a positive definite, absolute homogeneous, subadditive function $p : X \to \mathbb{R}$, where $p(x)$ is called the *norm* of $x \in X$. In particular, it can be shown that the function $\|\ \| : \mathbb{R}^n \to \mathbb{R}$ given by

$$\|x\| := \sqrt{\langle x, y \rangle} = \sqrt{x_1^2 + \cdots + x_n^2}. \tag{2.1.2}$$

gives a *norm* on $\mathbb{R}^n$ (Exercise 2.2). Therefore, $\mathbb{R}^n$ is a normed space, where $\|x\|$ and their norms is called the norm of a vector x induced by the inner product (2.1.1). An interesting relation between the inner product of two points $x, y \in \mathbb{R}^n$ is the *Cauchy-Schwartz inequality* given by

$$|\langle x, y \rangle| \le \|x\|\|y\|, \quad \text{for} \quad x, y \in \mathbb{R}^n. \tag{2.1.3}$$

A yet another interesting relation for the case when $n = 3$ is the identity given by

$$\|a \times b\|^2 = \|a\|^2 \|b\|^2 - \langle a, b \rangle^2, \quad \text{for} \quad a, b \in \mathbb{R}^3, \tag{2.1.4}$$

which finds many important applications, where the *cross product* $a \times b$ of vectors $a = (a_1, a_2, a_3)$ and $b = (b_1, b_2, b_3)$ is a vector in $\mathbb{R}^3$ given by

$$a \times b := (a_2 b_3 - a_3 b_2, \; a_3 b_1 - a_1 b_3, \; a_1 b_2 - a_2 b_1). \tag{2.1.5}$$

Notice that, for any vector $x \in \mathbb{R}^n$, the norm $\|x\|$ as in (2.1.2) gives the (radial) distance of the point $x \in \mathbb{R}^n$ from the origin $0 \in \mathbb{R}^n$. More generally, the function $d_2(x, y)$ defined using the usual *Euclidean distance* between two points $x, y \in \mathbb{R}^n$:

$$d_2(x, y) := \|x - y\| = \sqrt{(x_1 - y_1)^2 + \cdots + (x_n - y_n)^2}, \tag{2.1.6}$$

gives $\mathbb{R}^n$ a *metric structure*, which facilitates the introduction of fundamental concept of *convergence* in $\mathbb{R}^n$.

Definition 2.1 A sequence $\langle x_k \rangle$ in the metric space $(\mathbb{R}^n, d_2)$ is said to *converge to a point* $a \in \mathbb{R}^n$ if for any $\varepsilon > 0$ there exists $K \in \mathbb{N}$ such that

$$k \geq K \quad \Rightarrow \quad \|x_k - a\| < \varepsilon.$$

In this case, we write $x_k \to a$ as $k \to \infty$ or simply $\lim_{k \to \infty} x_k = a$.

Clearly, every convergent sequence $\langle x_k \rangle$ is bounded. That is, there exists a positive number $M > 0$ such that $\|x_k\| \leq M$, for all $k \in \mathbb{N}$. In fact, $\langle x_k \rangle$ is a *Cauchy sequence* in the sense that *almost all* terms of $\langle x_k \rangle$ are very close to each other. More precisely, for any $\varepsilon > 0$ there exists $K \in \mathbb{N}$ such that

$$j, k \geq K \quad \Rightarrow \quad \|x_j - x_k\| < \varepsilon.$$

Some of the fundamental theorems related to the convergence of (bounded) sequences in $\mathbb{R}^n$ are as stated below.

Theorem 2.1 (Bolzano–Weierstrass Theorem) *Every bounded sequence of points in $\mathbb{R}^n$ contains a convergent subsequence.*

Notice that, for $x_k = \left(x_1^k, \ldots, x_n^k \right)$ and $a = \left(a_1, \ldots, a_n \right)$ in $\mathbb{R}^n$, we have

$$\lim_{k \to \infty} x_k = a \quad \Leftrightarrow \quad \lim_{k \to \infty} x_i^k = a_i, \quad \text{for all } 1 \leq i \leq n. \tag{2.1.7}$$

Therefore, a sequence $\langle x_k \rangle$ in $\mathbb{R}^n$ is Cauchy with respect to metric d_2 if and only if it is a convergent sequence. That is, $(\mathbb{R}^n, d_2)$ is a complete metric space. Hence, $\mathbb{R}^n$ is a *Hilbert space* with respect to the inner product (2.1.1), and so $\mathbb{R}^n$ is also a Banach space with respect to the norm (2.1.2). Therefore, we can extend the concept of convergence of an infinite series to $\mathbb{R}^n$.

Definition 2.2 For an infinite sequence $\langle x_k \rangle$ in $\mathbb{R}^n$, we say the series $\sum_{k=1}^{\infty} x_k$ converges if the *sequence of partial sums* $\langle s_p \rangle$ given by

$$s_p = \sum_{k=1}^{p} x_k = x_1 + \cdots + x_p$$

is convergent, and the limit $s = \lim_{p \to \infty} s_p$ gives the sum of the series.

We know that, in general, a normed space X is a Banach space if and only if every infinite series $\sum_{k=1}^{\infty} x_k$ in X, with $\|x_k\| \leq 2^{-k}$, is convergent. As in the case when $n = 1$, an infinite series $\sum_{k=1}^{\infty} x_k$ is *absolutely convergent* if the series $\sum_{k=1}^{\infty} \|x_k\|$ converges. It follows easily that a normed space X is a Banach space if and only if every absolutely convergent infinite series in X converges.

Definition 2.3 With respect to the inner product (2.1.1), two vectors $x, y \in \mathbb{R}^n$ are said to be *orthogonal* if $\langle x, y \rangle = 0$. Further, we say $x \in \mathbb{R}^n$ is a *normalised vector* (or a *unit vector*) if $\|x\| = 1$.

For $i = 1, \ldots, n$, the standard ith unit vectors in $\mathbb{R}^n$ is given by

$$e_i := (0, \ldots, 0, 1, 0, \ldots, 0),$$

where 1 appears at the ith coordinate. The set of n unit vectors $e_1, \ldots, e_n$ forms an *orthonormal basis* of $\mathbb{R}^n$ in the sense that each $x \in \mathbb{R}^n$ can be written uniquely as

$$x = \langle x, e_1 \rangle e_1 + \cdots + \langle x, e_n \rangle e_n. \tag{2.1.8}$$

In particular, $\mathbb{R}^n$ is a *finite dimensional* Hilbert space. Together with the standard *orthonormal basis* as given above, $\mathbb{R}^n$ is the most suitable *model space* for many applications in science and engineering.

However, for the topics we intend to discuss in this book, we also have to use certain other important Hilbert spaces such as consisting of real-valued functions defined over an *open set* $\Omega \subseteq \mathbb{R}^n$ (Definition 2.4). In actual applications, the topology of an open set Ω plays a significant role while analysing a practical problem.

Euclidean Topology

For any $a \in \mathbb{R}^n$, a *basic neighbourhood* of the point a with respect the Euclidean metric is the set $B(a; r)$ given by

$$B(a; r) := \{ x \in \mathbb{R}^n : \| x - a \| < r \}, \quad \text{where } r > 0.$$

The set $B(a; r) \subset \mathbb{R}^n$ is called an n-dimensional ball of radius r centred at point a (or simply an r-ball at point a). For example, an r-ball at a point $a \in \mathbb{R}$ is an open interval of the form $(a - r, a + r)$; an r-ball at a point $(a, b) \in \mathbb{R}^2$ is an *open disk* written as

$$D^2 := \{ (x, y) \in \mathbb{R}^2 : \| (x, y) - (a, b) \| < r \};$$

and so on. The topology on $\mathbb{R}^n$ defined by such a system of neighbourhoods is known as the *Euclidean topology* (or simply the *usual topology*). For any set $S \subseteq \mathbb{R}^n$, and any $x \in \mathbb{R}^n$, we say

1. the element x is an *interior point* of the set S if we can find an r-ball $B(x; r)$ such that $B(x; r) \subseteq S$;
2. the element x is an *exterior point* of S if we can finds an r-ball $B(x; r)$ such that $B(x; r) \subset \mathbb{R}^n \setminus S$;
3. the element x is a *boundary point* of S if every r-ball $B(x; r)$ intersects both S and the complement $\mathbb{R}^n \setminus S$.

We may write the set of *interior points* of a set $S \subseteq \mathbb{R}^n$ as int(S) (or simply as S^o), and the set of *boundary points* of the set S is written as ∂S. For example,

$$S_r^{n-1}(a) := \partial B(a; r) = \{ x \in \mathbb{R}^n : \| x - a \| = r \}$$

is the $(n - 1)$-dimensional sphere of radius r centred at a. We usually write $\mathbb{S}^{n-1}$ for the $(n - 1)$-dimensional sphere of radius 1 centred at the origin $\mathbf{0}$, which is known as the *unit sphere* in the space $\mathbb{R}^n$, for $n \geq 2$. Notice that $\mathbb{S}^0 = \{-1, 1\}$.

Definition 2.4 A set $U \subseteq \mathbb{R}^n$ is said to be an *open set* if each $a \in U$ is an *interior point* of U. That is, for each $a \in U$, we can find some $r > 0$ such that

$$\|x - a\| < r \quad \Rightarrow \quad x \in U.$$

A set $F \subseteq \mathbb{R}^n$ is called a *closed set* if the complement $U = \mathbb{R}^n \setminus F$ is an open set.

For $x \in \mathbb{R}^n$, each open set containing x is a *neighbourhood* of the point x. For example, every open ball $B(x; r)$ is a neighbourhood of the point x. In particular, it follows that points in $\mathbb{R}^n$ can have arbitrary small neighbourhoods. Also, a point $x \in \mathbb{R}^n$ belongs to a closed F if and only if $x \in \partial U$ or x is an exterior point of the set $U = \mathbb{R}^n \setminus F$. That is, each ball $B(x; r)$ intersects both $F = \mathbb{R}^n \setminus U$ and U or it lies inside the set F.

Definition 2.5 Let $S \subset \mathbb{R}^n$. We say $x \in \mathbb{R}^n$ is a *limit point* of the set S if, for any $r > 0$, there exists $y \in S$ such that $0 < \|x - y\| < r$. That is, every ball $B(x; r)$ intersects S in a point distinct from x, in case $x \in S$. We write S' for the set of limit points of set S, which is called the *derived set* of the set S. The set $\overline{S} = S \bigcup S'$ is called the *closure* of the set S.

Clearly, $F \subseteq \mathbb{R}^n$ is a closed set if and only if $\overline{F} = F$. Said differently, a closed set in $\mathbb{R}^n$ is precisely the set that contains all its limit points. Also, $x \in \mathbb{R}^n$ is a limit point of a set S implies, for each $k \in \mathbb{N}$, we can find

$$x_k \in B(x; 1/k) \bigcap S \quad \Rightarrow \quad x_k \to x \text{ as } k \to \infty.$$

Conversely, if there exists a sequence $\langle x_k \rangle$ in a set S with $x_k \neq x$, for each $k \in \mathbb{N}$, such that $x_k \to x$ as $k \to \infty$, then we have $x \in S'$. An element $x \in S \setminus S'$ is called an *isolated point* of the set S. A closed set with no isolated points is called a *perfect set*.

Definition 2.6 A set $K \subset \mathbb{R}^n$ is called a *bounded set* if, for some $r > 0$, we have $K \subseteq B(\mathbf{0}; r)$.

According to celebrated *Heine-Borel Theorem*, a set $K \subset \mathbb{R}^n$ is compact if and only if it is a closed and bounded set. In particular, for any $x \in \mathbb{R}^n$, every closed r-ball $\overline{B}(x; r) = B(x; r) \cup S(x; r)$ is a compact set. Recall that $K \subset \mathbb{R}^n$ is said to be a *compact set* if every open cover of K contains a finite subcover. Or, equivalently, every infinite sequence in K contains a convergent subsequence, with the limit in the set K only. The latter assertion is also known as the *Bolzano-Weierstrass Theorem*. Another interesting equivalent condition for a set $K \subset \mathbb{R}^n$ to be compact is that every family $\mathscr{F}$ of closed sets in K, with *finite intersection property*, has nonempty intersection. According to *Tychonoff's Theorem*, product of every family of compact sets is compact.

Definition 2.7 We say $C \subseteq \mathbb{R}^n$ is a *connected set* if there do not exists any pair of nonempty open sets U and V in $\mathbb{R}^n$ such that

$$C = (U \cap C) \cup (V \cap C),$$

Further, C is *path-connected* if, for every pair of points $x, y \in C$, there is some continuous function $\gamma : [a, b] \to C$ such that $\gamma(a) = x$ and $\gamma(b) = y$.

Clearly, each path-connected set is connected, but the converse may not hold in general. However, each open connected set $U \subset \mathbb{R}^n$ is path-connected. If for a path-connected set $C \subseteq \mathbb{R}^n$ it is always possible to take γ as given by

$$\gamma(t) = (1 - t)x + t\,y, \quad \text{for } t \in [0, 1],$$

then C is a *convex set*. In general, a set $C \subseteq \mathbb{R}^n$ is said to be *convex* if the line joining any two points of C is contained in the set C itself. For example, for any $x \in \mathbb{R}^n$, each r-ball $B(x; r)$ is a convex set. In certain applications, compact convex subsets of $\mathbb{R}^n$ play a significant role.

Continuity and Differentiability

Let $\Omega \subseteq \mathbb{R}^n$ be an open set, and $\overline{\Omega} = \Omega \bigcup \partial\Omega$. As said earlier, to study the existence and uniqueness of a solution of an *initial-boundary value problem* for an ordinary or a partial differential equation, we must know certain analytical/geometrical properties for a function of the form $u = u(x, t)$, where $x = (x_1, \ldots, x_n) \in \Omega$ represents the *spatial variables*, and $t \geq 0$ represents time. We begin the discussion with the next definition.

Definition 2.8 Let $\varphi : \Omega \to \mathbb{R}$ be a function, $a \in \overline{\Omega}$, and $\ell \in \mathbb{R}$. We say ℓ is a *limit of function* φ at the point a if for any $\varepsilon > 0$ there is some $\delta = \delta(a, \varepsilon) > 0$ such that we have

$$x \in \Omega, \quad 0 < \|x - a\| < \delta \quad \Rightarrow \quad |\varphi(x) - \ell| < \varepsilon. \tag{2.1.9}$$

In this case, we write $\lim_{x \to a} \varphi(x) = \ell$ or $\varphi(x) \to \ell$ as $x \to a$.

Notice that the point a in the above definition may not belong to Ω. However, as $\overline{\Omega}$ is a closed set, the assumption $a \in \overline{\Omega}$ ensures that, for any $\delta > 0$, there are enough points $x \in \Omega$ to meet the requirement that $\|x - a\| < \delta$. In general, *limit* of a function at a point may not exist. However, it is unique, whenever it exists. It is also clear that the limiting value ℓ is independent of the value $\varphi(a)$, even when it is defined. In actual practice, we find it more convenient to use the condition as given in the next theorem.

Theorem 2.2 (Sequential Limit) *Let* $\varphi : \Omega \to \mathbb{R}$ *be a function, and* $a \in \overline{\Omega}$. *Then, we have*

$$\lim_{x \to a} \varphi(x) = l \Leftrightarrow \lim_{n \to \infty} \varphi(x_n) = l,$$

for every sequence $\langle x_n \rangle$ *in* Ω, *with* $x_n \neq a$, *such that* $\lim_{n \to \infty} x_n = a$.

Example 2.1 Consider the function $\varphi : \mathbb{R}^2 \to \mathbb{R}$ given by

$$\varphi(x) = \begin{cases} \dfrac{xy}{x^2 + y^2}, & \text{for } x = (x, y) \neq 0 \\ 0, & \text{for } x = 0 \end{cases}$$

Then, along the *line* $y = m\,x$, we have

$$\varphi(x) = \frac{xy}{x^2 + y^2} = \frac{m}{1 + m^2}.$$

Therefore, as $x \to 0$, the values $\varphi(x)$ approaches different values. Hence, $\lim_{x \to 0} \varphi(x)$ does not exist.

Example 2.2 Consider the function $\varphi : \mathbb{R}^2 \to \mathbb{R}$ given by

$$\varphi(x) = \begin{cases} \dfrac{x^2 y}{x^4 + y^2}, & \text{for } x = (x, y) \neq 0 \\ 0, & \text{for } x = 0 \end{cases}$$

In this case, along the *parabola* $y = m\,x^2$, we have

$$\varphi(x) = \frac{x^2 y}{x^4 + y^2} = \frac{m}{1 + m^2}.$$

So, as $x \to 0$, the function $\varphi(x)$ approach to different values. Hence, $\lim_{x \to 0} \varphi(x)$ does not exist.

More generally, consider a vector function $f : \Omega \to \mathbb{R}^m$, with

$$f(x) = \big(f_1(x), \ldots, f_m(x)\big), \qquad \text{for } x \in \Omega, \tag{2.1.10}$$

where $f_i : \Omega \to \mathbb{R}$ are the m *component functions* of f. It is easy to show that, for $l_1, \ldots, l_m \in \mathbb{R}$, we have

$$\lim_{x \to a} f(x) = (l_1, \ldots, l_m) \qquad \Leftrightarrow \qquad \lim_{x \to a} f_i(x) = l_i, \tag{2.1.11}$$

for each $i = 1, \ldots, m$. Therefore, f has the limit $l = (l_1, \ldots, l_m) \in \mathbb{R}^m$ at a point $a \in \Omega$ if, for every $\varepsilon > 0$, there exists $\delta = \delta(a, \varepsilon) > 0$, such that

$$0 < \|x - a\| < \delta \quad \Rightarrow \quad \big|f_i(x) - l_i\big| < \varepsilon, \quad \text{for each } i = 1, \ldots, m. \tag{2.1.12}$$

Definition 2.9 A function $\varphi : \Omega \to \mathbb{R}$ is said to be *continuous at* a point $a \in \Omega$ if for each $\varepsilon > 0$ there exists $\delta = \delta(a, \varepsilon) > 0$ such that

$$\|x - a\| < \delta \quad \Rightarrow \quad \big|\varphi(x) - \varphi(a)\big| < \varepsilon. \tag{2.1.13}$$

Said differently, a function $\varphi = \varphi(x)$ is continuous at a point a if $\lim_{x \to a} \varphi(x) = \varphi(a)$.

A function $\varphi : \Omega \to \mathbb{R}$ is said to be continuous over Ω if it is continuous at each point $x \in \Omega$. We may write $C(\Omega)$ for the set of continuous functions defined over an open set Ω. The next theorem follows directly from Theorem 2.2.

Theorem 2.3 (Sequential Continuity) *A function $\varphi : \Omega \to \mathbb{R}$ is continuous at a point $a \in \Omega$ if and only if for every sequence $\langle x_n \rangle$ in Ω, with $x_n \neq a$ and $\lim_{n \to \infty} x_n = a$, we have $\lim_{n \to \infty} \varphi(x_n) = \varphi(a)$.*

Notice that the two functions $\varphi : \mathbb{R}^2 \to \mathbb{R}$ as defined in Example 2.1 and Example 2.2 are not continuous at $x = 0 \in \mathbb{R}^2$. The next two examples illustrates the continuity of some other types of *nonlinear* functions $\varphi : \Omega \to \mathbb{R}$, with $\Omega \subseteq \mathbb{R}^2$.

Example 2.3 Consider a function $\varphi : \mathbb{R}^2 \to \mathbb{R}$ given by

$$\varphi(x) = \begin{cases} \dfrac{x^6 - 2y^4}{x^2 + y^2}, & \text{for } x = (x, y) \neq 0 \\ 0, & \text{for } x = 0 \end{cases}$$

Since $\|x\|^2 = x^2 + y^2$, we have $|x|^2 \leq \|x\|^2$, and so $|x| \leq \|x\|$. Similarly, $|y| \leq \|x\|$. For $0 < \|x\| < 1$, we have

$$\begin{aligned} \left| \varphi(x) - \varphi(0) \right| &\leq \frac{|x|^6 + 2|y|^4}{\|x\|^2} \\ &\leq \frac{\|x\|^6 + 2\|x\|^4}{\|x\|^2} \\ &= \|x\|^2 (\|x\|^2 + 2) \\ &< 3\|x\|^2 < 3\|x\|. \end{aligned}$$

Therefore, given any $\varepsilon > 0$, we may choose $\delta = \min \left\{ 1, \varepsilon/3 \right\}$ so that

$$0 < \|x - 0\| < \delta \quad \Rightarrow \quad \left| \varphi(x) - 0 \right| \leq 3\|x\| < 3\delta \leq \varepsilon.$$

Hence, we have $\lim_{x \to 0} \varphi(x) = 0 = f(0)$. That is, the function φ is continuous at $x = 0$. Clearly, φ is continuous for all $x \neq 0$.

Example 2.4 Consider the function $\varphi : \mathbb{R}^2 \to \mathbb{R}$ given by $\varphi(x) = 3x + 2y^2$, where $x = (x, y) \in \mathbb{R}^2$. As in the previous example, since $\|x\|^2 = x^2 + y^2$, we have $|x| \leq \|x\|$ and $|y| \leq \|x\|$. Using these, for $\|x\| < 1$, we have

$$|\varphi(x) - 0| \leq 3|x| + 2|y|^2 \leq 3\|x\| + 2\|x\|^2 \leq 5\|x\|.$$

For given $\varepsilon > 0$, choose $\delta = \min\{1, \varepsilon/5\}$. Then, for all x with $|x - 0| < \delta$, we have

$$|\varphi(x) - 0| \le 5|x| < 5\delta \le \varepsilon.$$

So, $\lim_{x \to 0} \varphi(x) = 0 = f(0)$. Hence, the function φ is continuous at $x = 0$. Clearly, φ is continuous for all $x \ne 0$.

The next two theorems state two fundamental properties of a continuous function defined on a compact set $K \subset \mathbb{R}^n$. Recall that a function $\varphi : \Omega \to \mathbb{R}$ is said to be *uniformly continuous* over an open set $\Omega \subseteq \mathbb{R}^n$ if for each $\varepsilon > 0$ there is some $\delta = \delta(\varepsilon) > 0$ such that, for any $x, y \in \Omega$, we have

$$\|x - y\| < \delta \quad \Rightarrow \quad |\varphi(x) - \varphi(y)| < \varepsilon. \tag{2.1.14}$$

Every uniformly continuous function is continuous, but the converse may not be true in general. The next theorem states that the converse holds when Ω is a compact set.

Theorem 2.4 (Heine–Cantor) *Let $K \subset \mathbb{R}^n$ be a compact set. If $\varphi : K \to \mathbb{R}$ be a continuous function, then φ is uniformly continuous.*

We need some preparation to state the next theorem. Recall that a scalar function $\varphi = \varphi(x)$ defined on an open set $\Omega \subseteq \mathbb{R}^n$ is said to have a *global maximum* over Ω if there is some $a \in \Omega$ such that

$$\varphi(x) \le \varphi(a), \qquad \text{for all} \ \ x \in \Omega. \tag{2.1.15}$$

In this case, we say $\varphi(a)$ is the *maximum value* of the function φ over Ω. Similarly, a function $\varphi = \varphi(x)$ has a *global minimum* over Ω if there is some $b \in \Omega$ such that

$$\varphi(b) \le \varphi(x), \qquad \text{for all} \ \ x \in \Omega. \tag{2.1.16}$$

In this case, we say $\varphi(b)$ is the *minimum value* of the function φ over Ω. It should be clear that a function $\varphi : \Omega \to \mathbb{R}$ may have the same maximum value (or the minimum value) at more than one points in Ω, if there exists at least one. The next theorem was proved by *Bernard Bolzano* in 1830. However, the formulation given below is due to *Karl Weierstrass*.

Theorem 2.5 (Extreme Value Theorem) *Let $K \subset \mathbb{R}^n$ be a compact set, and $\varphi \in C(K)$. Then, for some $a, b \in K$, we have*

$$\varphi(a) = \inf \{\varphi(x) : x \in K\} = \inf \varphi(K);$$
$$\varphi(b) = \sup \{\varphi(x) : x \in K\} = \sup \varphi(K).$$

That is, over the set K, $\varphi(a)$ is the minimum value *and $\varphi(b)$ is the* maximum value, *of the function φ.*

More generally, a vector-valued function $f : \Omega \to \mathbb{R}^m$ is continuous at a point $a \in \Omega$ if each component function $f_i : \Omega \to \mathbb{R}$ of f is continuous at a, for $i = 1, \ldots, m$.

As before, we say f is continuous over Ω if all m functions f_i are continuous over Ω. An interesting class of continuous functions $f : \Omega \to \mathbb{R}^m$ is given by the linear transformations between linear spaces. We need some preparation with certain simple concepts of linear algebra. Let V, W be two linear spaces over $\mathbb{R}$. We say a function $T : V \to W$ is a *linear transformation* if we have

$$T(a u + b v) = a T(u) + b T(v), \quad \text{for any } u, v \in V \text{ and } a, b \in \mathbb{R}.$$

We may write $\mathscr{L}(V, W)$ for the set of all linear transformations $T : V \to W$. It is easy to see that $\mathscr{L}(V, W)$ is a linear space over $\mathbb{R}$ with respect to pointwise addition and scalar multiplication. In particular, $V^* = \mathscr{L}(V, \mathbb{R})$ is called the *dual space* of the space V. The *operator norm* $\|T\|$ of a linear transformation $T : \mathbb{R}^n \to \mathbb{R}^m$ is given by

$$\|T\| := \sup\left\{\|T(x)\| : x \in \mathbb{R}^n\right\} = \sup\left\{\|T(u)\| : u \in \mathbb{S}^{n-1}\right\}, \quad (2.1.17)$$

where $\mathbb{S}^{n-1} \subset \mathbb{R}^n$ is the *unit sphere*. We say T is a *bounded operator* if $\|T\| < \infty$. Notice that, if the standard basis for both $\mathbb{R}^n$ and $\mathbb{R}^m$ are written as $e_1, e_2, \ldots$, and we have

$$T(e_j) = \sum_{i=1}^{m} a_{ij} e_i, \quad \text{with } a_{ij} \in \mathbb{R}, \quad \text{for } 1 \leq i \leq m \text{ and } 1 \leq j \leq n,$$

it follows from the Cauchy–Schwartz inequality (2.1.3) that

$$\|T(x)\| \leq \|T\|\|x\|, \quad \text{for all } x \in \mathbb{R}^n, \quad \text{with } \|T\| = \sqrt{\sum_{i,j=1}^{m,n} a_{ij}^2}.$$

Definition 2.10 A vector function $f : \Omega \to \mathbb{R}^m$ is said to be *Lipschitz continuous* over an open set $\Omega \subseteq \mathbb{R}^n$ if, for some $M > 0$, we have

$$\|f(x) - f(y)\| \leq M \|x - y\|, \quad \text{for all } x, y \in \Omega.$$

The next theorem proves that every linear transformation $T : \mathbb{R}^n \to \mathbb{R}^m$ is Lipschitz continuous and hence uniformly continuous.

Theorem 2.6 *Every linear map $T : \mathbb{R}^n \to \mathbb{R}^m$ is Lipschitz continuous.*

Proof To prove the assertion, we write

$$\alpha := \max\left\{\|T(e_i)\| : 1 \leq i \leq n\right\}.$$

Now, since $x \in \mathbb{R}^n$ can be written uniquely as

$$x = \langle x, e_1 \rangle e_1 + \cdots + \langle x, e_n \rangle e_n,$$

it follows that
$$T(x) = \langle x, e_1 \rangle T(e_1) + \cdots + \langle x, e_n \rangle T(e_n),$$

and so we have
$$\|T(x)\| = \|\langle x, e_1 \rangle T(e_1) + \cdots + \langle x, e_n \rangle T(e_n)\|$$
$$\leq |\langle x, e_1 \rangle| \|T(e_1)\| + \cdots + |\langle x, e_n \rangle| \|T(e_n)\|$$
$$\leq \left(|\langle x, e_1 \rangle| + \cdots + |\langle x, e_n \rangle| \right) \alpha$$
$$= \alpha \|x\|_1,$$

where $\|x\|_1$ is given by

$$\|x\|_1 := \max \left\{ |\langle x, e_i \rangle| : 1 \leq i \leq n \right\}.$$

Since $\|x\|_1 \leq \|x\|$, the proof is complete.

In particular, it follows that the *coordinate function* $x^i : \mathbb{R}^n \to \mathbb{R}$ given by

$$x^i(x) := \langle x, e_i \rangle, \quad \text{for} \quad x \in \mathbb{R}^n, \tag{2.1.18}$$

are continuous functions, for $i = 1, \ldots, n$. A linear transformation $T : \mathbb{R}^n \to \mathbb{R}^n$ is called an **isometry** if $\|T(x)\| = \|x\|$, for all $x \in \mathbb{R}^n$. Clearly, every *isometry* $T : \mathbb{R}^n \to \mathbb{R}^m$ is a continuous function. In particular, as $T(0) = 0$, it follows that every *rotation* is a linear transformation. However, the *translation maps* $x \mapsto x + a :$ $\mathbb{R}^n \to \mathbb{R}^n$ by a nonzero vector $a \in \mathbb{R}^n$ is a *nonlinear* continuous function.

Remark 2.1 More generally, if V and W are two normed spaces, then a linear map $T : V \to W$ is *continuous* if and only if T is continuous at $0 \in V$. In fact, this statement holds even when the term *continuous* is replaced by some other terms such as *bounded, Lipschitz continuous, uniformly continuous*, and also *continuous at a point*. Notice that, if $\{b_1, \ldots, b_n\}$ is any *orthonormal basis* of an n-dimension inner product space V, then the *coordinate function* $x^i : V \to \mathbb{R}$ as given in (2.1.18) provide a canonical basis for the dual space V^*. Further, when W is an n-dimensional inner product space, Theorem 2.6 can be extended for $T \in \mathscr{L}(V, W)$.

We now introduce some notations. Let $\Omega \subseteq \mathbb{R}^n$ be an open set. We may write $C_m(\Omega)$ for the set of continuous functions $f : \Omega \to \mathbb{R}^m$, which is a linear space over $\mathbb{R}$ with respect to *pointwise* addition and scalar multiplication. Notice that the space $C_1(\Omega)$ is same the space $C(\Omega)$, which also contains all the n-variable polynomial functions of any finite *degree*. Therefore, in particular, the linear space $C_m(\Omega)$ is an infinite dimensional linear space, for $m \geq 1$.

Differentiability and Some Important Theorems

In actual applications, Theorem 2.5 applies only when a function $\varphi \in C(K)$ has certain nice properties such as given in terms of the derivatives of the function φ at a point in K. Let us start the discussion with the next definition. As before, $\Omega \subseteq \mathbb{R}^n$ is an open set.

Definition 2.11 For $1 \leq i \leq n$, the ith *partial derivative* $D_i \varphi(a)$ of a function $\varphi \colon \Omega \to \mathbb{R}$ at a point $a = (a_1, \ldots, a_n) \in \Omega$ is given by

$$D_i \varphi(a) := \lim_{t \to 0} \frac{\varphi(a_1, \ldots, a_i + t, \ldots, a_n) - \varphi(a_1, \ldots, a_n)}{t}, \qquad (2.1.19)$$

provided the limit exists. We also write $D_i \varphi(a)$ as $\partial \varphi / \partial x_i \big|_a$.

In terms of the standard basis $\{e_1, \ldots, e_n\}$ of $\mathbb{R}^n$, the condition (2.1.19) can be written as

$$D_i \varphi(a) = \lim_{t \to 0} \frac{\varphi(a + t\,u) - \varphi(a)}{t}, \qquad \text{where} \quad u = e_i. \qquad (2.1.20)$$

Let a function $\varphi_i : \mathbb{R} \to \mathbb{R}$ be given by

$$\varphi_i = \varphi(a + te_i), \qquad \text{for } t \in \mathbb{R}.$$

Then, the ith partial derivative $D_i \varphi(a)$ of the function φ exists at point $a \in \Omega$ if the (ordinary) derivative of the function φ_i exist at $t = 0$. In particular, it follows that all the usual rules of (ordinary) differentiation hold for the partial derivatives.

For $k \geq 2$, a kth order partial derivative of a function $\varphi : \Omega \to \mathbb{R}$ at a point $a \in \Omega$ is defined by applying (2.1.20) to the $(k-1)$th order partial derivatives of the function φ at point a. In particular, for $1 \leq i, j \leq n$, the *second order* partial derivatives of a function φ at point a is given by

$$D_j D_i \varphi(a) := D_{ij}\varphi(a) = \frac{\partial^2 \varphi}{\partial x_i \partial x_j}(a);$$

$$D_i D_j \varphi(a) := D_{ji}\varphi(a) = \frac{\partial^2 \varphi}{\partial x_j \partial x_i}(a).$$

It is easy to find examples to show that the *mixed partial derivatives* as defined above may not be the same in general. For example, consider the function $\varphi : \mathbb{R}^2 \to \mathbb{R}$ given by

$$\varphi(x) = \begin{cases} \dfrac{x^3 y - x y^3}{x^2 + y^2}, & \text{for } x = (x, y) \neq 0 \\ 0, & \text{for } x = 0 \end{cases}$$

The next theorem gives a sufficient condition for the equality of the second order mixed partial derivatives of a function φ at point a.

Theorem 2.7 *Let $\Omega \subset \mathbb{R}^n$ be an open set. Suppose $\varphi : \Omega \to \mathbb{R}$ has partial derivatives of order ≤ 2 and all these derivatives are continuous over $\overline{\Omega} = \Omega \bigcup \partial\Omega$. Then*

$$D_j D_i(\varphi) = D_i D_j(\varphi), \quad \text{for all} \ \ 1 \leq i, j \leq n.$$

Recall that $C(\Omega)$ is the linear space of all continuous functions $\varphi : \Omega \to \mathbb{R}$. More generally, for $k \geq 1$, we may write $C^k(\Omega)$ for the space of functions $\varphi : \Omega \to \mathbb{R}$ such that all the kth order partial derivatives exist and belong to the space $C^{k-1}(\Omega)$. In this case, we say that φ is a C^k-*smooth function*. A function $\varphi \in C^1(\Omega)$ is also known as a *continuously differentiable* function. A function $\varphi : \Omega \to \mathbb{R}$ is called a C^∞-smooth (or simply a *smooth function*) if

$$\varphi \in C^\infty(\Omega) = \bigcap_{k \geq 1} C^k(\Omega).$$

In view of inductive nature of the above definition, we have

$$C(\Omega) \supset C^1(\Omega) \supset C^2(\Omega) \supset C^3(\Omega) \supset \cdots,$$

where all the inclusions are *proper*. For example, the function f given by

$$f(x) = \begin{cases} 0, & \text{for } x \leq 0 \\ x, & \text{for } x > 0 \end{cases}$$

lies in the collection $C(\mathbb{R}) \setminus C^1(\mathbb{R})$, and the function g given by

$$g(x) = \begin{cases} 1+, & \text{for } x \leq 0 \\ 1 + x^2, & \text{for } x > 0 \end{cases}$$

lies in the collection $C^1(\mathbb{R}) \setminus C^2(\mathbb{R})$ (see Fig. 2.1). We may write the partial derivatives of a function $\varphi = \varphi(x, y, \dots)$ as

$$\varphi_x = \partial_x \varphi = \frac{\partial \varphi}{\partial x}, \quad \varphi_y = \partial_y \varphi = \frac{\partial \varphi}{\partial y}, \quad \varphi_{xy} = \partial_{xy} \varphi = \frac{\partial^2 \varphi}{\partial x \partial y}, \quad \text{etc.}$$

A natural generalisation of the partial derivatives of a function $\varphi = \varphi(x)$ at a point $a \in \Omega$ is the concept of the *directional derivatives* of φ at a in *any given direction* $v \in \mathbb{R}^n$. It is obtained from the relation (2.1.20) by using the vector v in place of the standard unit vector e_i.

Definition 2.12 The *directional derivative* of a function $\varphi : \Omega \to \mathbb{R}$ at a point $a \in \mathbb{R}^n$ in the direction of a vector $v = (v_1, \dots, v_n) \in \mathbb{R}^n$, denoted by $D_v \varphi(a)$, is given by

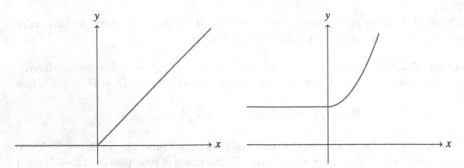

Fig. 2.1 Inclusions $C^2(\mathbb{R}) \subset C^1(\mathbb{R}) \subset C(\mathbb{R})$ are proper

$$D_v\varphi(a) := \lim_{\lambda \to 0} \frac{\varphi(a + \lambda v) - \varphi(a)}{\lambda}$$

$$= \sum_{i=1}^{n} v_i \frac{\partial \varphi}{\partial x_i}(a), \qquad (2.1.21)$$

if the limit exists.

For any function $\varphi \in C(\Omega)$ representing some scalar quantity, and any vector $v \in \Omega$, the directional derivatives $D_v\varphi$ gives the *variations in values* of the function φ in the direction v. For example, let $\Omega \subset \mathbb{R}^3$ represents a region of a city, and suppose a function $\varphi \in C(\Omega)$ is such that $\varphi(x)$ represents the steady-state temperature at $x \in \Omega$. Then $D_v\varphi(x)$ gives the *temperature variations* at the location x along the vector v, which is depicted by changing colour hues on the surface given by the graph Γ_φ of the function φ.

Definition 2.13 Let $\Omega \subseteq \mathbb{R}^n$ be a *domain*, and $x \in \Omega$. A *scalar field* defined over the region Ω is any function $\varphi \in C(\Omega)$ such that each $\varphi(x) \in \mathbb{R}$ represents a scalar quantity. A *vector field* defined over the region Ω is any function $\varphi \in C_n(\Omega)$ such that each $f(x) \in \mathbb{R}^n$ represents a vector quantity.

Example 2.5 In actual applications, a scalar field φ at a spatial position $x \in \Omega$ may represent physical quantities such as the mass density, pressure distribution of a fluid, temperature across a body, electric potential in electrostatic field, and gravitational force in scalar theory of gravitation. On the other hand, a vector field f at a spatial position $x \in \Omega$ may represent physical quantities such as the velocity of a fluid flow or the field of gravitational/electromagnetic forces.

More generally, a sufficiently smooth function F is said to define a **field** over a region $\Omega \subseteq \mathbb{R}^n$ if each pair $(x, F(x))$ represents a physical quantity at the *spatial position* $x \in \Omega$. In terms of the physical interpretation, as illustrated above, the directional derivatives $D_v F$ of a field F in the direction v is a fundamental tool to study many different types of problems in science and engineering. Notice that, in actual practice, we need high-end machines to *interpolate* a huge amount of real-time data to approximates a field F over a region Ω.

Next, let F be a *field* defined over a region $\Omega \subseteq \mathbb{R}^n$. We discuss the fundamental concept of the *differential* of F at point $a \in \Omega$. To start with, suppose $\Omega = I$ is an open subinterval of $\mathbb{R}$. In terms of Carathéodory formulation, a function $f : I \to \mathbb{R}$ is differentiable at a point $a \in I$, with the derivative $f'(a)$, if the function g given by

$$g(x) = \begin{cases} \dfrac{f(x) - f(a)}{x - a}, & \text{for } x \neq a \\ f'(a), & \text{for } x = a \end{cases},$$

is continuous over I such that the following relation holds:

$$f(x) - f(a) = g(x)(x - a), \qquad \text{for all } x \in I. \tag{2.1.22}$$

The above statement is known as the *Carathéodory theorem* when the derivative $f'(a)$ exists in usual *limit argument* sense. The function g as above is called the *increment function* of f. As $f'(a) = g(a)$, and

$$y - f(a) = m\,(x - a)$$

is the equation of the line passing through the point $(a, f(a))$, with *slope m*, it follows that a function f is differentiable at point $a \in I$ if and only if $m = f'(a) = g(a)$.

More generally, according to Carathéodory, a function $\varphi : \Omega \to \mathbb{R}$ is said to be *differentiable* at a point $a \in \Omega$ if there exists a continuous function $G : \Omega \to \mathscr{L}(\mathbb{R}^n, \mathbb{R})$ such that

$$\varphi(x) - \varphi(a) = G(x)(x - a), \qquad \text{for all } x \in \Omega. \tag{2.1.23}$$

We write $\lim_{x \to a} G(x) = D\varphi(a)$, which is called the *derivative* of the function φ at the point a. Recall that we can view $\mathscr{L}(\mathbb{R}^n, \mathbb{R})$ as the space $\mathbb{R}^{n+1}$. The above formulation of the concept of differentiability of a scalar function at a point is more appropriate while dealing with Banach spaces. However, in this text, we shall use the formulation due to Frechét, as given in the next definition.

Definition 2.14 Let $\Omega \subseteq \mathbb{R}^n$ be an open set, and consider a function $\varphi : \Omega \to \mathbb{R}$. We say φ is **differentiable** at a point $a \in \Omega$ if there exists $L_a \in \mathscr{L}(\mathbb{R}^n, \mathbb{R})$ such that

$$\lim_{h \to 0} \frac{\varphi(a + h) - \varphi(a) - L_a(h)}{\|h\|} = 0. \tag{2.1.24}$$

which is called the **differential** of φ at point a. We write $L_a = D\varphi(a)$ for the *derivative* of a function φ at a point $a \in \Omega$. We say φ is differentiable over Ω if it is differentiable at each point $x \in \Omega$, and we write $D(\Omega)$ for the linear space of functions $\varphi : \Omega \to \mathbb{R}$ differentiable over Ω.

Example 2.6 For $b \in \mathbb{R}^n$ and $\alpha \in \mathbb{R}$, the *affine map* $\varphi_b : \mathbb{R}^n \to \mathbb{R}$ given by

$$\varphi_b(x) = \langle x, b \rangle + \alpha, \quad x \in \mathbb{R}^n,$$

is differentiable, with $L_a = b$. In particular, every linear map $\varphi : \mathbb{R}^n \to \mathbb{R}$ is differentiable, with $L_a = \varphi$, for all $a \in \mathbb{R}^n$. Also, the function $\varphi : \Omega \to \mathbb{R}$ given by $\varphi(x) = \langle x, x \rangle = \|x\|^2$ is differentiable, with $L_a = 2a$.

With notations as in Definition 2.14, the *translation* $\Omega - a$ of the open set $\Omega \subset \mathbb{R}^n$ is an open set that contains the origin $0 \in \mathbb{R}^n$. Therefore, it makes sense to take the limit $h \to 0$ within $\Omega - a$. More precisely, by using openness of $\Omega - a \subset \mathbb{R}^n$, we can find $\delta > 0$ such that $B(a; \delta) \subseteq \Omega$. Hence, in topological terms, we can reformulate the condition (2.1.24) as

$$h \in B(0; \delta) \quad \Rightarrow \quad \varphi(a + h) - \varphi(a) = L_a(h) + e(h), \qquad (2.1.25)$$

where the *error function* $e : B(0; \delta) \to \mathbb{R}$ satisfies the property

$$\|h\| \to 0 \quad \Rightarrow \quad e(h) \to 0. \qquad (2.1.26)$$

In most applications, it is safe to assume that a function $\varphi \in D(\Omega)$ is defined and continuous on the boundary surface $\Gamma = \partial \Omega$ of a domain Ω. The *Leibniz rule* as stated below holds for any two functions $\varphi, \phi \in D(\Omega)$:

$$D\langle \varphi, \phi \rangle = \langle D\varphi, \phi \rangle + \langle \varphi, D\phi \rangle, \qquad (2.1.27)$$

where the inner product $\langle \varphi, \phi \rangle$ is defined in some appropriate way. Moreover, the three important statements as given in the next theorem follow directly from the condition (2.1.25). By using the mapping $x \mapsto L_x : \Omega \to \mathscr{L}(\mathbb{R}^n, \mathbb{R})$, it also proves that the equivalence of the two formulations due to Carathéodory and Frechét.

Theorem 2.8 *Let $\Omega \subset \mathbb{R}^n$ be an open set, and $a \in \Omega$. Suppose a function $\varphi : \Omega \to \mathbb{R}$ is differentiable at a, with differential $L_a = D\varphi(a)$. Then we have*

1. *$L_a : \mathbb{R}^n \to \mathbb{R}$ is unique.*
2. *φ is continuous, i.e. $C(\Omega) \supset D(\Omega)$.*
3. *$D_v f(a)$ exists for every $v \in \mathbb{R}^n$, and we have*

$$D_v\varphi(a) = L_a(v) = \sum_{i=1}^{n} v_i \frac{\partial \varphi}{\partial x_i}(a). \qquad (2.1.28)$$

Definition 2.15 Let $\Omega \subset \mathbb{R}^n$ be open set, $a \in \Omega$, and $\varphi : \Omega \to \mathbb{R}$ be a function differentiable at point $a \in \Omega$. For $n > 1$, the *tangent map* $\tau_\varphi = \tau_\varphi(u)$ of the function φ at point a is given by

$$\tau_\varphi(u) = D_u\varphi(a), \quad \text{for} \quad u \in \mathbb{S}^{n-1},$$

where $\mathbb{S}^{n-1} = \{v \in \mathbb{R}^n : \|v\| = 1\}$.

Notice that, by (2.1.28), $\tau_\varphi(u) = L_u\varphi$, for all $u \in \mathbb{S}^{n-1}$. Furthermore, if the *direction v* as used in Definition 2.12 is an arbitrary vector, and $0 \neq \alpha \in \mathbb{R}$, then the relation

$$D_{\alpha u}\varphi(a) = \lim_{t \to 0} \frac{\varphi(a + (t\alpha)u) - \varphi(a)}{t} = \alpha\, D_u\varphi(a)$$

implies that the directional derivative of φ exists in the direction of the vector $\alpha\, u$. Therefore, in geometrical terms, we say that the collection K of all directions $v \in \mathbb{R}^n$ such that $D_v\varphi(a)$ exists forms a **cone** with vertex at the point a. Hence, the tangent map τ_φ is defined over a cone $K \subset \mathbb{R}^n$, and the *graph surface* of τ_φ given by

$$\Gamma_{\tau_a} = \big\{(u, \alpha) \in K \times \mathbb{R} : D_u\varphi(a) = \alpha\big\}$$

is a *plane* if and only if τ_φ is a linear map. Notice that, however, this later condition may not hold in general (Exercise 2.27).

Remark 2.2 The significance of the above discussion becomes clearer while using the *method of characteristics* to solve a first order quasilinear partial differential equation given by

$$f(x, y, z)\partial_x\varphi + g(x, y, z)\partial_y\varphi = h(x, y, z),$$

where $z = \varphi(x, y) \in C^1(\Omega)$ is an unknown function, and the coefficients f, g, h are some sufficiently smooth functions defined over a region $\Omega \times \mathbb{R}$, with $\Omega \subseteq \mathbb{R}^2$. We shall study such types of equations in Chap. 6.

Definition 2.16 Let $\Omega \subseteq \mathbb{R}^n$ be an open set, and $a \in \Omega$. A vector function $f : \Omega \to \mathbb{R}^m$ is said to be *differentiable* at point a if each component function $f_j : \Omega \to \mathbb{R}$ of f is differentiable at a, for $j = 1, \dots, m$. As before, f is differentiable over Ω if all the m functions f_i are differentiable over Ω. The derivative of a vector function f at a point $a \in \Omega$ is written as $Df(a)$.

To continue with the discussion, we also need to use a *matrix formulation* of the differential $L_a : \mathbb{R}^n \to \mathbb{R}^m$ at a, where f is a vector function differentiable at a point $a \in \Omega$. In general, let $L : \mathbb{R}^n \to \mathbb{R}^m$ be a linear map with competent functions $\ell_j : \mathbb{R}^n \to \mathbb{R}$, for $j = 1, \dots, m$. Then, by *Riesz theorem*, there is a vector $v_j = (a_{j1}, \dots, a_{jn}) \in \mathbb{R}^n$ such that we can write $\ell_j(x) = \langle v_j, x \rangle$, for any $x \in \mathbb{R}^n$ and for each $1 \leq j \leq m$. Notice that, by taking

$$a_{ij} := \langle L(e_j), e_i \rangle, \quad \text{for } 1 \leq i \leq m \text{ and } 1 \leq j \leq n,$$

where e_j's and e_i's are, respectively, the standard basis of $\mathbb{R}^n$ and $\mathbb{R}^m$, and $x = x_1 e_1 + \cdots + x_n e_n \in \mathbb{R}^n$, we have

$$
\begin{aligned}
L(x) = L\Big(\sum_{j=1}^{n} x_j e_j \Big) &= \sum_{j=1}^{n} x_j L(e_j) \\
&= \sum_{j=1}^{n} x_j \Big(\sum_{i=1}^{m} \langle L(e_j), e_i \rangle e_i \Big) \\
&= \sum_{i=1}^{m} \Big(\sum_{j=1}^{n} x_j \langle L(e_j), e_i \rangle \Big) e_i \\
&= \sum_{i=1}^{m} \Big(\sum_{j=1}^{n} a_{ij} x_j \Big) e_i \\
&= \Big(\sum_{j=1}^{n} a_{1j} x_j \Big) e_1 + \cdots + \Big(\sum_{j=1}^{n} a_{mj} x_j \Big) e_m \\
&= \ell_1(x) e_1 + \cdots + \ell_m(x) e_m.
\end{aligned}
$$

Therefore, for any column vector $x \in \mathbb{R}^n$, we have the relation $L(x) = Mx$, where M is the $m \times n$ matrix with v_i as the ith row. We usually write $M := M_L$. Conversely, every $m \times n$ matrix A over $\mathbb{R}$ defines a linear map $L_A : \mathbb{R}^n \to \mathbb{R}^m$ given by $L_A(x) = Ax$, where $x \in \mathbb{R}^n$ is a column vector. We write $\mathcal{M}_{m,n}(\mathbb{R})$ for the space of all $m \times n$ matrices over the field $\mathbb{R}$.

Definition 2.17 [Matrix Form] Let $\Omega \subseteq \mathbb{R}^n$ be an open set, and $a \in \Omega$. A function $f : \Omega \to \mathbb{R}^m$ is differentiable at a if there exists a matrix $B := B_a \in \mathcal{M}_{m,n}(\mathbb{R})$ such that

$$
\lim_{h \to 0} \frac{f(a+h) - f(a) - Bh}{\|h\|} = 0. \tag{2.1.29}
$$

In this case, we say the matrix B is the derivative of f at a, and we may write $Df(a) = B_a$.

Notice that the product Bh as given in (2.1.29) makes sense only when $h \in \mathbb{R}^n$ is a column vector. Also, if $f_j : \Omega \to \mathbb{R}$ are the m component functions of f, then it follows from the relation (2.1.28) that, for $1 \le i \le n$ and $1 \le j \le m$, we have

$$
L_a(e_i) = D_i f_j(a) = \frac{\partial f_j}{\partial x_i}\Big|_a \quad \Rightarrow \quad Df_j(a) = \big(D_1 f_j(a), \ldots, D_n f_j(a) \big)
$$

where $D_i f_j(a)$ is as usual the ith partial derivative of the function $f_j : \Omega \to \mathbb{R}$ at a. Therefore, for a function $f = (f_1, \ldots, f_m)^t$ that is differentiable at a, we can write the linear map $Df_j(a) : \mathbb{R}^n \to \mathbb{R}$ uniquely as

$$
Df_j(a)(x) = \langle v_j, x \rangle, \quad \text{for any} \ \ x \in \mathbb{R}^m,
$$

where the vector v_j is given by

$$v_j = \left(D_1 f_j(a), \ldots, D_n f_j(a)\right), \quad \text{for} \ \ 1 \leq j \leq m. \tag{2.1.30}$$

Definition 2.18 Let $\Omega \subseteq \mathbb{R}^n$ be an open set, $a \in \Omega$, and $\varphi : \Omega \to \mathbb{R}$ be a function differentiable at a. Then the vector

$$\text{grad } \varphi(a) := \left(D_1 \varphi(a), \ldots, D_n \varphi(a)\right), \tag{2.1.31}$$

is called the **gradient** of the function φ at the point a. We may also write grad $\varphi(a) := \nabla \varphi(a)$, where

$$\nabla := \left(\frac{\partial}{\partial x_1}, \ldots, \frac{\partial}{\partial x_n}\right) \tag{2.1.32}$$

is called the n-dimensional *del operator* in n variables $x = (x_1, \ldots, x_n)$.

Let $\Omega \subseteq \mathbb{R}^n$ be an open set. A function $\varphi \in C^2(\Omega)$ is said to be **harmonic** if φ satisfies the Laplace equation $\nabla^2 \varphi = 0$ over Ω. That is,

$$\frac{\partial^2 \varphi}{\partial x_1^2} + \cdots + \frac{\partial^2 \varphi}{\partial x_n^2} = 0, \quad \text{for all} \ \ x \in \Omega.$$

It follows by direct computation that the functions given below are harmonic:

$$x^2 - y^2, \quad e^x \cos y, \quad \sin x \cosh y, \quad \text{and} \quad \ln(x^2 + y^2),$$

Notice that the first three functions are harmonic over $\mathbb{R}^2$, whereas the fourth is a *harmonic function* over the region $\mathbb{R}^2 \setminus \{0\}$. In general, as we shall see, such types of functions play a significant role in solving problems concerning vector fields of the form $f = \nabla \varphi$, for some $\varphi \in C^2(\Omega)$. The related phenomena are governed by the *potential theory*.

Since we can write each $f_j \in D(\Omega)$ uniquely in terms of the partial derivatives over Ω, it follows from the next theorem that the matrix B_a of the differential $L_a(f)$: $\mathbb{R}^n \to \mathbb{R}^m$ with respect to standard basis is given by the $m \times n$ matrix $\left(\partial f_j / \partial x_i\right)_a$, which is called the **Jacobi matrix** of the function f at the point a, and is denoted by $J_f(a)$. In particular, when $m = n$, the determinant det $\left(J_f(a)\right)$ is called the **Jacobian** of a function f at the point a.

Theorem 2.9 *Let $\Omega \subset \mathbb{R}^n$ be an open set, and $f : \Omega \to \mathbb{R}^m$ be a function, with $f = (f_1, \ldots, f_m)$, which is differentiable at a point $a \in \Omega$. Then, for each $1 \leq i \leq n$ and $1 \leq j \leq m$, the partial derivatives $D_i f_j(a)$ exist. Also, matrix $J_f(a)$ of the differential $L_a = Df(a)$ at a with respect to standard basis is given by*

$$J_f(a) = \big(Df_1(a),\, Df_2(a),\, \dots,\, Df_m(a)\big)^t$$

$$= \begin{pmatrix} D_1 f_1(a) & D_2 f_1(a) & \cdots & D_n f_1(a) \\ D_1 f_2(a) & D_2 f_2(a) & \cdots & D_n f_2(a) \\ \vdots & \vdots & \vdots & \vdots \\ D_1 f_m(a) & D_2 f_m(a) & \cdots & D_n f_m(a) \end{pmatrix}.$$

Let $\Omega \subseteq \mathbb{R}^n$ be an open set. The concept of a *local maximum* (or a *local minimum*) of a function $\varphi : \Omega \to \mathbb{R}$ at a point $a \in \Omega$ is defined by restricting the condition (2.1.15) (or (2.1.16)) to a ball $B(a; r) \subset \Omega$, for some $r > 0$. It should be clear that the *isolated* local maxima (or a local minima) are always strict, i.e., we have $\varphi(x) < \varphi(a)$ (or $\varphi(x) > \varphi(a)$) if $x \neq a$ in the ball $B(a; r)$. In the latter case, we say $\varphi(a)$ is a *strict local maxima* (or a *strict local minima*). We add here a comment on the problem of finding a *local maxima* or a *local minima* of a differentiable function $\varphi : \Omega \to \mathbb{R}$. Let us start with the next definition.

Definition 2.19 Let $\Omega \subseteq \mathbb{R}^n$ be an open set, and $\varphi : \Omega \to \mathbb{R}$ be a function differentiable at a point $a \in \Omega$. We say a is a *critical point* (or a *stationary point*) of φ if $\nabla\varphi(a)$ is not defined or it is zero. In this case, $\varphi(a)$ is called a *critical value* of the function φ. A point $a \in \Omega$ is said to be a *regular point* of a function φ if $\nabla\varphi(a) \neq 0$.

For a function φ differentiable at a point a, with $D\varphi(a)$ defined, a critical point a is the point of *local maxima* or a *local minima* or a *saddle point*[1] The nature of a critical point a of a function $\varphi \in C^2(\Omega)$ is analysed by using the eigenvalues of the symmetric matrix of order n given by *Critical point!of a scalar field*

$$H_\varphi(a) := \begin{pmatrix} \frac{\partial^2 \varphi}{\partial x_1^2}(a) & \frac{\partial^2 \varphi}{\partial x_1 \partial x_2}(a) & \cdots & \frac{\partial^2 \varphi}{\partial x_1 \partial x_n}(a) \\ \frac{\partial^2 \varphi}{\partial x_2 \partial x_1}(a) & \frac{\partial^2 \varphi}{\partial x_2^2}(a) & \cdots & \frac{\partial^2 \varphi}{\partial x_2 \partial x_n}(a) \\ \vdots & \vdots & \vdots & \vdots \\ \frac{\partial^2 \varphi}{\partial x_n \partial x_1}(a) & \frac{\partial^2 \varphi}{\partial x_n \partial x_2}(a) & \cdots & \frac{\partial^2 \varphi}{\partial x_n^2}(a) \end{pmatrix} \tag{2.1.33}$$

which is known as the *Hessian matrix* of the function φ at point a. The concept was developed by *Ludwig Hesse* (1811–1874), who used the terminology *functional determinants*. Notice that the Hessian matrix H_φ is *symmetric* because the mixed partial derivatives of a C^2-function φ are always equal.

More generally, in terms of the *Jacobi matrix* of a function $f : \Omega \to \mathbb{R}^m$, a point $a \in \Omega$ is a critical point if the differential $L_a f : \mathbb{R}^n \to \mathbb{R}^m$ is not defined or else the matrix $J_f(a)$ is not of the *full rank*. In particular, for the case when $m = n$, $a \in \Omega$ is a *critical point* of a vector field $f : \Omega \to \mathbb{R}^n$ if the *Jacobian* $\det\big(J_f(a)\big) = 0$. Or, equivalently, the matrix $J_f(a)$ is singular. In this latter situation, $f(a)$ is called a *critical value* of the function f. For a function $\varphi \in C^2(\Omega)$, we have

[1] A critical point a of a function φ is called a **saddle point** if $\varphi(a)$ is a local maxima and a local maxima along two different paths passing through the point a.

$$H_\varphi(x) = J_{\nabla\varphi}(x), \qquad \text{for all } \; x \in \Omega.$$

For any vector $v = (v_1, \ldots, v_n)$, we have

$$D_v\varphi(a) = \sum_{i=1}^{n} v_i D_i\varphi(a) \qquad \text{and} \qquad D_v^2\varphi(a) = v\, H_\varphi(a)\, v^t.$$

Theorem 2.10 *Let Ω be an open set, and $\varphi \in C^1(\Omega)$. If $a \in \Omega$ is a critical point of the function φ, then we have*

1. *φ has a strict local maxima at a if $D_v^2\varphi(a) < 0$, for all $v \neq 0$.*
2. *φ has a strict local maxima at a if $D_v^2\varphi(a) > 0$, for all $v \neq 0$.*
3. *φ has a saddle point at a if there exists two directions u and u such that $D_u^2\varphi(a) < 0 < D_v^2\varphi(a)$.*

The above theorem serves the purpose in most applications if basic concepts of linear algebra such as eigenvalues and eigenvectors are known and also the method of *Lagrange multipliers*. For further details, the reader is referred to the text ([1], Chap. 6).

To continue with the general discussion, notice that the converse of Theorem 2.7 may not hold in general. The next theorem states that the converse holds for continuously differentiable functions, i.e., we have the inclusion $C^1(\Omega) \subset D(\Omega)$, for any open set $\Omega \subset \mathbb{R}^n$.

Theorem 2.11 *Let $\Omega \subset \mathbb{R}^n$ be an open set, and $a \in \Omega$. If the partial derivatives of each component function of $f : \Omega \to \mathbb{R}^m$ exist, and are continuous in some open ball $B(a; \varepsilon) \subset \Omega$, then $Df(a)$ exists.*

The next theorem is used frequently in the sequel.

Theorem 2.12 (Chain Rule) *Let $U \subset \mathbb{R}^n$ is an open set, and a function $f : U \to \mathbb{R}^m$ be differentiable such that $V = f(U) \subset \mathbb{R}^m$ is an open set. If $g : V \to \mathbb{R}^p$ is a differentiable function, then $g \circ f : U \to \mathbb{R}^p$ is differentiable, and*

$$D(g \circ f)(a) = Dg(f(a))Df(a).$$

Further, if $f \in C^r(U)$ and $g \in C^2(V)$ then $g \circ f \in C^r(U)$.

The next theorem gives a formula to obtain the derivative of the inverse of a differentiable functions, provided it is known that the inverse too is differentiable.

Theorem 2.13 *Let $U \subset \mathbb{R}^n$ is an open set, and a function $f : U \to \mathbb{R}^n$ be differentiable at $a \in U$. Let V be an open set containing the point $b = f(a)$, and $g : V \to \mathbb{R}^n$ be such that $g(f(x)) = x$ for all x in some neighbourhood of a contained in U. If g is differentiable at b then $Dg(b) = [Df(a)]^{-1}$.*

Proof Since the derivative of the identity function is the identity matrix, and $g(f(x)) = x$, chain rule shows that $Dg(b)Df(a) = I$, which proves the assertion.

The *inverse function theorem*, as stated below, says the converse of Theorem 2.13 holds, but only *locally*. Recall that, for any $n \times n$ matrix A, the linear transformation $L_A : \mathbb{R}^n \to \mathbb{R}^n$ given by $L_A : x \mapsto Ax$ defines a *linear system* of equations

$$Ax = b, \quad \text{for any } b \in \mathbb{R}^n.$$

And, we know that this system has a unique solution if and only if A is *invertible*, i.e. $\det(A) \neq 0$. Theorem 2.14 states that a *nonlinear system* of the form $f(a) = b$ can be solved locally at a provided $J_f(a)$ is nonsingular. More precisely, it proves that the *local isomorphism* implies the *local diffeomorphism*. We say a bijective smooth function $f : U \to V$ between two open sets $U, V \subset \mathbb{R}^n$ is a *diffeomorphism* if the inverse $f^{-1} : V \to U$ is also a smooth function. Notice that, by Theorem 2.13, the Jacobi matrix $J_f(a)$ of every diffeomorphism $f : U \to V$ is nonsingular. That is, $Df : \mathbb{R}^n \to \mathbb{R}^n$ is an isomorphism.

Theorem 2.14 (Inverse Function Theorem) *Let $U, V \subset \mathbb{R}^n$ be open sets, and $f : U \to V$ be a C^r-function, for some $r \geq 1$, such that $J_f(a)$ is nonsingular, for some $a \in U$. Then, there exists an open neighbourhood $B_a \ni a$ in U such that $g = f|_{B_a} : B_a \to f(B_a)$ is a diffeomorphism, and $Dg(b) = [Df(a)]^{-1}$, with $b = f(a)$.*

Remark 2.3 The main purpose to study a *diffeomorphism* between two open sets in the sense of Theorem 2.14 is to find some *suitable coordinates* so that the geometry of the map f is simplified. More precisely, given a smooth function $f : U \to V$, we find diffeomorphisms $\Phi : U \to U_1$ and $\Psi : V \to V_1$, for some open sets $U_1, V_1 \subseteq \mathbb{R}^n$, such that the *transformation* given by the function

$$f_1 = \Psi \circ f \circ \Phi^{-1} : U_1 \to V_1$$

provides a simpler geometric description of the function f. For example, applying the diffeomorphisms $\tau_1 : x \mapsto x - a$ and $\tau_2 : x \mapsto x - f(a)$, we may take $U_1 = \tau_1(U)$ and $V_1 = \tau_2(V)$. In the new coordinates system, the point a is replaced by the origin $0 \in \mathbb{R}^m$ and f_1 satisfies the condition $f_1(0) = 0$.

The next theorem is equivalent to Theorem 2.14.

Theorem 2.15 (Implicit Function Theorem) *For $n = k + m$, let $\Omega \subset \mathbb{R}^n$ be an open sets, and $a = (a_1, a_2) \in \Omega$, where*

$$a_1 = (a_1, \ldots, a_k) \quad \text{and} \quad a_2 = (a_{k+1}, \ldots, a_n,).$$

Let $f : \Omega \to \mathbb{R}^m$ be a differentiable function, with $f(a_1, a_2) = b$, such that the Jacobi matrix $J_f(a)$ has the rank m. Then there exists some open set $U \subset \mathbb{R}^k$ containing the point a_1, a differentiable function $g : U \to \mathbb{R}^m$ such that $g(a_1) = a_2$, and an open set $V \subset \Omega$ containing a such that

$$f^{-1}(\{b\}) \bigcap V = \Gamma_g := \{(x, g(x)) : x \in U\}, \tag{2.1.34}$$

For a simpler description of the above theorem, suppose a function $f : \Omega \to \mathbb{R}^m$ is continuously differentiable over an open set $\Omega \subset \mathbb{R}^{k+m}$. Let $a_1 \in \mathbb{R}^k$ and $a_2 \in \mathbb{R}^m$, with $(a_1, a_2) \in \Omega$, are such that we have $f(a_1, a_2) = 0$. Now, if the Jacobi matrix $J_f(a_1, a_2)$ is invertible, i.e. the Jacobian $\det\left(J_f(a_1, a_2)\right) \neq 0$, for some small ball $B(a_1; \varepsilon) \subset \mathbb{R}^k$ and a continuous function $g : B(a_1; \varepsilon) \to \mathbb{R}^m$, we have $g(a_1) = a_2$ and $f(x, g(x)) = 0$, for every $x \in B(a_1; \varepsilon)$.

A study of the *local growth* (or rate of change) of a smooth function $f : \Omega \to \mathbb{R}^m$ is important for many applications in science and engineering. In general, there are mainly two types of *mean value theorems*, where each gives an estimate about by how much the global growth (or the rate) *is bounded by* the local growth (or the rate), in some sense. We give here statements of the mean value theorems *for derivatives*,[2] i.e. in terms of *rate of change*. We start with the next theorem, for the case when $n = 1$, that suggests *how far* the classical mean value theorem holds for the vector-valued function $\gamma : I \to \mathbb{R}^m$, where $I \subseteq \mathbb{R}$ is an interval.

Theorem 2.16 *If $\gamma : [a, b] \to \mathbb{R}^m$ is continuous on $[a, b]$ and differentiable on (a, b), then there is a $c \in (a, b)$ such that*

$$\|\gamma(b) - \gamma(a)\| \leq (b - a)\|\gamma'(c)\|.$$

The following version holds for a function $f : \Omega \to \mathbb{R}$.

Theorem 2.17 (Mean Value Theorem) *Let $\Omega \subset \mathbb{R}^n$ be an open set, containing the line segment $L := \{(1 - t)a + tb : 0 \leq t \leq 1\}$ joining the points $a, b \in \Omega$, and $\varphi \in C^1(\Omega)$. Then there exists an interior point c of the line L such that*

$$\varphi(b) - \varphi(a) = \nabla\varphi(c) \cdot (b - a).$$

Proof The proof follows easily by applying single variable mean value theorem, and the chain rule, to the function $f : \mathbb{R} \to \mathbb{R}$ given by $f(t) := \varphi[(1 - t)a + tb]$, for $t \in [0, 1]$.

In the general case, the inequality given in the next theorem holds.

Theorem 2.18 (Mean Value Inequality) *Suppose $\Omega \subset \mathbb{R}^n$ is an open set, and let $f : \Omega \to \mathbb{R}^m$ be a differentiable function. Let $a, b \in \Omega$ be such that the line segment $L := \{(1 - t)a + tb : 0 \leq t \leq 1\}$ lies inside Ω. Then, there exists an interior point x_0 of the line L such that*

$$\|f(a) - f(b)\| \leq \|Df(x_0)\|\|a - b\|.$$

The inequality given in Theorem 2.18 is used to study *locally* some reasonably well-behaved functions $\varphi : \Omega \to \mathbb{R}^m$. As before, by applying a suitable translation, we only have to focus our study at the point $a = 0$.

---

[2] The mean value theorems *for integrals* (in terms of *averages*) are discussed in the later part of the text.

For the proofs of theorems stated above, the reader is referred to the companion volume [2] or any other standard text such as [3].

2.2 Classical Theory of Surfaces and Curves

Let $\Omega \subseteq \mathbb{R}^n$ be a domain, and consider a C^1-function $\varphi : \Omega \to \mathbb{R}$, with $\nabla\varphi \neq 0$ over Ω. We may write Γ_φ for the *graph* of φ as given by

$$\Gamma_\varphi = \left\{ \left(x, \varphi(x)\right) \in \Omega \times \varphi(\Omega) : x \in \Omega \right\} \subset \mathbb{R}^{n+1}.$$

In the particular case, when $\Omega = \mathbb{R}^n$ and $\varphi : \mathbb{R}^n \to \mathbb{R}$ is a linear map, the graph $\Gamma_\varphi \subset \mathbb{R}^{n+1}$ is a linear subspace of dimension n with a basis given by the vectors

$$\left(e_1, \varphi(e_1)\right), \ \ldots, \left(e_n, \varphi(e_n)\right),$$

where $e_1, \ldots, e_n$ is the standard basis of the space $\mathbb{R}^n$. In the general case, when $n = 1, 2$, a function φ can be visualised easily in terms of its graph. For example, if D is a domain in $\mathbb{R}^2$ and $f \in C^1(D)$, then the geometry of the *graph surface* Γ_f can be identified as a *family of curves* obtained as the intersection of Γ_f with planes parallel to coordinates planes. A more interesting situation corresponds to the case when such types of curves are *sections* of Γ_f formed by using the planes $z = c$, for $c \in f(D)$. We call these as the *level curves* of surface Γ_f, provided $\nabla f(c) \neq 0$.

Example 2.7 For the function $f_1 : \mathbb{R}^2 \to \mathbb{R}$ given by

$$f_1(x, y) = 4x^2 + 9y^2, \quad \text{with} \quad (x, y) \in \mathbb{R}^2 \setminus \{(0, 0)\},$$

the level curves are *ellipses* given by the equation $4x^2 + 9y^2 = f_1(a, b)$, for some $(a, b) \neq (0, 0)$. Similarly, for the function $f_2(x, y) = xy$, $(x, y) \in \mathbb{R}^2 \setminus \{(0, 0)\}$, the level curves are *hyperbolas* of the type $xy = f_2(a, b)$, for some $(a, b) \neq (0, 0)$. Also, for the function $f_3(x, y) = x^2 + y^2 - 1$, $(x, y) \in \mathbb{R}^2 \setminus \{(0, 0)\}$, the level curves are *circles* given by $x^2 + y^2 - 1 = f_3(a, b)$, for some $(a, b) \neq (0, 0)$.

When $\Omega \subseteq \mathbb{R}^3$ is a domain and $F \in C^1(\Omega)$, with $\nabla F \neq 0$ over Ω, a *projected surface curve* given by $\Gamma_F \bigcap \mathbb{R}^3$ is called a *contour map* if the function F remains constant along the curve. In general, *contour map* in lower dimensions is obtained by keeping fixed one of the independent variables, which provides a visually intuitive way to see the *level surfaces* of the graph Γ_F. For example, taking $x = a$, the contour map is the intersection of the graph Γ_F with the plane $x = a$. A *level surface* of the function F is the contour map obtained from the intersection of Γ_F with the plane $z = c$. In what follows, we reserve the term level set for the surface obtained from Γ_φ by slicing it with the plane $x_{n+1} = \varphi(x_0)$, for some $x_0 \in \Omega$.

Definition 2.20 Let $\varphi \in C^1(\Omega)$, where $\Omega \subseteq \mathbb{R}^n$ be a domain. For any point $c \in \varphi(\Omega)$ such that $\nabla\varphi(c) \neq 0$, the set $\varphi^{-1}(c)$ given by

$$\varphi^{-1}(c) := \{x \in \Omega : \varphi(x) = c\}$$

is called the *level set* of the function φ, at the *level c*.

In the above situation, we also say that $\varphi(x) = c$ define a single-parameter *family of surfaces* such that through each point on Γ_φ passes a particular member of the family, for any given c. The *gradient condition* $\nabla\varphi(c) \neq 0$ ensures that each level set $\varphi^{-1}(c)$ indeed represents a *surface* in $\mathbb{R}^{n+1}$. As said above, the level sets help understand the geometry of the graph surface Γ_φ.

Let D be a plain domain, and $(x_0, y_0) \in D$. Suppose a function $f \in C^1(D)$ satisfies the gradient condition at point $(x_0, y_0) \in D$. Then, for $F(x, y, z) = f(x, y) - z$, the level curve passing through the point $(x_0, y_0, z_0) \in \Gamma_f$ is given by the equation

$$F(x, y, z) = F(x_0, y_0, z_0) = 0, \quad \text{for any } (x, y, z) \in \Omega_2 \times \mathbb{R},$$

Further, if $\Omega \subseteq \mathbb{R}^3$ is a domain and $F \in C^1(\Omega_3)$ is such that the gradient $\nabla F(x) \neq 0$ at a point $x_0 = (x_0, y_0, z_0) \in \Omega$, then the equation

$$F(x) = F(x_0), \quad \text{for } x \in \Omega$$

is the level surface of the function F passing through the point x_0, which we may write as Γ_{x_0}. Notice that Γ_{x_0} may not be the *graph surface* of any function $z = f(x, y) \in C^1(D)$. Also, since each point on the graph surface Γ_F lies on exactly one level surface $F^{-1}(c)$, for some $c \in F(\Omega)$, and

$$x, y \in \Gamma_F \bigcap F^{-1}(c) \iff F(x) = F(y) = c,$$

we can view the graph Γ_F as *lamination* of level surfaces $F^{-1}(c)$. Therefore, $F(x) = c$ represents a 1-parameter family of surfaces on Γ_F.

Remark 2.4 In practical applications, if a scalar field $\varphi \in C^1(\Omega)$ represents the *potential energy* of some force field $f : \Omega \to \mathbb{R}^3$, then the level surfaces represent the layers of constant potential, which are also known as the *equipotential surfaces*. In particular, if φ represents the temperature distribution on the boundary surface $\partial\Omega_3$ of a (bounded) domain Ω, then $F(x, y, z) = c$ represents the *isothermal surfaces*. The concepts such as the above play a significant role while interpreting a solution of a *differential equation model* of a phenomenon.

Example 2.8 For the function $F(x, y, z) = 4x^2 + y^2 - z$, we have $\nabla F = (8x, 2y, -1)$, and so we may take $\Omega = \mathbb{R}^3$. For any $x_0 \in \mathbb{R}^3$ such that $F(x) = 0$ the level surface $z = f(x, y) = 4x^2 + y^2$ is the bowl-shaped *elliptical paraboloid*, with bottom at the origin. For $z = 0$, it has a degenerate level curve. Next, for the function φ given by

$$\varphi(x, y, z) = \sqrt{x^2 + y^2 + z^2}, \qquad \text{for } x^2 + y^2 + z^2 > 0,$$

the level surface $\varphi(x, y, z) = c$ is the sphere of radius r centred at the origin. In this case, the level curves are the *equators*, i.e. the circles of latitude. Also, for the function

$$\phi(x, y, z) = x^2 + y^2 - z^2, \qquad \text{for } (x, y, z) \in \mathbb{R}^3 \setminus \{(0, 0, 0)\},$$

three particularly important level surfaces are a given below: A *two-sheeted hyperboloid* for $\phi(x, y, z) = -1$; a *double cone* for $\phi(x, y, z) = 0$; and, a *single-sheeted hyperboloid* for $\phi(x, y, z) = 1$.

By Theorem 2.20, normal to the surface Γ_{x_0} is given by

$$\mathbf{n}(x_0) := \nabla F \Big|_{x_0} = \big(F_x(x_0), F_y(x_0), F_z(x_0)\big).$$

Therefore, the tangent plane T_{x_0} to the surface Γ_{x_0} at point x_0 is given by

$$T_{x_0} = \big\{(x, y, z) \in \mathbb{R}^3 : (x - x_0) F_x(x_0) + (y - y_0) F_y(x_0) + (z - z_0) F_z(x_0) = 0\big\}.$$

More generally, if $f_1, \ldots, f_m : \Omega \to \mathbb{R}$ are the components of a vector function $f : \Omega \to \mathbb{R}^m$, then we may write $f = \sum_{j=1}^m f_j e_j$, where e_j's are the standard basis of $\mathbb{R}^m$. In this case, the graph surface Γ_f of the function f is a subspace of $\mathbb{R}^{n+m}$. For a regular point $c = (c_1, \ldots, c_m) \in \mathbb{R}^m$ of f, as

$$f^{-1}(\{c\}) = \bigcap_{j=1}^m \{x \in \Omega : f_j(x) = c_j\} = \bigcap_{j=1}^m f_j^{-1}(\{c\}),$$

it follows that the surface Γ_f is given in terms of the level sets of the functions $f_j : \Omega \to \mathbb{R}$, for $j = 1, \ldots, m$.

Example 2.9 Let $f : \mathbb{R}^2 \to \mathbb{R}^2$ be a function with two components given by functions

$$f_1(x, y, z) = x^2 + y^2 + z^1 - 1 \qquad \text{and} \qquad f_2(x, y, z) = 4(x^2 + y^2) - 1.$$

Clearly, $f_1^{-1}(0)$ is the sphere $\mathbb{S}^2$ and $C = f_2^{-1}(0)$ is the right circular cylinder about the z-axis with the guiding circle $x^2 + y^2 = (0.5)^2$. Therefore, the level surface $f^{-1}((0, 0))$ is given by the intersection of the sphere $\mathbb{S}^2$ and the cylinder C. Notice that, in this case, the equations $f_1 = f_2 = 0$ can be solved easily, and we find that $f^{-1}((0, 0))$ consists of two circles on $\mathbb{S}^2$.

Tangent Space of Level Sets

Let $f : \Omega \to \mathbb{R}^m$ and $g : \Omega \times \mathbb{R}^m \to \mathbb{R}^m$ be given by

$$g(x, y) = f(x) - y, \quad \text{for } x \in \Omega \text{ and } y \in \mathbb{R}^m.$$

Clearly, for $0 \in \mathbb{R}^m$, we have

$$(x, y) \in g^{-1}(\{0\}) \quad \Longleftrightarrow \quad (x, y) \in \Gamma_f.$$

Therefore, the graph surface Γ_f of every function $f : \Omega \to \mathbb{R}^m$ is a level set. However, the converse may not hold in general. We may apply the *Implicit Function Theorem* to show that each (regular) level surface is the graph surface of some function, at least *locally*. The main argument uses the geometry of the differential $L_a : \mathbb{R}^n \to \mathbb{R}^m$, considering the fact that near the point a we can write

$$f(x) \approx f(a) + L_a(x - a),$$

provided the Jacobi matrix $J_f(a)$ has the *full rank*. Notice that, in view of the condition (2.1.26), the associated *approximation error* is negligible in comparison to the distance $\|x - a\|$. Taking $b = f(a)$, we have

$$
\begin{aligned}
f^{-1}(b) &= \{x : f(x) = b\} \\
&\approx \{x : f(a) + L_a(x - a) = b\} \\
&= \{x : L_a(x - a) = 0\} \\
&= \{x : \langle \nabla f_j(a), x - a \rangle = 0, \text{ for } 1 \le j \le m\}
\end{aligned}
$$

Hence, the best *linear approximation* to the level set $f^{-1}(b)$ near a is given by

$$
\begin{aligned}
T_f(a) &:= \{x \in \mathbb{R}^n : L_a(x - a) = 0\} \\
&= a + \{x \in \mathbb{R}^n : L_a(x) = 0\},
\end{aligned}
\tag{2.2.1}
$$

which is called the *tangent space* to the level set $f^{-1}(b)$ near the point a. A vector $x = (x_1, \ldots, x_n) \in \mathbb{R}^n$ satisfies the equation $L_a(x) = 0$ if and only if $(x_1, \ldots, x_n)$ is the solution of the system of homogeneous linear equations given by

$$
\frac{\partial f_1}{\partial x_1}\bigg|_a x_1 + \cdots + \frac{\partial f_1}{\partial x_n}\bigg|_a x_n = 0
$$

$$
\vdots
\tag{2.2.2}
$$

$$
\frac{\partial f_m}{\partial x_1}\bigg|_a x_1 + \cdots + \frac{\partial f_m}{\partial x_n}\bigg|_a x_n = 0
$$

The coefficients matrix of the above system is the Jacobi matrix $J_f(a)$, which is known to have the *rank m* if and only if the dimension of the tangent space $T_f(a)$ is given by $n - m$. Therefore, by choosing m linearly independent columns, the remaining $n - m$ columns correspond to a complete set of independent variables that define a function having the graph surface $T_f(a)$ at point a. Further, the equality

$$L_a(x) = \langle \nabla f_j(a), x \rangle$$

implies that the *tangent space* $T_f(a)$ *shifted to the point* a consists of vectors that are orthogonal to the gradient $\nabla f_j(a)$. Accordingly, the *normal space* to the level set $f^{-1}(b)$ near a is given by

$$N_f(a) := a + \left\{ \sum_{j=1}^{m} a_j \nabla f_j(a) : a_j \in \mathbb{R} \right\}. \tag{2.2.3}$$

The fundamental idea of identifying an *infinitesimal curved part* around a point on a regular surface with the tangent space at the point, as described above, facilitates applications of algebraic tools to study the geometry of a smooth surfaces embedded in the Euclidean space $\mathbb{R}^{n+m}$. In particular, by using Theorem 10.2.2, it is possible to study the geometry of the graph surface Γ_φ of a function $\varphi \in D(\Omega)$ in terms of the (shifted) tangent space $T_\varphi(a)$ of the surface Γ_φ at point $(a, \varphi(a))$, which is defined by the *differential* $L_a : \mathbb{R}^n \to \mathbb{R}$, with $\det(L_a) \neq 0$.

Example 2.10 Let $\Omega \subseteq \mathbb{R}^2$ be a domain, and Γ_φ be the graph surface of a function $\varphi : \Omega \to \mathbb{R}$, with $\nabla \varphi \not\equiv 0$ over Ω. Then, the first order Taylor approximation of Γ_φ at a point $a = (x_0, y_0) \in \Omega$ is given by

$$D\varphi(x, y) = L_a((x, y))$$
$$= \varphi(a) + \left(\partial_x(a) \ \partial_y(a) \right) \begin{pmatrix} x - x_0 \\ y - y_0 \end{pmatrix}$$
$$= \varphi(a) + (x - x_0)\partial_x(a) + (y - y_0)\partial_y(a).$$

In general, for $F(x, y, z) = \varphi(x, y) - z$ and $x_0 = (x_0, y_0) \in \Omega$, the equation of the tangent plane to the surface Γ_F at the point (x_0, y_0, z_0) (with $z_0 = \varphi(x_0)$) is given by

$$(x - x_0)\partial_x F(x_0) + (y - y_0)\partial_y F(x_0) + (z - z_0)\partial_z F(x_0) = 0.$$

Remark 2.5 By using algebraic approach to the geometry, as suggested by the above discussion, it is possible to linearise a nonlinear problem involving a function $\varphi \in D(\Omega)$ locally. The tangent space $T_\varphi(a)$ is generated by the tangent vectors to the curves obtained from the surface Γ_φ taking its sections with the orthogonal planes (see (2.3.4)). In the special case, when the planes are parallel to the coordinate planes, the partial derivatives of φ constitute generators for the tangent space (see also Definition 2.15).

The parametric approach to basic geometric objects such as curves and surfaces helps in general to use *coordinate free* geometric arguments. In rest of the section, we review some basic concept related to *parametrisation* of curves and surfaces.

Parametrised Curves

Let I be a subinterval of $\mathbb{R}$. An n-dimensional *parametric function* (or simply a *parametrisation*) is a vector function $\gamma : I \to \mathbb{R}^n$ given by

$$\gamma(t) = \big(\gamma_1(t), \ldots, \gamma_n(t)\big), \quad \text{for } t \in I,$$

where $\gamma_i \in C(I)$, for $i = 1, \ldots, n$, and the independent variable t is called a *parameter*. We say a parametrisation γ is *sufficiently smooth* if each $\gamma_i \in C^1(I)$, for $i = 1, \ldots, n$. The interval I in any particular situation is determined by considering the intervals where the n component functions γ_i are well-defined. The next example illustrates the point.

Example 2.11 If the component functions γ_i of a function $\gamma : I \to \mathbb{R}^3$ are given by

$$\gamma_1(t) = a\cos t, \quad \gamma_2(t) = a\sin t, \quad \gamma_3(t) = bt,$$

then we may take a *parameter* $t \in I = \mathbb{R}$ because each $\gamma_i(t)$ is defined for every $t \in \mathbb{R}$. Therefore, we can write the parametrisation γ as

$$\gamma(t) = \big(a\cos t, a\sin t, bt\big), \quad \text{for } t \in \mathbb{R}, \quad \text{and} \quad a, b > 0.$$

However, to ensure the validity of the component function γ_1 of the parametrisation γ given by $\gamma(t) = (3\tan t, 4\sec t, 5t)$, we must take *parameter* t so that

$$t \in \mathbb{R} \setminus \big\{(2n+1)\pi/2 : n \in \mathbb{Z}\big\}.$$

The most common approach to describe a curve in $\mathbb{R}^3$ is by using a *parametric representation*. We start with the next definition.

Definition 2.21 The trace (or trajectory) of a parametrisation $\gamma = \gamma(t) : I \to \mathbb{R}^n$ given by the (connected) set

$$C_\gamma := \big\{x \in \mathbb{R}^n : x = \gamma(t), \text{ for some } t \in I\big\}$$

is called a *parametrised curve* in $\mathbb{R}^n$. We say C_γ is *simple curve* if it has no self-intersections; i.e., γ is an injective function. Further, C_γ is said to be a *closed curve* if $\gamma(a) = \gamma(b)$, for $I = [a, b]$.

Clearly, for any continuous function $y = f(t) : I \to \mathbb{R}$, the graph Γ_f of f given by

$$\Gamma_f = \big\{(t, f(t)) \in \mathbb{R}^2 : t \in I\big\}$$

is a *simple curve*, parametrised naturally by the independent variable t. Notice that no vertical line intersects the curve Γ_f in more than one point. However, in general, this latter property does not hold simply because not every curve in $\mathbb{R}^n$ is necessarily the graph of some function.

Example 2.12 The parametrised curve $C_\gamma \subset \mathbb{R}^3$ traced by the point

$$\gamma(t) = (a\cos t,\ a\sin t,\ bt), \qquad \text{for}\ \ t \in \mathbb{R},$$

is a *double helix* on a right circular cylinder of base radius a, which winds around the z-axis at the rate b/a, with rise per revolution; i.e., the *pitch* given by $2\pi b$ and the *total height* attained is given by $\ell\sqrt{a^2 + b^2}$, for $t \in [0, \ell]$. In this case, the parameter t measures the angle between the x-axis and the line joining the origin to the projection of the point $\gamma(t)$ onto the xy-plane.

In general, to visualise a parametrised curve $C_\gamma \subset \mathbb{R}^3$, we may use the geometry of its *projections* on any of the three coordinate planes $x = 0$, $y = 0$, or $z = 0$. In particular, we find it more interesting to consider the parametrised curve in $\mathbb{R}^2$ obtained from the first two coordinate functions of $\gamma : I \to \mathbb{R}^3$.

Example 2.13 For the parametrised curve $C_\gamma \subset \mathbb{R}^3$ given by

$$\gamma(t) = (4\cos t,\ 3\sin t,\ 6t), \qquad \text{for}\ \ t \in \mathbb{R},$$

the first two coordinate functions given by $\gamma_1(t) = 4\cos t$ and $\gamma_2(t) = 3\sin t$ satisfy the equation of an ellipse as given below:

$$\frac{x^2(t)}{4^2} + \frac{y^2(t)}{3^2} = 1, \qquad \text{for}\ \ t \in \mathbb{R}.$$

Therefore, it follows that the locus of the point $\gamma(t)$ is an *elliptical helix*. By similar argument, for a parametrisation $\gamma : \mathbb{R} \to \mathbb{R}^3$ given by

$$\gamma(t) = (at\cos t,\ at\sin t,\ bt), \qquad \text{for}\ \ t \in \mathbb{R},$$

the coordinate functions of parametrised curve C_γ satisfy the equation of a cone given by $b^2(x^2 + y^2) = a^2 z^2$. Therefore, in this case, C_γ is a *circular conic helix*. On the other hand, for a parametrisation $\gamma : \mathbb{R} \to \mathbb{R}$ given by

$$\gamma(t) = (5\cos t,\ 3\sin t,\ 4\sin t), \qquad \text{for}\ \ t \in \mathbb{R},$$

the parametrised curve C_γ *lies on sphere* because $\|\gamma(t)\| = 5$, for all $t \in \mathbb{R}$.

Definition 2.22 Let $\Omega \subset \mathbb{R}^n$ be an open set, and $n \geq 1$. A 1-dimensional connected open set $C \subset \Omega$ is called a *parametrised curve* if it is the *trajectory* of some sufficiently smooth parametrisation $\gamma : I \to \mathbb{R}^n$. In this case, we write $C = C_\gamma$, and call C a *smooth curve*.

Example 2.14 The *Viviani curve* C obtained as the intersection of the sphere $x^2 + y^2 + z^2 = a^2$ with the cylinder $x^2 + y^2 = ax$ has a parametrisation given by the differentiable function

$$\gamma(\theta) := \left(a\cos^2\theta,\ a\cos\theta\sin\theta,\ a\sin\theta\right), \qquad \theta \in [0, 2\pi].$$

In general, a parametric curve C_γ can be parametrised in infinitely many ways, where the parameter t can be changed by using any differentiable function $\theta : I \to \mathbb{R}$, with $\theta'(t) > 0$ for all t. For, taking $J = \theta(I)$, the differentiable function given by $\gamma \circ \theta^{-1} : J \to \mathbb{R}^n$ gives another parametrisation, which is called a *re-parametrisation* of the curve C_γ. Notice that a re-parametrisation do not change the *shape* of the parametric curve C_γ. It only changes the *speed* of the particle $\gamma(t)$ as it moves along the curve (see Exercise 2.29).

Example 2.15 Every parametrised curve C_γ such as given in Example 2.13 and Example 2.14 can be re-parametrised by using any one of the following three functions:

$$t \mapsto t^2; \qquad t \mapsto at + b; \qquad \text{and } t \mapsto e^t, \text{ for } t > 0.$$

We discuss next some useful geometric concepts associated with a smooth parametrised curve $C_\gamma \subset \mathbb{R}^n$, with $\gamma \in C^1([a, b])$. Let $\delta\gamma$ represents an infinitesimal change in the position of a point $\gamma(t)$ along the curve C_γ corresponding to an instantaneous change in (time) variable from t_0 to $t_0 + \delta t$. Then the *tangent vector* $\gamma'(t)$ to the curve C_γ at a point $P = \gamma(t_0)$ is given by

$$\begin{aligned}
\gamma'(t) &= \frac{d\gamma(t)}{dt}\Big|_{t=t_0} \\
&= \lim_{\delta t \to 0} \frac{\gamma(t_0 + \delta t) - \gamma(t_0)}{\delta t} \\
&= \left(\gamma_1'(t_0), \ldots, \gamma_n'(t_0)\right),
\end{aligned} \qquad (2.2.4)$$

provided the limit exists.

Definition 2.23 A parametrised curve $C_\gamma \subset \mathbb{R}^n$ is called a *regular curve* if $\gamma : I \to \mathbb{R}^n$ is a continuously differentiable function such that $\gamma'(t) \neq 0$, for any $t \in I$.

In general, the n components $\gamma_i'(t)$ of the *velocity vector* $v(t) = \gamma'(t)$ at a point $P = P(t)$ in the direction of a vector $\mathbf{u} \in \mathbb{R}^n$ are given by

$$\gamma'(t)\Big|_P \cdot \frac{\mathbf{u}}{|\mathbf{u}|}\Big|_P.$$

For example, the tangent vector $\gamma'(t)$ of a parametrised curve C_γ traced by the function

$$\gamma(t) = \left(t, t - \sin t, 1 - \cos t\right), \qquad \text{for } t \in [0, 1],$$

is a unit vector at the origin.

Definition 2.24 The *arc length* $s(t)$ of a parametrised curve C_γ, with $\gamma \in C^1(I)$, from $\gamma(t_0)$ to the point $\gamma(t)$ is given by

$$s(t) = \int_{t_0}^{t} \|\gamma'(t)\| \, dt = \int_{t_0}^{t_1} \sqrt{[\gamma_1']^2 + \cdots + [\gamma_n']^2} \, dt. \qquad (2.2.5)$$

Recall that one *radian* on the unit circle refers to the arc length of unit length. It follows using *chain rule* that the arc length of a parametrised curve C_γ is independent of parametrisation γ. Further, by the *Fundamental Theorem*, we have

$$s'(t) = \|\gamma'(t)\|, \qquad \text{for all} \quad t \in [a, b]. \qquad (2.2.6)$$

In particular, the arc length s of the graph of a function $f \in C^1([a, b])$ is given by

$$s = \int_{a}^{b} \|f'(x)\| \, dx = \int_{a}^{b} \sqrt{1 + [f'(x)]^2} \, dx. \qquad (2.2.7)$$

A regular C_γ parametrised by the arc length s is a *unit-speed curve* because the point $\gamma(s)$ moves along the curve with unit speed. In general, if C_γ is a regular curve so that we have $v(t) = \gamma'(t) \neq 0$, for any $t \in (a, b]$. To obtain the arc length re-parametrisation of C_γ, we need to produce a *formula* $t = t(s)$. For, notice that we can invert the arc length function $s = s(t)$ *locally*, by using the regularity condition and the *Inverse Function Theorem*. We have thus proven the next theorem.

Theorem 2.19 *Every regular curve can be re-parametrisation by its arc length function.*

Example 2.16 The trajectory of the function $\gamma : \mathbb{R} \to \mathbb{R}$ given by

$$\gamma(t) = \left(a \cos t, a \sin t, bt \right), \qquad \text{for} \quad t \in \mathbb{R}.$$

is a *circular helix* C_γ. In view of Exercise 2.28, the function given by

$$\varphi(s) = \left(a \cos \frac{s}{\sqrt{a^2 + b^2}}, a \sin \frac{s}{\sqrt{a^2 + b^2}}, \frac{bs}{\sqrt{a^2 + b^2}} \right),$$

gives the arc length re-parametrisation of C_γ, which is obtained by using the substitution $t = s/\sqrt{a^2 + b^2}$. Since $\|\varphi'(s)\| = 1$, for all s, the curve C_φ is a *unit-speed curve*.

As shown in Example 2.9, curves in $\mathbb{R}^3$ can also be obtained as the intersection of two nondegenerate surfaces. Let $\varphi, \psi \in C^1(\Omega_3)$ be such that none of the two vector fields $\nabla\varphi$ and $\nabla\psi$ vanishing over the domain Ω_3. Further, assume that

$$\nabla\varphi \times \nabla\psi \neq 0, \qquad \text{at every point in} \quad \Omega_3. \qquad (2.2.8)$$

In this case, we say functions φ, ψ are *functionally independent*. Notice that, geometrically, condition (2.2.8) means at no point in Ω_3 the vectors $\nabla\varphi$ and $\nabla\psi$ are parallel. Then, for each pair of constants (c_1, c_2), the intersection of the level surfaces given by

$$\varphi(x, y, z) = c_1 \quad \text{and} \quad \psi(x, y, z) = c_2, \qquad (2.2.9)$$

if nonempty, gives a curve. Since we have

$$\nabla\varphi \times \nabla\psi = \Big(J[\varphi, \psi; y, z], \ J[\varphi, \psi; z, x], \ J[\varphi, \psi; x, y] \Big), \qquad (2.2.10)$$

where the Jacobian $J[\varphi, \psi; z, y]$ of functions φ and ψ with respect to variables z and y is given by

$$J[\varphi, \psi; y, z] := \frac{\partial(\varphi, \psi)}{\partial(y, z)} = \det\begin{pmatrix} \varphi_y & \varphi_z \\ \psi_y & \psi_z \end{pmatrix}, \qquad (2.2.11)$$

and the Jacobians $J[\varphi, \psi; z, x]$ and $J[\varphi, \psi; x, y]$ are defined similarly, it follows that at least one of these Jacobians is nonzero. Therefore, by implicit function theorem, it is possible to obtain locally a parametrised curve near the point x_0, where one of the independent variables x, y, z could be used as a parameter. Hence, we obtain a 2-parameter family of curves, for different values of the pair (c_1, c_2).

Formation of 2-*parameter family of curves* by using the procedure as described above is used later in Chap. 5 to find the general solution of a first order quasilinear differential equation by using Lagrange's method. We will also need the concept of the *first integral* of a vector field.

Definition 2.25 A scalar function $\varphi \in C^1(\Omega)$, with $\Omega \subseteq \mathbb{R}^n$ a domain, is called a *first integral* of a C^1 vector field $f : \Omega \to \mathbb{R}^n$ if $f \cdot \nabla\varphi = 0$, over Ω.

The next lemma proves that *integral curves* of a vector field f over a domain $\Omega \subseteq \mathbb{R}^3$ are given by two functionally independent first integrals of the field f. Recall that a curve C contained in Ω is an *integral curve* of a field $f : \Omega \to \mathbb{R}^3$ if the tangent to the curve C at each point is the vector $f(x)$, for $x \in \Omega$. In practical terms, integral curves of a vector field represent the *lines of force* (also known as the *field lines*) such as lines of flow of a velocity field of a fluid.

Lemma 2.1 *Let $f = (f_1, f_2, f_3)$ be a vector field defined over a domain $\Omega \subseteq \mathbb{R}^3$, and functionally independent functions φ, $\psi \in C^1(\Omega)$ be first integrals of the field f. Then the integral curves of f over Ω are given by the level surfaces such as (2.2.9), for some appropriate values $(c_1, c_2) \in \mathbb{R}^2$. However, not necessarily uniquely.*

Proof As functions φ, ψ satisfy the condition (2.2.8), and we have

$$f \cdot \nabla\varphi = 0 \quad \text{and} \quad f \cdot \nabla\psi = 0, \quad \text{over} \quad \Omega, \qquad (2.2.12)$$

each member $C(c_1, c_2)$ of 2-parameter family of curves represented by two relations given in (2.2.9) is an integral curve of the field f. On the other hand, if C_f be an integral curve of the field f, then we must show that $C_f = C(c_1, c_2)$, for some suitable values of (c_1, c_2). For, let $x_0 = (x_0, y_0, z_0)$ be a point on C, and consider a curve C' obtained from (2.2.9), with

$$c_1 = \varphi((x_0, y_0, z_0)) \quad \text{and} \quad c_2 = \varphi((x_0, y_0, z_0)).$$

Therefore, both C and C' are integral curves of field f, passing through the point $((x_0, y_0, z_0))$. Hence, as we shall prove in Chap. 4, it follows that $C = C'$ (Theorem 3.19).

It follows from Lemma 2.1 that the collection of all first integrals of a vector field f over a domain in $\mathbb{R}^3$ is a 2-parameter family of curves. The next lemma proves that it is possible to form infinitely many first integrals of a vector field f from any given first integral of f.

Lemma 2.2 *If $\varphi \in C^1(\Omega \times \mathbb{R})$ is a first integral of the vector field $f = (a, b, c)$, then $\psi = G(\varphi)$ is also a first integral of f, for any single-variable C^1-function defined on a suitable subinterval of $\mathbb{R}$. Furthermore, if $\varphi, \psi \in C^1(\Omega \times \mathbb{R})$ are first integrals of f, then $G(\varphi, \psi) = 0$ is also a first integral of f, for any two-variable C^1-function defined on a suitable domain in $\mathbb{R}^2$.*

Proof Differentiating $\varphi = G(\psi)$ with respect to x and y, we have

$$\varphi_x + \varphi_z\, p = g'(\psi)(\psi_x + \psi_z\, p);$$
$$\varphi_y + \varphi_z\, q = g'(\psi)(\psi_y + \psi_z\, q).$$

Eliminating the function $g'(\psi)$, we obtain

$$\det \begin{pmatrix} \varphi_x + \varphi_z\, p & \psi_x + \psi_z\, p \\ \varphi_y + \varphi_z\, q & \psi_y + \psi_z\, q \end{pmatrix} = 0.$$

Therefore, the conclusion follows from the condition $f \cdot \nabla \psi = 0$. A proof of the second assertion is a part of the proof of Theorem 6.3.

Parametrised Surfaces

We recall some basic concepts related to *regular surfaces*. Informally, an $(n + 1)$-dimensional *regular surfaces* is the trace of a scalar-valued function $\varphi \in C^2(\Omega)$, where $\Omega \subset \mathbb{R}^n$ is an open (connected) set. In particular, the concept of a *parametrised surface* in $\mathbb{R}^3$ is a 2-dimensional analogue of parametrised curves.

Definition 2.26 Let I and J be any types of intervals, and take $\Omega = I \times J$. For a continuous function $\mathbf{r} : I \times J \to \mathbb{R}^3$, let

$$S_{\mathbf{r}} = \left\{ x \in \mathbb{R}^3 : x = \mathbf{r}(u, v), \text{ for some } (u, v) \in I \times J \right\}.$$

Then $S_{\mathbf{r}}$ is called a *parametrised surface* in $\mathbb{R}^3$, with $(u, v) \in I \times J$ being the *parameters*, and $\mathbf{r}$ a parametrisation. We say a parametrised surface $S_{\mathbf{r}}$ is *regular* if the parametrisation $\mathbf{r}$ is a continuously differentiable function.

For $(u, v) \in I \times J$, the three *coordinate functions* $x_i : I \times J \to \mathbb{R}$ for a point $\mathbf{r}(u, v) \in S_{\mathbf{r}}$ can be written as

$$x_1 = x_1(u, v), \quad x_2 = x_2(u, v), \quad x_3 = x_3(u, v).$$

Clearly, the idea introduced above can be extended to define a two-parameter *hypersurface* in $\mathbb{R}^{n+2}$. In particular, the graph Γ_φ of any (differentiable) scalar-valued function $z = \varphi(x, y)$ given on a region $\Omega \subseteq \mathbb{R}^2$ gives a two-parameters (regular) surface, where $(x, y) \in \Omega$ are the *natural parameters* for the parametrised surface Γ_φ. In this case, we may write

$$\Gamma_\varphi(x, y, z) : \quad \varphi(x, y) - z = 0.$$

Similarly, a differentiable scalar-valued function $w = \varphi(x, y, z)$ gives 3-dimensional regular surface $\Gamma_\varphi \subset \mathbb{R}^4$ with respect to the natural parameters (x, y, z). In general, a *regular surface* in $\mathbb{R}^3$ is specified by the conditions as specified in the next definition.

Definition 2.27 A connected set $S \subset \mathbb{R}^3$ is called a *piecewise regular surface* if for each point $P \in S$ there is an open ball $V_P := B(P; \varepsilon) \subset \mathbb{R}^3$ and a continuously differentiable injective function $f : V_P \to \mathbb{R}^3$ that maps the open set $W_P := S \cap V_P$ onto an open set $U \subset \mathbb{R}^2$ (taken as a uv-plane) in $\mathbb{R}^3$. If $\mathbf{r} = \left(f\big|_{W_P} \right)^{-1}$ then $\mathbf{r}(U) = W_P$, and the function $\mathbf{r} \in C_3^1(U)$ is given by

$$\mathbf{r}(u, v) = \big(x(u, v), \, y(u, v), \, z(u, v) \big),$$

which is called a *regular local representation* of the surface S. We also say S is a parametrised surface, and $\mathbf{r}$ is called a *parametrisation* of S. A surface S is said to be of class C^r if the coordinate function

$$x = x(u, v), \quad y = y(u, v), \quad z = z(u, v), \quad \text{for } (u, v) \in U,$$

of $\mathbf{r}$ are so. And, S called a *regular surface*, if $\mathbf{r}_u \times \mathbf{r}_v \neq 0$ or, equivalently, rank of the Jacobi matrix $J_{\mathbf{r}}$ is everywhere two.

Some simple surfaces $S \subset \mathbb{R}^3$ can be parametrised by a single function $\varphi \in C^1(U)$, where $U \subset \mathbb{R}^2$ is an open set.

Example 2.17 We consider first a plane $\Gamma_a \subset \mathbb{R}^3$ passing through a point $a = (a_1, a_2, a_3)$, and parallel to vectors $\mathbf{p} = (p_1, p_2, p_3)$ and $\mathbf{q} = (q_1, q_2, q_3)$. We can parametrise Γ_a by a function $\varphi : \mathbb{R}^2 \to \Gamma_a$ given by

$$\varphi(u, v) = a + u\mathbf{p} + v\mathbf{q}$$
$$= (a_1 + up_1 + vq_1, a_3 + up_2 + vq_2, a_3 + up_3 + vq_3),$$

for $(u, v) \in \mathbb{R}^2$. And, the sphere $\mathbb{S}^2 \subset \mathbb{R}^3$ at the origin of radius $r > 0$ can be parametrised in terms of *spherical coordinates* (u, v), where the coordination functions of a parametrisation $\mathbf{r}(u, v)$ are given by

$$x(u, v) = r \sin v \cos u, \quad y(u, v) = r \sin v \sin u, \quad z(u, v) = r \cos v.$$

Notice that $u \in [0, 2\pi)$ is the called the *longitudinal angle*, and $v \in [0, \pi]$ is called the *latitudinal angle*. Next, recall that a *helicoid* is a (ruled) surface like a spiral staircase in a building, described by a straight line that rotates at a constant angular rate around a fixed axis, and intersects the axis at a constant angle, say θ. For $(u, v) \in \mathbb{R}^2$, the coordination functions of a parametrisation $\mathbf{r}(u, v)$ of the helicoid are given by

$$x(u, v) = u \cos(v\theta), \quad y(u, v) = u \sin(v\theta), \quad z(u, v) = v.$$

Theorem 2.20 *Let $\Omega \subseteq \mathbb{R}^n$ be a domain, and $\varphi \in C^1(\Omega)$. A normal to the graph surface given by*

$$\Gamma_\varphi := \{(x, \varphi(x)) \in \mathbb{R}^{n+1} : x \in \Omega\},$$

at any point $a \in \Omega$, is given by the gradient $\nabla\varphi(a)$.

Proof Suppose $\gamma : I \to \mathbb{R}^n$ is a parametric curve passing through the point a that lies on the level set $\varphi \equiv c$, where $c \in \mathbb{R}$. Now, since a point on C_γ is given by

$$\gamma(t) = (x_1(t), x_2(t), \dots, x_n(t)), \quad t \in I,$$

we have $\varphi(\gamma(t)) = c$. Therefore, it follows that

$$\frac{d}{dt}\varphi(\gamma(t)) = 0 \quad \Rightarrow \quad \nabla\varphi \cdot \gamma'(t) = 0,$$

by the chain rule. Hence, the gradient $\nabla\varphi(a)$ of φ at a point $a \in \Omega$ is orthogonal to the tangent vector $\gamma'(t)$, provided it is nonzero at a. Notice that this latter condition is no restriction because, by the differentiability of φ, it can be assumed that C_γ is a *unit-speed* curve so that $\gamma' \not\equiv 0$. That is, the vector $\nabla\varphi(a)$ is normal to the surface Γ_φ at a.

In general, if the angle between the gradient vector $\nabla\varphi(x)$ at a point $x \in \Omega$ and a vector $\mathbf{v}$ is given by θ, then we have

$$D_\mathbf{v}\varphi = \nabla\varphi \cdot \frac{\mathbf{v}}{|\mathbf{v}|} = |\nabla\varphi| \cos\theta. \tag{2.2.13}$$

Therefore, when $\cos\theta = 1$, i.e. when $\theta = 0$, it follows that the directional derivative $D_v\varphi$ attains its *maximum value* given by $|\nabla\varphi|$. Said differently, $D_v\varphi$ attains the *maximum* in the direction v given by the vector $\nabla\varphi(x)$. This simple observation, together with Theorem 2.20, explains the importance of the concept of *normal derivative* of a scalar-valued continuously differentiable function φ given on a region $\Omega \subseteq \mathbb{R}^n$, as given in the next definition.

Definition 2.28 Let $x \in \Omega$. The *normal derivative* of φ, denoted by $\partial_n\varphi = \partial\varphi/\partial\mathbf{n}$, is the directional derivative $D_n\varphi(x)$ taken in the direction of the (outward) normal $\mathbf{n}$ to the surface Γ_φ at the point x.

The normal derivative $\partial_n\varphi$ at a point $\big(x, \varphi(x)\big) \in \Gamma_\varphi$ is also given by the relation

$$\partial_n\varphi = D_{\nabla\varphi}\varphi = \nabla\varphi \cdot \mathbf{n} = |\nabla\varphi|, \quad \text{where} \quad \mathbf{n} = \frac{\nabla\varphi}{|\nabla\varphi|},$$

which is the *maximum value* of the gradient $\nabla\varphi$. In physical terms, we say $\partial_n\varphi$ is the rate of change in values of the quantity $\varphi(x)$ as we move *orthogonally* across the boundary $\partial\Omega$ of a surface Ω.

2.3 Vector Calculus

The development of vector analysis is primarily due to English mathematician *Oliver Heaviside* (1850–1925), and independently by American mathematician *Josiah Gibbs* (1839–1903). Heaviside published his work in 1893 as part of the book "*The Elements of Vectorial Algebra and Analysis*", whereas Gibbs' work first appeared as the book "*Elements of Vector Analysis*", published in 1901 as the compilation of the his lectures delivered in 1881 at Yale University. Heaviside applied vector analysis tools to reformulate the twelve of twenty equations related to *electromagnetic radiations* in *vector form*, which were originally proposed by Scottish mathematician and scientist *James Maxwell* (1831–1879) during 1861–62. The twelve equations are recognised in modern physics as the Maxwell's four fundamental equations (see Appendix A.2 for details). Most notations and terminology introduced in this section are due to Gibbs.

In this section, we discuss the three fundamental theorems due to Gauss, Stokes, and Helmholtz that are applied in the next two chapters to derive *differential equation models* for some important practical problems related to physical phenomena such as *fluid flow, heat conduction, mechanical vibrations, and electromagnetic waves*. In all that follows, the 3-dimensional *del operator* as introduced earlier plays the lead role. We shall use shorthand operator notations as given below:

$$\partial_x \equiv \frac{\partial}{\partial x}, \quad \partial_y \equiv \frac{\partial}{\partial y}, \quad \partial_{xx} \equiv \frac{\partial^2}{\partial x^2}, \quad \partial_{yx} \equiv \frac{\partial^2}{\partial y \partial x}, \quad \text{etc.}$$

Let $\Omega \subseteq \mathbb{R}^n$ be a domain. Recall that a C^1-function $\varphi : \Omega \to \mathbb{R}$ is called a scalar field, and, a vector field is a C^1-function $f : \Omega \to \mathbb{R}^n$. Clearly, each coordinate function $f_i : \Omega \to \mathbb{R}$ of a vector field f is a C^1-scalar field, for $i = 1, \ldots, n$. More generally, for $k \geq 1$, a vector field f is a C^k-function if and only if each coordinate function $f_i \in C^k(\Omega)$. That is, for $k \geq 1$,

$$f = (f_1, \ldots, f_n) \in C^k(\Omega) \quad \Leftrightarrow \quad f_i \in C^k(\Omega), \quad \text{for all } 1 \leq i \leq n.$$

Therefore, f is a C^∞-function if and only if each coordinate function $f_i \in C^\infty(\Omega)$. In latter case, we also say that f is a *smooth* vector field. A similar modification holds for other notations and terminology applicable to the vector fields. We are mainly dealing with the case when $n = 2$ or $n = 3$.

Definition 2.29 For $I \subseteq \mathbb{R}$, and a domain $U \subseteq \mathbb{R}^n$, let $\Omega = I \times U \subseteq \mathbb{R}^{n+1}$, and $f : U \to \mathbb{R}^n$ be a C^1 vector field. For any $(t_0, x_0) \in \Omega$, let $\delta > 0, \varepsilon > 0$ be such that $J = [t_0 - \delta, t_0 + \varepsilon] \subset I$. A C^1-function $x : [t_0 - \delta, t_0 + \varepsilon] \to \Omega$ is called an *integral curve* of the field f if

$$x'(t) = f(x(t)), \quad \text{for all } t \in J, \quad \text{with } x(t_0) = x_0. \tag{2.3.1}$$

Example 2.18 The vector field $f = (f_1, f_2)$ given by coordinate functions

$$f_1(x, y) = -y \quad \text{and} \quad f_2(x, y) = x, \quad \text{for } (x, y) \in \mathbb{R}^2,$$

represents 2-dimensional *counter-clockwise* circular motions of some types of particles about the origin, which follows easily by looking at the *integral curves* of the field f. By similar reasoning, it follows that the vector field $f = (f_1, f_2, 0)$ given by coordinate functions

$$f_1(x, y, z) = -\frac{y}{x^2 + y^2}, \quad f_2(x, y, z) = \frac{x}{x^2 + y^2}, \quad f_3(x, y, z) \equiv 0, \tag{2.3.2}$$

represents 3-dimensional velocity field of some types of particles over the region $\Omega = \mathbb{R}^3 \setminus \{(0, 0, z)\}$. Further, the velocity of the projectile defines a vector field along its trajectory, a time-dependent *spatial change* of a physical entity in a magnetic or gravitational field is represented a vector field, and so on.

The linear spaces of scalar and vector fields defined over a domain Ω are written, respectively, as $S(\Omega)$ and $V(\Omega)$. For any scalar field $\varphi \in C^1(\Omega)$, the gradient

$$f = \nabla \varphi := (\partial_x \varphi, \partial_y \varphi, \partial_z \varphi). \tag{2.3.3}$$

is a vector field that gives at each point $x \in \Omega$ 3-dimensional spatial changes of the scalar quantity $\varphi(x)$. It follows easily that the *gradient operator* grad : $S(\Omega) \to V(\Omega)$ satisfies all usual properties of a differential operator, where grad φ points in the direction of steepest increase in the values $\varphi(x)$, whereas the *negative gradient*

$(-\operatorname{grad} \varphi(a))$ points in the direction of steepest decrease in the values $\varphi(x)$, for $x \in \Omega$.

Example 2.19 Suppose P is the pressure of a gas contained in a volume V. Then the force f it exerts on any *volume element* δV is given by $f = -\nabla P \delta V$. Also, according to Fourier law, the *heat flux* Q due to variations of temperature $T(r)$ in a body of thermal conductivity k is given by $Q = -k \nabla T$.

Further, for a C^1 scalar field $\varphi \in S(\Omega)$, if the directional derivative $D_v \varphi(x)$ of φ at a point $x \in \Omega$ in the direction v exists, then we have

$$D_v \varphi(x) = \nabla \varphi(x) \cdot \widehat{v}, \qquad \text{where} \quad \widehat{v} := \frac{v}{|v|}. \qquad (2.3.4)$$

The above relation can also be written in *operator notation* using the operator

$$D_v := v_1 \frac{\partial}{\partial x_i} + \cdots + v_n \frac{\partial}{\partial x_n}$$
$$= v \cdot \left(\frac{\partial}{\partial x_i}, \ldots, \frac{\partial}{\partial x_n} \right) = v \cdot \nabla. \qquad (2.3.5)$$

In practical terms, as a point x moves by a small length $d\ell$ in the direction v, the quantity $D_v \varphi(x) \, d\ell$ gives an infinitesimal change in the scalar quantity $\varphi(x)$ that we write as $d\varphi$. Therefore, for any scalar field $\varphi \in C^1(\Omega)$, the directional derivative $D_v \varphi$ in the direction v gives the *spatial changes* in the values $\varphi(x)$ as we move away from the point x in the direction v.

Our main concern here is to discuss some important theorems related to the *del operator* ∇ in 3 variables as in (2.3.3). In general, the operator ∇ is fundamental to many applications in physics and engineering concerning phenomena such as mechanical vibrations, heat propagation, fluid flow, and electromagnetic waves, mainly due to the fact that the vector fields that are gradients of some C^1 scalar field are important objects of study. Notice that the expression for the operator ∇ changes with a change in coordinates and so will be the case with all other operators defined in terms of ∇. The four fundamental operators introduced below are very significant for our discussion about the topics as discussed in subsequent chapters.

1. The **divergence** of a (smooth) vector field $f : \Omega \to \mathbb{R}^3$ with $f = (f_1, f_2, f_3)$, denoted by $\nabla \cdot f$, is a scalar field given by

$$\nabla \cdot f := \partial_x f_1 + \partial_y f_2 + \partial_z f_3, \qquad (2.3.6)$$

The divergence of a vector field at point gives the magnitude of a *source* (or a *sink*) (see also Definition 2.35).

2. The **curl** of a vector field $f : \mathbb{R}^3 \to \mathbb{R}^3$ with $f = (f_1, f_2, f_3)$, denoted by $\nabla \times f$, is given by

$$\nabla \times f := \left(\partial_y f_3 - \partial_z f_2, \ \partial_z f_1 - \partial_x f_3, \ \partial_x f_2 - \partial_y f_1 \right). \qquad (2.3.7)$$

The curl of a vector field at point helps in identifying *vortices* of the field f (see also Definition 2.36). Recall that the velocity field f of a rotating rigid body at any point is given by $\mathbf{V} = \omega \times \mathbf{r}$, where ω is the angular velocity of the body.

3. As said earlier, the (scalar) differential operator of the form

$$\nabla^2 := \partial_{x_1 x_1} + \cdots + \partial_{x_n x_n}$$

is called the *scalar Laplacian* (or simply the **Laplacian**) in n variables, which we also write as Δ. For any C^2 scalar field $\varphi = \varphi(x, y, z)$, the Laplacian $\Delta\varphi$ is the divergence of the gradient of φ that gives a measure of the rate at which the average value of the field φ changes over spheres centred at P as its radius shrinks to zero.

4. It follows easily that, for any C^2 vector field $f = (f_1, f_2, f_3)$, we have

$$\nabla^2 f := \nabla (\nabla \cdot f) - \nabla \times (\nabla \times f),$$

which is called the *vector Laplacian* of the field f. In actual practice, to compute the component functions of the vector $\nabla^2 f$, we use the expression given by

$$\nabla \times \nabla f = \left(\nabla^2 \partial_x f_1, \nabla^2 \partial_y f_2, \nabla^2 \partial_z f_3\right). \tag{2.3.8}$$

Example 2.20 Let $\mathbf{r} = (x, y, z)$ be an arbitrary point on a curve in $\mathbb{R}^3$, with $r = |\mathbf{r}|$, and f be any differentiable function defined over a domain in $\mathbb{R}^3$ containing the curve. Then, we have

$$
\begin{aligned}
\nabla^2 x f(r) &= \frac{\partial^2 (x f(r))}{\partial x^2} + \frac{\partial^2 (x f(r))}{\partial y^2} + \frac{\partial^2 (x f(r))}{\partial z^2} \\
&= \frac{\partial}{\partial x}\left[f(r) + x \frac{\partial f(r)}{\partial x} \right] + x \frac{\partial^2 (f(r))}{\partial y^2} + x \frac{\partial^2 (f(r))}{\partial z^2} \\
&= 2 \frac{\partial f(r)}{\partial x} + x \nabla^2 f(r) = 2 \frac{\partial f(r)}{\partial x} + f''(r) + \frac{2}{r} f'(r) \\
&= \frac{2x}{r} f'(r) + f''(r) + \frac{2}{r} f'(r) = f''(r) + \frac{2x+2}{r} f'(r).
\end{aligned}
$$

Similarly, we obtain

$$\nabla^2 y f(r) = f''(r) + \frac{2y+2}{r} f'(r) \quad \text{and} \quad \nabla^2 z f(r) = f''(r) + \frac{2z+2}{r} f'(r).$$

Putting together the above three expressions, by equation (2.3.8), we obtain

$$\nabla^2 f(r)\mathbf{r} = \left(\nabla^2 x f(r), \nabla^2 y f(r), \nabla^2 z f(r)\right)$$

$$= \left(f''(r) + \frac{2x+2}{r}f'(r), f''(r) + \frac{2y+2}{r}f'(r), f''(r) + \frac{2z+2}{r}f'(r)\right)$$

$$= \left[f''(r) + \frac{2}{r}f'(r)\right]\mathbf{v} + 2f'(r)\widehat{\mathbf{r}} = \nabla^2 f(r)\mathbf{v} + 2f'(r)\widehat{\mathbf{r}},$$

where $\mathbf{v} = (1, 1, 1)$. In particular, when $f(r) = r^n$, it follows that for $n \geq 1$ we have

$$\nabla^2 r^n \mathbf{r} = n(n+1)r^{n-2}\mathbf{v} + 2nr^{n-2}\mathbf{r} = n(n+3)r^{n-2}\mathbf{r}.$$

The next theorem states some useful properties of curl and divergence operators.

Theorem 2.21 *With notions as above, we have*

1. Both the operators curl *and* div *are **linear**:*

$$\nabla \cdot (\alpha \, \mathbf{f}_1 + \beta \, \mathbf{f}_2) = \alpha \nabla \cdot \mathbf{f}_1 + \beta \nabla \cdot \mathbf{f}_2; \qquad (2.3.9a)$$

$$\nabla \times (\alpha \, \mathbf{f}_1 + \beta \, \mathbf{f}_2) = \alpha \nabla \times \mathbf{f}_1 + \beta \nabla \times \mathbf{f}_2. \qquad (2.3.9b)$$

2. Both the operators curl *and* div *are **derivations**:*

$$\nabla \cdot (\varphi \, \mathbf{f}) = \text{grad}\,(\varphi) \cdot \mathbf{f} + \varphi \nabla \cdot \mathbf{f}; \qquad (2.3.10a)$$

$$\nabla \times (\varphi \, \mathbf{f}) = \text{grad}\,(\varphi) \times \mathbf{f} + \varphi \nabla \times \mathbf{f}. \qquad (2.3.10b)$$

3. Both the operators curl *and* div *obey basic **product laws**:*

$$\nabla \cdot (\mathbf{f} \times \mathbf{g}) = \mathbf{g}(\nabla \times \mathbf{f}) - (\nabla \times \mathbf{g})\mathbf{f}; \qquad (2.3.11a)$$

$$\nabla \times (\mathbf{f} \times \mathbf{g}) = (\mathbf{g} \cdot \nabla)\mathbf{f} - (\nabla \cdot \mathbf{f})\mathbf{g} + (\nabla \cdot \mathbf{g})\mathbf{f} - (\mathbf{f} \cdot \nabla)\mathbf{g}. \qquad (2.3.11b)$$

Proof Left for the reader as an exercise.

It follows from properties such as 2.3.9 and Exercise 2.33 that all the three differential operators in the *two-link* chain given below are linear operators:

$$S(\Omega) \xrightarrow{\text{grad}} V(\Omega) \xrightarrow{\text{curl}} V(\Omega) \xrightarrow{\text{div}} S(\Omega). \qquad (2.3.12)$$

The two related vector identities as given in the next theorem are significant to many applications in physics and engineering (see also Theorem 2.28).

Theorem 2.22 *For sufficiently smooth $\varphi \in S(\Omega)$ and $\mathbf{f} \in V(\Omega)$, we have*

$$(\text{curl} \circ \text{grad})\varphi = 0 \quad \text{and} \quad (\text{div} \circ \text{curl})\mathbf{f} = 0. \qquad (2.3.13)$$

Proof Left for the reader as an exercise.

Definition 2.30 A C^1 vector field f is said to be **conservative** (or an *irrotational field*) if curl $(f) = 0$, and it is said to be an *incompressible field* (or **solenoidal**) if div $(f) = 0$.

In general, by Theorem 2.22, the field $f = \text{grad } \varphi$ is conservative, for every C^1 function $\varphi \in S(\Omega)$. Also, the field $g = \text{curl } f$ is incompressible, for every C^1 function $f \in V(\Omega)$. The *integral definitions* of the above two concepts are given in later part of the section.

Example 2.21 The vector field f given by

$$f(x, y, z) = (y + z, \ x + z, \ x + y), \quad \text{for } (x, y, z) \in \mathbb{R}^3, \quad (2.3.14)$$

is both irrotational and solenoidal. However, the vector field f given by

$$f(x, y, z) = (2xyz^3, \ x^2z^3, \ 3x^2yz^2), \quad \text{for } (x, y, z) \in \mathbb{R}^3, \quad (2.3.15)$$

is irrotational, but not solenoidal. Let f represent the electrostatic force of attraction or repulsion that electric charges exert on each other. If $r = |\mathbf{r} - \mathbf{r}'|$ stands for the *radial distance* of a test charge at the source $q = q(\mathbf{r}')$, then by Coulomb's law we have

$$f = \frac{1}{4\pi\varepsilon_0} \frac{q}{r^2} \widehat{\mathbf{r} - \mathbf{r}'}.$$

In this case, the electrostatic field $\mathbf{E}$ defined by f is given by $\mathbf{E} = f/q$, in a deleted neighbourhood of the point charge q. Such an electrostatic field in a charge-free region is both irrotational and solenoidal, whereas in a charged region, it is irrotational, but not solenoidal. Further, a steady magnetic field in a current carrying conductor is solenoidal, but not irrotational.

Definition 2.31 A C^1 function $\varphi \in S(\Omega)$ is called a *scalar potential* of a vector field $f \in V(\Omega)$ if $f = \nabla\varphi$.

Clearly, if a C^1 vector field f has a scalar potential, then it is necessarily conservative, i.e. $\nabla \times f = 0$. It easy to see that potential function of a field is unique up to a constant. In some cases, it is convenient to find a potential function φ of a conservative field f, by using the relation $d\varphi = f_1 \, dx + f_2 \, dy + f_3 \, dz$, where f_1, f_2, f_3 are the coordinate functions of the field f.

Example 2.22 Let f represents the *gravitational field*, with coordinate functions given by

$$f_1(x, y, z) = \frac{x}{(x^2 + y^2 + z^2)^{3/2}};$$

$$f_1(x, y, z) = \frac{y}{(x^2 + y^2 + z^2)^{3/2}};$$

$$f_1(x, y, z) = \frac{z}{(x^2 + y^2 + z^2)^{3/2}}.$$

It follows easily that f is conservative. Therefore, if $\varphi = \varphi(x, y, z)$ is a *gravitational potential* of f, then we have

$$d\varphi = f_1\, dx + f_2\, dy + f_3\, dz$$

$$= \frac{x}{(x^2 + y^2 + z^2)^{3/2}}\, dx + \frac{y}{(x^2 + y^2 + z^2)^{3/2}}\, dy + \frac{z}{(x^2 + y^2 + z^2)^{3/2}}\, dz$$

$$= d\left(-\frac{1}{\sqrt{x^2 + y^2 + z^2}}\right)$$

$$\Rightarrow \qquad \varphi = -\frac{1}{\sqrt{x^2 + y^2 + z^2}}$$

Hence, $f = \nabla\varphi$. Further, as the velocity field f given in (2.3.14) is conservative, a potential function φ of the vector field f can also be computed by applying the *integration method*. That is, we follow the procedure as given below. First, integrating with respect to x, we have

$$\frac{\partial\varphi}{\partial x} = y + z \qquad \Rightarrow \qquad \varphi = yx + zx + g(y, z),$$

where g is some two-variable C^1-function. Next, by using the above expression of φ and integrating with respect to y, we obtain

$$\frac{\partial\varphi}{\partial y} = x + z \quad \Rightarrow \qquad x + \frac{\partial g}{\partial y} = x + z$$

$$\Rightarrow \qquad g(y, z) = yz + h(z),$$

where h is some single variable C^1-function. Finally, by using the function $g(y, z)$ as above, and also φ as obtained at the first step, it follows that

$$\frac{\partial\varphi}{\partial z} = x + y \quad \Rightarrow \qquad x + y + h'(z) = x + y$$

$$\Rightarrow \qquad h(z) = \alpha, \quad \text{where } \alpha \text{ is some constant.}$$

Hence, the potential function of f is given by

$$\varphi(x, y, z) = xy + xz + yz + \alpha,$$

for some constant α.

In general, existence of *potential function* for a vector field $f \in V(\Omega)$ depends on the type of the field we have and also on the topology of the domain Ω. Notice that if a vector function

$$\mathbf{r}(t) = \big(x(t),\, y(t),\, z(t)\big), \qquad \text{for } t \in I,$$

defines the *flow lines* of a vector field $f = \nabla\varphi$, then we have

$$\frac{d}{dt}(\varphi \circ \mathbf{r}) = \nabla\varphi(\mathbf{r}) \cdot \mathbf{r} = |\mathbf{r}|^2 \geq 0$$

implies that the function $\varphi \circ \mathbf{r} : I \to \mathbb{R}$ is an *increasing function*. It thus follows that the *flow lines of a conservative field cannot be loops*.

Example 2.23 The velocity field f of moving fluid particles given by the coordinate functions

$$f_1(x, y, z) = -\frac{y}{x^2 + y^2} \qquad f_2(x, y, z) = \frac{x}{x^2 + y^2}, \qquad f_3(x, y, z) \equiv 0,$$

defined over the domain $\Omega = \mathbb{R}^3 \setminus \{(0, 0, z)\}$ has circular flow lines around the z-axis, and hence it cannot be conservative. Further, by using the relations

$$\cos\theta = \frac{x}{x^2 + y^2} \qquad \sin\theta = \frac{y}{x^2 + y^2},$$

where θ is the angle xz-plane makes with the plane passing through the point $P(\mathbf{r})$ and z-axis, it follows that $d\theta = d\tan^{-1}(y/x) = D$, say, which is the total differential of the function $u = \theta$. So, for any closed curve C in xy-plane around z-axis oriented positively with respect to θ, we have

$$\int_C D = \int_0^{2\pi} d\theta = 2\pi \neq 0.$$

Line Integral

To start with, suppose a force field $f = f(\mathbf{r})$ acts along a smooth curve $C := C_\gamma \subset \mathbb{R}^n$, say parametrised by a smooth function $\gamma : [a, b] \to \mathbb{R}^n$, where the position vector $\mathbf{r} = \mathbf{r}(t)$ of a moving point along C is given by

$$\mathbf{r}(t) := (\gamma_1(t), \ldots, \gamma_n(t)), \quad t \in I = [a, b],$$

such that $\|\mathbf{r}'(t)\| > 0$ for all t. The *work done* by the force f in displacing a particle through a *line element* $d\ell$ along the curve C is given by $f \cdot d\ell$. So, the *total work done* W by the force f in moving along the curve C is given by the *line integral*

$$W := \int_C f \cdot d\ell = \int_a^b (f \cdot \mathbf{r}') \, dt.$$

The expression $f \cdot d\ell$ is also known as a *1-form field*. Clearly, the value of the line integral depends on the path taken to move between points (Theorem 2.23), and also on the preferred *orientation* of the underlying parametrised curve. A similar remark holds for the other two types of line integrals as given below.

$$\int_C \varphi \, d\ell \quad \text{and} \quad \int_C F \times d\ell.$$

The *orientation*[3] of a parametrised curve also plays a significant role while finding a *potential function* of a vector field.

Definition 2.32 A regular curve $C_\gamma \subset \mathbb{R}^2$ parametrised by arc length s, so that $\gamma'(s)$ and $\gamma''(s)$ are nonzero, is said to be *positively oriented* if orientation of the orthonormal basis $\{\gamma'(s), \gamma''(s)\}$ is same as the standard basis $\{e_1, e_2\}$. More generally, two unit-speed curves C_{γ_1} and C_{γ_2} in $\mathbb{R}^n$ are said to have the *opposite orientations* if γ_1 and γ_2 are related as $\gamma_1(s) = \gamma_2(\ell - s)$, where ℓ is the length of the curve C.

Notice that the problem highlighted in Example 2.23 arises mainly because the region $\Omega = \mathbb{R}^3 \setminus \{(0, 0, z)\}$ is not *simply connected* because no loop around the z-axis cannot be shrunk continuously to a point. For example, $\mathbb{R}^2 \setminus \{0\}$ is not simply connected because no simple closed curve in $\mathbb{R}^2$ around the origin 0 can be continuously shrunk to a point.

Definition 2.33 A *circulation* of a vector field f around a simple closed curve (or a *loop*) C is the line integral of f over C. A vector field $f \in V(\Omega)$ is said to be *conservative* if its circulation around every loop is zero.

In general, Theorem 2.23 proves that a vector field $f \in C_3^1(\Omega)$ over a simply connected region Ω is *conservative* (or a *vortex field*), i.e. $f = \text{grad}(\varphi)$, for some nonzero scalar field $\varphi \in C^1(\Omega)$. In particular, if Ω is a *simply connected* domain, and f is the force field of a *potential function* $\varphi = \varphi(x)$ defined over Ω so that we can write $f = \nabla\varphi$, then

$$d\varphi = \nabla\varphi \cdot d\ell = f \cdot d\ell \quad \Rightarrow \quad f \text{ is conservative.}$$

Conversely, given any conservative field f defined over a simply connected domain Ω, we may take

$$\varphi(\mathbf{r}) := \int_0^{\mathbf{r}} f \cdot d\ell, \tag{2.3.16}$$

which is well-defined by our assumption, and the origin 0 is the chosen as the reference point. Therefore,

$$d\varphi = f \cdot d\ell \quad \Rightarrow \quad \nabla\varphi \cdot d\ell = f \cdot d\ell.$$

---

[3] The concept of *oriented manifolds*, in general, play an important role in many applications.

As the above relation holds for every line element $d\ell$, it follows that $f = \nabla\varphi$. Hence, the relation (2.3.16) implies that every vector field f that satisfies the condition $\nabla \times f = 0$ over Ω is conservative. We apply the differential operator

$$\text{grad} : S(\Omega) \to V(\Omega) \quad \text{given by} \quad \text{grad } \varphi := \nabla\varphi,$$

to prove that most typical vector fields $f \in C^1(\Omega)$ that one may come across in classical mechanics, hydrodynamics, electrodynamics, etc., arise in this way only.

Theorem 2.23 *Let $\Omega \subset \mathbb{R}^3$ be a simply connected region. Then, for any field $f \in C^1(\Omega)$, the following are equivalent:*

1. *There exists some $\varphi \in C^1(\Omega)$ such that $f = \nabla\varphi$;*
2. *For any closed curve C lying inside Ω,*

$$\oint_C f \cdot d\ell = 0.$$

   Or, equivalently, $\nabla \times f = 0$, i.e. f, is conservative;
3. *(Gradient Theorem) The line integral of the field f is path independent; i.e., the line integral of vector field f around any closed curve C depends only on the end points.*

Proof By (2.3.16), every vector field f that satisfies the condition $\nabla \times f = 0$ over a simply connected region Ω has a scalar potential. That is, (2) $\Rightarrow$ (1). The converse holds by Theorem 2.22. Next, let $f(r) = \nabla\varphi(r)$, and consider a curve C parametrised by $\gamma : [a, b] \to \Omega$ such that $r(t) = (\gamma_1(t), \dots, \gamma_n(t))$, with $r_0 = \gamma(a)$ and $r_1 = \gamma(b)$. Then

$$\int_C f \cdot d\ell = \int_C \nabla\varphi(r) \cdot d\ell$$

$$= \int_a^b \varphi'(\gamma(t))\gamma'(t)dt = \int_a^b \frac{d}{dt}\Big[\varphi(\gamma(t))\Big]dt$$

$$= \varphi(r_1) - \varphi(r_0),$$

which implies (3), using (2.3.16). The last implication is also known as the *fundamental theorem for line integral*, which follows directly from the Stokes' theorem, as we will prove shortly. Next, suppose (3) holds for a field $f \equiv (f_1, f_2, f_3)$. As seen in worked examples above, $f = \nabla\varphi$ is equivalent to the statement that

$$f \cdot dx = \nabla\varphi(x) \cdot dx$$
$$= \varphi_x(x)dx + \varphi_y(x)dy + \varphi_z(x)dz$$
$$= d\varphi.$$

So, assuming that (3) holds over some smooth parametrised curve C_γ with end points $\gamma(a) = \mathbf{r}_0$ and $\gamma(b) = \mathbf{r}_1$, proof of the implication (3) $\Rightarrow$ (1) completes if we show that $d\varphi = f_1\, dx + f_2\, dy + f_3\, dz$. Without loss of generality, we may take Ω to be path-connected, where paths can be assumed to be smooth. So, fixing a point $\mathbf{r}_0 \in \Omega$, we may take

$$\varphi(\mathbf{r}) = \int_{\mathbf{r}_0}^{\mathbf{r}} f(\mathbf{u}) \cdot d\ell.$$

Then, along the line $\mathbf{r}' = \mathbf{r} + (\delta x, 0, 0)$, we obtain

$$\varphi(\mathbf{r}') - \varphi(\mathbf{r}) = \int_{\mathbf{r}_0}^{\mathbf{r}'} f(\mathbf{u}) \cdot d\ell - \int_{\mathbf{r}_0}^{\mathbf{r}} f(\mathbf{u}) \cdot d\ell$$

$$= \int_{x}^{x+\delta x} f_1(u, y, z)dt = f_1(\mathbf{r})\delta x$$

$$\Rightarrow \quad \varphi_x(\mathbf{r}) = f_1(\mathbf{r}).$$

Similarly, we have $\varphi_y(\mathbf{r}) = f_2(\mathbf{r})$ and $\varphi_z(\mathbf{r}) = f_3(\mathbf{r})$. $\square$

Flux and Divergence Theorem

As it has been for the line integral of a vector field along a smooth curve, the *surface integral* (or the *volume integral*) of a vector field over a regular surface S also depends on the *orientation* of the surface. To introduce the concept, let $\mathbf{r} = \mathbf{r}(u, v) : \Omega \to \mathbb{R}^3$ be a parametrisation of S. In general, we study the geometry of S at a point $a = (u_0, v_0) \in \Omega$ by using the two (orthogonal) curves given by

$$\mathbf{r}_1(u) = \mathbf{r}(u, v_0) \quad \text{and} \quad \mathbf{r}_2(v) = \mathbf{r}(u_0, v),$$

respectively, called the *u-curve* and *v-curve*. Notice that the derivatives $\mathbf{r}_u = \mathbf{r}'(u)$ and $\mathbf{r}_v = \mathbf{r}'(v)$ are, respectively, the *tangent vectors* to the two curves $\mathbf{r}_1$ and $\mathbf{r}_2$ on S. Also, by the vector identity

$$\left| \mathbf{r}_u \times \mathbf{r}_v \right|^2 = (\mathbf{r}_u \times \mathbf{r}_v) \cdot (\mathbf{r}_u \times \mathbf{r}_v) = \det \begin{pmatrix} \mathbf{r}_u \cdot \mathbf{r}_u & \mathbf{r}_u \cdot \mathbf{r}_v \\ \mathbf{r}_v \cdot \mathbf{r}_u & \mathbf{r}_v \cdot \mathbf{r}_v \end{pmatrix},$$

it follows that the *regularity condition* as given in Definition 2.27 is equivalent to the condition that the vectors $\mathbf{r}_u$, $\mathbf{r}_v$ are linearly independent. Therefore, there is a unique (shifted) tangent plane $\Pi(a)$ at $\mathbf{r}(a) \in S$ spanned by the tangent vectors $\mathbf{r}_u$ and $\mathbf{r}_v$. In fact, the two vectors form a natural basis for the tangent plane $\Pi(x)$ in the sense we explain shortly. In particular, if $\varphi \in C^1(\Omega)$, then for the regular surface

$$\Gamma_\varphi(x, y, z) : \quad \varphi(x, y) - z = 0,$$

we have $\mathbf{r}_x = (1, 0, \varphi_x)$ and $\mathbf{r}_y = (0, 1, \varphi_y)$, which implies that

$$\mathbf{r}_x \times \mathbf{r}_y = (-\varphi_x, -\varphi_y, 1),$$

and the equation of the tangent plane at a point $a = (x_0, y_0, z_0)$ is given by

$$\varphi_x(x - x_0) + \varphi_y(y - y_0) - (z - z_0) = 0, \qquad \text{where} \quad z_0 = \varphi(x_0, y_0).$$

Therefore, for any $x \in \Gamma_\varphi$, we have

$$\mathbf{n}(x) = \pm \frac{\mathbf{r}_u \times \mathbf{r}_v}{\|\mathbf{r}_u \times \mathbf{r}_v\|} = \frac{(-\varphi_x, -\varphi_y, 1)}{\sqrt{\varphi_x^2 + \varphi_y^2 + 1}}.$$

define **unit normals** to the surface Γ_φ at the point $(x, \varphi(x))$, $x \in \Omega$, which is clearly orthogonal to the tangent plane $\Pi(x)$. We say the surface Γ_φ is *orientable* if it is possible to assign an *orientation* to S by choosing one of the two sides of the plane $\Pi(x)$, $x \in \Omega$. That is, by taking one of the two possible *signs* for the normal $\mathbf{n}$ as *positive*.

Definition 2.34 A regular (closed) surface $S \subset \mathbb{R}^3$ is said to be *positively oriented* if it is possible to orient the tangent plane $\Pi(x)$ at each point $x \in S$ such that the normal $\mathbf{n}(x)$ always points *outward*.

Example 2.24 The unit sphere $\mathbb{S}^2 \subset \mathbb{R}^3$ is an orientable surface, and so is the *torus* $\mathbb{T} := \mathbb{S}^1 \times \mathbb{S}^1$ that has a parametrisation $\mathbf{r} : [0, 2\pi) \times [0, 2\pi) \to \mathbb{T}$ given by

$$\begin{aligned}
x(u, v) &= (1 + \cos v) \cos u \\
y(u, v) &= (1 + \cos v) \sin u \qquad\qquad (2.3.17) \\
z(u, v) &= \sin v.
\end{aligned}$$

A classical example of a *non-orientable* surface $S \subset \mathbb{R}^3$ is provided by the trace of the parametrisation $\mathbf{r} : (-1, 1) \times [0, 2\pi) \to \mathbb{R}^3$ given by

$$\begin{aligned}
x(u, v) &= (1 + u \cos(v/2)) \cos v, \\
y(u, v) &= (1 + u \cos(v/2)) \sin v, \qquad\qquad (2.3.18) \\
z(u, v) &= u \sin(v/2),
\end{aligned}$$

which is known as the *Möbius strip*. However, the *severed surface* obtained from a Möbius strip by restricting the function γ to the region $\Omega = (-1, 1) \times (0, 2\pi)$ is indeed orientable.

Remark 2.6 More generally, notice that a *closed surface* S that is the boundary of a region $\Omega \subset \mathbb{R}^3$ can be oriented even when it is not a connected set. For, at each point $x \in S$, it is possible to figure out difference between continuously varying normals pointing inwards or outwards with respect to various parts of Ω. One may choose the orientations for these parts arbitrarily, depending upon whether the coordinate system is *right-handed* or *left-handed*. Accordingly, we say S is *oriented with respect to Ω*. Of course, connected surfaces have *coherent orientation*, i.e., an orientation at one point defines orientation everywhere.

We now introduce the concept of *surface integral* of a vector field over a regular surface S. For a field $f = f(x)$ such as the *velocity field* of a fluid flow through a region $\Omega \subset \mathbb{R}^3$ with a *closed boundary surface* S, the scalar quantity $f \cdot \mathbf{n}\, da$ defines the (outward) *flux element* of the field f through the infinitesimal *surface element* $d\mathbf{a}$, say between two end faces of an infinitesimal cubical box. So, the total flux, i.e. the *volume flow rate* of the field f over the surface S, is given by the **surface integral**

$$\text{Flux} := \oint_S f \cdot \mathbf{n}\, da. \tag{2.3.19}$$

Notice that $f \cdot \mathbf{n}$ gives the component of f in the direction of the normal $\mathbf{n}$ so that $f \cdot \mathbf{n}\, da$ represents the Flux across the surface element $d\mathbf{a}$. Suppose a regular surface S such as the graph Γ_φ of a function $\varphi \in C^1(\Omega)$ is parametrised (naturally) by the parameters $(x, y) \in [a, b] \times [c, d]$ so that the speed of streamline flow in a rectangular pipe, say of cross section area A, is $|f| = f(x, y)$. Then

$$\text{Flux} := \iint_S f(x, y)d\mathbf{a} = \int_c^d \left(\int_a^b f(x, y)dx \right) dy.$$

Notice that, for a circular cross sections such as for a cylindrical pipe, it is more convenient to use the polar coordinates $(1, \theta)$ so that (almost) rectangular surface element is given by $d\mathbf{a} = r\, d\theta\, dr$, and hence

$$\text{Flux} := \iint_S f(x, y)d\mathbf{a} = \int_0^1 \int_0^{2\pi} f(1, \theta)r\, d\theta\, dr.$$

In general, if a surface S is parametrised by the parameters (u, v) over some open set $U \subset \mathbb{R}^2$, then an infinitesimal *surface element* $d\mathbf{a}$ is given by $d\mathbf{a} = \mathbf{r}_u \times \mathbf{r}_v\, dudv$, and hence

$$\text{Flux} := \oint_S f \cdot \mathbf{n}\, da = \oint_S (f \cdot \mathbf{r}_u \times \mathbf{r}_v)dudv. \tag{2.3.20}$$

The expression $f \cdot \mathbf{r}_u \times \mathbf{r}_v$ is also known as a *2-form field*. Notice that in exactly the same manner we can deal with the following other three types of surface integrals:

$$\oint_S \varphi \, da, \qquad \oint_S \varphi \, \mathbf{n} \, da, \qquad \text{and} \qquad \oint_S \mathbf{f} \times \mathbf{n} \, da,$$

where φ is a scalar field. For example, by using polar coordinates, the surface integral of the scalar field $\varphi(x, y) = e^{-x^2 - y^2}$ over $S = \mathbb{R}^2$ is easily found to equal π. In general, all integrals are evaluated taking due care of positivity of the normal $\mathbf{n}$. Again, it is important to find the class of vector fields for which the surface integral is independent of the surface, and is determined wholly by the boundary curve.

Theorem 2.24 *Let $\Omega \subset \mathbb{R}^3$ be a domain, star-shaped with respect to a point $x_0 \in \Omega$. For any field $\mathbf{f} \in C^1(\Omega)$, the following are equivalent:*

1. *The field $\mathbf{f}$ is solenoidal, i.e. $\nabla \cdot \mathbf{f} = 0$ everywhere.*
2. *The integral $\int \mathbf{f} \cdot d\mathbf{a}$ is surface independent, for any boundary curve.*
3. *The integral $\oint \mathbf{f} \cdot d\mathbf{a} = 0$, for any closed surface.*
4. *There exists some field $\mathbf{g} \in C^1(\Omega)$ such that $\mathbf{f} = \nabla \times \mathbf{g}$.*

Proof The reader is encouraged to complete the proof by supplying the details for each one of simple implications as given below:

$$(4) \Rightarrow (1); \qquad (1) \Leftrightarrow (3); \qquad (3) \Leftrightarrow (2).$$

The implication $(1) \Rightarrow (4)$ is a nontrivial part of the proof. Suppose (1) holds, and let φ_1, φ_2 be two *functionally independent* first integrals of the field $\mathbf{f}$ defined over a neighbourhood Ω_1 of a point $x_0 = (x_0, y_0, z_0) \in \Omega$ (Definition 2.29). Since $\mathbf{f}$ is parallel to the vector $\nabla \varphi_1 \times \nabla \varphi_2$, at each point of Ω_1, we may write

$$\mathbf{f}(x) = \lambda(x)(\nabla \varphi_1 \times \nabla \varphi_2), \qquad \text{for } x \in \Omega_1, \tag{2.3.21}$$

for some $\lambda : \Omega_1 \to \mathbb{R}$. Notice that, as both $\varphi_1, \varphi_2 : \Omega_1 \to \mathbb{R}$ are actually C^2-functions, we have

$$\lambda = \frac{\mathbf{f} \cdot (\nabla \varphi_1 \times \nabla \varphi_2)}{|\nabla \varphi_1 \times \nabla \varphi_2|} \in C^1(\Omega_1).$$

As both divergence and curl are derivations (Theorem 2.21), it follows from the first relation in Theorem 2.22 that

$$\begin{aligned}
0 &= \div(\mathbf{f}) \\
&= \nabla \lambda \cdot (\nabla \varphi_1 \times \nabla \varphi_2) + \lambda \big[(\nabla \times \nabla \varphi_1) \cdot \nabla \varphi_2 - (\nabla \times \nabla \varphi_2) \cdot \nabla \varphi_1 \big] \\
&= \nabla \lambda \cdot (\nabla \varphi_1 \times \nabla \varphi_2), \tag{2.3.22}
\end{aligned}$$

which shows that $\nabla \lambda$ is perpendicular to the vector $\nabla \varphi_1 \times \nabla \varphi_2$, at each point of Ω_1. Therefore, we have

$$f_1 \lambda_x + f_2 \lambda_z + f_3 \lambda_z = 0, \qquad \text{over } \Omega_1. \tag{2.3.23}$$

As we shall see in Chap. 3, in this case, it is possible to write λ as a function of φ_1 and φ_2. More specifically, by Lemma 2.2, there is a neighbourhood $\Omega_0 \subset \Omega_1$ of point x_0, and a C^1-function $F = F(\varphi_1, \varphi_2)$ such that

$$\lambda(x, y, z) = F(\varphi_1(x, y, z), \varphi_2(x, y, z)), \quad \text{for} \quad (x, y, z) \in \Omega_0. \quad (2.3.24)$$

We may take a function $G = G(\varphi_1, \varphi_2)$ given by

$$F(\varphi_1, \varphi_2) := \frac{\partial G}{\partial \varphi_1}(\varphi_1, \varphi_2). \quad (2.3.25)$$

It follows from (2.3.21), (2.3.24), and (2.3.25) that, over Ω_0, we have

$$f = \left(\frac{\partial G}{\partial \varphi_1} \nabla \varphi_1\right) \times \nabla \varphi_2 = \nabla \varphi_1 \times \nabla \varphi_2 \quad (2.3.26)$$

Therefore, to complete the proof, it suffices to show that (2.3.26) can be written as $f = \nabla \times (G \nabla \varphi_2)$, because the rest will be taken care of by the first relation in Theorem 2.22 and the identity (2.3.11b). For, taking $g = G \nabla \varphi_2$, it follows that (4) holds.

The vector field g in (4) above is called a *vector potential* of (necessarily *solenoidal*) field f. If g_1 and g_2 are two vector potentials of a vector field f, then

$$\nabla \times (g_1 - g_2) = 0 \quad \Rightarrow \quad g_1 - g_2 = \nabla \varphi,$$

for some $\varphi \in C^1(\Omega)$. Therefore, vector potential of a vector field is unique, up to an additive *gradient factor*.

Replacing *surface element da* in the above discussion about surface integral by a *volume element $dv \subset \Omega$*, we obtain the concept of a *volume integral* of a scalar field $\varphi = \varphi(x) \in C^1(\Omega)$ given by

$$\text{Vol} := \int_V \varphi \, dv. \quad (2.3.27)$$

For example, if $\varphi = \varphi(x)$ is the density of a fluid so that $dm := \varphi \, dv$ gives the infinitesimal mass of the fluid contained in the volume element dv, then Vol gives the total mass M of the fluid inside the volume Ω; in particular, for $\varphi \equiv 1$, Vol gives the actual volume of a bounded region. In exactly the same manner, as above, the volume integral of a vector field $f \in C^1(\Omega)$, with $f \equiv (f_1, f_2, f_3)$, is given by

$$\text{Vol} := \int_V f dv = \left(\int_V f_1 dv, \int_V f_2 dv, \int_V f_3 dv\right).$$

In each of other cases, the volume integral of a field is given by the triple integral, while taking care in setting the limits of integration and also the order in which such integrals are evaluated.

Definition 2.35 Let $\Omega \subset \mathbb{R}^3$ be an open set. The *divergence* of a vector field f : $\Omega \to \mathbb{R}^3$ at a point $x \in \Omega$ is the integral

$$\text{div } f := \lim_{\delta v} \frac{1}{\delta v} \oint_{\delta a} f \cdot \mathbf{n} \, da, \tag{2.3.28}$$

where δv is a volume element around x with surface element δa.

By the *integral definition* of the divergence as above, $\text{div}(f)$ is the ratio of the flux of the vector field f to the volume δv, which explains why divergence at a point gives a *local spread* of f (also see Lemma 2.26). The condition $\text{div}(f)(x) > 0$ implies existence of a *source* (or a *faucet*) at a point x, and $\text{div}(f)(x) < 0$ correspond to a *sink point* (or a *drain*). When f is given in terms of component functions using Cartesian coordinates, the above expression coincides with the earlier definition, i.e. $\text{div } f = \nabla \cdot f$.

A celebrated result of Gauss, as stated in the next theorem, shows that the *total spread* of a continuously differentiable vector field within a volume equals its *out flux* from the boundary surface of the volume. This is a mathematical form of a typical *conservation law* of a physical system. For example, if $\mathbf{v}$ is the velocity field of an *incompressible fluid* enclosed in a region Ω then the total amount of fluid flowing out of the surface per unit time, given by Flux of $\mathbf{v}$, must equal the amount of fluid pouring out through the boundaries of the region, given by $\text{div}(\mathbf{v})$.

Theorem 2.25 (Divergence Theorem) *Let $f \in C^1(\Omega)$ be a field, and S denote the closed surface forming the boundary of the volume enclosed by a region $\Omega \subset \mathbb{R}^3$. Then*

$$\int_{\Omega} \nabla \cdot f \, dv = \oint_S f \cdot \mathbf{n} \, da. \tag{2.3.29}$$

Proof At an infinitesimal level, (2.3.36) is equivalent to

$$\nabla \cdot f \approx \frac{1}{\delta V_i} \oint_{\delta S_i} f \cdot \mathbf{n} \, da, \tag{2.3.30}$$

so that we have

$$\sum_i (\nabla \cdot f) \delta V_i \approx \sum_i \oint_{\delta S_i} f \cdot \mathbf{n} \, da. \tag{2.3.31}$$

Taking limit $\delta V_i \to 0$, by definition, the left side of the last equation gives the left side of the equation (2.3.29). For the right side, consider the two adjacent volume element δV_1 and δV_2, with normals $\mathbf{n}_1$ and $\mathbf{n}_2$ of the corresponding surface elements δS_1 and δS_2. Since at the common faces, $\mathbf{n}_1 = -\mathbf{n}_2$, therefore the contributions from the interior of the volume V to the right side sum of (2.3.31) is zero. So, only the exterior surface contributes to sum. This completes the proof.

Furthermore, taking a suitable field f, there are some other interesting consequences that follow directly from Theorem 2.25. The reader are encouraged to supply a proof for each one of the following important facts:

1. Take $f = a\varphi$, where $\varphi \in C^1(\Omega)$ and $a \in \mathbb{R}^3$. Then

$$\int_\Omega \nabla\varphi \, dv = \oint_S \varphi \mathbf{n} \, da, \qquad (2.3.32)$$

where $S = \partial\Omega$.

2. Take $f = a \times \mathbf{G}$, where $\mathbf{G} \in C^1(\Omega)$ and $a \in \mathbb{R}^3$. Then

$$\int_\Omega -(\nabla \times \mathbf{G}) dv = \oint_S \mathbf{G} \times \mathbf{n} \, da. \qquad (2.3.33)$$

3. (*Green's First Identity*) Take $f = \psi\nabla\varphi$, where $\psi, \varphi \in C^1(\Omega)$. Then

$$\int_\Omega (\nabla\psi \cdot \nabla\varphi + \psi\nabla^2\varphi) dv = \oint_S \psi\nabla\varphi \cdot \mathbf{n} \, da. \qquad (2.3.34)$$

4. (*Green's Second Identity*) Take $f = \psi\nabla\varphi - \varphi\nabla\psi$, where $\psi, \varphi \in C^1(\Omega)$. Then

$$\int_\Omega (\psi\nabla^2\varphi - \varphi\nabla^2\psi) dv = \oint_S [\psi\nabla\varphi - \varphi\nabla\psi] \cdot \mathbf{n} \, da. \qquad (2.3.35)$$

Theorem 2.26 (Localisation Theorem) *Let* $\varphi \in C^1(\Omega)$. *For* $x_0 \in \Omega$, *and* $\varepsilon > 0$, *let* S_ε *be the sphere of radius* ε *at* x_0, *with volume* V_ε. *Then*

$$\varphi(x_0) = \lim_{\varepsilon \to 0} \frac{1}{V_\varepsilon} \oint_{S_\varepsilon} \varphi(x) \, dv.$$

In particular, $\varphi \equiv 0$ if $\int_K \varphi \, dV = 0$ for every closed set $K \subset \Omega$.

Proof By a well known theorem in integral calculus, we have

$$\left| \varphi(\boldsymbol{x}_0) - \frac{1}{V_\varepsilon} \oint_{\mathbb{S}_\varepsilon} \varphi(\boldsymbol{x}) \, d\mathbf{v} \right|$$

$$\leq \frac{1}{V_\varepsilon} \oint_{\mathbb{S}_\varepsilon} |\varphi(\boldsymbol{x}_0) - \varphi(\boldsymbol{x})| d\mathbf{v}$$

$$\leq \frac{1}{V_\varepsilon} \oint_{\mathbb{S}_\varepsilon} \sup_{\boldsymbol{x} \in \Omega} |\varphi(\boldsymbol{x}_0) - \varphi(\boldsymbol{x})| d\mathbf{v}$$

$$= \max_{\boldsymbol{x} \in \Omega} |\varphi(\boldsymbol{x}_0) - \varphi(\boldsymbol{x})|,$$

using *maximum principle*. The assertion follows by continuity of φ as $\varepsilon \to 0$. The second part follows because $\mathbb{S}_\varepsilon \subset \Omega$ is a closed set.

As an interesting application of Theorem 2.25, we prove the following simple fact that helps to prove uniqueness of solution of a *Laplace equation* $u_{xx} + u_{yy} + u_{zz} = 0$, where $u = u(x, y, z)$ is a sufficiently smooth scalar field defined over a region $\Omega \subset \mathbb{R}^3$.

Lemma 2.3 *Let φ be a sufficiently smooth scalar field defined over a domain $\Omega \subset \mathbb{R}^3$ such that $\varphi = 0$ over the closed surface $S = \partial\Omega$, and $\nabla^2\varphi = 0$. Then $\varphi = 0$ everywhere over Ω.*

Proof By derivation property of the operator ∇, we have

$$\nabla^2 \varphi = 0 \quad \Rightarrow \quad \nabla \cdot (\varphi \nabla \varphi) - \nabla\varphi \cdot \nabla\varphi = 0.$$

Integrating over Ω and applying divergence theorem, we obtain

$$\oint_S \varphi \nabla\varphi \cdot \mathbf{n} da - \int_\Omega |\nabla\varphi|^2 d\mathbf{v} = 0.$$

By the assumption that $\varphi = 0$ on $S = \partial\Omega$, the first integral is zero, and so we conclude $\nabla\varphi = 0$ over Ω, because $|\nabla\varphi|^2 \geq 0$, always. Therefore, φ is constant over Ω. But then, since $\varphi = 0$ on $S = \partial\Omega$, this constant must be zero, using connectedness of Ω.

Circulations and Stokes' Theorem

Here, we prove Stokes' *curl theorem*. Notice that, since Theorem 2.26 holds when $d\mathbf{v}$ is replaced by $d\mathbf{a}$, while making respective changes in the statement, the *integral definition* of curl $\boldsymbol{f} = \nabla \times \boldsymbol{f}$ is obtained as given in the next definition.

Definition 2.36 Let $\boldsymbol{f} \in C^1(\Omega)$ be a 3-dimensional field. The component of $\nabla \times \boldsymbol{f}$ in the direction of the normal $\mathbf{n}$ is the integral given by

$$\mathbf{n} \cdot (\nabla \times \boldsymbol{f}) := \lim_{\delta S} \frac{1}{\delta S} \oint_{\delta C} \boldsymbol{f} \cdot \mathbf{r}, \tag{2.3.36}$$

where δS is a surface element orthogonal to normal $\mathbf{n}$, and δC is the *positively oriented*[4] boundary of the surface element δS.

In this case, $\nabla \times f$ is the ratio of the work done by the field f while moving around the loop δC to the area of the surface element δS, which explains why curl measures how much the field f *swirls locally*. So, $\mathrm{curl}(f)(x) \neq 0$ gives a region of *whirlpool* of positive or negative curvature, and $\mathrm{curl}(f)(x) = 0$ correspond to a the point of *circulation-free* motions. Expressing component functions of f in terms of Cartesian coordinates and using standard basis $\mathbf{e}_i$ for $\mathbf{n}$, this definition gives earlier definition of curl as

$$\mathrm{curl}\ f = (\mathbf{e}_1 \cdot \nabla \times f, \mathbf{e}_2 \cdot \nabla \times f, \mathbf{e}_3 \cdot \nabla \times f).$$

Theorem 2.27 (Stokes Theorem) *Let f be a continuously differentiable vector field defined over a surface S, with a closed boundary curve C. Then*

$$\int_S \nabla \times f \cdot \mathbf{n}\, da = \oint_C f \cdot d\mathbf{r}, \qquad (2.3.37)$$

where C is positively oriented with respect to the normal $\mathbf{n}$ in the sense as described earlier in a footnote remark.

Proof Notice that, at infinitesimal level, (2.3.36) gives

$$\nabla \times f \cdot \mathbf{n} \approx \frac{1}{\delta S_i} \oint_{\delta C_i} f \cdot \ell, \qquad (2.3.38)$$

where the surface element δS_i is orthogonal to $\mathbf{n}$, and has δC_i as the posively oriented closed boundary curve. Adding contributions over all infinitesimal surface elements, we obtain

$$\sum_i [\nabla \times f \cdot \mathbf{n}]\delta S_i \approx \sum_i \oint_{\delta C_i} f \cdot \mathbf{r}. \qquad (2.3.39)$$

As $\delta S_i \to 0$, by Definition 2.36, the first sum gives the surface integral as in the statement of the theorem, whereas the second sum gives the asserted line integral, because the right side sum contributes nothing to the integral from the common portions of the bounding curves of any two infinitesimal surface elements.

Notice that the left side of equation (2.3.37) equals zero for any *closed* surface S. For, if there is a small ball $B(a; r) \subset S$ then by the above theorem the surface integral of $\nabla \times f \cdot \mathbf{n}$ over $\overline{B}(a; r)$ equals the line integral of $f \cdot d\mathbf{r}$ over the boundary circle $C(a; r) = \overline{B}(a; r) \setminus B(a; r)$. So, taking $r \to 0$, assertion follows.

---

[4] That is, as we move along the closed boundary δC anticlockwise, the region δS lies on our left. Or, equivalently, motion is anticlockwise with regard to positive $\mathbf{n}$.

Also, taking a suitable field f defined over a surface S with a closed boundary curve C, there are some other interesting consequences that follow directly from Theorem 2.27. The reader are encouraged to supply a proof for each one of the following important facts:

1. Take $f = a\varphi$, where φ is a continuously differentiable scalar field defined over the surface S, and $a \in \mathbb{R}^3$. Then

$$\int_S -\nabla\varphi \times \mathbf{n}\,da = \oint_C \varphi\,d\ell. \tag{2.3.40}$$

2. Take $f = a \times \mathbf{G}$, where $\mathbf{G}$ is a continuously differentiable vector field defined over the surface S, and $a \in \mathbb{R}^3$. Then

$$\int_S -(\mathbf{n} \times \nabla) \times \mathbf{G}\,da = \oint_C \mathbf{G} \times d\ell. \tag{2.3.41}$$

3. (*Green's Theorem*) Let $S \subset \mathbb{R}^2$ be an open set. Take $f = (\varphi, \psi, 0)$, where $\psi, \varphi \in C^1(S)$. Then

$$\int_S (\psi_x - \varphi_y)\,dxdy = \oint_C [\varphi dx + \psi dy]. \tag{2.3.42}$$

Helmholtz Decomposition Theorem

We conclude the chapter with a discussion about the fundamental theorem of vector calculus due to *Hermann von Helmholtz* (1821–1894): *Every sufficiently well-behaved vector field f defined over a simply connected domain $\Omega \to \mathbb{R}^3$, with a piecewise smooth boundary, can be expressed as the sum of two suitably chosen vector fields, where the one is curl-free and the other divergence-free*. It is also known as the *Helmholtz's decomposition theorem*, which has numerous applications in physics and engineering, especially to problems related to electromagnetism. The theorem was known to Stokes since 1849, who published the related work in 1856.

Recall that a conservative field $f = (f_1, f_2, f_3)$ can be written as

$$f = -\nabla\varphi \quad \Longleftrightarrow \quad f_1 = \varphi_x, \;\; f_2 = \varphi_y, \;\; f_3 = \varphi_z,$$

where $\varphi \in C^1(\Omega)$ is called a scalar potential of the field f (Theorem 2.23). If f represents the velocity field of a conservative fluid flow, then the level curves of φ are known as the *potential lines* of the flow. Therefore, to solve a system of differential equations for the function f, it suffices to solve the relate differential equations for the function φ. In most such cases, we are led to solve a Laplace equation of the form

$$u_{xx} + u_{yy} + u_{zz} = 0, \quad \text{for some} \;\; u = u(x, y, z) \in C^2(\Omega).$$

Also, since it is known by Maxwell law that a magnetic field **B** do not diverge from anything, it only *curls around*, i.e.,

B is the curl of some vector field f, called a vector potential, and so it is always solenoidal (see Appendix A.2 for details). Also, for the Newton's vector field

$$f(x) = -c \, \frac{x - x_1}{\|x - x_1\|^3},$$

defined over the *star-shaped* region

$$\Omega = \mathbb{R}^3 \setminus \{x_1 + u(x_1 - x_0) : u \geq 0\}, \quad \text{for } x_0 \neq x_1,$$

with respect to the point x_0, the vector potential $\mathbf{w} = \mathbf{w}(x)$ is given by

$$-c \, \frac{(x_0 - x_1) \times (x - x_1)}{\|x_0 - x_1\| \, \|x - x_1\|^2 + \big((x_0 - x_1) \times (x - x_1)\big)\|x - x_1\|}.$$

In general, when a vector field f is *solenoidal*, i.e., $\nabla \cdot f = 0$, we can write $f = \nabla \times \mathbf{w}$, for some vector field $\mathbf{w}$ (Theorem 2.24). The vector field $\mathbf{w}$ is called a *vector potential* for the field f.

Example 2.25 Consider the field $f(x, y, z) = (x, y, -2z)$ defined over $\mathbb{R}^3$, which is viewed as a star-shaped region with respect to the point $x_0 = \mathbf{0}$. Clearly, $\nabla \cdot f = 0$. Now, we have

$$\mathbf{w} = \int_0^1 t \, f(x_0 + t(x - x_0)) \times (x - x_0) dt$$

$$= \int_0^1 t \, f(tx, ty, -2tz) \times (x, y, z) dt$$

$$= \int_0^1 t^2 f(3yz, -3xz, 0) dt = (yz, -xz, 0),$$

which satisfies the condition $f = \nabla \times \mathbf{w}$. That is, the field $\mathbf{w}$ is a *vector potential* for f.

More generally, let $f \in C^1(\Omega)$ has field $\mathbf{u}$ as a *vector potential*, and suppose $\psi = \nabla \cdot \mathbf{u} \neq 0$. Now, if $f_1 = \mathbf{u} + \nabla\varphi$ has to be a solenoidal vector potential for the field f, then we must have

$$0 = \nabla \cdot f_1(\mathbf{r}) = \nabla \cdot \mathbf{u}(\mathbf{r}) + \nabla^2 \varphi(\mathbf{r}) = \psi(\mathbf{r}) + \nabla^2 \varphi(\mathbf{r}).$$

Therefore, for a (necessarily solenoidal) field f to have a solenoidal vector potential of the form $f_1 = \mathbf{u} + \nabla\varphi$, we must solve the *Poisson equation* given by

$$\nabla^2 \varphi(\mathbf{r}) = -f(\mathbf{r}).$$

That is, we need to find an appropriate function φ that is unique up to to an additive harmonic function. Moreover, such a harmonic function can be assumed to be zero subject to assigned boundary conditions for the function φ, thereby ensuring uniqueness of the function φ. A similar approach also helps to find a conservative field $f_2 = \nabla\varphi$, with $\nabla \cdot f_2 = \psi$. Notice that

$$\psi(\mathbf{r}) = \nabla \cdot f_2(\mathbf{r}) = \nabla^2 \varphi,$$

which is again a *Poisson equation*. Finally, to find a field f with specified curl and divergence, respectively, given by

$$\mathbf{v}(\mathbf{r}) = \nabla \times f(\mathbf{r}) \quad \text{and} \quad \varphi(\mathbf{r}) = \nabla \cdot f(\mathbf{r}), \tag{2.3.43}$$

then we may take $f = f_1 + f_2$, and uniqueness of the decomposition may be subject to certain suitable boundary conditions. Therefore, if scalar and vector potentials exist, and also the related Poisson equations have unique solutions, then a fairly general situation given by the vector field f can be expressed as a sum of a conservative field f_2 and a solenoidal field f_1.

1. Suppose $f = \nabla \times \mathbf{u}$. Find solenoidal field f_1 as above, and write

$$f = f_1 + (f - f_1).$$

Then, the field f_1 is solenoidal, and $f - f_1$ is conservative; alternatively,

2. Suppose $f = \nabla \cdot f$. Find conservative field f_2 as above, and write

$$f = (f - f_2) + f_2.$$

Then, in this case, $f - f_2$ is solenoidal, and f_2 is conservative.

It is also possible obtain a type of *Helmholtz decomposition* of the form

$$\nabla^2 f = \nabla(\nabla \cdot f) + \nabla \times (\nabla \times (-f)). \qquad \text{(see Exercise 2.43)}$$

Theorem 2.28 *Let $\Omega \subseteq \mathbb{R}^3$ be a domain, and $f \in V(\Omega)$ be such that $f(\mathbf{r}) \to 0$ as $r = \|\mathbf{r}\| \to \infty$. Suppose both $\varphi(\mathbf{r}) = \nabla \cdot f(\mathbf{r})$ and $v(\mathbf{r}) = \nabla \times f(\mathbf{r})$ tend to zero faster than the function $1/r^2$ as $r \to \infty$. Then, the field f can be expressed uniquely as*

$$f = -\nabla u + \nabla \times \mathbf{u}, \tag{2.3.44}$$

where the scalar potential $u = u(r)$ and the vector potential $\boldsymbol{u} = U(r)$ of the field $\boldsymbol{f} = F(r)$ are, respectively, given by the relations as follow:

$$u(r) = \frac{1}{4\pi} \int_{\Omega} \frac{v(r)}{\|r - s\|} dv; \qquad (2.3.45)$$

$$U(r) = \frac{1}{4\pi} \int_{\Omega} \frac{\varphi(r)}{\|r - s\|} dv, \qquad (2.3.46)$$

where dv is the differential volume element.

Proof Let $\varphi(\mathbf{r})$ and $\mathbf{v}(\mathbf{r})$ be as in (2.3.43), both specified over a simply-connected region $\Omega \subseteq \mathbb{R}^3$, with a nice boundary surface. We may assume that both vector functions approach to zero faster than the function $1/r^2$ as $r \to \infty$, where $r = \|\mathbf{r}\|$. By Theorem 2.22, we know that the field $\mathbf{v} = \mathrm{curl}(f)$ satisfies $\mathrm{div}\,\mathbf{v} = 0$; i.e., it is a solenoidal field. In physical terms, $\mathbf{v}$ is said to satisfy the basic laws of fluid mechanics such as conservation of mass and momentum. If (2.3.44) holds, we have

$$\varphi(\mathbf{r}) = \nabla \cdot \boldsymbol{f}(\mathbf{r}) = -\nabla^2 u(\mathbf{r}) = -\frac{1}{4\pi} \int_{\Omega} d\nabla^2 \left(\frac{1}{\|\mathbf{r} - \mathbf{s}\|} \right) dv;$$

$$\mathbf{v} = \nabla \times \boldsymbol{f} = -\nabla^2 \mathbf{u} + \nabla(\nabla \cdot \mathbf{u}). \quad \text{(Exercise 2.43)}$$

Further, notice that we have

$$-\nabla^2 \mathbf{u} = -\frac{1}{4\pi} \int_{\Omega} \varphi \nabla^2 \left(\frac{1}{\|\mathbf{r} - \mathbf{s}\|} \right) dv = \int_{\Omega} \varphi(\mathbf{s}) \delta^3 (\mathbf{r} - \mathbf{s}) dv = \varphi(\mathbf{r}).$$

Therefore, it suffices to show that $\nabla(\nabla \cdot \mathbf{u}) \equiv 0$. For, by using the integration by parts, the relation given by

$$\int_{\Omega} \varphi(\nabla \cdot \mathbf{u}) dv = - \int_{\Omega} \mathbf{u} \cdot (\nabla \varphi) dv + \int_{\Gamma} \varphi \cdot \mathbf{u} \cdot ds,$$

and also the fact that the derivatives of $\|\mathbf{r} - \mathbf{s}\|$ with respect to $\mathbf{s}$ differ by a sign from those with respect to $\mathbf{r}$, the desired conclusion follows.

Clearly, to gain from the above theorem, it is important to know how to solve the partial differential equations such as the *Poisson equations*. The above theorem also helps in dealing with a vector field $\boldsymbol{f} = F(\mathbf{r}) \in V(\Omega)$ such that it is neither conservative nor solenoidal. For example, consider the problems related to potential of an electric field $\mathbf{E} = \mathbf{E}(t)$ in a charged medium, with time varying magnetic field $\mathbf{H} = \mathbf{H}(t)$. As proved above, under some mild conditions, every field $\boldsymbol{f}$ can be written as the sum of a conservative and solenoidal fields.

Remark 2.7 As there are nontrivial fields that are both *conservative* and *solenoidal*, it is not possible to have uniqueness of Helmholtz decomposition. For example, exploring possibility to express a velocity field with a solenoidal flow as a Helmholtz decomposition with a nonzero u does not amount to expressing the velocity field as a different Helmholtz decomposition in which $u = 0$. However, under certain mild limiting conditions, the decomposition is unique.

Notice that, in most applications, an unknown field f is given by natural sources, and so we are required to find the component functions of f analytically or numerically using the assigned data. Also, since differential equation models are formulated using data sources, every relation among fields is studied in terms of solutions of these differential equations. For example, four fundamental Maxwell equations discussed later in Appendix A.2 are used to study electric and magnetic fields in terms of distributions of charges and currents.

Exercises 2

2.1. For $x, y \in \mathbb{R}^n$, let $\langle x, y \rangle$ be as in (2.1.1). Show that, for any $x, y \in \mathbb{R}^n$ and $a \in \mathbb{R}$, we have

    a. $\langle x + y, z \rangle = \langle x, z \rangle + \langle y, z \rangle$;
    b. $\langle a x, y \rangle = a \langle x, y \rangle$;
    c. $\langle x, y \rangle = \langle y, x \rangle$;
    d. $\langle x, x \rangle \geq 0$, with $\langle x, x \rangle = 0$ only if $x = 0$.

2.2. For $x \in \mathbb{R}^n$, let $\|x\|$ be as in (2.1.2). Show that, for any $x, y \in \mathbb{R}^n$ and $a \in \mathbb{R}$, we have

    a. $\|x + y\| \leq \|x\| + \|y\|$;
    b. $\|a x\| = |a| \|x\|$;
    c. $\|x\| = 0 \implies x = 0$.

2.3. Show that *parallelogram law* holds in $\mathbb{R}^n$. That is, for any $x, y \in \mathbb{R}^n$, we have

$$\|x + y\|^2 + \|x - y\|^2 = 2\|x\|^2 + 2\|y\|^2.$$

2.4. Let $a_1, \ldots, a_n$ be positive real numbers. Use the inequality (2.1.3) to show that

$$\left(\sum_{k=1}^n a_k \right) \left(\sum_{k=1}^n \frac{1}{a_k} \right) \geq n^2.$$

2.5. Prove the identity (2.1.4).
2.6. Let $\langle a_k \rangle$ be sequence in $\mathbb{R}$. Show that if the series $\sum_{k=1}^\infty a_k^2$ converges then the series $\sum_{k=1}^\infty (1/k) a_k$ converges absolutely.
2.7. Let $f(x, y) = 5x^2 y/(x^2 + y^2)$ for $x = (x, y) \neq (0, 0)$. Find the limit of $f(x)$ as $x \to (0, 0)$, if it exists.
2.8. Test the continuity of $f : \mathbb{R}^2 \to \mathbb{R}$ defined by $f(x) = x^4 y/(x^8 + y^2)$ where $x = (x, y)$ and $f(0, 0) = 0$.

2.9. (Peano) Let $\varphi(x, y) = (x^2 - y^2)/(x^2 + y^2)$, and $f(x, y) = xt\varphi(y, x)$. Show that $D_1 D_2 f(0, 0) \neq D_2 D_1 f(0, 0)$.

2.10. Let $1 \leq k \leq n$ be fixed. Is the kth projection function $\pi_k : \mathbb{R}^n \to \mathbb{R}$ given by $\pi_k(x_1, \ldots, x_m) = x_k$, differentiable? If yes, what is the derivative?

2.11. Find the derivative of the function $f : \mathbb{R}^n \to \mathbb{R}$ given by $f(x) = x^t Ax$, where $x \in \mathbb{R}^n$ is viewed as a column vector.

2.12. Let $y \in \mathbb{R}^n$ be a fixed vector. Is the function $f : \mathbb{R}^n \to \mathbb{R}$ given by $f(x) = \langle x, y \rangle$ differentiable? Justify your answer.

2.13. Show that the derivative of $f : \mathbb{R}^n \to \mathbb{R}$ given by $f(x) = \|x\|$ is $x/\|x\|$, for any $0 \neq x \in \mathbb{R}^n$.

2.14. Is the function $f : \mathbb{R}^n \to \mathbb{R}^n$ given by $f(x) = x\|x\|$ differentiable?

2.15. Suppose a function $\varphi : \mathbb{R}^n \to \mathbb{R}$ satisfies the condition

$$|\varphi(x)| \leq \|x\|^2, \quad \text{for all } x \in \mathbb{R}^n.$$

Show that φ is differentiable at the origin $0 \in \mathbb{R}^n$.

2.16. Let $f : \mathbb{R} \to \mathbb{R}$ be an even functions, and $\varphi : \mathbb{R}^n \to \mathbb{R}$ be defined by $\varphi(x) = f(\|x\|)$, for $x \in \mathbb{R}^n$. Show that if f is differentiable at 0 then the function φ is differentiable at the origin $0 \in \mathbb{R}^n$.

2.17. Let $E \subseteq \mathbb{R}^m$ be an open connected set, and $\varphi : E \to \mathbb{R}^n$ be a differentiable function such that $Df(x) = 0$, for all $x \in E$. Prove that f is constant.

2.18. Let $\varphi(x, y) = e^{xy}$, for $(x, y) \in \mathbb{R}^2$. Write the equation of the tangent plane to the graph Γ_φ at the origin.

2.19. Let $E \subseteq \mathbb{R}^m$ be an open connected set, $\varphi : E \to \mathbb{R}$ be a differentiable function such that $\text{grad}\,\varphi(x) \neq 0$, and $u \in E$ be a unit vector. Show that $D_u\varphi(x)$ is *maximum* when u is in the direction of the vector $\nabla\varphi(x)$, and $D_u\varphi(x)$ is *minimum* when u is in the opposite direction, i.e. along the vector $(-\nabla\varphi(x))$.

2.20. Let $E \subseteq \mathbb{R}^m$ be an open connected set, and $\varphi : E \to \mathbb{R}$ be a differentiable function, with $\varphi(a) = c$. If $\gamma : \mathbb{R} \to \mathbb{R}^m$ is a curve lying entirely in the level set $\varphi(x) = c$, with $\gamma(t_0) = a$, show that $\text{grad}\,\varphi(a)$ is orthogonal to the tangent vector $\gamma'(t)$ at $t = t_0$.

2.21. Let $E \subseteq \mathbb{R}^m$ be an open connected set, and $\varphi : E \to \mathbb{R}$ be a differentiable function. Show that, for any unit vector $u \in E$, $D_u\varphi(x) = \nabla\varphi(x) \cdot u$, for all $x \in E$. [**Hint:** Apply the chain rule to the function $\gamma(t) = a + t\,u$, $t \in \mathbb{R}$.]

2.22. Suppose that $E \subset \mathbb{R}^m$ is an open set, and let $f : E \to \mathbb{R}^n$ be a differentiable function. Let $a, b \in E$ be such that the line segment $L[a; b]$ with end points a and b lies inside E. Show that

$$\|f(a) - f(b)\| = (b - a) \int_0^1 J_f[(1 - t)a + tb]dt.$$

2.23. If $f : \mathbb{R}^m \to \mathbb{R}^n$ is differentiable at a, prove that there exists $\delta > 0$ and $M > 0$ such that

$$x \in B(a; \delta), \ x \neq a, \quad \Rightarrow \quad \frac{\| f(x) - f(a) \|}{\| x - a \|} < M.$$

Hence, prove Theorem 2.12.

2.24. Let $I \subseteq \mathbb{R}$ be an interval, and $f : I \to \mathbb{R}$ be a differentiable function such that $f(t, f(t)) = 0$, for all $t \in I$ and some differentiable function $f = f(x, y)$. Show that $f'(t) = -\left(\partial_x f(t, f(t)) / \partial_y f(t, f(t)) \right)$, wherever the denominator is nonzero. How about the formula if $f, g : I \to \mathbb{R}$ are two differentiable functions such that

$$\begin{cases} f_1(t, f(t), g(t)) = 0 \\ f_2(t, f(t), g(t)) = 0 \end{cases}, \quad \text{for } t \in I,$$

and $f_1 = f_1(x, y, z)$, $f_2 = f_2(x, y, z)$ are some differentiable functions.

2.25. (*Euler Theorem*) Let $E \subset \mathbb{R}^m \setminus \{0\}$ be an open set, and $\varphi \in C^1(E)$. For $a \in \mathbb{R}$, we say φ is a *homogeneous function* of degree a (or simply a-*homogeneous*) over E if $\varphi(tx) = t^a \varphi(x)$, for all $x \in E$ and $t > 0$. Show that φ is a-homogeneous if and only if $\nabla \varphi(x) \cdot x = a \varphi(x)$, for all $x \in E$.

2.26. Let $E \subset \mathbb{R}^m$ be an open set, $a \in E$, and $f : E \to \mathbb{R}^n$ be differentiable at a. Prove that

a. For any $\varepsilon > 0$, there exists a ball $B(a; \varepsilon)$ such that

$$\| f(x) - f(a) \| \leq \left(\| Df(a) \| + \varepsilon \right) \| x - a \|, \quad \text{for all} \ \ x \in B(a; \varepsilon).$$

b. For any $\varepsilon > 0$, there exists a ball $B(a; \varepsilon)$ such that

$$\| f(x) - f(y) \| \leq \left(\| Df(a) \| + \varepsilon \right) \| x - y \|, \quad \text{for all} \ \ x, y \in B(a; \varepsilon).$$

2.27. Let $\varphi(x, y) = x^2 y / (x^2 + y^2)$ for $x = (x, y) \neq (0, 0)$, with $\varphi(0, 0) = 0$. Show that, at $(0, 0)$, φ is continuous and all its directional derivatives vanish, but the tangent map $\tau_{(0,0)}$ is not a linear map.

2.28. Show that the arc length function $s(t)$ of the right circular helix $h(t) = (a \cos t, a \sin t, bt)$ over the interval $[0, 2\pi]$ is given by the formula $s(t) = \sqrt{a^2 + b^2} \, t$.

2.29. Consider the following two parametrisations of a space curve $C \subset \mathbb{R}^3$:

$$f_1(t) = (\sin t^2, \cos t^2, t^2) \quad \text{and} \quad f_2(t) = (\sin t, \cos t, t).$$

Show that the velocity $v(t)$ at $t = 1$ of a moving particle with position vector $f_2(t)$ is half than with position vector f_1.

2.30. For each of the following situations, calculate $f'(t)$, and find the equation of the tangent line at $f(0)$: (i) $f(t) = (t + 1, t^2 + 1, t^3 + 1)$; (ii) $f(t) = (e^t + 1, e^{2t} + 1, e^t + 1)$.

2.31. Let $\mathbf{f}(t) = \left(\dfrac{\cos t}{\sqrt{1 + a^2 t^2}}, \dfrac{\sin t}{\sqrt{1 + a^2 t^2}}, \dfrac{-at}{\sqrt{1 + a^2 t^2}} \right)$, with $a \neq 0$. Show that
(a) $\|\mathbf{f}(t)\| = 1$, for all t; and that (b) $\mathbf{f}'(t) \cdot \mathbf{f}(t) = \mathbf{0}$, for all t.

2.32. The highly scary helter skelter at the fair is a cylindrical tower of height 20 m and circumference 8 m. The slide is wound around the tower exactly 5 times. What's the total length of the slide? [**Hint:** Write the circular helix, and find the arc length]

2.33. Let φ_1, φ_2 be scalar fields, and $a, b \in \mathbb{R}$. Show that

a. $\operatorname{grad}(a\varphi_1 + b\varphi_2) = a \operatorname{grad}(\varphi_1) + b \operatorname{grad}(\varphi_2)$.
b. $\operatorname{grad}(\varphi_1 \cdot \varphi_2) = \varphi_1 \cdot \operatorname{grad}(\varphi_2) + \varphi_2 \cdot \operatorname{grad}(\varphi_1)$.
c. $\operatorname{grad}\left(\dfrac{\varphi_1}{\varphi_2} \right) = \dfrac{\varphi_2 \cdot \operatorname{grad}(\varphi_1) - \varphi_1 \cdot \operatorname{grad}(\varphi_2)}{\varphi_2^2}$.

2.34. Let $\varphi : \mathbb{R}^3 \to \mathbb{R}$ be scalar field. Show that $\operatorname{grad} \varphi$ gives the *rate* and the *direction* of the change in the scalar field φ.

2.35. Show that for a scalar fied $\varphi = \varphi(r, \theta)$,

$$\operatorname{grad}(\varphi) \equiv \frac{\partial \varphi}{\partial r} \widehat{e}_r + \frac{1}{r} \frac{\partial \varphi}{\partial \theta} \widehat{e}_\theta,$$

where $\widehat{e}_r$ and $\widehat{e}_\theta$ are the unit vectors along the two orthogonal axes.

2.36. Let $\varphi(x, y, z) = x^2 yz + 4xz^2$, $\mathbf{v} = 2\widehat{i} - \widehat{j} - 2\widehat{k}$, and $P = (1, -2, -1)$. Compute the value of the directional derivative $D_{\mathbf{v}} \varphi(P)$. What maximum value $D_{\mathbf{v}} \varphi(P)$ can possibly take at the point P?

2.37. Let $\varphi(x, y, z) = x^2 y^2 z^2$ and $P = (1, 1, -1)$. Compute the directional derivative $D_{\mathbf{v}} \varphi(P)$ in the direction of the tangent to the curve

$$\mathbf{u}(t) = e^t \widehat{i} + 2 \sin t \widehat{j} + (t - \cos t) \widehat{k},$$

at the point given by $t = 0$.

2.38. Compute the directional derivative of the surface $\varphi(x, y, z) = \nabla \cdot (\nabla f)$, where $f(x, y, z) = 2x^3 y^2 z^4$, at point $P = (1, -2, 1)$ in the direction of the normal to the surface $\varphi(x, y, z) = xy^2 z - z^2 - 3x$.

2.39. Find the equation of the tangent plane and normal line to the surface $\varphi(x, y, z) = x^2 yz + 6$ at the point $P = (1, -2, 3)$.

2.40. Suppose the two surfaces $\varphi_1(x, y, z) = ax^2 - byz - (a + 2)x$ and $\varphi_2(x, y, z) = 4x^2 y + z^3 - 4$ intersect orthogonally at the point $P = (1, -1, 2)$. Find the values of the constants a, b.

2.41. Let $\mathbf{r} = (x, y, z)$ be a point on a space curve with $r = |\mathbf{r}|$, and f be any differentiable function over any domain containing the curve. Show that $\nabla f(r) = f'(r) \widehat{\mathbf{r}}$.

2.42. Let $\mathbf{r} = (x, y, z)$ be a point on a space curve with $r = |\mathbf{r}|$, and f be any differentiable function over any domain containing the curve. Show that $\operatorname{div}(f(r)\mathbf{r}) = rf'(r) + 3f(r)$ and $\operatorname{curl}(f(r)\mathbf{r}) = 0$.

2.43. Show that, for any C^2 vector field $f = (f_1, f_2, f_3) \in V(\Omega)$, we have

$$\nabla \times (\nabla \times f) := \text{grad div } f - \nabla^2 f. \tag{2.3.47}$$

2.44. Find scalar potential of irrotational vector field $\mathbf{v} = r^{-3}\mathbf{r}$.

2.45. A particle moves on a circular path with an angular velocity $\mathbf{w} = (w_1, w_2, w_3)$, and suppose the velocity vector at a point $P(\mathbf{r}) = P(x, y, z)$ is given by $\mathbf{v} = (v_1, v_2, v_3)$. Show that $\text{curl}(\mathbf{v}) = 2\mathbf{w}$.

2.46. Let $\mathbf{r} = (x, y, z)$ be a point on a space curve with $r = |\mathbf{r}|$, and f be any differentiable function over any domain containing the curve. Show that

$$\nabla^2 f(r) = f''(r) + \frac{2}{r} f'(r).$$

2.47. Prove the identities (2.3.32) through (2.3.35).

2.48. Prove the identities (2.3.40) through (2.3.42).

2.49. (*Fundamental Lemma of Variational Calculus*) Let $f \in C(a, b)$ be such that

$$\int_a^b f(x)\varphi(x)\,dx = 0, \qquad \text{for all } \varphi \in C^1(a, b).$$

Show that $f \equiv 0$ over the interval (a, b).

References

1. Shifrin, T. (2005). *Multivariable mathematics—Linear algebra, multivariable calculus, and manifolds*. Wiley.
2. Razdan, A., & Ravichandran, V., (in press). *Fundamentals of analysis with applications*. Springer.
3. Munkres, J. R. (1991). *Analysis on manifolds*. Advanced Book Program, Addison-Wesley Publishing Company.

Chapter 3
Ordinary Differential Equations

Further, the dignity of the science itself seems to require that every possible means be explored for the solution of a problem so elegant and so celebrated.

Carl Friedrich Gauss (1777–1855)

In this chapter, we review some aspects of ordinary differential equations that are relevant to solve some important *initial-boundary value problems* related to multivariate phenomena modelled by linear partial differential equations. The method of *characteristic curves* discussed later in Chap. 6 depends on the method of solving first order system of initial value problems for ordinary differential equations. The method of *separation of variables* discussed later in Chaps. 7 and 8 depends on the method of solving boundary value problems for a second order linear ordinary differential equations, with Sturm–Liouville types of boundary conditions. We also discuss *differential equation models* for some interesting univariate phenomena to demonstrate connection such types of equations has with the physical world.

In Sect. 3.1, we recall the general concepts and fix notations. In Sect. 3.2, we review some solution methods for simple *integrable* first order differential equations and prove the fundamental Picard–Lindelöf theorem about the *existence* and *uniqueness* of a local solution of an initial value problem for a general first order differential equation. In Sect. 3.3, linear differential equations of order ≥ 2 are discussed. The method of *power series solution* for second order equation is also briefly described. In Sect. 3.4, we discuss boundary value problems for second order linear differential equations, focusing mainly on the *Sturm–Liouville eigenvalue problem* for regular type of boundary conditions. The same is used to develop the concept of *generalised Fourier series*. In Sect. 3.5, we discuss general first order system of differential equations and prove the *existence* and *uniqueness* theorem related to initial value problem for such a system.

The reader may refer the standard texts such as [1–7], for any missing details and further readings on the topic. For the analysis prerequisites, we recommend [8].

3.1 Introduction

Let I be a subinterval of $\mathbb{R}$. An *ordinary differential equation* (or simply a *differential equation*) for a sufficiently smooth univariate unknown function $x = x(t) : I \to \mathbb{R}$ is an equation involving x, and some of its derivatives. A more formal definition is as given below.

Definition 3.1 For open sets $U \subseteq \mathbb{R}^{n+1}$, $V \subseteq \mathbb{R}^k$ and $\Omega \subseteq I \times U \times V$, let $F : \Omega \to \mathbb{R}$ be a sufficiently smooth *nonconstant* function such that we have

$$F\left(t;\ x;\ x',\ldots,x^{(n)}; b_1,\ldots,b_k\right) = 0, \quad \text{for some function } x \in C^n(I). \quad (3.1.1)$$

A general nth order **ordinary differential equation** for the unknown function $x = x(t)$ is an implicit equation of the form (3.1.1), where functions $b_1,\ldots,b_k \in C(I)$ are called *parameters* of the equation. The interval I is called the *domain interval* of the differential equation.

The differential equation (3.1.1) is said to be of **order** n because the order of the highest derivative of function $x = x(t)$ it involves is n. Further, in practical terms, the functions b_j's represent the *physical constraints* of the physical system under study, which appear as coefficients of terms containing the variables x, x', ..., $x^{(n)}$ or as a function representing the *external force* applied to the system.

Definition 3.2 An Eq. (3.1.1) is called a *linear differential equation* if the function F is linear with respect to the variables x, x', ..., $x^{(n)}$. And, it is said to be a *nonlinear differential equation* if it is not linear in the above sense.

Said differently, a linear differential equation is the one that do not involve any power of x or its derivates or products of the form $x_i x'_j$, etc. In practical terms, a physical process is *linear* if every small change in values of $t \in I$, near a point of equilibrium, produces only proportional change in the values of the function $x = x(t)$. Therefore, there is no *blow-up*, and solutions for such types of differential equations stabilises automatically.

Example 3.1 In this example, we give some simple differential equations for an unknown function $x = x(t)$ to illustrate the above defined terminologies.

1. The equation $x'(t) = t^2 + \sin t$ is a first order linear differential equation, with coefficient $b_1 = 1$, and the *forcing function* given by $b(t) = t^2 + \sin t$.
2. The equation $3x''(t) - 2x'(t) + x(t) = t + 3t + e^{2t}$ is a second order linear differential equation, with *coefficients* $b_1 = 3$, $b_2 = -2$, $b_3 = 1$, and the *forcing function* given by $b(t) = t + 3t + e^{2t}$.

3. The equation $\big(x'(t) - x(t)\big)\big(x'(t) - 2x(t)\big) = 0$ is a nonlinear equation of first order.
4. The equation $\big[x''(t)\big]^2 + \big[x(t)\big]^2 = 0$ is a nonlinear equation of second order.

Notice that the domain interval in each case is some interval $I \subseteq \mathbb{R}$.

As rightly said by Newton, it is useful to solve differential equations. Clearly, we first need to understand what we mean by a *solution* of a differential equation.

Definition 3.3 A **solution** of an nth order differential equation defined over an interval I is a function $\varphi = \varphi(t)$ such that (a) $\varphi \in C^n(J)$; (b) $\big(t, \varphi(t)\big) \in U$; and, (c) $F\big(t; \varphi(t), \varphi'(t), \dots, \varphi^{(n)}(t)\big) = 0$, where J is some subinterval of the domain interval I. We say φ is a *global solution* of the differential equation, when $J = I$.

The three significant questions that we need to ponder on are as given below:

1. *When a differential equation has a solution?*
2. *How to find the one, if it exists?*
3. *Is a solution of a differential equation always unique?*

The answer to the first question depends on several factors. In general, a general differential equation of the form (3.1.1) may not have any nontrivial solution. For example, the second order nonlinear equation as in (4) of Example 3.1 has only the trivial solution $\varphi \equiv 0$. Further, the *general solution* may not be an *n-parameter family of solutions*. For example, the first order nonlinear equation as in (3) of Example 3.1 has a solution of the form

$$\big(x(t) - a\,e^t\big)\big(x(t) - b\,e^{2t}\big) = 0, \quad \text{for } a, b \in \mathbb{R}.$$

In some simpler situations, the *order* and *linearity type* of a differential equation help prove the *existence* of a solution, by applying some known analytical or geometric ideas. Next, it is difficult to answer the second question for a general differential equation is of the form (3.1.1). However, for

$$\big(t; \boldsymbol{x}; \boldsymbol{b}\big) = \big(t; x; x', \dots, x^{(n)}; b_1, \dots, b_k\big) \in \Omega,$$

when the function $F = F(t; \boldsymbol{x}; \boldsymbol{b})$ is known to be sufficiently smooth in some neighborhood of a point $(t_0; \boldsymbol{x}_0) \in I \times U$ such that

$$\frac{\partial F}{\partial (t; \boldsymbol{x})_k}\big[(t_0; \boldsymbol{x}_0)\big] \neq 0,$$

we can invoke *implicit function theorem* to write a general equation (3.1.1) *locally* in the *Normal form* as given below:

$$\frac{d^n x}{dt^n} = G\big(t; x, x', \dots, x^{(n-1)}; b_1, \dots, b_k\big), \quad \text{for } t \in I, \tag{3.1.2}$$

where G is some nice function. As we shall see in the sequel, it is possible to develop a general theory for nth order differential equations of the form (3.1.2). Moreover, in this case, the general solution is given by an *n-parameter* family of functions in $C^n(J)$, for some interval $J \subseteq I$.

We consider next the third important question: *When a solution of a differential equation is unique?* By definition, we say a solution $\varphi : J \to \mathbb{R}$ of a differential equation is *unique* if, for any other solution $\phi : K \to \mathbb{R}$, we have $\varphi \equiv \phi$ over the interval $J \cap K$, where both J and K are contained in the domain interval I. In practical terms, the (global) uniqueness of a solution in above sense is important to interpret analytical or geometrical properties of a solution in physical terms. More generally, to obtain a *particular solution* from an n-parameter family of solutions of a differential equation of the form (3.1.2), it should also be clear that at least n equations involving such parameters must be known in advance. Such *auxiliary equations* are usually obtained by using certain types of predefined conditions that the unknown function $x = x(t)$ satisfies at certain point(s) of the domain interval I.

Uniqueness of a solution is always subject to assigned *auxiliary conditions*. In any particular situation, the values assigned to the involved unknown function $\varphi : I \to \mathbb{R}^n$, and some of its derivatives, at a point $t_0 \in I$ are called the initial values. Further, for a *boundary value problem* related to an ordinary differential equation, various types of boundary conditions are specified as the sum of values of $\varphi, \varphi', \ldots$ at the two end points of the interval I. In practical terms, specification of *initial values* is related to the *state* of the system, and *boundary conditions* takes into consideration the physical environment of the problem. Therefore, specification of above two types of *initial data* for the involved unknown function is a part of the modeling process of any univariate phenomenon.

Definition 3.4 Given an nth order differential equation for an unknown function $x : I \to \mathbb{R}$, say of the form (3.1.2), the values of the functions $x, x', \ldots, x^{(n-1)}$ specified at a single-point $t_0 \in I$ are known as a set of *initial values* of the differential equation; and, a system of n equations involving values of the function $x, x', \ldots, x^{(n-1)}$ specified at the two end points of I are known as the *boundary conditions* of the differential equation.

In actual practice, the initial values and boundary conditions are specified by considering some relevant *physical constraints* of the dynamical system under investigation. The importance of initial values is related to the study of the *state*[1] of a dynamical system, whereas the assigned boundary conditions concern factors influencing the *environment* of the system. Therefore, we are led to the following two types of fundamental problems related to various types of differential equations.

1. An *initial-value problem* (or simply IVP) for an nth order differential equation (3.1.2) refers to the problem of finding a function $\varphi \in C^n(I)$ satisfying the equation, and also the assigned initial values at some $t = t_0 \in I$. We say an IVP is

---

[1] The terminology **state** of a dynamical system refers to the mathematical data that fully describes the system at any time $t > t_0$, provided the same is known at some initial time $t = t_0$. In some situations, a physical theory postulates how the state of a dynamical system be determined.

homogeneous if all the assigned initial values are zero. We come across initial value problems, also known as a *Cauchy problems*, while studying the time-evolution of a dynamical system in terms of *data* given at some initial time $t_0 \in I$.

2. A *boundary value problem* (or simply BVP) for an nth order differential equation (3.1.2) is about finding a function $\varphi \in C^n(I)$ that satisfies the equation, and also certain boundary conditions involving the functions $x, x', \ldots, x^{(n-1)}$ assigned at the two end points of the interval I. We may write n linear boundary conditions $B_i[x]$ for a function $x = x(t)$ as

$$B_i[x]: \quad \sum_{j=1}^{n} a_{ij} x^{(j-1)} + \sum_{j=1}^{n} b_{ij} x^{(j-1)} = c_i, \quad \text{for } i = 1, \ldots, n, \quad (3.1.3)$$

where $a_{ij}, b_{ij}, c_i \in \mathbb{R}$. In general, we may come across more complex types of boundary conditions. A BVP is said to be *homogeneous* if the values c_i are zero.

Further, a *solution* φ of an *initial/boundary value problem* is a solution in the sense of Definition 3.3 such that φ also satisfies the assigned initial and/or boundary conditions. For some practical problems related to a differential equation model of a dynamical system under investigation, it is enough to know the analytic/geometric properties of a solution of the underlying *initial/boundary value problems*. In all such cases, we do not seek a solution explicitly.

However, in many other cases, we have to find a suitable analytical or numerical method to determine a unique solution of a given initial/boundary value problem, and also study the *maximal interval* of the solution validity. Therefore, practical significance of a differential equation model for a dynamical system is determined in terms of how *well-posed* the associated *initial-boundary value problem* is.

Definition 3.5 *(Hadamard)* An IVP (or a BVP) is said to be **well-posed** if

1. (*Existence*) the problem has <u>at least one</u> solution;
2. (*Uniqueness*) the problem has <u>at most one</u> solution;
3. (*Continuity*) if (1) and (2) hold, the solution <u>depends continuously</u> on the assigned *initial values* (or, respectively, *boundary conditions*).

Since the condition (3) pertains to the *stability* of the dynamical system under investigation, it is most significant among the three. We say an IVP (or a BVP) is **ill-posed** if any of the above three conditions fails to hold.

In general, the *existence* of a solution $\varphi = \varphi(t)$ of an IVP (or a BVP) is about specifying necessary and/or sufficient condition(s) that a possible solution $\varphi : J \to \mathbb{R}$ of the problem must satisfy, where the subinterval $J \subseteq I$ is also known as a the *solution interval*. Therefore, the existence of a solution of an IVP (or a BVP) is a subject matter of abstract analysis. In the sequel, we prove *existence theorems* for some important initial value problems, and also for certain special types of boundary value problems. The issues related to *uniqueness* of a solution of an IVP (or a BVP) are always subject to assigned initial values (or boundary conditions). As said earlier, we need to consider such issues to facilitate *physical interpretations* of a solution in terms of its analytical and/or geometric properties.

In Sect. 3.2, our main focus is to prove the *existence* and *uniqueness* of a local solution of an IVP for a general first order differential equation. A generalisation of the same for a *first order system* is proved in Sect. 3.5. In Sect. 3.4, we develop the method of *eigenfunctions expansion* and use the same to solve some important types of *boundary value problems* for second order linear differential equation. In all that follows, we review methods of solving an IVP (or a BVP) for some simple differential equations of the form (3.1.2).

3.2 First Order Differential Equations

In this section, our aim is to prove the *existence* and *uniqueness* theorems for an *initial value problem* of the form

$$\frac{dx}{dt} = f(t; x; b), \quad \text{with } x(t_0) = x_0, \tag{3.2.1}$$

where $f = f(t, x)$ is a continuous function defined on a domain $D \subseteq I \times \mathbb{R}$, and the vector $b \in \mathbb{R}^k$ ($k \geq 1$) of *parameters* represents the *physical constraints* of the underlying dynamical system. For brevity, we may write the function f on the right side of (3.2.1) simply as $f = f(t, x)$. In geometrical terms, the function f defines a *direction field* as the collection of *line elements*[2] at points $(t, x) \in D$, and so a solution of Eq. (3.2.1) over a subinterval $J \subseteq \mathbb{R}$ is a function $\varphi : J \to \mathbb{R}$ such that

$$\frac{d\varphi}{dt} = f(t, \varphi(t)), \quad \text{for } (t, \varphi(t)) \in D,$$

which is *unique* subject to the *initial condition* $\varphi(t_0) = x_0$ (Theorem 3.2). Therefore, a general solution of a first order differential equation is given by a *family of curves* that covers a subdomain $J \times \mathbb{R}$ (Theorem 3.1).

Notice that, if $\gamma : J \to D$ is a parametrisation of a unique solution φ given by $t \mapsto (t, \varphi(t))$, then the curve

$$C_\gamma(t_0) = \{(t, \varphi(t)) : t \in I\} \subset D$$

is such that $\gamma(t_0) = (t_0, x_0)$, and the *gradient* $\nabla\gamma = \varphi'$ at a point $(t, \varphi(t)) \in D$ is the function $f = f(t, x)$. In this case, we say C_γ is an *integral curve* of Eq. (3.2.1). In general, a mathematical study of the geometry of integral curves of a *differential equation model* of the type (3.2.1) is fundamental to many different types of practical problems concerning some univariate dynamical systems [1, 9].

For example, most problems related to the growth and decay of some specie in a competitive environment leads to such types of initial value problems, where

---

[2] A numerical triplet (t, x, p) is called a **line element** at a point $(t, x) \in D$ defined by an initial value problem (3.2.1) if $p = x'(t) = f(t, x)$.

$f(t, x)$ represents the difference between the *birth* and *death rates*. In such situations, the parameters are defined considering the effects of environmental factors such as pollution, technological advancements, market competition, and growth fluctuations across species of diverse social fabric. We discuss a prototypical *growth model* in the next example.

Example 3.2 *(Growth Models)* Suppose $p(t)$ denote the number of units of a population at time t, with the *initial population* being given by $p(t_0) = p_0$. We can take the function $p(t)$ to be nice enough if the population size is assumed to be sufficiently large. Further, it is assumed that the population under investigation is *isolated* so that the possibility of an immigration or emigration is ruled out. The *logistic law*[3] of *population growth* states that

$$p'(t) = \alpha\, p - \beta\, p^2, \quad \text{with } \beta > 0 \text{ and } p(t_0) = p_0. \qquad (3.2.2)$$

The *parameters* $\beta \ll \alpha$ play a vital role in predicting the future growth in a competitive environment, where α is the (relative) growth rate and the ratio α/β is the *maximum occupation density*. The above differential equation model was first published in 1837 by the Belgian mathematician-biologists *Pierre Verhulst* (1804–1849). The involved first order *nonlinear equation* is commonly known as the Bernoulli's equation. It follows easily that, under certain suitable conditions, $p(t) \to \alpha/\beta$ as $t \to \infty$, regardless of the value assigned to p_0. Notice that, in biological terms, the ration α/β is called the *carrying capacity*. Clearly then, the expression given by

$$\frac{p'}{p} = \beta\left(\frac{\alpha}{\beta} - p\right) \qquad (3.2.3)$$

represents the *per capita rate* of decrease in population. The particular case when

$$f(t, p) = \rho\, p(t) \qquad (3.2.4)$$

gives a simplest type of *growth model*, which is also known as the *Malthusian law* of unrestricted growth, where the constant ρ represents the available resource for an early stage colonisation. However, this particular model serves well only when the time period of the study is small or the population size is not too large. In general, we need to make a number of *simplifying assumptions* in order to obtain a *linear growth models*. For example, as expected, no population is isolated *per se*. Also, in general, *parameters* α and β in Eq. (3.2.2) are functions of time t.

As a warm-up, we first focus on the cases when a specific form of $f = f(t, x)$ makes a given differential equation an *integrable system*. That is, it would be possible to write a general solution in explicit form.

---

[3] The terminology is justified on the basis of the fact that, for $p_0 < \alpha/2\beta$, the solution is an *S-shaped* curve.

3.2.1 Integrable Forms

To start with, we consider the case of a first order differential equation given in a symmetric form. That is, when

$$f(t, x) = -\frac{q(t, x)}{p(t, x)}, \quad \text{for some } 0 \neq p, \ q \in C(D),$$

which we can also write as

$$p\frac{dx}{dt} + q = 0 \quad \text{or} \quad p\,dx + q\,dt = 0. \tag{3.2.5}$$

Clearly, to avoid trivial situations, we must have $p^2 + q^2 > 0$ over D.

Definition 3.6 A differential equation of the form (3.2.5) is said to be **exact** over a domain $D \subseteq I \times \mathbb{R}$ if the pair (q, p) is a *gradient field*, i.e. there exists a function $F = F(t, x) \in C^1(D)$ such that

$$q(t, x) = \frac{\partial F}{\partial t} \quad \text{and} \quad p(t, x) = \frac{\partial F}{\partial x}. \tag{3.2.6}$$

The function F is called a *potential function* for the field (q, p).

Therefore, in this case, we have

$$dF = p(t, x)\frac{dx}{dt} + q(t, x) = 0 \quad \Longrightarrow \quad F(t, x) = \text{constant},$$

which gives a general solution of (3.2.5).

Lemma 3.1 *A necessary condition for (3.2.5) to be exact over a domain $D \subseteq I \times \mathbb{R}$ is that we have $\partial p/\partial t, \ \partial q/\partial x \in C(D)$, and*

$$\frac{\partial p(t, x)}{\partial t} = \frac{\partial q(t, x)}{\partial x}. \tag{3.2.7}$$

Conversely, if D is a simply connected[4] *domain, and the two functions $p, q \in C^1(D)$ satisfy the condition (3.2.7), then Eq. (3.2.5) is exact.*

Proof Suppose (3.2.5) is exact, where functions p, q satisfy the given condition. Then, for a potential function $F = F(t, x) \in C^1(D)$, the two equations in (3.2.6) hold. Therefore, we have

$$\frac{\partial p(t, x)}{\partial t} = \frac{\partial^2 F(t, x)}{\partial x \partial t} = \frac{\partial^2 F(t, x)}{\partial t \partial x} = \frac{\partial q(t, x)}{\partial x}. \tag{3.2.8}$$

---

[4] Recall that a domain D is said to be **simply connected** if every simple closed curve in D can be *shrunk continuously* to a point. In geometrical terms, we say the domain D is without any *hole*.

Conversely, suppose D is a simply connected domain and for some $p, q \in C^1(D)$, Eq. (3.2.7) holds. Then, it can be shown that the C^1-function $F = F(t, x)$ given by

$$F(t, x) = \int^t p(u, x)du + \int^x \left[q(t, y) - \frac{\partial}{\partial y} \int^t p(u, y)du\right]dy \qquad (3.2.9)$$

is a potential function for the field (q, p). More specifically, the function $F = F(t, x)$ is obtained as the line integral of the 1-form $pdx + qdt$ along a C^1-path from a fixed point (t_0, x_0) to the point (t, x), and the condition (3.2.7) ensures that F is independent of the path (Theorem 2.23).   □

Notice that, since $p^2 + q^2 > 0$, it follows from (3.2.6) that

$$\left(\frac{\partial F}{\partial t}\right)^2 + \left(\frac{\partial F}{\partial x}\right)^2 > 0 \quad \text{over the domain } D.$$

Let $(t_0, x_0) \in D$ be such that $F(t_0, x_0) = c$, for some constant c. Now, for example, if $\partial F/\partial x \neq 0$ then by *implicit function theorem* $F(t, x) = c$ has a unique C^1-solution $x = x(t)$ in some neighbourhood of the point (t_0, x_0) so that $F(t, x(t)) = c$ is a solution of Eq. (3.2.5).

Example 3.3 Let $p(t, x) = 2t^2 + x^2$ and $q(t, x) = 4tx$. Clearly then, with these p and q, Eq. (3.2.5) is exact. We also have

$$F(t, x) = \int^t p(u, x)du + \int^x \left[q(t, y) - \frac{\partial}{\partial y} \int^t p(u, y)du\right]dy$$

$$= \int^t [2u^2 + x^2]du + \int^x \left[4ty - \frac{\partial}{\partial y} \int^t [2u^2 + y^2]du\right]dy$$

$$= (2/3)t^3 + tx^2 + \int^x \left[4ty - \frac{\partial}{\partial y}((2/3)u^3 + ty^2)\right]dy$$

$$= (2/3)t^3 + tx^2 + \int^x [4ty - 2ty]dy$$

$$= (2/3)t^3 + 2tx^2.$$

Hence, a general solution of the exact equation

$$(2t^2 + x^2)\frac{dx}{dt} + (4tx) = 0$$

is given by a potential function $F(t, x) = (2/3)t^3 + 2tx^2 = c$, where c is a constant.

We consider next the case when a differential equation (3.2.5) is not exact. In this case, we find a suitable function $\mu = \mu(t, x)$ such that the equation given by

$$\mu(t, x)p(t, x)\frac{dx}{dt} + \mu(t, x)q(t, x) = 0 \qquad (3.2.10)$$

is an *exact equation*. That is, we have

$$\frac{\partial\big(\mu(t, x)p(t, x)\big)}{\partial t} = \frac{\partial\big(\mu(t, x)q(t, x)\big)}{\partial x}. \qquad (3.2.11)$$

Definition 3.7 A function $\mu = \mu(t, x)$ satisfying the condition (3.2.11) is called an *integrating factor* (or *Euler multiplier*) for Eq. (3.2.5).

If D is a simply connected domain, it follows from (3.2.11), and above lemma, that a function $\mu = \mu(t, x) \in C^1(D)$ is an integrating factor for Eq. (3.2.5) if and only if μ is a solution of the first order partial differential equation given by

$$q(t, x)\frac{\partial\mu}{\partial t} - p(t, x)\frac{\partial\mu}{\partial x} + \Big(\frac{\partial q}{\partial t} - \frac{\partial p}{\partial x}\Big)\mu(t, x) = 0. \qquad (3.2.12)$$

Therefore, in general, the problem of finding an integrating factor for Eq. (3.2.5) is as hard as solving the equation itself (also see Remark 3.9).

However, it is possible in some particular cases to find a formula for $\mu = \mu(t, x)$ that works just fine. For a ready reference, we list some explicit formulas of $\mu = \mu(t, x)$ that are helpful to solve some specific types of equations of the form (3.2.1). Each one of the formulas given below is subject to a specified condition that the given coefficient functions p and q must satisfy.

1. Suppose $\mu(t, x) = g(t)$, for some function $g \in C(I)$. In this case, it follows from (3.2.11) that

$$\frac{dg}{g} = \frac{\partial p/\partial x - \partial q/\partial t}{q}\, dt.$$

Therefore, if the functions p and q satisfy the condition

$$\frac{\partial p/\partial x - \partial q/\partial t}{q} = f(t),$$

then the integrating factor $\mu(t, x)$ is given by

$$\mu(t, x) = g(t) = e^{\int^t f(u)du}.$$

For example, if

$$p(t, x) = 3x^2 + t \quad \text{and} \quad q(t, x) = t^2 + 6tx + 2,$$

then we have $f(t) = -2t$, and hence $\mu(t, x) = e^{-t^2}$ is an integrating factor.

2. Suppose $\mu(t, x) = h(x)$, for some function $g \in C(I)$. In this case, it follows from (3.2.11) that

$$\frac{dh}{h} = \frac{\partial q/\partial t - \partial p/\partial x}{p} dx.$$

Therefore, if the functions p and q satisfy the condition

$$\frac{\partial q/\partial t - \partial p/\partial x}{p} = k(x),$$

then the integrating factor $\mu(t, x)$ is given by

$$\mu(t, x) = h(x) = e^{\int^x k(u)du}.$$

3. Suppose $\mu(t, x) = g(tx)$, for some function $g \in C(I)$. In this case, it follows from (3.2.11) that

$$\frac{dg(u)}{g(u)} = \frac{\partial p/\partial x - \partial q/\partial t}{xq - tp} du, \quad \text{with } u = tx.$$

Therefore, if the functions p and q satisfy the condition

$$\frac{\partial p/\partial x - \partial q/\partial t}{xq - tp} = f(u),$$

then the integrating factor $\mu(t, x)$ is given by

$$\mu(t, x) = g(u) = e^{\int^u f(y)dy}.$$

4. Suppose $\mu(t, x) = g(t/x)$, for some function $g \in C(I)$. In this case, it follows from (3.2.11) that

$$\frac{dg(u)}{g(u)} = \frac{x^2(\partial p/\partial x - \partial q/\partial t)}{tp + xq} du, \quad \text{with } u = \frac{t}{x}.$$

Therefore, if the functions p and q satisfy the condition

$$\frac{x^2(\partial p/\partial x - \partial q/\partial t)}{tp + xq} = f(u),$$

then the integrating factor $\mu(t, x)$ is given by

$$\mu(t, x) = g(u) = e^{\int^u f(y)dy}.$$

5. Suppose $\mu(t, x) = g(x/t)$, for some function $g \in C(I)$. In this case, it follows from (3.2.11) that

$$\frac{dg(u)}{g(u)} = \frac{t^2(\partial q/\partial t - \partial p/\partial x)}{tp + xq} \, du, \quad \text{with } u = \frac{x}{t}.$$

Therefore, if the functions p and q satisfy the condition

$$\frac{t^2(\partial q/\partial t - \partial p/\partial x)}{tp + xq} = f(u),$$

then the integrating factor $\mu(t, x)$ is given by

$$\mu(t, x) = g(u) = e^{\int^u f(y) dy}.$$

In addition, if

$$p(t, x) = t\left(A\, t^a x^b + B\, t^c x^d\right) \quad \text{and} \quad q(t, x) = t\left(C\, t^a x^b + D\, t^c x^d\right),$$

then one may use $\mu(t, x) = t^\alpha x^\beta$ as a trial integrating factor, and then try to determine suitable values for constants α and β, by actual substitutions.

Variables Separable Form

Suppose the function $f = f(t, x)$ is of *variables separable* form. That is, it is possible to express the function $f = f(t, x)$ as

$$f(t, x) = g(t)h(x), \quad \text{for some } g \in C(I) \quad \text{and} \quad h \in C(J), \qquad (3.2.13)$$

with $h(x) \in \mathbb{R} \setminus \{0\}$, for all $x \in J$. We first consider the particular case when $f(t, x)$ is independent of the variable x. Notice that, in this case, the direction field is defined on the strip $I \times \mathbb{R}$, which is *translation invariant* with respect to x-axis. Therefore, it suffices to find the integral curve of the differential equation (3.2.1) in terms of the *line elements* for the points in the set $I \times \{0\}$. For, let $t_0 \in I$ be fixed, and define $G : I \to \mathbb{R}$ by

$$G(t) := \int_{t_0}^{t} g(s) \, ds, \quad \text{with } G(t_0) = 0.$$

Clearly then, since $G' = g$ by the fundamental theorem of integral calculus, it follows that the function G is a solution. In general, we can obtain an integral curve passing through the point $(\tau, \xi) \in I \times \mathbb{R}$ as given by

$$x(t; \tau, \xi) = G(t) + (\xi - G(\xi)),$$

which is a *global solution*.

Example 3.4 Consider the first order equation $x'(t) = t^2 + \sin t$ so that $g(t) = t^2 + \sin t$, with $x(0) = 0$. Therefore, in this case, the global solution is given by

$$x(t; \tau, \xi) = \frac{t^3}{3} - \cos t + \left(\xi - \frac{\xi^3}{3} + \cos \xi \right).$$

Next, we consider the particular case when $f(t, x)$ in (3.2.13) is independent of the variable t. That is, we have an *autonomous* first order equation of the form

$$\frac{dx}{dt} = h(x), \quad \text{for } x \in J.$$

In this case, the direction field is defined on the strip $\mathbb{R} \times J$, which is *translation invariant* with respect to t-axis. Therefore, it suffice to know the integral curve in terms of the line elements for the points in the set $\{0\} \times I$. By using the continuity of h, we may also assume that $h > 0$ over the interval J. For $x_0 \in J$ fixed, define the primitive of $1/h$ over J by a differentiable function $H : J \to \mathbb{R}$ given by

$$H(x) := \int_{x_0}^{x} \frac{1}{h(y)} \, dy, \quad \text{for } x \in J.$$

It thus follows from the condition $h > 0$ that the function H is monotonically increasing, and hence injective. Also, $H^{-1} : H(J) \to J$ is differentiable, by *inverse function theorem*. Indeed, we have that $H(J)$ is an interval containing 0. Now, for $H(x) = t$, we have

$$\frac{d}{dt}(H(x)) = \frac{dH}{dx}\frac{dx}{dt} = \frac{1}{h(x)}\frac{dx}{dt} = 1.$$

So, the function $x(t) = H^{-1}(t)$ provides a *particular solution* of the equation $x'(t) = h(x)$, with $x(0) = x_0$. Thus, in this case, the curve $x = x(t)$ passes through the point $(0, x_0)$, which is not a global solution. We show that an arbitrary solution is given by

$$x(t) = H^{-1}(t - c), \quad \text{for } t \in c + H(J).$$

For, suppose $\varphi = \varphi(t)$ is a solution given over some interval I_φ. Then, $\varphi'(t) = h(\varphi(t))$. For $t_0 \in I_\varphi$, with $\varphi(t_0) = y_0$, we have

$$\int_{t_0}^{t} \frac{\varphi'(u)}{h(\varphi(u))} \, du = \int_{t_0}^{t} 1 \, du = t - t_0,$$

which gives

$$H(\varphi) - \int_{x_0}^{y_0} \frac{1}{h(y)} \, dy = t - t_0 \quad \text{i.e. } H(\varphi) - H(y_0) = t - t_0.$$

Therefore, it follows that $t - t_0 + H(y_0) \in H(J)$. Hence, for any $x \in t_0 - H(y_0) + H(J)$, we can write the general solution as

$$\varphi(t) = H^{-1}(t - t_0 + H(y_0)).$$

In this case, the integral curve passes through every point of the domain $\mathbb{R} \times J$.

Example 3.5 Consider an *autonomous* first order equation given by $x'(t) = -x^2$. Here, we have $h(x) = -x^2$. It follows from the above discussion that the one-parameter family of solutions is given by $\varphi(x) = (x + c)^{-1}$. Also, $\varphi \equiv 0$ over $\mathbb{R}$ is a solution. Further, for an autonomous first order equation given by $x'(t) = 1 + x^2$, we have $h(x) = 1 + x^2$. In this case, the general solution is given by $\varphi(x) = \tan(x + c)$.

Finally, in the general case, a solution is obtained by putting together solutions of the two auxiliary first order equations given by

$$\frac{dx}{du} = h(x) \quad \text{and} \quad \frac{du}{dt} = g(t). \tag{3.2.14}$$

The main argument is based on the assertion that, for $(t_1, x_1) \in I \times J$, a solution $x = x(t)$ satisfying the condition $x(t_1) = x_1$ is given by

$$x(t) = H^{-1}(G(t) - G(t_1) + H(x_1)), \tag{3.2.15}$$

where the functions G, H are as defined in above cases. For, there exists a unique $u_1 \in H(J)$ such that $H^{-1}(u_1) = x_1$. The solution of the first equation in (3.2.14) satisfying the condition $x(u_1) = x_1$ is given by

$$x(u) = H^{-1}(u - u_1 + H(x_1)), \quad \text{for } u \in u_1 - H(x_1) + H(J). \tag{3.2.16}$$

Also, solution of the second equation in (3.2.14) satisfying the condition $u(t_1) = u_1$ is given by

$$u(t; t_1, u_1) = G(t) + (u_1 - G(t_1)) \cdot G(t) \in G(t_1) - H(x_1) + H(J). \tag{3.2.17}$$

By combining together the last two equations, the assertion is proven. Notice that the formula makes sense for $t \in I$ such that

$$G(t) - (u_1 - G(t_1)) \in u_1 - H(x_1) + H(J) \quad \text{i.e. if } G(t) \in G(t_1) - H(x_1) + H(J).$$

It can be shown that the set

$$\{t \in I : G(t) \in G(t_1) - H(x_1) + H(J)\}$$

is a nondegenerate interval. Hence, the general solution of (3.2.13) is given by

$$x(t) = H^{-1}(G(t) - c). \qquad (3.2.18)$$

Example 3.6 Consider the first order differential equation $x' = e^x \sin t$. By separation of variables, the single-parameter general solution is given by

$$x(t; c) = -\log(\cos t + c), \quad \text{with } \cos t + c > 0.$$

Notice that, in this case, the direction field is periodic of period 2π with respect to the variable t, and it is symmetric about the x-axis. Also, the above solution behaves differently depending on the value of the parameter c. Clearly, for $c > 1$, solutions exist over $\mathbb{R}$, and all are bounded. However, for $-1 < c \le 1$, solutions exist only over bounded intervals, and each increases without bounds. For example, with $x(0) = \alpha$, a unique solution of the initial value problem is given by

$$x(t; e^{-\alpha} - 1) = -\log(\cos t + e^{-\alpha} - 1),$$

so that, when $\alpha = -\log 2$, it follows that the solution exists only over the interval $(-\pi, \pi)$, and it approaches ∞ as $t \to \pm\pi$. As said earlier, for $\alpha < -\log 2$, the solutions exist over $\mathbb{R}$, and all are bounded; whereas, for $\alpha > -\log 2$, the solutions exist only over the interval $|t| < \cos^{-1}(1 - e^{-\alpha})$. Notice that, in the latter case, the length of interval of validity approaches 0 as $\alpha \to \infty$.

Linear Differential Equation

A first order differential equation (3.2.1) is called a *linear equation* if the function $f = f(t, x)$ is of the form

$$f(t, x) = -p(t)x(t) + q(t), \quad \text{for some } p, q \in C(I).$$

In this case, we need to solve a given initial value problem for the equation

$$\mathscr{L}x(t) = q(t), \quad \text{where } \mathscr{L} \equiv \frac{d}{dt} + p(t). \qquad (3.2.19)$$

As said earlier, we say (3.2.19) is a *homogeneous equation* if $q \equiv 0$ over I. Clearly then, if φ_1 and φ_2 are any two solutions of the equation $\mathscr{L}x = 0$, the linearity of the first order *differential operator* L implies that, for any scalar c, both $\varphi_1 + \varphi_2$ and $c\,\varphi_1$ are also solution of the equation $\mathscr{L}x = 0$. Recall that this property is called the *principle of superposition*.

Therefore, the set of all solutions of a homogeneous equation $\mathscr{L}x = 0$ is a subspace of the linear space $C(I)$. Further, it should also be clear that any two solutions of the *nonhomogeneous equation* $\mathscr{L}x = q$ differ only by a solution of the corresponding homogeneous equation $\mathscr{L}x = 0$. Hence, a general solution of a linear differential equation (3.2.19) is given by

$$\varphi = \varphi_h + \varphi_p, \quad \text{with } \mathscr{L}\varphi_h = 0 \text{ and } \mathscr{L}\varphi_p = q. \qquad (3.2.20)$$

We call φ_h *complementary function* and φ_p a *particular solution* of Eq. (3.2.19). The procedure described below shows how to obtain the two function φ_h and φ_p for any given linear equation of first order.

For, let $t_0 \in I$ be arbitrary fixed point. Recall that, by variable separable method, a single-parameter complementary function φ_h is given by

$$\varphi(t; c) = c\, e^{\int_{t_0}^{t} p(s)\, ds}, \quad \text{for } t \in I \quad \text{and} \quad c = \varphi(t_0). \tag{3.2.21}$$

Next, we find a particular solution φ_p by well known technique, called the *variation of parameters*. That is, we find a suitable function $c : I \to \mathbb{R}$ such that the function

$$\varphi_p(t) = c(t)e^{\int_{t_0}^{t} p(s)\, ds} \tag{3.2.22}$$

is a solution of Eq. (3.2.19). In this case, by a simple computation, it follows that

$$c(t) = c(t_0) + \int_{t_0}^{t} \left(\int_{t_0}^{u} p(s)\, ds \right) q(u)\, du. \tag{3.2.23}$$

Notice that $\varphi_p(t_0) = c(t_0)$. Hence, in this case, a unique solution of the initial value problem (3.2.1) is given by

$$\varphi(t) = \varphi_h(t) + \varphi_p(t)$$

$$= \left[\varphi_p(t_0) + \int_{t_0}^{t} \left(\int_{t_0}^{u} p(s)\, ds \right) q(u)\, du \right] e^{-\int_{t_0}^{t} p(s)\, ds}. \tag{3.2.24}$$

Example 3.7   Consider the initial value problem

$$x'(t) - 2x(t) = 3t^2 - t + 4, \quad \text{with } x(0) = 4.$$

Here, we have $p(t) = -2$ and $q(t) = 3t^2 - t + 4$. Therefore, by using the formula (3.2.24), we obtain

$$\varphi(t) = \left[\varphi_p(t_0) + \int_{t_0}^{t} \left(\int_{t_0}^{u} p(s)\, ds \right) q(u)\, du \right] e^{-\int_{t_0}^{t} p(s)\, ds}$$

$$= \left[4 + \int_{0}^{t} [3u^2 - u + 4]e^{-2u}\, du \right] e^{2t}$$

Alternatively, we may also apply the method of *integrating factor* to find a particular solution φ_p. Here, the basic idea is to find a suitable function $\mu(t)$, called an *integrating factor*, such that multiplying Eq. (3.2.19) by $\mu(t)$ results into an equation having the left side as an *exact differential*. That is, the equation

$$\mu(t)x'(t) + \mu(t)p(t)x(t) = 0$$

is an *exact equation*. In this case, it can be seen easily that the desired function μ is given by the formula

$$\mu(t) := e^{\int^t p(s)\,ds}, \quad \text{for } t \in I, \tag{3.2.25}$$

so that we can write Eq. (3.2.19) as

$$\frac{d}{dt}\left(e^{\int^t p(s)\,ds}x(t)\right) = q(t)e^{\int p\,dt}.$$

Once again, we obtain the formula (3.2.24) for a unique solution of the initial value problem (3.2.1), where $\varphi_p(t_0) = x(t_0)$.

Example 3.8 With $p = \alpha$ and $q = e^{-\beta t}$, we have

$$\frac{d}{dt}\left(e^{\int^t \alpha\,ds}x(t)\right) = e^{-\beta t} \times e^{\int^t \alpha\,ds} = e^{(\alpha-\beta)t},$$

and so the initial condition $x(0) = 0$ gives

$$x(t) = \frac{e^{-(\gamma-\alpha)t}}{\gamma}\left[1 - e^{-\gamma t}\right], \quad \text{with } \gamma = \alpha - \beta. \tag{3.2.26}$$

Transformation
It is possible to solve *initial value problems* for some other types of first order differential equations of the form (3.2.1), by using a simple *transformation* involving the variables t and x. In each case, the basic idea is to apply a suitable transformation to change the given equation to a known forms such as discussed above. We start with the case when the function $f = f(t, x)$ in (3.2.1) is *homogeneous*.

Definition 3.8 A function $f = f(t, x)$ defined on a domain $D \subseteq I \times \mathbb{R}$ is called a *homogeneous function* of degree k if

$$f(u\,t, u\,x) = u^k f(t, x), \quad \text{for all } (t, x) \in D.$$

Equivalently, a function $f(t, x)$ is said to be *homogeneous* if we can write $f(t, x) = g(x/t)$, for some function g, with $t \neq 0$. In this case, we use the transformation $y = x/t$ so that we obtain

$$\frac{dy}{dt} = -\frac{x}{t^2} + \frac{1}{t}\frac{dy}{dt} = \frac{g(y) - y}{t},$$

which is an equation in *separable form*. More generally, if

$$f(t, x) = f\left(\frac{ax + bt + c}{\alpha x + \beta t + \gamma}\right),$$

and $a\beta - b\alpha \neq 0$, we use a transformations given by $u = x - x_0$ and $v = t - t_0$, where (x_0, t_0) are obtained as the unique solution of the linear (algebraic) equations

$$ax + bt + c = 0 \quad \text{and} \quad \alpha x + \beta t + \gamma = 0.$$

It then follows that we need to solve the homogeneous equation of the form

$$\frac{du}{dt} = G\left(\frac{au + bv}{\alpha u + \beta v}\right), \quad \text{for some function } G.$$

Notice that, if $a\beta - \alpha b = 0$, then we have

$$\frac{dx}{dt} = G(ax + bt), \quad \text{for some function } G,$$

which transforms easily to an equation of the form (3.2.19), by using $u = ax + bt$.

Example 3.9 Consider a first order differential equation of the form

$$x'(t) = \frac{x + 1}{t + 2} - \exp\left(\frac{x + 1}{t + 2}\right).$$

Here, we have $t_0 = -2$ and $x_0 = -1$, and so the differential equation for u is

$$\frac{du}{dv} = \frac{u}{v} - \exp\left(\frac{u}{v}\right).$$

Therefore, for $w = u/v$, we obtain the differential equation $vw' = -e^w$, which gives

$$e^{-w} = \log|v| + c = \log C|v|, \quad \text{where } \log C = c > 0.$$

Hence, as long as $C|v| > 1$, we have

$$w = \log\left(\log C|v|\right).$$

That is, a general solution of the differential equation is given by

$$x(t) = -1 - (t + 2)\log\left(\log C|t + 2|\right), \quad \text{for } C|t + 2| > 1.$$

Notice that the solution passing through the origin is obtained for the value $C = (1/2)\exp\left(e^{-1/2}\right)$, which exists for $t > (1/c) - 2 \approx -0.9095$.

A first order differential equation of the form

$$x'(t) = p(t)x + q(t)x^n, \quad \text{for } p, q \in C(I) \text{ and } n \neq 0, 1, \tag{3.2.27}$$

is called a *Bernoulli equation*. In this case, by using the transformation $u = x^{1-n}$, we obtain the linear equation

$$\frac{du}{dt} = (1 - n)p(t)u(t) + (1 - n)q(t).$$

Further, a first order differential equation of the form

$$x'(t) = p(t)x + q(t)x^2 + r(t), \quad \text{with } p, q, r \in C(I), \tag{3.2.28}$$

is known as a *Riccati equation*, which is named after the Italian mathematician *Jacopo Francesco Riccati* (1676–1754). Except in some simple cases, it is not possible in general to obtain a closed form solution of this equation. However, if one solution $x_1(t)$ is known, we can use the transformation $u = 1/(x - x_1)$ to obtain the first order differential equation

$$\frac{du}{dt} = -\big(p(t) + 2x_1(t)q(t)\big)u(t) - q(t),$$

which is a linear equation. The details are left for the reader as an exercise.

3.2.2 Picard–Lindelöf Theorem

We now consider the general case of the *initial value problem* as in (3.2.1). By the fundamental theorem of integral calculus, the case when $f(t, x)$ is a function of only x variable, say given by $g(x)$, has a unique (global) solution given by the integral curve passing through the point (t_0, x_0) if and only if the function g is continuous. In this case, a unique solution is given by

$$\varphi(t) = x_0 + \int_{t_0}^{t} g(s)\,ds, \quad \text{for } t \in I. \tag{3.2.29}$$

Next, if $\varphi \in C(I)$ is a solution in the general case so that we have

$$\frac{d\varphi}{dt} = f\big(t, \varphi(t)\big), \quad \text{for all } t \in I, \quad \text{with } \varphi(t_0) = x_0, \tag{3.2.30}$$

it then follows that

$$\varphi(t) = x_0 + \int_{t_0}^{t} f\big(s, \varphi(s)\big)\, ds, \quad \text{for all } t \in I, \tag{3.2.31}$$

using the fact that the continuity of $\varphi(t)$ implies the continuity of $f\big(t, \varphi(t)\big)$. Therefore, by using the initial condition $\varphi(t_0) = x_0$, we see that φ is a solution of the integral equation given by

$$x(t) = x_0 + \int_{t_0}^{t} f\big(s, x(s)\big)\, ds, \quad \text{for all } t \in I. \tag{3.2.32}$$

Conversely, if a function $\varphi \in C(I)$ is a solution of the integral equation (3.2.41), then by the second fundamental theorem of integral calculus, it follows that φ is a solution of the initial value problem (3.2.1). This simple observation plays an important role in all that follows in this section.

Example 3.10 If $f(t, x) = t\,x$, with $x(0) = 1$, then a solution φ is found as a uniform limit of the sequence of iterates $\langle \varphi_n \rangle$ obtained by using the formula (3.2.31). Here, for $\varphi_0(t) = 1$, we have

$$\varphi_1(t) = 1 + \frac{t^2}{2}, \quad \varphi_2(t) = 1 + \frac{t^2}{2} + \frac{t^4}{8}, \ \ldots, \quad \varphi_n(t) = \sum_{k=0}^{n} \frac{1}{k!}\Big[\frac{t^2}{2}\Big]^2.$$

Hence, we obtain $\varphi(t) = e^{t^2/2}$, for $t \in \mathbb{R}$.

For the general case, we first prove Peano's classical *existence theorem*. The formulation given in the next theorem is due to *Émile Picard* (1856–1914).

Theorem 3.1 (Peano, 1890) *For some suitably chosen constants a and b, let $R \subset \mathbb{R} \times \mathbb{R}$ be a rectangle such that*

$$(t, x) \in R \iff t_0 \le t \le t_0 + a \ \text{ and } \ |x - x_0| \le b.$$

Suppose f and f_x are real-valued functions, both continuous on R, with

$$M := \max_{(t,x)\in R} \big|f(t, x)\big| \ \text{ and } \ \alpha = \min\big\{a, b/M\big\}. \tag{3.2.33}$$

Then the initial value problem (3.2.1) has at least one solution $\varphi = \varphi(t)$ defined on the interval $\big[t_0, t_0 + \alpha\big]$. A similar assertion holds for the case when $t < t_0$.

Proof As said above, the idea is to construct a sequence of function $\langle \varphi_n \rangle$ by *successive approximation* by using the *iteration formula* given by (3.2.41) such that $\varphi_n \to \varphi$ uniformly over the interval $\big[t_0, t_0 + \alpha\big]$, and φ is a solution of Eq. (3.2.1). We may start with the value $x(t_0) = x_0$ as the *1st approximate* given by $\varphi_1(t) = x_0$, for all t. Next, in view of (3.2.41), the function given by

$$\varphi_2(t) = x_0 + \int_{t_0}^{t} f(s, \varphi_0) \, ds,$$

gives the *2nd approximate*, and so on. Proceeding in this manner, we arrive at the general Picard's *iteration formula* given by

$$\varphi_{n+1}(t) = x_0 + \int_{t_0}^{t} f(s, \varphi_n(s)) \, ds, \quad \text{for } n \geq 1. \tag{3.2.34}$$

By what is said above, we cannot expect the *convergence for all t*. However, it is a remarkable fact that, on some interval J contained in $[t_0, t_0 + a]$, the sequence $\langle \varphi_n \rangle$ is *uniformly bounded*. That is, for some fixed constant K, we have $|\varphi_n(t)| \leq K$, for all $t \in J$. For, we may choose constants a and b such that

$$R := \{(t, x) : t_0 \leq t \leq t_0 + a \quad \text{and} \quad |x - x_0| \leq b\},$$

and also the constants M and α as in (3.2.33). Then, by using the iteration scheme (3.2.34), a simple inductive argument shows that

$$|\varphi_n(t) - x_0| \leq M(t - t_0), \quad \text{for } t_0 \leq t \leq t_0 + \alpha. \tag{3.2.35}$$

Notice that the above equation implies that, for $t_0 \leq t \leq t_0 + \alpha$, the graph of the each function φ_n is enclosed within the lines

$$x = x_0 + M(t - t_0) \quad \text{and} \quad x = x_0 - M(t - t_0).$$

Also, since the two lines leave the rectangle R at $t = t_0 + a$, if $a \leq b/M$; and, at $t = t_0 + b/M$, if $b/M < a$, our choice for the constant α is apt to meet the convergence requirement of the sequence $\langle \varphi_n \rangle$. In view of our assumption that the partial derivative f_x exists, and is continuous, this part turns out be simpler affair. For,

$$\varphi_n = x_0 + (\varphi_1 - x_0) + (\varphi_2 - \varphi_1) + \cdots + (\varphi_n - \varphi_{n-1})$$

implies that

$$\lim_{n \to \infty} \varphi_n(t) = x_0 + \sum_{n=1}^{\infty} (\varphi_n(t) - \varphi_{n-1}(t)).$$

We also have

$$\left| \varphi_n(t) - \varphi_{n-1}(t) \right| = \left| \int_{t_0}^{t} \left[f\left(s, \varphi_{n-1}(s)\right) - f\left(s, \varphi_{n-2}(s)\right) \right] ds \right|$$

$$\leq \int_{t_0}^{t} \left| f\left(s, \varphi_{n-1}(s)\right) - f\left(s, \varphi_{n-2}(s)\right) \right| ds$$

$$= \int_{t_0}^{t} \left| f_x(s, \eta(s)) \right| \left| \varphi_{n-1}(s) - \varphi_{n-2}(s) \right| ds,$$

for some $\eta(s) \in \left[\varphi_{n-1}(s), \varphi_{n-2}(s) \right]$, given by the *mean value theorem*. By (3.2.35), we have that all the points $\left(s, \eta(s) \right)$ lie in the rectangle R, for $s \leq t_0 + \alpha$. Therefore, taking $L := \max_{(t,x) \in R} \left| f_x(t, x) \right|$, it follows that

$$\left| \varphi_n(t) - \varphi_{n-1}(t) \right| \leq L \int_{t_0}^{t} \left| \varphi_{n-1}(s) - \varphi_{n-2}(s) \right| ds, \quad \text{for } t_0 \leq t \leq t_0 + \alpha. \quad (3.2.36)$$

Hence, proceeding inductively, we obtain

$$\left| \varphi_n(t) - \varphi_{n-1}(t) \right| \leq \frac{M L^{n-1} (t - t_0)^n}{n!}, \quad \text{for } t_0 \leq t \leq t_0 + \alpha, \quad (3.2.37)$$

which proves that

$$\sum_{n=1}^{\infty} \left| \varphi_n(t) - \varphi_{n-1}(t) \right| \leq \frac{M}{L} \left[\alpha L + \frac{(\alpha L)^2}{2!} + \cdots \right] = \frac{M}{L} \left(e^{\alpha L} - 1 \right) < \infty. \quad (3.2.38)$$

Consequently, the sequence of Picard iterates $\langle \varphi_n \rangle$ converges uniformly over the interval $\left[t_0, t_0 + \alpha \right]$, by using the *Weierstrass M-test*. A similar argument shows that $\langle \varphi_n \rangle$ converges uniformly over the interval $\left[t_0 - \beta, t_0 \right]$, where we may take $\beta = \min \left\{ a, b/N \right\}$, with

$$N := \max \left\{ |f(t, x)| : (t, x) \in [t_0 - a, t_0] \times [x_0 - b, x_0 + b] \right\}.$$

Finally, we show that $\varphi(t) = \lim_{n \to \infty} \varphi_n(t)$ is a solution of the initial value problem (3.2.1). For, as said earlier, it suffices to prove that the integral equation (3.2.41) holds for φ. However, this follows directly from (3.2.34) by taking the limit as $n \to \infty$, and by using

$$\left| \int_{t_0}^{t} f(s, \varphi(s)) - \int_{t_0}^{t} f(s, \varphi_n(s)) \right| \longrightarrow 0, \quad \text{as } n \to \infty. \tag{3.2.39}$$

Notice that the above assertion follows from the inequality

$$\left| \int_{t_0}^{t} f(s, \varphi(s)) - \int_{t_0}^{t} f(s, \varphi_n(s)) \right| \le M\alpha \sum_{k=n+1}^{\infty} \frac{(\alpha L)^k}{k!},$$

which in turn is given by the inequality as in (3.2.37), and the scalar L is as defined above. Clearly, the above series approaches to zero as $n \to \infty$. To complete the proof, it remains to be shown that the limit φ is continuous. For, let $\varepsilon > 0$, we may choose a number N large enough so that

$$\frac{M}{L} \sum_{k=N+1}^{\infty} \frac{(\alpha L)^k}{k!} < \frac{\varepsilon}{3}.$$

It then follows from the inequality

$$\left| \varphi_n(s) - \varphi_{n-1}(s) \right| \le \frac{M}{L} \sum_{k=n+1}^{\infty} \frac{(\alpha L)^k}{k!},$$

that, for $t < t_0 + \alpha$ and $t + h < t_0 + \alpha$, we have

$$\left| \varphi(t+h) - \varphi_N(t+h) \right| < \frac{\varepsilon}{3} \quad \text{and} \quad \left| \varphi_N(t) - \varphi(t) \right| < \frac{\varepsilon}{3}.$$

Now, as $\varphi_N(t)$ is continuous, there exists $\delta > 0$ such that

$$\left| \varphi_N(t+h) - \varphi_N(t) \right| < \frac{\varepsilon}{3}, \quad \text{for } |h| < \delta.$$

Therefore, φ is continuous. This completes the proof. □

Unfortunately, in general, there may not be any necessary and sufficient condition for *uniqueness* of a *local solution* of an *initial value problem* such as (3.2.1) [10]. However, there are a large number of sufficient conditions to guarantee the *uniqueness* [11]. We prove here an important *uniqueness theorem* due to *Rudolf Lipschitz* (1832–1903). We first need to introduce some terminology. Let us start with the next definition.

Definition 3.9 Let $\Omega := I \times U \subset \mathbb{R}^2$ be a *domain*, and $(t_0, x_0) \in \Omega$. We say a function $f : \Omega \to \mathbb{R}$ satisfies the *Lipschitz condition* in the second variable, uniformly with respect to the first variable, if we have

$$\left|f(t, x_1) - f(t, x_1)\right| \leq L\left|x_1 - x_2\right|, \quad \text{for all } (t, x_i) \in \Omega, \tag{3.2.40}$$

for some $L > 0$, called a *Lipschitz constant*. Sometimes, we also say that f is *Lipschitz continuous* over Ω.

Example 3.11 Consider the function $f(t, x) = 3x^{2/3}$ defined over a domain $\Omega \subseteq \mathbb{R}^2$, with $(0, 0) \in \Omega$. Then, in this case

$$\left|f(t, 0) - f(t, x)\right| = \left|0 - 2x^{2/3}\right| \leq L\left|0 - x\right|$$

implies that f is does not satisfy the *Lipschitz condition* in any strip given by $|x| \leq a$, for any $a > 0$. However, it does satisfy the *Lipschitz condition* (3.2.40) in every strip given by $|x| \geq a$, with $L = 3a^{-1/3}$. More generally, it is known that every *compactly supported* function satisfies the Lipschitz condition (by the case $n = 1$ of Lemma 3.3).

The proof of (existence and) uniqueness of a *local solution* passing through the point (t_0, x_0), as given in the next theorem, uses sequence of Picard's iterates $\langle \varphi_n \rangle$ given by (3.2.34), and the *Banach contraction principle* (Theorem A.1). For, once again, consider a rectangular region

$$\Omega' = \left[t_0 - a, t_0 + a\right] \times \left[x_0 - b, x_0 + b\right] \subset \Omega, \quad \text{for } a > 0 \quad \text{and} \quad b > 0.$$

By using the continuity of f, there exists some $M > 0$ such that

$$\left|f(t, x)\right| \leq M, \quad \text{for all } (t, x) \in \Omega'.$$

As before, we may take $\alpha = \min\left\{a, b/M\right\} \in \mathbb{R}$ so that

$$\left[t_0 - \alpha, t_0 + \alpha\right] \times \left[x_0 - \alpha M, x_0 + \alpha M\right] \subset \Omega'.$$

Let $J = \left[t_0 - \alpha, t_0 + \alpha\right]$. The space $C(J)$ of real-valued continuous functions on J is a *complete metric space* with respect to the *supremum metric* d_∞ given by

$$d_\infty(g, h) = \sup\left\{\left|g(t) - h(t)\right| : t \in J\right\}.$$

In what follows, we may view x_0 as the (continuous) constant function $h(t) = x_0$, for $t \in J$.

Theorem 3.2 (Lipschitz, 1876) *Let Y be the subspace of the complete metric space* $\left(C(J), d_\infty\right)$ *consisting of functions g such that* $d_\infty(g, x_0) < \alpha M$. *Then, the function* $T : Y \to Y$ *given by*

$$T\left(g(t)\right) = x_0 + \int_{t_0}^{t} f\left(s, g(s)\right) ds, \quad \text{for } t \in J, \tag{3.2.41}$$

is a contraction, for $\alpha L < 1$, and so a unique solution of the initial value problem (3.2.1) is given by the unique fixed-point of of the map T.

Proof Recall that a function $\varphi \in C(J)$ is a *solution* of Eq. (3.2.1) if and only if $T\varphi = \varphi$ over J. We also have $T\varphi(t_0) = x_0$, and $T\varphi$ is a continuous function. Now, for $t \in J$, we have

$$\left| T\varphi(t) - x_0 \right| = \left| \int_{t_0}^{t} f(s, \varphi(s)) \, ds \right|$$

$$\leq \int_{t_0}^{t} \left| f(s, \varphi(s)) \right| ds$$

$$\leq \alpha M,$$

Therefore, it follows that $T\varphi \in Y$. Finally, for $\varphi_1, \varphi_2 \in Y$, we have

$$\left| T\varphi_1(t) - T\varphi_1(t) \right| = \left| \int_{t_0}^{t} \left[f(s, \varphi_1(s)) - f(s, \varphi_2(s)) \right] ds \right|$$

$$\leq \int_{t_0}^{t} \left| f(s, \varphi_1(s)) - f(s, \varphi_2(s)) \right| ds$$

$$\leq L \int_{t_0}^{t} \left| \varphi_1(s) - \varphi_2(s) \right| ds$$

$$\leq (L \alpha) \, d_\infty(\varphi_1, \varphi_2)$$

That is,

$$d_\infty(T\varphi_1, T\varphi_1) \leq (\alpha L) d_\infty(\varphi_1, \varphi_2),$$

so that T is a contraction on the complete metric space Y. Therefore, by Theorem A.1, there exists a unique $\varphi \in Y$ such that $T(\varphi) = \varphi$, which is indeed a *unique solution* of (3.2.1). □

Example 3.12 In particular, for $T\varphi(t) = -t\,\varphi(t)$, we have

$$\varphi'(t) = \frac{d}{dt} T_a \varphi(t) = -t\,\varphi(t).$$

Therefore, $\varphi(0) = a \Rightarrow \varphi(t) = ae^{-t^2/2}$, for $t \in [-c, c]$. Notice that it is possible to extend the solution $\varphi(t) = ae^{-t^2/2}$ as obtained above to $\mathbb{R}$. More generally, to extend the local solution across Ω', we may write

$$J_1 = J, \quad b_1 = b, \quad t_1 = t_0 + b, \quad \text{and} \quad x_1 = \varphi(t_1),$$

and then by applying Theorem 3.2 to the pair (t_1, x_1) we obtain $J_2, b_2, (t_2, x_2)$, and the solutions $\varphi_1 = \varphi$ on J_1 and φ' on J_2, both agreeing on an interval, so that φ' gives the unique solution on $J_1 \cup J_2$, by connectedness argument. Proceeding inductively, we find a sequence $\langle t_n \rangle$, with $t_{n+1} > t_n$, and points $(t_n, x_n) \in \Omega'$ having properties as stated above. So, if Ω' is assumed to bounded, then

$$a_n = d_2((t_n, x_n), \partial \Omega') \longrightarrow 0, \quad \text{as} \quad n \to \infty.$$

Thus, if the sequence $\langle b_n \rangle$ is chosen to satisfy

$$b_n = \min \left\{ \frac{a_n}{B^2 + 1}, \frac{1}{2M} \right\},$$

then since the series $\sum_{n=1}^{\infty} b_n$ is convergent we conclude

$$\sum_{n=1}^{\infty} a_n < \infty \quad \text{and} \quad \lim_{n \to \infty} a_n = 0.$$

3.3   Higher Order Linear Differential Equations

An nth order differential equation for an unknown function $x \in C^n(I)$ of the form

$$\frac{d^n x}{dt^n} = G\big(t; x, x', \ldots, x^{(n-1)}; b\big), \quad \text{for } t \in I, \tag{3.3.1}$$

arise naturally as a *mathematical model* for many phenomena in science and engineering. In majority of the cases, we have $n = 2$. That is, a *differential equation model* related to many different kinds of applications is a second order differential equation given in *normal form* as

$$\frac{d^2 x}{dt^2} = f\big(t; x, x'; b\big), \quad \text{for } t \in I, \tag{3.3.2}$$

where the function f is usually assumed to be continuous. In physical terms, f represents the *total force* acting on the dynamical system under investigation. A practical situations in physics illustrated in the next example is a typical equation of the form (3.3.2).

Example 3.13 *(Newton's Momentum Law)* Let $x : \mathbb{R} \to \mathbb{R}$ be a twice-differentiable function such that the point $\big(t, x(t)\big)$ represents the *position* of a particle of mass m in $\mathbb{R}^2$ moving in a force field given by a function $f : \Omega \to \mathbb{R}$, with $\Omega \subseteq \mathbb{R}^2$ being an open set. It is assumed that $x(t) \in \Omega$, for all $t \in \mathbb{R}$. Then, by Newton's *momentum law*, we have

$$m\frac{d^2x}{dt^2} = f(t; x, x'; \boldsymbol{b}), \quad \text{for } t \in I, \tag{3.3.3}$$

where the vector $\boldsymbol{b} \in \mathbb{R}^k$ ($k \geq 1$) represents the *parameters* influencing the motion of the particle. The above equation is a case of second order differential equation for the unknown function $x : \mathbb{R} \to \mathbb{R}$.

Linear Differential Equations

A differential equation for an unknown function $x = x(t) \in C^n(I)$ of the form

$$a_0\frac{d^nx}{dt^n} + a_1\frac{d^{n-1}x}{dt^{n-1}} + \cdots + a_{n-1}\frac{dx}{dt} + a_nx(t) = g(t), \quad t \in I, \tag{3.3.4}$$

is called a general *linear differential equation* of order n, where the $(n+1)$ coefficients $a_0, \ldots, a_n \in C(I)$, with $a_0 \not\equiv 0$ over I. The (known) function g may not be continuous in general, and it is called a *forcing function* (or an *equilibrium function*) of the dynamical system under investigation.

Definition 3.10 A function $\varphi \in C^n(J)$ that satisfies Eq. (3.3.4) identically over a subinterval $J \subseteq I$ is called a *solution*. Also, for a given *initial-boundary value problem* related to Eq. (3.3.4), a solution φ is also required to satisfy the specified *initial-boundary conditions*.

When $g \equiv 0$ over I, we say that Eq. (3.3.4) is *homogeneous*. Otherwise, it is called a *nonhomogeneous equation*. Also, Eq. (3.3.4) is *linear* in the sense that the *differential operator* $\mathscr{L} : C^n(I) \to C^k(I)$ ($k \leq n - 1$) given by

$$\mathscr{L} \equiv a_0\frac{d^n}{dt^n} + a_1\frac{d^{n-1}}{dt^{n-1}} + \cdots + a_{n-1}\frac{d}{dt} + a_n, \tag{3.3.5}$$

is a *linear transformation*. That is, for any $\varphi, \psi \in C^n(I)$ and scalars $c, d \in \mathbb{R}$, we have

$$\mathscr{L}(c\varphi + d\psi) = c\mathscr{L}(\varphi) + d\mathscr{L}(\psi).$$

It is common to write Eq. (3.3.4) in *operator notation* as

$$\mathscr{L}x(t) = g(t), \quad \text{for } t \in I. \tag{3.3.6}$$

The linearity of $\mathscr{L}$ is equivalent to the *principle of superposition*.[5]

Theorem 3.3 (Principle of Superposition) *If functions φ_1 and φ_2 satisfy the homogeneous equation $\mathscr{L}x = 0$, then the linear combination $c_1\varphi_1 + c_2\varphi_2$, for any scalars $c_1, c_2 \in \mathbb{R}$, also satisfies it.*

---

[5] This principle explains why it is easier to deal with dynamical systems modelled by a system of linear differential equations. More so, when the coefficients a_i in (3.3.4) are *scalars*.

Proof By assumption, we have

$$\mathscr{L}\varphi_1 = 0 \quad \text{and} \quad \mathscr{L}\varphi_2 = 0,$$

and so, using linearity of the operator $\mathscr{L}$, it follows that

$$\mathscr{L}(c_1\varphi_1 + c_2\varphi_2) = c_1\mathscr{L}\varphi_1 + c_2\mathscr{L}\varphi_2 = 0,$$

which proves that the function $c_1\varphi_1 + c_2\varphi_2$ also satisfies the equation $\mathscr{L}x = 0$. $\square$

The case when $n = 2$ is a perfect prototype of the general case. So, we first consider a second order linear differential equation for an unknown function $x \in C^2(I)$ given by

$$a_0(t)\frac{\mathrm{d}^2 x}{\mathrm{d}t^2} + a_1(t)\frac{\mathrm{d}x}{\mathrm{d}t} + a_2(t)x(t) = g(t), \quad \text{for } t \in I, \tag{3.3.7}$$

where $0 \neq a_0, a_1, a_2 \in C(I)$, and $g : I \to \mathbb{R}$ is some known function. As said above, a function $\varphi \in C^2(J)$ is a *solution* of Eq. (3.3.7) if we have

$$a_0(t)\frac{\mathrm{d}^2 \varphi}{\mathrm{d}t^2} + a_1(t)\frac{\mathrm{d}\varphi}{\mathrm{d}t} + a_2(t)\varphi(t) = g(t), \quad \text{for all } t \in J \subseteq I. \tag{3.3.8}$$

To continue with our discussion, we need some basic concepts of linear algebra. Recall that two functions $\varphi_1, \varphi_2 : I \to \mathbb{R}$ are *linearly independent* over a subinterval $I \subseteq \mathbb{R}$ if, for any $c_1, c_2 \in \mathbb{R}$, we have

$$c_1\varphi_1 + c_2\varphi_2 = 0 \quad \Longrightarrow \quad c_1 = c_2 = 0.$$

Otherwise, φ_1, φ_2 are called *linearly dependent*. In general, a simple tool to check the *linear independence* of any $(n-1)$-times differentiable functions $\varphi_i : I \to \mathbb{R}$, for $i = 1, \ldots, n$, is provided by the concept of *Wronskian of functions* $\varphi_1, \ldots, \varphi_n$, denoted by $W = W(\varphi_1, \ldots, \varphi_n)$.

Definition 3.11 Let $\varphi_1, \ldots, \varphi_n \in C^{n-1}(I)$. The **Wronskian** $W = W(\varphi_1, \ldots, \varphi_n)$ is a function $W : I \to \mathbb{R}$ given by

$$W(t) := \det \begin{pmatrix} \varphi_1(t) & \varphi_2(t) & \cdots & \varphi_n(t) \\ \varphi_1'(t) & \varphi_2'(t) & \cdots & \varphi_n'(t) \\ \vdots & \vdots & \ddots & \vdots \\ \varphi_1^{(n-1)}(t) & \varphi_2^{(n-1)}(t) & \cdots & \varphi_n^{(n-1)}(t) \end{pmatrix}, \quad \text{for } t \in I. \tag{3.3.9}$$

Notice that, in particular, if functions $\varphi_1, \varphi_2 \in C^2(J)$ are solutions of a homogeneous linear differential equation given by

$$a_0(t)\frac{\mathrm{d}^2 x}{\mathrm{d}t^2} + a_1(t)\frac{\mathrm{d}x}{\mathrm{d}t} + a_2(t)x(t) = 0, \quad \text{for } t \in I, \tag{3.3.10}$$

then, for any c_1, $c_2 \in \mathbb{R}$, $c_1\varphi_1 + c_2\varphi_2 = 0$ implies that $c_1\varphi_1' + c_2\varphi_2' = 0$. Therefore, if $W(\varphi_1, \varphi_2)(t_0) \neq 0$ for some $t_0 \in I$, it follows from the two linear (algebraic) equations given by

$$c_1\varphi_1 + c_2\varphi_2 = 0 \quad \text{and} \quad c_1\varphi_1' + c_2\varphi_2' = 0,$$

that the unknowns c_1 and c_2 are both zero. Hence, two solutions φ_1 and φ_2 of the homogeneous equation (3.3.10) are *linearly independent* over the interval I. Conversely, suppose for two solutions φ_1, $\varphi_2 \in C^2(J)$ of Eq. (3.3.10), we have

$$W(\varphi_1, \varphi_2)(t_0) = 0, \quad \text{for some } t_0 \in I. \tag{3.3.11}$$

Clearly then, the two linear algebraic equations in unknowns c_1 and c_2 given by

$$c_1\varphi_1(t_0) + c_2\varphi_2(t_0) = 0 \quad \text{and} \quad c_1\varphi_1'(t_0) + c_2\varphi_2'(t_0) = 0.$$

has a *nontrivial* solution, say given by $c, d \in \mathbb{R}$. Therefore, by Theorem 3.3, the function $\varphi \in C^2(J)$ given by

$$\varphi(t) := c\,\varphi_1(t) + d\,\varphi_2(t), \quad \text{for } t \in J \quad \text{and} \quad c, d \in \mathbb{R},$$

is also a solution of Eq. (3.3.10), and so of the *initial value problem* given by

$$a_0 \frac{d^2\varphi}{dt^2} + a_1 \frac{d\varphi}{dt} + a_2\varphi(t) = 0, \quad \text{with } \varphi(t_0) = \varphi'(t_0) = 0.$$

But then, by the uniqueness part of Theorem 3.7, it follows that $\varphi \equiv 0$ over J. That is, $c\,\varphi_1(t) + d\,\varphi_2(t) = 0$, for all $t \in J$. Hence, if φ_1, φ_2 are two linearly independent solutions of (3.3.10) over J, then (3.3.11) must be false. That is, we must have

$$W(\varphi_1, \varphi_2)(t_0) \neq 0, \quad \text{for all } t_0 \in J. \tag{3.3.12}$$

A simple corollary of the above argument is that, if φ_1, $\varphi_2 \in C^2(I)$ are two solutions of (3.3.10) over an interval J, then $W(\varphi_1, \varphi_2) \equiv 0$ over J or $W(\varphi_1, \varphi_2)(t) \neq 0$, for all $t \in J$. This is known as the *Abel's identity*. Now, the next theorem can be proven easily.

Theorem 3.4 *If φ_1, $\varphi_2 \in C^2(J)$ are two linearly independent solutions of the homogeneous equation (3.3.10) over an interval J, then the general solution of the equation is given by*

$$\varphi_h(t) = c_1\,\varphi_1(t) + c_2\,\varphi_2(t), \quad \text{for } t \in J \tag{3.3.13}$$

where c_1, $c_2 \in \mathbb{R}$.

Proof Left to the reader as an exercise. $\qquad\qquad\qquad\qquad\qquad\qquad\qquad\qquad\square$

Every pair of functions φ_1, $\varphi_2 \in C^2(J)$ as in Theorem 3.4 is called the *fundamental solutions* of the linear equation (3.3.7), and the function φ_h is called the *complementary function*.

Definition 3.12 A solution φ_p of the linear equation (3.3.7) is called a *particular solution* if it can be obtained from a function φ_h as in (3.3.13), for some suitable values of *parameters* c_1 and c_2. Also, we say a function $\varphi \in C^2(J)$ is a *singular solution*[6] of (3.3.7) if $\varphi \neq \varphi_p$, for any choice of c_1 and c_2.

The method of *variation of parameters* says that, if φ_1 and φ_2 are two linearly independent solutions of (3.3.10), then a particular solution φ_p of Eq. (3.3.7) is given by

$$\varphi_p(t) = c_1(t)\varphi_1(t) + c_2(t)\varphi_2(t), \quad \text{for } t \in J, \tag{3.3.14}$$

for some c_1, $c_2(t) \in C^2(J)$. The next theorem gives a formula to compute the functions c_1 and c_2 explicitly, provided the functions φ_1, φ_2 are known.

Theorem 3.5 (Variation of Parameters) *Suppose all coefficient functions a_i and the (forcing) function g as in (3.3.7) are continuous. Then the functions c_1 and c_2 as in (3.3.14) are given by the integrals*

$$c_1(t) = \int_J \frac{\widehat{W}_1(\varphi_1, \varphi_2)(s)}{W(\varphi_1, \varphi_2)(s)} \, ds;$$

$$c_2(t) = \int_J \frac{\widehat{W}_2(\varphi_1, \varphi_2)(s)}{W(\varphi_1, \varphi_2)(s)} \, ds, \tag{3.3.15}$$

where the modified Wronskians $\widehat{W}_2$ and $\widehat{W}_2$ are obtained by using the column vector $= (0 \ g/a_0)^t$ respectively as the 1st and 2nd column of the Wronskian W.

Proof Left to the reader as an exercise. □

Notice that Theorem 3.5 is helpful only when the two linearly independent solutions φ_1 and φ_2 as in (3.3.13) are known in advance.

Remark 3.1 If the forcing function g in a given situation is of the form

$$g(t) = g_1(t) + \cdots + g_m(t), \quad \text{for } t \in I,$$

and $\varphi_p^k \in C^2(J)$ is a *particular solution* of the second order linear equation

$$a_0(t)\frac{d^2x}{dt^2} + a_1(t)\frac{dx}{dt} + a_2(t)x(t) = g_k(t), \quad \text{for } k = 1, \ldots, m, \tag{3.3.16}$$

it then follows easily that the function given by the sum

---

[6] For example, that is precisely the case for the nonlinear equation (4) as in Example 3.1.

$$\varphi_p(t) = \varphi_p^1(t) + \cdots + \varphi_p^m(t), \quad \text{for } t \in J,$$

is a solution of the equation

$$a_0(t)\frac{d^2x}{dt^2} + a_1(t)\frac{dx}{dt} + a_2(t)x(t) = g_1(t) + \cdots + g_m(t).$$

Therefore, we need only to focus attention on solving an equation of the form (3.3.16).

Finally, let $\varphi \in C^2(J)$ be *any* solution of the linear equation (3.3.7). Then

$$L(\varphi - \varphi_p) = L\varphi - L\varphi_p = g - g = 0,$$

implies that the function $\varphi - \varphi_p$ is a solution of the homogeneous equation (3.3.10). Therefore, if $\varphi - \varphi_p = \varphi_h$, we may write a *unique solution* of the linear differential equation (3.3.7) as

$$\varphi(t) = c_1\,\varphi_1(t) + c_2\,\varphi_2(t) + \varphi_p(t), \quad \text{for } t \in J, \tag{3.3.17}$$

where values of the scalars $c_1, c_2 \in \mathbb{R}$ are determined by using the assigned *initial* or *boundary conditions*. Hence, the existence and uniqueness of a solution of an *initial-boundary value problem* related to a second order linear differential equation is established. Notice that, in this case, we can write a unique solution explicitly provided it is known how to find two linearly independent solutions φ_1 and φ_2 as in (3.3.13). Therefore, the main problem boils down to solving a homogeneous linear equation of the form (3.3.7).

Remark 3.2 The only known *solution method* that can be used to deal with the general case is the method of *power series solutions*, as describe in the next part of the section.

It should be clear that the concepts we discussed above for the case $n = 2$ extend in straight forward way to the general case of an nth order linear differential equation. Recall that a function $\varphi \in C^n(I)$ is a *solution* of Eq. (3.3.4) if

$$\frac{d^n\varphi}{dt^n} + a_1\frac{d^{n-1}\varphi}{dt^{n-1}} + \cdots + a_{n-1}\frac{d\varphi}{dt} + a_n\varphi(t) = g(t), \quad \text{for } t \in I. \tag{3.3.18}$$

As before, if the n functions $\varphi_1, \ldots, \varphi_n \in C^n(I)$ are *linearly independent*[7] solutions of the homogeneous equation $\mathscr{L}[x(t)] = 0$, then it follows from Theorem 3.3 that the *fundamental solution* $\varphi_h = \varphi_h(t)$ of Eq. (3.3.4) is given by

$$\varphi_h(t) = c_1\varphi_1 + \cdots + c_n\varphi_n, \quad \text{with } c_i \in \mathbb{R}. \tag{3.3.19}$$

---

[7] Notice that this property is equivalent to the condition that **Wronskian** $W(\varphi_1, \ldots, \varphi_n) \neq 0$, for some $t \in I$.

Recall that we also call φ_h the *complementary solution* of the differential equation. Sometimes, we may also write φ_c for φ_h. A *Particular solution* $\varphi_p = \varphi_p(t)$ of the differential equation is a solution obtained from Eq. (3.3.19) by assigning some particular values to the *parameters* $c_1, \ldots, c_n$, which could be determined by using the given initial/boundary conditions. In general, according to the method of *variation of parameters*, if

$$\varphi_p(t) = c_1(t)\varphi_1(t) + \cdots + c_n(t)\varphi_n(t), \quad \text{for } t \in J, \tag{3.3.20}$$

for some $c_1, \ldots, c_n \in C^n(J)$, then the n functions c_i can be computed explicitly by using the formula as given in the next theorem.

Theorem 3.6 (Variation of Parameters) *With notations as above, let* $\varphi_1, \ldots, \varphi_n \in C^n(I)$ *be* n *linearly independent solutions of the homogeneous equation* $\mathscr{L}x = 0$. *If a function* φ_p *as in* (3.3.20) *is solution of Eq. (3.3.4), for some* $c_1, \ldots, c_n \in C^n(J)$, *then we have*

$$c_i(t) = \int^t \frac{\widehat{W}_i(\varphi_1, \ldots, \widehat{\varphi}_i, \ldots, \varphi_n)(s)}{W(\varphi_1, \ldots, \varphi_n)(s)} \, ds, \quad \text{for } i = 1, \ldots, n, \tag{3.3.21}$$

where the modified Wronskian $\widehat{W}_i$ *is obtained by using the column vector* $e_n = \begin{pmatrix} 0 & 0 & \cdots & g/a_0 \end{pmatrix}^t$ *in place of the* ith *column of the Wronskian* W.

Proof By using the assumption that φ_p is a solution of Eq. (3.3.4), a simple algebraic manipulation gives the system of equations

$$\begin{pmatrix} \varphi_1(t) & \varphi_2(t) & \cdots & \varphi_n(t) \\ \varphi_1'(t) & \varphi_2'(t) & \cdots & \varphi_n'(t) \\ \vdots & \vdots & \ddots & \vdots \\ \varphi_1^{(n-1)}(t) & \varphi_2^{(n-1)}(t) & \cdots & \varphi_n^{(n-1)}(t) \end{pmatrix} \begin{pmatrix} c_1'(t) \\ c_2'(t) \\ \vdots \\ c_n'(t) \end{pmatrix} = \begin{pmatrix} 0 \\ 0 \\ \vdots \\ g(t)/a_0(t) \end{pmatrix} \tag{3.3.22}$$

Therefore, the parameters $c_i = c_i(t)$ are obtained as in (3.3.21). $\square$

Finally, the *complete solution* of an nth order differential equation of the form (3.3.4) is given by

$$\varphi(t) = c_1\varphi_1 + \cdots + c_n\varphi_n + \varphi_p(t), \quad \text{for } t \in J \quad \text{and} \quad c_i \in \mathbb{R}. \tag{3.3.23}$$

We thus have proven the next fundamental theorem.

Theorem 3.7 (Existence—Uniqueness Theorem) *Suppose the functions* a_i *and* g *in Eq. (3.3.4) are continuous on the interval* I. *Then, for any* $t_0 \in I$ *and arbitrary scalars* $\alpha_1, \ldots, \alpha_n \in \mathbb{R}$, *the initial value problem*

$$\begin{aligned} \mathscr{L}x(t) &= g(t), \quad \text{for } t \in I; \\ x^{(k-1)}(t_0) &= \alpha_k, \quad \text{for } k = 1, 2, \ldots, n, \end{aligned} \tag{3.3.24}$$

has a unique solution.

3.3.1 The Case of Constant Coefficients

When all the coefficients $a_0, \ldots, a_n$ in Eq. (3.3.4) are *constants*, it is possible to produce an explicit formula for the solution guaranteed by Theorem 3.7. In this case, we call the equation as a linear differential equations *with constant coefficients*. As our first application, we discuss a *differential equation* model of a *mass-spring-dashpot system* as in the next example, which is given by a second order linear differential equation with constant coefficients of the form

$$a_0 \frac{d^2 x}{dt^2} + a_1 \frac{dx}{dt} + a_2\, x(t) = g(t), \quad \text{with } a_0 \neq 0, \tag{3.3.25}$$

where $x \in C^2(I)$ is an unknown function, and the *forcing function* $g \in C(I)$ is known. In general, such types of *models* are also used to study some more complex phenomena in science and engineering.

Example 3.14 *(Spring-Mass-Dashpot System)* Consider a mass–spring–dashpot system, as shown in Fig. 3.1, and let $x = x(t)$ represent the motion of a particle of mass m attached to free-end of a spring, moving back and forth about a fixed equilibrium position. We say the particle $x(t)$ has a free undamped motion (or *simple harmonic motion*) if the resistance and damping factors of the medium are negligible. In general, by Newton's momentum law, the *inertial force* $F = ma$ equals the sum of the harmonic driving force F_d, say with angular frequency ω and amplitude α_0; the *spring force* F_s, which is linearly proportional to (vertical or horizontal) displacement given by the function $x = x(t)$; and, the *frictional force* F_f, which is linearly proportional to the velocity $v = x'$. Considering the directions in which the above

Fig. 3.1 Spring–mass oscillator system

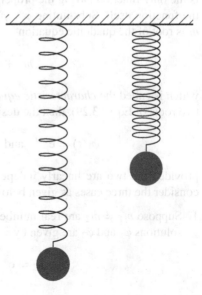

forces act, we obtain the equation

$$m\,x''(t) + c\,x'(t) + k\,x(t) = \alpha_0 \sin \omega t, \qquad (3.3.26)$$

which is a *differential equation model* of the system of the type (3.3.25). In practical terms, the mass m, constants c and k, and α_0 are *adjustable parameters*. Two additional parameters that we need to have for the uniqueness of a solution are given by the assigned initial conditions, say $x(t_0) = x_0$ and $x'(t_0) = v(t_0) = v_0$. Notice that each dynamical system modelled by an equation of the form (3.3.26) involves some *damping* due to factors like *friction, air resistance*, or due to a damper called a *dashpot*.

As in the general case, the situation when $n = 2$ is a perfect prototype of an arbitrary nth order linear differential equation with *constant coefficients*, with $a_0 \neq 0$. So, we may first consider a second order equation of the form

$$\frac{d^2x}{dt^2} + p\,\frac{dx}{dt} + q\,x(t) = f(t), \quad \text{with } p = \frac{a_1}{a_0} \text{ and } q = \frac{a_2}{a_0}, \qquad (3.3.27)$$

where $x \in C^2(I)$ is an unknown function, and $f : I \to \mathbb{R}$ is some known function. In view of our previous discussion, we only need to know how to find two linearly independent solutions of a homogeneous differential equation given by

$$x'' + p\,x' + q\,x = 0, \quad \text{where } p \text{ and } q \text{ are some scalars.} \qquad (3.3.28)$$

It is plausible to assume that a function $\varphi : J \to \mathbb{R}$ given by $\varphi(t) = e^{mt}$, where m is a real or complex number, is a solution of Eq. (3.3.28) because the *exponential function* is the only function having the property that every derivative is a scalar multiple of the function itself. Substituting φ'', φ' and φ back into the equation, it follows that m is root of the quadratic equation

$$m^2 + p\,m + q = 0, \qquad (3.3.29)$$

which is called the *characteristic equation* of (3.3.28). Therefore, if m_1 and m_2 are two roots of Eq. (3.3.29), then the desired solutions are given by

$$\varphi_1(t) = e^{m_1 t} \quad \text{and} \quad \varphi_2(t) = e^{m_2 t}, \quad \text{for } t \in J,$$

provided the two are linearly independent over a subinterval $J \subseteq I$. We need to consider the three cases as given below.

1. Suppose $m_1 \neq m_2$ are real numbers. In this case, the two linearly independent solutions φ_1 and φ_2 are given by

$$\varphi_1(t) = e^{m_1 t} \quad \text{and} \quad \varphi_2(t) = e^{m_2 t}.$$

2. Suppose $m_1 = m_2 = m$, say. In this case, it can be shown that the two linearly independent solutions φ_1 and φ_2 are given by

$$\varphi_1(t) = e^{mt} \quad \text{and} \quad \varphi_2(t) = te^{mt}.$$

3. Suppose m_1 and m_2 are *complex conjugates*. That is, $m_1 = a + ib$ and $m_2 = a - ib$, for some $a, b \in \mathbb{R}$. In this case, it follows from Theorem 3.3 that the two linearly independent solutions φ_1 and φ_2 given by

$$\varphi_1(t) = e^{at} \cos(bt) \quad \text{and} \quad \varphi_2(t) = e^{at} \sin(bt).$$

In all above three cases, the linear independence of φ_1 and φ_2 follows easily.

Example 3.15 We solve the equation $x'' + 4x = 1$. The *characteristic equation* is $m^2 + 4 = 0$, and so $x_1(t) = \cos 2t$ and $x_2(t) = \sin 2t$ are two linearly independent solutions of the homogeneous equation $x'' + 4x = 0$. In this case, the Wronskian $W(x_1, x_2) = 2$. Also, by (3.3.15), we have

$$c_1(t) = -\frac{1}{2} \int^t \sin 2s \, ds = \frac{1}{4} \cos 2t;$$

$$c_2(t) = \frac{1}{2} \int^t \cos 2s \sec(s) \, ds = \frac{1}{4} \sin 2t.$$

Therefore, by (3.3.14), a particular solution is given by

$$x_p(t) = \frac{1}{4}\left[\cos 2t + \sin 2t \right].$$

Hence, a general solution of the equation is given by

$$x(t) = A \cos 2t + B \sin 2t + \frac{1}{4}\left[\cos 2t + \sin 2t \right],$$

where A, B are constants.

Example 3.16 We solve the equation $x''(t) + x(t) = \sec t$. In this case, the *characteristic equation* is $m^2 + 1 = 0$, and so $\varphi_1(t) = \cos t$ and $\varphi_2(t) = \sin t$ are two linearly independent solutions of the homogeneous equation $x'' + x = 0$. The Wronskian $W(\varphi_1, \varphi_2) = 1$. Also, by (3.3.15), we have

$$c_1(t) = -\int^t \sin(s) \sec(s) \, ds = \ln|\cos t|;$$

$$c_2(t) = \int^t \cos(s) \sec(s) \, ds = t.$$

Therefore, by (3.3.14), a particular solution is given by

$$\varphi_p(t) = \cos t \, \ln |\cos t| + t \, \sin t.$$

Hence, a general solution of the equation is given by

$$\varphi(t) = c_1 \cos t + c_2 \sin t + \cos t \, \ln |\cos t| + t \, \sin t,$$

where $c_1, c_2 \in \mathbb{R}$.

The next example illustrates the procedure for an interesting practical situation.

Example 3.17 Consider the linear equation (3.3.26) so that we have

$$p = \frac{c}{m}, \quad q = \frac{k}{m}, \quad \text{and} \quad f(t) = \frac{\alpha_0 \sin \omega t}{m}.$$

Therefore, in this case, the characteristic equation (3.3.29) has

1. two real roots if

$$p^2 - 4q = \frac{1}{m^2} [c^2 - 4km] > 0 \quad \text{i.e. } c^2 > 4\,km.$$

   In physical terms, this case corresponds to *over-damping*, i.e. there are no oscillations.
2. equal roots if $c^2 = 4\,km$. In physical terms, this case corresponds to *critical damping*, i.e. there are no oscillations, but the system oscillates if the damping is reduced even by a little.
3. complex conjugate roots if $c^2 < 4\,km$. In physical terms, this case corresponds to *under-damping*, i.e. there are oscillations, with decreasing amplitude over time.

Notice that, for all the three damped systems, we have

$$\varphi(t) = c_1 \varphi_1(t) + c_2 \varphi_2(t) \to 0 \quad \text{as} \quad t \to \infty,$$

and so it would always attains the equilibrium position over a period of time.

As said earlier, each one of the above discussed concepts holds for a general nth order linear differential equations with *constant coefficients*. Hence on, we may assume that all coefficients $a_0, \ldots, a_n$ of the operator $\mathcal{L}$ as given in (3.3.5) are *constants*. For sake of completeness, we shall repeat here all statements of theorems given earlier for the case $n = 2$. In general, the *complementary function* φ_c of a homogeneous differential equation $\mathcal{L}x = 0$ is given by the n linearly independent functions $\varphi_i(t) = e^{m_i t}$, for $1 \leq i \leq n$, where m_i is a root of the *characteristic equation*

$$m^n + a_1 m^{n-1} + \cdots + a_{n-1} m + a_n = 0, \tag{3.3.30}$$

of the differential equation (3.3.18). After some rearrangements, we can write the n roots as $m_1, \ldots, m_n$.

1. Assuming that the first r roots $m_1, \ldots, m_r$ are distinct and real, then the first r linearly independent solutions of the equation $\mathscr{L}x = 0$ are given by

$$\varphi_1(t) = e^{m_1 t}, \quad \ldots \quad , \; \varphi_r(t) = e^{m_r t}; \tag{3.3.31}$$

2. Assuming that the next $r_1 + \cdots + r_k$ roots are given by

$$m_{r+1} = \cdots = m_{r+r_1} = d_1;$$
$$\vdots \; ; \tag{3.3.32}$$
$$m_{r+\sum_{j=1}^{k-1} r_j + 1} = \cdots = m_{r+r_1+\ldots+r_k} = d_k,$$

then the next $r_1 + \cdots + r_k$ linearly independent solutions are given by

$$\varphi_{r+1}(t) = e^{d_1 t},$$
$$\varphi_{r+2}(t) = (1+t)e^{d_1 t},$$
$$\vdots \; ,$$
$$\varphi_{r+r_1}(t) = (1+t+\ldots+t^{d_1-1})e^{d_1 t};$$
$$\vdots \tag{3.3.33}$$
$$\varphi_{r+\sum_{j=1}^{k-1} r_j}(t) = e^{d_k t},$$
$$\vdots$$
$$\varphi_{r+\sum_{j=1}^{k} r_j}(t) = (1+t+\ldots+t^{d_k-1})e^{d_k t}$$

3. Let $s = r + \sum_{j=1}^{k} r_j$, and assume that the next set of $2u$ distinct complex conjugates roots are given by

$$\alpha_{s+1} \pm i\beta_{s+1}, \quad \ldots \quad , \; \alpha_{s+u} \pm i\beta_{s\pm u}. \tag{3.3.34}$$

Then, in this case, the *pair* of linearly independent solutions are given by

$$\varphi_{s+1}(t) = e^{\alpha_{s+1} t} \cos \beta_{s+1} t, \qquad \phi_{s+1}(t) = e^{\alpha_{s+1} t} \sin \beta_{s+1} t$$
$$\vdots \tag{3.3.35}$$
$$\varphi_{s+u}(t) = e^{\alpha_{s+u} t} \cos \beta_{s+u} t, \qquad \phi_{s+u}(t) = e^{\alpha_{s+u} t} \sin \beta_{s+u} t$$

4. Let $n' = n - 2u$, and assume that the next set of roots are $2u$ distinct complex conjugates given by

$$m_{n'+1} = \cdots = m_{n'+2u_1} = \gamma_1 \pm i\delta_1; \cdots;$$

$$\vdots \tag{3.3.36}$$

$$m_{n'+2\sum_{j=1}^{\ell-1} u_j+1} = \cdots = m_{n'+2\sum_{j=1}^{\ell} u_j} = \gamma_\ell \pm i\delta_\ell,$$

Accordingly, the associated linearly independent solutions are given by

$$\varphi_{n'+1}(t) = e^{\gamma_{n'+1}t} \cos \delta_{n'+1}t, \quad \xi_{n'+1}(t) = e^{\gamma_{n'+1}t} \sin \delta_{n'+1}t \tag{3.3.37a}$$

$$\vdots$$

$$\varphi_{n'+2}(t) \doteq (1+t)e^{\gamma_{n'+1}t} \cos \delta_{n'+1}t, \quad z_{n'+2}(t) = (1+t)e^{\gamma_{n'+1}t} \sin \delta_{n'+1}t \tag{3.3.37b}$$

$$\vdots$$

$$\varphi_{u_1}(t) = (1+t+\ldots+t^{u_1-1})e^{\gamma_{n'+1}t} \cos \delta_{n'+1}t;$$

$$\xi_{u_1}(t) = (1+t+\ldots+t^{u_1-1})e^{\gamma_{n'+1}t} \sin \delta_{n'+1}t$$

$$\vdots$$

$$\varphi_{m_n}(t) = (1+t+\ldots+t^{\ell-1})e^{\gamma_\ell t} \cos \delta_\ell t, \tag{3.3.37c}$$

$$\xi_{m_n}(t) = (1+t+\ldots+t^{\ell-1})e^{\gamma_\ell t} \sin \delta_\ell t. \tag{3.3.37d}$$

The next example illustrates an interesting case.

Example 3.18 Consider the differential equation $x^{(4)}(t) - x(t) = 0$. Here, the four roots of the characteristic equation are $m = \pm 1$ and $\pm i$. Thus, by using (3.3.31) and (3.3.35), the fundamental solutions are given by

$$\varphi_1(t) = e^t, \quad \varphi_2(t) = e^{-t}, \quad \varphi_3(t) = \cos t, \quad \text{and} \quad \varphi_4(t) = \sin t.$$

Therefore, the general function is given by

$$\varphi(t) = c_1 e^t + c_2 e^{-t} + c_3 \cos t + c_4 \sin t, \quad \text{with } c_i \in \mathbb{R}.$$

In general, some constants c_i may be zero for certain given types of initial conditions.

Theorem 3.8 *Suppose the coefficients a_i in Eq. (3.3.4) are constants, and $g \in C(I)$. Then, for any $t_0 \in I$, and arbitrary choice of scalars $\alpha_1, \ldots, \alpha_n \in \mathbb{R}$, the initial value problem*

$$\mathscr{L}x(t) = g(t), \quad \text{for } t \in I;$$
$$x^{(k-1)}(t_0) = \alpha_k, \quad \text{for } k = 1, 2, \ldots, n, \tag{3.3.38}$$

has a unique solution, which can be found explicitly.

Method of Undetermined Coefficients

The method of *undetermined coefficients*, also known as the *annihilator method*, provides a convenient procedure to find a particular solution φ_p of a linear differential equations with constant coefficients. In some cases, it helps to avoid integral based method as given by Theorem 3.6. The method applies when the (forcing) function g is a linear combination of products of the functions of the type

$$P(t), \quad , e^{at}, \quad \cos(bt), \quad \text{or} \quad \sin(bt), \quad \text{with } a, b \in \mathbb{R}, \tag{3.3.39}$$

where $P(t)$ is a polynomial, provided none of the above functions is a solution of the homogeneous equation. In the case of failure, we may multiply the trial particular solution φ_p by t^k, for some suitable $k \in \mathbb{Z}^+$. For a second order differential equation, the following procedure is used to find a particular solution φ_p:

1. If g is a second degree polynomial, we try a solution of the form $\varphi_p(t) = At^2 + Bt + C$.
2. If g is a trigonometric function of the form $a \sin(bt)$ or $a \cos(bt)$, we try a solution of the form $\varphi_p(t) = A \sin(\beta t) + B \cos(\beta t)$.
3. If $g(t) = e^{kt}$, we try a solution of the form $\varphi_p(t) = Ae^{kt}$, for $k \neq 2$; and, otherwise, a solution of the form $\varphi_p(t) = Ate^{2t}$.
4. It should be clear how to choose a trial particular solution $\varphi_p(t)$ in other cases such as a linear combination of products of the functions in (3.3.39). Notice that Remark 3.1 applies for the case when g is a linear combination of such types of functions.

We summarise the above discussion as the statement of the next theorem.

Theorem 3.9 *Consider an nth order linear differential equation with constant coefficients $\mathscr{L}x(t) = g(t)$, where $\mathscr{L}$ is given by (3.3.5), and the (forcing) function g is given by*

$$g(t) = P(t)e^{at} \cos(bt) + Q(t)e^{at} \sin(bt), \quad \text{for } a, b \in \mathbb{R},$$

where P, Q are polynomials. Suppose $n = \max\{\deg(P), \deg(Q)\}$, and $k \in \mathbb{Z}^+$ is the smallest such that $t^k e^{at} \cos(bt)$ is not a solution of the homogeneous equation $\mathscr{L}x = 0$. Then for some constants $A_0, A_1, \ldots, A_n; B_0, B_1, \ldots, B_n$, we have that

$$\varphi_p(t) = t^k [A_0 t^n + \cdots + A_{n-1}t + A_n]e^{at} \cos(bt)$$
$$+ t^k [B_0 t^n + \cdots + B_{n-1}t + B_n]e^{at} \sin(bt)$$

is a particular solution of the equation $\mathscr{L}x(t) = g(t)$.

Example 3.19 For the linear differential equation given by

$$x'' - 2x' - 3x = 5\cos 2t + 4t^2 + e^{3t},$$

the functions $\varphi_1(t) = e^{3t}$ and $\varphi_2(t) = e^{-t}$ are the linearly independent solutions of the homogeneous equation. Therefore, we may write a particular solution φ_p of the given equation as a sum $\varphi_p = \phi_1 + \phi_2 + \phi_3$, where ϕ_1, ϕ_2, ϕ_3 are respectively trial particular solutions of the three equations as given below:

$$x'' - 2x' - 3x = 5 \cos 2t;$$
$$x'' - 2x' - 3x = 4t^2;$$
$$x'' - 2x' - 3x = e^{3t}$$

It is only in the case of third equation that we have to take a trial particular solution of the form $\phi_3(t) = Cte^{3t}$, for some constant C, because e^{3t} is a solution of the corresponding homogeneous equation. It follows easily that we have

$$\varphi_p(t) = -\frac{1}{13}\left[7 \cos 2t + 4 \sin 2t\right] - \frac{1}{27}\left[36t^2 - 48t + 56\right] + t^{3t}.$$

Cauchy-Euler Equation

As seen earlier, some first order differential equations can solved by transforming the original equation. Similarly, some special types of higher order equations with *variable coefficients* can also be transformed to an equation with *constant coefficients*. Such types of differential equation are known as the *Cauchy-Euler equation*. In particular, a second order Cauchy-Euler equation is given by

$$t^2 x''(t) + a t x'(t) + b x(t) = g(t), \quad \text{where } a \text{ and } b \text{ are constants.} \qquad (3.3.40)$$

In general, the transformation of the form $t = e^s \Leftrightarrow s = \ln|t|$ will do the trick. For example, by using the relations

$$\frac{dx}{dt} = \frac{1}{t}\frac{dx}{ds};$$
$$\frac{d^2 x}{dt^2} = -\frac{1}{t^2}\frac{dx}{ds} + \frac{1}{t^2}\frac{d^x}{ds^2},$$

Equation (3.3.40) transforms to the equation of the form

$$x''(s) + (a - 1) x'(s) + b x(s) = g(e^s), \qquad (3.3.41)$$

which is a second order equations with *constant coefficients*. Clearly, the same argument applies for Cauchy–Euler equations of higher order.

3.3.2 Power Series Solution

The *series solution method*, due to *Lazarus Fuchs* (1833–1902) and *Georg Frobenius* (1849–1971), provides a procedure of finding the two linearly independent *ana-*

lytic solutions of an arbitrary second order homogeneous linear differential equations of the form (3.3.10). The related concepts are widely used in solving many interesting problems in science and engineering.

We start with a simple illustration as given in the next example.

Example 3.20 Suppose a homogeneous differential equation given by

$$x''(t) + 3t\, x'(t) + 3\, x(t) = 0$$

has an analytic solution of the form

$$\varphi(t) = c_0 + c_1\, t + c_2\, t^2 + c_3\, t^3 + \cdots, \quad \text{for } c_k \in \mathbb{R},$$

which is valid in some interval about the point $t = 0$. By substituting in the above equation, the expressions for φ'', φ' and φ, we obtain

$$\sum_{k=0}^{\infty} k(k-1)c_k\, t^{k-2} + \sum_{k=0}^{\infty} \big[3k+3\big]c_k\, t^k = 0$$

Now, to have the summation over the powers t^{n-2} in the second term (same as in the first term), we apply *re-indexing*.[8] For, let $k = n - 2$ so that

$$\sum_{k=0}^{\infty} \big[3k+3\big]c_k\, t^k = \sum_{n=2}^{\infty} \big[3n-3\big]c_{n-2}\, t^{n-2},$$

implies that we can write the last equation as

$$\sum_{k=0}^{\infty} k(k-1)c_k\, t^{k-2} + \sum_{k=2}^{\infty} \big[3k-3\big]c_{k-2}\, t^{k-2} = 0.$$

Therefore, by using the fact that a power series is *identically zero* in the interval of convergence if and only if the sequence of coefficients $\langle c_n \rangle$ is zero, we obtain a second order *recurrence relation* given by

$$k(k-1)c_k + \big[3k-3\big]c_{k-2} = 0, \quad \text{for } k \geq 2.$$

So, the coefficients c_2, c_3, c_4, ... can be determined in terms of the coefficients c_0 and c_1. Hence, the general solution of the given differential equation can be written as

$$\varphi(t) = c_0\, \varphi_1(t) + c_1\, \varphi_2(t).$$

---

[8] It is important to do *re-indexing* wherever necessary.

In most cases, it can be shown that the analytic solutions φ_1 and φ_2 are linearly independent, say by using the Wronskian $W(\varphi_1, \varphi_2)(0)$. Notice that, for the given equation, we have $c_k = -(3/n)c_{k-2}$ so that

$$c_{2k} = -\frac{3}{2k}c_{2k-2} \quad \text{and} \quad c_{2k+1} = -\frac{3}{2k+1}c_{2k-1},$$

implies that we can write

$$c_{2k} = \frac{(-1)^k 3^k}{2^k k!} c_0 \quad \text{and} \quad c_{2k+1} = \frac{(-1)^k 3^k}{3 \cdot 7 \cdots (2k+1)} c_1, \quad \text{for } k \geq 1.$$

In turn, it follows that we have

$$\varphi_1(t) = 1 + \sum_{k=1}^{\infty} \frac{(-3)^k}{2^k k!} t^{2k} \quad \text{and} \quad \varphi_2 = t + \sum_{k=1}^{\infty} \frac{(-3)^k 2^k k!}{(2k+1)!} t^{2k+1}.$$

Ratio test proves that the above two power series are valid for each $t \in \mathbb{R}$.

For a general discussion, we first consider the case when $a_0(t)$ is not zero for any t in the domain interval I, and write

$$p(t) = \frac{a_1(t)}{a_0(t)} \quad \text{and} \quad q(t) = \frac{a_2(t)}{a_0(t)}, \quad \text{for } t \in I. \tag{3.3.42}$$

We say an interior point $t_0 \in I$ is an *Ordinary point* of the equation

$$x''(t) + p(t)x(t) + q(t)x(t) = 0, \quad \text{for } t \in I. \tag{3.3.43}$$

if both p and q are *analytic functions* at t_0. Therefore, $p, q \in C^\infty(I)$, and we can write the two functions as

$$p(t) = \sum_{n=0}^{\infty} \frac{p^{(n)}(t_0)}{n!}(t - t_0)^n \quad \text{and} \quad q(t) = \sum_{n=0}^{\infty} \frac{q^{(n)}(t_0)}{n!}(t - t_0)^n, \tag{3.3.44}$$

say with *radii of convergence* given by $\rho_1 > 0$ and $\rho_2 > 0$. That is, the two series in (3.3.44) respectively converge uniformly in intervals $(t_0 - \rho_1, t_0 + \rho_1)$ and $(t_0 - \rho_2, t_0 + \rho_2)$, both contained in the interval I. In this case, the next theorem ensures existence of a series solution of Eq. (3.3.43) valid in the common *intervals of convergence* for the power series of p and q about the point t_0.

Theorem 3.10 *Suppose the functions p and q are sufficiently nice so that they have convergent power series expansions of the form*

$$p(t) = \sum_{n=0}^{\infty} p_n (t - t_0)^n \quad \text{and} \quad q(t) = \sum_{n=0}^{\infty} q_n (t - t_0)^n, \tag{3.3.45}$$

valid in some intervals $(t_0 - \rho, t_0 + \rho) \subseteq I$. *Then there exists a unique solution of the initial value problem*

$$Lx(t) = x''(t) + p(t)x(t) + q(t)x(t) = 0, \quad for \ t \in I;$$
$$x(t_0) = \alpha, \qquad x'(t_0) = \beta, \quad with \ \alpha, \ \beta \in \mathbb{R}, \tag{3.3.46}$$

given by a power series

$$\varphi(t) = \sum_{n=0}^{\infty} c_n (t - t_0)^n, \quad for \ t \in (t_0 - \rho, t_0 + \rho). \tag{3.3.47}$$

Further, the same assertion holds for the nonhomogeneous equation $Lx = f$, *provided the function* f *can be expressed by a power series about the point* t_0, *valid in the same intervals* $(t_0 - \rho, t_0 + \rho)$.

Proof Left for the reader as an exercise. $\qquad\qquad\qquad\qquad\qquad\qquad\qquad$ $\square$

Notice that, in general, it can be assumed that $t_0 = 0$. By above theorem, we have that a solution $\varphi = \varphi(t)$ of Eq. (3.3.43) is also *analytic* at $t_0 = 0$ such that the power series

$$\varphi(t) = \sum_{k=0}^{\infty} c_k t^k = c_0 + c_1 t + c_2 t^2 + \cdots, \quad \text{with } c_i \in \mathbb{R}, \tag{3.3.48}$$

converges uniformly in a subinterval of some interval $(-\delta, \varepsilon)$. Then, by substitution, we obtain certain *recurrence relation* for the sequence $\langle c_n \rangle$, which helps in writing the coefficients $c_2, c_3, c_4, \ldots$ in terms of the coefficients c_0 and c_1. Finally, we obtain

$$\varphi(t) = c_0 \, \varphi_1(t) + c_1 \, \varphi_2(t), \tag{3.3.49}$$

where φ_1 and φ_2 are two linearly independent solutions of (3.3.43). The next example illustrates the above procedure.

Example 3.21 The origin is an ordinary point of the *Legendre equation* given by

$$(1 - t^2)x'' - 2t \, x' + \lambda(\lambda + 1)x = 0, \quad for \ -1 < t < 1 \quad and \quad \lambda \in \mathbb{R}, \tag{3.3.50}$$

because, for $|t| < 1$, we can write

$$p(t) = -\frac{2t}{1 - t^2} = -2 \sum_{k=0}^{\infty} t^{2k+1};$$

$$q(t) = \frac{\lambda(\lambda + 1)}{1 - t^2} = \lambda(\lambda + 1) \sum_{k=0}^{\infty} t^{2k}.$$

Let a series solution near the origin be given by (3.3.48) so that by substituting for φ'', φ' and φ in Eq. (3.3.50), and simplification, we obtain a *recurrence relation* given by

$$c_{k+2} = -\frac{(\lambda - k)(\lambda + k + 1)}{(k + 1)(k + 2)} c_k, \quad \text{for } k \geq 0. \tag{3.3.51}$$

This above recurrence relation helps to write the coefficients $c_2, c_3, c_4, \ldots$ in terms of the coefficients c_0 and c_1:

$$c_{2k} = \frac{(-1)^k \lambda (\lambda - 2) \cdots (\lambda - 2k + 2)(\lambda + 1)(\lambda + 3) \cdots (\lambda + 2k + 1)}{(2k)!} c_0;$$

$$c_{2k+1} = \frac{(-1)^k (\lambda - 1)(\lambda - 3) \cdots (\lambda - 2k + 1)(\lambda + 2)(\lambda + 4) \cdots (\lambda + 2k)}{(2k + 1)!} c_1.$$

and hence we obtain

$$\varphi(t) = c_0 \, \varphi_\lambda(t) + c_1 \, \phi_\lambda(t), \tag{3.3.52}$$

where φ_λ and ϕ_λ are two linearly independent solutions of (3.3.43). Notice that, when $\lambda = n \in \mathbb{Z}^+$, it follows from (3.3.51) that $c_{n+2} = c_{n+4} = \cdots = 0$ so that φ_n is a polynomial of degree n, if n is even. Similarly, if n is odd, then ϕ_n is a polynomial of degree n. Therefore, in the general solution (3.3.52), we can assume that φ_n is a polynomial of degree n, called the *Legendre polynomial* of the first kind, and ϕ_n is an infinite series, called the *Legendre function* of the second kind. A normalised Legendre function $L(t)$ is given by

$$L(t) = \begin{cases} \varphi_n(1)\phi_n(t), & \text{for } n \text{ even} \\ -\phi_n(1)\varphi_n(t), & \text{for } n \text{ odd} \end{cases}, \quad \text{for } |t| < 1. \tag{3.3.53}$$

The *Legendre polynomial* is usually denoted by $P_n(t)$. To obtain a compact expression for the polynomial $P_n(t)$, a convenient way is to take

$$c_n = \frac{(2n)!}{2^n (n!)^2}, \quad \text{so that we have } \varphi_n(1) = 1. \tag{3.3.54}$$

Then, if we rewrite the relation (3.3.51) as

$$c_{n-2} = -\frac{(n - 1)n}{2(2n - 1)} c_n, \tag{3.3.55}$$

then by using (3.3.54) it follows that

$$c_{n-2} = -\frac{(2n - 4)!}{2^n (n - 1)!(n - 2)!} \quad \text{and} \quad c_{n-4} = -\frac{(2n - 2)!}{2^n 2!(n - 2)!(n - 4)!}$$

Hence, by simple inductive argument, we obtain

$$c_{n-2k} = \frac{(-1)^k (2n - 2k)!}{2^n k!(n - k)!(n - 2k)!}, \quad \text{for } k \geq 1.$$

This implies the *Legendre polynomial* $P_n(t)$ is given by

$$P_n(t) = \sum_{k=0}^{n} \frac{(-1)^k (2n - 2k)!}{2^n k!(n - k)!(n - 2k)!} t^{n-2k}. \tag{3.3.56}$$

However, a simpler way is to use the *Rodrigues' formula* given by

$$P_n(t) = \frac{1}{2^n n!} \frac{d^n}{dt^n} (t^2 - 1)^n, \quad \text{for } n \in \mathbb{Z}^+. \tag{3.3.57}$$

Definition 3.13 A point $t_0 \in I$ is called a **singular point** of Eq. (3.3.43) if the function p or q is *not analytic* at t_0. A singular point $t_0 \in I$ is said to be a **regular singular point** if the function $(t - t_0)^2 p(t)$ and $(t - t_0)q(t)$ are analytic at t_0. Otherwise, we say t_0 is an *irregular singular point* of Eq. (3.3.43).

As before, we may assume that $t_0 = 0$ is a *regular singular point* of the second order differential equation given by

$$Lx(t) = 0, \quad \text{where } L \equiv t^2 D^2 + t p(t)D + q, \quad \text{for } t \in I, \tag{3.3.58}$$

where power series expansions of the functions p and q are as given in (3.3.45) (with $t_0 = 0$). In this case, according to *Frobenius method*, we take a series solution for Eq. (3.3.58) of the form

$$\varphi(t) = \sum_{n=0}^{\infty} c_n t^{n+s}, \quad \text{for } 0 < t < \rho. \tag{3.3.59}$$

We can assume that $c_0 \neq 0$. It then follows that we have

$$(n + s)(n + s - 1)c_n + \sum_{k=0}^{n} \left[p_{n-k}(k + s) + q_{n-k} \right]c_k = 0, \quad \text{for } n \in \mathbb{Z}^+. \tag{3.3.60}$$

As $c_0 \neq 0$, we obtain the quadratic equation

$$h(s) = 0, \quad \text{where } h(s) = s(s - 1) + p_0 s + q_0, \tag{3.3.61}$$

is the *indicial polynomial* of the differential equation (3.3.58). The two roots, say s_1 and s_2, are called the *characteristic exponents* of Eq. (3.3.58). Notice that we have

$$h(s_1 + n) = n[n + (s_1 - s_2)]. \tag{3.3.62}$$

Also, since we have

$$h(s + n) = (s + n)(s + n - 1) + p_0(s + n) + q_0,$$

we can rewrite (3.3.60) as

$$h(s + n)c_n + \sum_{k=0}^{n-1} [p_{n-k}(k + s) + q_{n-k}]c_k = 0, \quad \text{for } n \geq 1. \tag{3.3.63}$$

This gives the recurrence relation of the form

$$h(s + n)c_n + g_n = 0, \quad \text{with } g_n = \sum_{k=0}^{n-1} [p_{n-k}(k + s) + q_{n-k}]c_k = 0, \quad \text{for } n \geq 1, \tag{3.3.64}$$

so that we can determine c_n, in terms of c_0 and s, by using

$$c_n(s) = -\frac{g_n}{h(s + n)}, \quad \text{for } n \geq 1, \tag{3.3.65}$$

provided $h(s + n) \neq 0$. Clearly then, we have

$$c_n(s) = -\frac{c_0 H_n(s)}{h(s + n)h(s + n - 1) \cdots h(s + 2)h(s + 1)}, \quad \text{for } n \geq 1, \tag{3.3.66}$$

where $H_n(s)$ is a polynomial in variable s. Now, if $\mathrm{Re}(s_1) \geq \mathrm{Re}(s_2)$, then $h(s_1 + n) \neq 0$, and so we have that $c_n(s_1)$ exists for all $n \geq 1$. In this case, assuming that $c_0(s_1) = 1$ and the related convergence, it follows that

$$\varphi_1(t) = t^{s_1} \sum_{n=0}^{\infty} c_n(s_1) t^n, \quad \text{for } 0 < t < \rho,$$

is a series solution of Eq. (3.3.58). Also, if $\mathrm{Re}(s_1) > \mathrm{Re}(s_2)$ and the difference $s_1 - s_2$ is not a positive integer, then

$$h(s_2 + n) = n[n - (s_1 - s_2)] \implies h(s_2 + n) \neq 0, \tag{3.3.67}$$

so that $c_n(s_2)$ exists for all $n \geq 1$. Once again, assuming that $c_0(s_2) = 1$ and the related convergence, it follows that

$$\varphi_2(t) = t^{s_2} \sum_{n=0}^{\infty} c_n(s_2) t^n, \quad \text{for } 0 < t < \rho,$$

is the second series solution of Eq. (3.3.58).

Remark 3.3 The above assertions remain valid for the case when $-\rho < t < 0$ if t^{s_1} is replaced by $|t|^{s_1}$ and t^{s_2} by $|t|^{s_1}$.

Theorem 3.11 (Fuchs Theorem) *Let $t_0 \in I$ be a regular singular point of Eq. (3.3.58), the functions p and q have power series expansions as given in (3.3.45) (with $t_0 = 0$), and s_1, s_2 be the roots of Eq. (3.3.61).*

1. *If $\mathrm{Re}(s_1) \geq \mathrm{Re}(s_2)$, for $0 < |t| < \rho$, then Eq. (3.3.58) has a (convergent) series solution given by*

$$\varphi_1(t) = |t|^{s_1} \sum_{n=0}^{\infty} \alpha_n t^n, \quad for \, |t| < \rho, \quad with \, \alpha_0(s_1) = 1. \tag{3.3.68}$$

2. *If $\mathrm{Re}(s_1) > \mathrm{Re}(s_2)$, and $s_1 - s_2$ is not a positive integer in $0 < |t| < \rho$, then the second linearly independent (convergent) series solution of Eq. (3.3.58) is given by*

$$\varphi_2(t) = |t|^{s_2} \sum_{n=0}^{\infty} \beta_n t^n, \quad for \, |t| < \rho, \quad with \, \beta_0(s_2) = 1. \tag{3.3.69}$$

The coefficients α_n and β_n are determined by direct substitution.

Notice that, when the two roots s_1 and s_2 of the *indicial equation* (3.3.61) are equal, we only have one (convergent) series solution φ_1 of Eq. (3.3.58) given by (3.3.68). In this case, we find the second linearly independent (convergent) series solution φ_2 by using the series

$$\varphi(t, s) = t^s \sum_{n=0}^{\infty} \alpha_n t^n, \quad for \, t > 0. \tag{3.3.70}$$

For, with operator L as in (3.3.58), we have

$$L[\varphi(t, s)] = \alpha_0 h(s) t^s + t^s \sum_{n=1}^{\infty} \left\{ h(n + s)\alpha_n + \sum_{k=0}^{n-1} [(k + s)p_{n-k} + q_{n-k}]\alpha_k \right\} t^n$$

$$= \alpha_0 h(s) t^s, \quad with \, \alpha_0 \neq 0. \tag{3.3.71}$$

Notice that the series in above vanishes because of relation (3.3.64). Differentiating the last equation with respect to s, we obtain

$$L\left[\frac{\partial}{\partial s} \varphi(t, s) \right] = \frac{\partial}{\partial s} L[\varphi(t, s)] = \alpha_0 [h'(s) + h(s) \ln t] t^s,$$

so that, for $s = s_1$, we have

$$\frac{\partial}{\partial s} L[\varphi(t, s)] \bigg|_{s=s_1} = \alpha_0 [h'(s_1) + h(s_1) \ln t] t^{s_1}.$$

However, since s_1 is a root of order two of the equation $h(s) = 0$, we have $h(s_1) = h'(s_1) = 0$. Therefore

$$L\left[\frac{\partial}{\partial s}\varphi(t, s)\right] = 0 \quad \Longrightarrow \quad \varphi_2(t) = \frac{\partial}{\partial s}\varphi(t, s)\Big|_{s=s_1}.$$

Hence, assuming the involved series is convergent, it follows from (3.3.70) that

$$\varphi_2(t) = t^{s_1}\ln t \sum_{n=0}^{\infty} \alpha_n(s_1)t^n + t^{s_1}\sum_{n=0}^{\infty}\alpha'_n(s_1)t^n.$$

We can also write this solution as

$$\varphi_2(t) = \varphi_1(t)\ln t + t^{s_1}\sum_{n=0}^{\infty}\alpha'_n(s_1)t^n, \quad \text{for } t > 0. \tag{3.3.72}$$

As before, for the case when $t < 0$, we can obtain the solutions by replacing t^{s_1} with $|t|^{s_1}$, and $\ln t$ by $\ln|t|$. Notice that $\alpha_0(s_1) = 1$ implies that $\alpha'_0(s_1) = 0$ so that, by using $h(s_1 + n) \neq 0$ and $\alpha_n(s) = -g_n(s)/h(s+n)$, we have that $\alpha'_n(s_1)$ exists for $n \geq 1$. Hence, the series in (3.3.72) is well-defined. That is, the second linearly independent (convergent) series solution φ_2 is given by

$$\varphi_2(t) = \varphi_1(t)\ln|t| + |t|^{s_1+1}\sum_{n=0}^{\infty}\beta_n t^n, \quad \text{for } 0 < |t| < \rho, \tag{3.3.73}$$

where the coefficient β_n is determined by direct substitution.

Finally, suppose the difference $s_1 - s_2 = m \in \mathbb{Z}^+$, and $t > 0$. Then, by (3.3.62), we have

$$h(s + m) = (s + m - s_1)(s + m - s_2) = (s - s_2)(s + m - s_2).$$

It thus follows from (3.3.66) that

$$\alpha_m(s) = \frac{\alpha_0 H_m(s)}{(s - s_2)(s + m - s_2)h(s + m - 1)\cdots h(s + 2)h(s + 1)}.$$

Notice that, in view of (3.3.67), $\alpha_n(s_2)$ exists for all $n \neq m$. Also, if $H_m(s_2) = 0$, then $\alpha_m(s_2)$ is defined. In this case, (convergent) series solution φ_2 is given by

$$\varphi_2(t) = t^{s_2}\sum_{n=0}^{\infty}\alpha_n(s_2)t^n, \quad \text{with } \alpha_0(s_2) = 1. \tag{3.3.74}$$

However, if $H_m(s_2) \neq 0$, we may choose $\alpha_0 - s - s_2$ so that

$$\alpha_m(s) = \frac{H_m(s)}{(s+m-s_2)h(s+m-1)\cdots h(s+2)h(s+1)},$$

which shows that $\alpha_n(s_2) < \infty$ for all n. As before, we define

$$\varphi(t,s) = t^s \sum_{n=0}^{\infty} \alpha_n(s) t^n, \quad \text{with } \alpha_0 = s - s_2.$$

By (3.3.71), we have

$$L[\varphi(t,s)] = (s - s_2) h(s) t^s, \qquad (3.3.75)$$

which gives $L[\varphi(t,s_2)] = 0$. Therefore, we obtain a solution by taking $\varphi_2(t) = \varphi(t,s_2)$. In view of (3.3.66), we can write

$$\alpha_n(s) = \frac{(s-s_2)H_n(s)}{h(s+n)h(s+n-1)\cdots h(s+2)h(s+1)},$$

and we also have

$$h(s_2+n) = n(n-m) \neq 0, \quad \text{for } n = 1, 2, \ldots, m-1,$$

so that

$$\alpha_1(s_2) = \alpha_2(s_2) = \cdots = \alpha_{m-1}(s_2) = 0. \qquad (3.3.76)$$

Hence, we have

$$\varphi_2(t) = t^{s_2} \sum_{n=m}^{\infty} \alpha_n(s_2) t^n$$

$$= t^{s_2+m} \sum_{n=0}^{\infty} \alpha_{m+n}(s_2) t^n$$

$$= t^{s_1} \sum_{n=0}^{\infty} \beta_n(s_2) t^n, \quad \text{with } s_1 = s_2 + m \text{ and } \beta_n \equiv \alpha_{m+n}. \qquad (3.3.77)$$

Clearly, $\varphi_2 = A\varphi_1$, for some scalar A.

Next, we find the second linearly independent (convergent) series solution $\varphi_2 = \varphi_2(t)$, for $t > 0$. For, differentiating (3.3.75) with respect to s, we obtain

$$\frac{\partial}{\partial s} L[\varphi(t,s)] = h(s)t^s + (s - s_2)[h'(s) + h(s) \ln t] t^s,$$

so that by letting $s = s_2$ it follows that $\varphi_2(t) = \partial \varphi(t,s)/\partial s \big|_{s=s_2}$ is a solution. That is,

$$\varphi_2(t) = \left(t^{s_2} \ln t\right) \sum_{n=0}^{\infty} \alpha_n(s_2)\, t^n + t^{s_2} \sum_{n=0}^{\infty} \alpha_n'(s_2)\, t^n.$$

However, by using (3.3.76) and (3.3.77), we can also write

$$\varphi_2(t) = A\, \varphi_1(t) \ln t + t^{s_2} \sum_{n=0}^{\infty} \beta_n\, t^n,$$

where $\beta_n = \alpha_n'(s_2)$. As before, for $t < 0$, we write

$$\varphi_2(t) = |t|^{s_1} \sum_{n=0}^{\infty} \alpha_n\, t^n;$$

$$\varphi_2(t) = A\, \varphi_1(t) \ln |t| + |t|^{s_2} \sum_{n=0}^{\infty} \beta_n\, t^n,$$

where the coefficients α_n and β_n are determined by direct substitution. The final conclusion in previous two cases is that one root s_1 of the indicial equation always gives a (convergent) series solution φ_1, and we can manage to obtain the other linearly independent series solution φ_2, which converges in the same interval of convergence ($|t| < \rho$). The next example illustrates the procedure.

Example 3.22 Consider *Bessel equations* of index $v > 0$ given by

$$t^2 x'' + t x' + \left(t^2 - v^2\right) x = 0, \quad \text{for } -\infty < t < \infty. \tag{3.3.78}$$

As $t = 0$ is a regular singular point, we may first take a series solution of the form

$$\varphi(t) = \sum_{n=0}^{\infty} a_n\, t^{s+n}, \quad \text{for } t > 0, \quad \text{with } a_0 \neq 0. \tag{3.3.79}$$

Substituting for φ'', φ' and φ in Eq. (3.3.78), and simplifying, we obtain

$$\left[s^2 - v^2\right] a_0 t^s + \left[(s+1)^2 - v^2\right] a_1 t^{s+1} + \sum_{n=2}^{\infty} b_n\, t^{s+n} = 0, \tag{3.3.80}$$

where the coefficients b_n are given by

$$b_n = \left[(s+n)^2 - v^2\right] + a_{n-2}, \quad \text{for } n \geq 2. \tag{3.3.81}$$

Here, we have $s_1 = v$ and $s_2 = -v$. Notice that the solution of the Bessel's equation for $s = s_1$ is trivial at $t = 0$, and for $s = s_2$ does not exists at $t = 0$. We first consider the case when $s = s_1 = v$. It thus follows from the second term in (3.3.80) that

$$\left[(s+1)^2 - v^2\right]a_1 = 0 \quad \Rightarrow \quad [2v+1]a_1 = 0 \quad \Rightarrow \quad a_1 = 0. \tag{3.3.82}$$

Also, (3.3.81) gives the second order recurrence relation

$$a_n = -\frac{a_{n-2}}{n(2v+n)}, \quad \text{for } n \geq 2. \tag{3.3.83}$$

Therefore, we have $a_n = 0$ for $n = 3, 5, 7, \ldots$. So, we may use

$$a_{2k} = \frac{(-1)^k a_0}{2^{2k} k! (v+k)(v+k-1) \cdots (v+1))}, \quad \text{for } k \geq 1. \tag{3.3.84}$$

A more convenient way to write the above relation is given by

$$a_{2k} = \frac{(-1)^k 2^v \Gamma(v+1) a_0}{2^{2k+v} k! \Gamma(v+k+1)}, \quad \text{for } k \geq 1, \tag{3.3.85}$$

where $\Gamma(a)$ is the *Gamma function* given by

$$\Gamma(a) := \int_0^\infty x^{a-1} e^{-x} \, dx, \quad \text{for } a \in \mathbb{R}. \tag{3.3.86}$$

Therefore, taking

$$a_0 := \frac{1}{2^v \Gamma(v+1)}, \tag{3.3.87}$$

the regular solution of Eq. (3.3.78) is written as

$$J_v(t) = \sum_{k=0}^\infty \frac{(-1)^k t^{2k+v}}{2^{2k+v} k! \Gamma(v+k+1)}, \tag{3.3.88}$$

which is called the *Bessel function* of the first kind of order v. Next, for $s = s_2 = -v$, we first consider the case when $v \neq 1/2$ so that (3.3.82) implies that $a_1 = 0$. Now, using the recurrence relation

$$a_n = -\frac{a_{n-2}}{n(n-2v)}, \quad \text{for } n \geq 2, \tag{3.3.89}$$

we obtain the *irregular solution* of Eq. (3.3.78) given by

$$J_{-v}(t) = \sum_{k=0}^\infty \frac{(-1)^k t^{2k-v}}{2^{2k-v} k! \Gamma(-v+k+1)}, \tag{3.3.90}$$

which is called the *Bessel function* of the first kind of order $-v$. It can be shown that functions J_v and J_{-v} are linearly independent (convergent) series solution, provided $v \notin \mathbb{Z}$. Therefore, the general solution of Bessel equation for the case when v is not an integer can be written as

$$\varphi(t) = c_1 J_v(t) + c_2 J_{-v}, \quad \text{for } c_1, c_2 \in \mathbb{R}. \tag{3.3.91}$$

It follows easily that the Bessel functions J_n and J_{-n} are linearly dependent for any $n \in \mathbb{Z}$. So, suppose $v \notin \mathbb{Z}$. We consider here an irregular solution of Eq. (3.3.78) due to Weber as given by

$$Y_v := \frac{\cos v\pi}{\sin v\pi} J_v(t) - \frac{1}{\sin v\pi} J_{-v}. \tag{3.3.92}$$

Clearly, for $v \notin \mathbb{Z}$, J_v and Y_v are linearly independent. Notice that, though Y_n is indeterminate for $v = n \in \mathbb{Z}^+$, but $Y_n(t) = \lim_{v \to n} Y_v(t)$ exists, and it is a solution of the Bessel equation. Also, Y_n and J_n are linearly independent. We call Y_v *Bessel function* of the second kind of order v. Hence, the general solution of the Bessel equation is given by

$$\varphi(t) = c_1 J_v(t) + c_2 Y_v, \quad \text{for } v \geq 0 \text{ and } c_1, c_2 \in \mathbb{R}. \tag{3.3.93}$$

The case when $v = 1/2$ is left for the reader as an exercise.

3.4   Boundary Value Problems

In most applications, the *boundary value problems* (or simply BVPs) for second order differential equations involve a space variable as the independent variable. Therefore, in this section, we may write unknown function as $y = y(x) \in C^2(I)$, where $I = [a, b]$ is some compact interval. However, at a later stage, we will also consider functions in the space $\mathscr{L}^2(J) \cap C^2(J)$, where $\mathscr{L}^2(J)$ is the Hilbert space of square integrable functions defined over a subinterval $J \subseteq \mathbb{R}$. It is well know that, when J is compact, we have

$$\mathscr{L}^2(J) \cap C^2(J) = C^2(J),$$

We will mainly discuss linear BVPs with *mixed types* of boundary conditions given by

$$\begin{aligned} c_1 y(a) + c_2 y'(a) = \alpha, \quad &\text{where } (c_1, c_2) \neq 0; \\ d_1 y(b) + d_2 y'(b) = \beta, \quad &\text{where } (d_1, d_2) \neq 0. \end{aligned} \tag{3.4.1}$$

The boundary conditions of above types are called *separated*, because each involves only one end point of the interval I. It is difficult to find an explicit solution of a BVP for a general second order differential equation of the form

$$\frac{d^2 y}{dx^2} = f(x; y, y'), \quad \text{for some nice function } f \in C(\mathbb{R}^3),$$

However, it is possible to prove the *existence* and *uniqueness* of a solution for such types of BVPs [12].

In what follows, we consider some important BVPs for a second order linear differential equation given by

$$Ly = f, \quad \text{where } L \equiv a_0 \frac{d^2}{dx^2} + a_1 \frac{d}{dx} + a_2, \quad (3.4.2)$$

and the functions a_0, a_1, a_2, f are assumed to be defined and continuous over the interval I, with $a_0 \neq 0$ over I. When $a_0(x) \neq 0$ for any $x \in I$, we may also write Eq. (3.4.2) as $L_1[y] = g$, where the operator L_1 is given by

$$L_1 \equiv \frac{d^2}{dx^2} + p \frac{d}{dx} + q, \quad \text{with } p = \frac{a_1}{a_0}, \ q = \frac{a_2}{a_0}, \ g = \frac{f}{a_0} \in C(I). \quad (3.4.3)$$

A BVP for a linear differential equation (3.4.2) is called **nonhomogeneous** if the assigned boundary conditions are of the type (3.4.1), with $(\alpha, \beta) \neq (0, 0)$. Also, a BVP for a homogeneous differential equation $Ly = 0$ is called **homogeneous** if the assigned boundary conditions are of the type

$$c_1 y(a) + c_2 y'(a) = 0, \quad (c_1, c_2) \neq 0;$$
$$d_1 y(b) + d_2 y'(b) = 0, \quad (d_1, d_2) \neq 0. \quad (3.4.4)$$

Many different types of important homogeneous BVPs for differential equations of the form (3.4.2) arise naturally while applying the method of *separation of variables* to solve *initial-boundary value problems* related to some fundamental practical problems in science and engineering.

Example 3.23 The homogeneous BVPs listed below are used later in Chap. 7, and also in some subsequent chapters.

1. The *Dirichlet problem* is a homogeneous BVP of the form

$$y'' + \lambda y = 0, \quad \text{for } 0 < x < \ell,$$
$$y(0) = y'(\ell) = 0. \quad (3.4.5)$$

2. The *Neumann problem* is a homogeneous BVP of the form

$$y'' + \lambda y = 0, \quad \text{for } 0 < x < \ell,$$
$$y'(0) = y'(\ell) = 0. \tag{3.4.6}$$

3. A BVP with *Robin type* of boundary conditions is given by

$$y'' + \lambda y = 0, \quad \text{for } 0 < x < \ell,$$
$$y(0) = 0, \quad y(\ell) + y'(\ell) = 0. \tag{3.4.7}$$

4. A homogeneous BVP for the *Legendre equation* is given by

$$(1 - x^2)y'' - 2xy' + \lambda y = 0, \quad \text{for } -1 < x < 1,$$
$$y(-1) = y'(1) = 0, \tag{3.4.8}$$

where it is assumed that both y and y' remain finite as $x \to \pm 1$. We come across such types of BVPs while solving *Laplace equation* over a spherical domain.

5. For a *Bessel equation* of index ν given in standard form as

$$x^2 y'' + xy' + (x^2 - \nu^2)y = 0, \quad \text{for } 0 < x \le \alpha. \tag{3.4.9}$$

a BVP is given by taking $y(\alpha) = 0$, with both y, $y' < \infty$ as $x \to 0^+$.

6. A homogeneous BVP for the *Chebyshev equation*, with *periodic boundary conditions*, is given by

$$(1 - x^2)y'' - x\, y' + n^2 y = 0, \quad \text{for } -1 < x < 1;$$
$$y(-1) = y(1) = 0. \tag{3.4.10}$$

Definition 3.14 A **solution** of a *nonhomogeneous BVP* (3.4.2) is a function $\varphi = \varphi(x) \in C^2(I)$ such that it satisfies the equation, and also the boundary conditions given by (3.4.1). Similarly, a function $\varphi \in C^2(I)$ is called a solution of a *homogeneous BVP* if it satisfies the equation $Ly = 0$, and also some homogeneous boundary conditions such as given in (3.4.4).

Example 3.24 The general solution of equation $y'' + \lambda\, y = 0$ is given by

$$y(x) = A \cosh mx + B \sinh mx, \quad \text{for } \lambda = -m^2 < 0;$$
$$y(x) = A + Bx, \quad \text{for } \lambda = 0;$$
$$y(x) = A \cos kx + B \sin kx, \quad \text{for } \lambda = k^2 > 0,$$

where A, B are some constants. Therefore, with $y(0) = 0$, we have $A = 0$, in each case. Taking $y'(\pi) = 0$, it follows that the corresponding homogeneous BVP has no nontrivial solution over the interval $I = (0, \pi)$. Notice that the same conclusion holds even when we take $y(\pi) - y'(\pi) = 0$ as the second boundary condition. Next, the nonhomogeneous BVP for differential equation $y'' + y = 1$ has a unique solution over the interval $[0, 1]$, provided we take boundary conditions as $y(0) = 0$

and $y(1) = 0$. However, the corresponding homogeneous boundary value problem has a trivial solution. Also, over the interval $I = (0, \pi)$, the equation $y'' + y = 0$ has infinitely many solution for the boundary conditions given by $y(0) = 0$ and $y(\pi) = 0$.

In general, unlike an initial value problem, even a simple looking BVP may not have any solution. And, when it does, solution may not be unique. It is a standard fact that *a nonhomogeneous BVP for a differential equation* (3.4.2) *has a unique solution, unless the associated homogeneous BVP has a nontrivial solution*. The main argument is that, if φ_1, φ_2 and φ_3 are respectively unique solutions of the initial value problems given by

$$Ly_1 = 0, \quad \text{with } y_1(a) = c_2 \quad \text{and} \quad y_1'(a) = -c_1; \tag{3.4.11a}$$

$$Ly_2 = 0, \quad \text{with } y_2(b) = d_2 \quad \text{and} \quad y_2'(b) = -d_1; \tag{3.4.11b}$$

$$Ly_3 = f, \quad \text{with } y_3(a) = y_3'(a) = 0, \tag{3.4.11c}$$

then the nonhomogeneous equation $Ly = f$ has a unique solution φ given by

$$\varphi(x) = A\,\varphi_1(x) + B\,\varphi_2(x) + \varphi_3(x), \quad \text{for } x \in I \quad \text{and} \quad A, B \in \mathbb{R}. \tag{3.4.12}$$

Therefore, when φ satisfies boundary conditions of the form (3.4.1), a simple algebraic manipulation implies that the first part of the assertion holds provided φ_1 and φ_2 are linearly independent over the interval $I = [a, b]$. Otherwise, for $\varphi_2 = \lambda\,\varphi_1$, with $\lambda \in \mathbb{R}$, it follows that the functions φ_1 and φ_2 satisfy the boundary conditions (3.4.4). Hence, in the latter situation, the second part of the assertion holds. We thus conclude that it suffices to focus on solving BVPs of the types as given below:

1. (*Nonhomogeneous BVP*) With functions p, q, and g be as in (3.4.3), solve a BVP for the equation $Ly = f$, where the assigned boundary conditions are as given in Eq. (3.4.1).
2. (*Homogeneous BVP*) With functions p and q be as in (3.4.3), solve a BVP for an equation $Ly = 0$, where the assigned boundary conditions are as given in Eq. (3.4.4).

Use of Green's function in solving nonhomogeneous BVPs is a standard procedure.

3.4.1 Green's Functions and Nonhomogeneous Problems

In 1828, the British mathematician and physicist *George Green* (1793–1841) published *An Essay on the Application of Mathematical Analysis to the Theories of Electricity and Magnetism* wherein he introduced an important function that Riemann called *Green's function*. As discussed in later chapter, the related procedure as described in the same paper sought solutions of Poisson's equation governing the electric potential inside a bounded open set $\Omega \subset \mathbb{R}^3$, considering certain specified boundary conditions on the surface $\Gamma = \partial\Omega$. In this part, we will derive the Green's function of some simple IVPs and BVPs for ordinary differential equations.

We start with a demonstration about how to use Green's function in solving an initial value problem for a nonhomogeneous ordinary differential equation such as

$$a_0(t)x''(t) + a_1(t)x'(t) + a_2(t)x(t) = g(t); \tag{3.4.13a}$$

$$x(0) = x_0 \quad x'(0), = v_0, \tag{3.4.13b}$$

where the coefficient a_i and the forcing function g are assumed to be continuous functions. As before, we may write Eq. (3.4.13a) as $L[x] = f$, where

$$L \equiv a_0(t)\frac{d^2}{dt^2} + a_1(t)\frac{d}{dt} + a_2(t),$$

so that the solution is formally given by $x = L^{-1}[g]$, which we write as

$$x(t) = \int G(t, u)\, g(u)\, du. \tag{3.4.14}$$

The kernel of the integral operator on the right side of the above equation, namely $G(t, u)$ is called the Green's function of the initial value problem.

We know that if x_1 and x_2 are two linearly independent solutions of the associated homogeneous ordinary differential equation then, according to Theorem 3.5, a particular solution of Eq. (3.4.13a) is given by

$$x_p(t) = x_2(t)\int_{t_1}^{t} \frac{g(u)x_1(u)}{a_0(u)W(u)}\, du - x_1(t)\int_{t_0}^{t} \frac{g(u)x_2(u)}{a_0(u)W(u)}\, du, \tag{3.4.15}$$

where W is the *Wronskian* of solutions $x_1(t)$ and $x_2(t)$. The goal is to obtain from the above equation the general solution of Eq. (3.4.13a) of the form

$$x(t) = c_1(t)x_1(t) + c_2(t)x_2(t) + \int_{0}^{t} G(t, u)\, g(u)\, du. \tag{3.4.16}$$

For, notice that it is possible to solve the given initial value problem by putting together the solutions of initial value problems given by

$$a_0(t)x_h''(t) + a_1(t)x_h'(t) + a_2(t)x_h(t) = 0, \quad x_h(0) = x_0 \text{ and } x_h'(t) = v_0; \tag{3.4.17a}$$

$$a_0(t)x_p''(t) + a_1(t)x_p'(t) + a_2(t)x_p(t) = g(t), \quad x_h(0) = 0 \text{ and } x_h'(t) = 0. \tag{3.4.17b}$$

By linearity, we conclude that a solution of initial value problem is given by

$$x(t) = x_h(t) + x_p(t), \tag{3.4.18}$$

which also satisfies the initial conditions:

$$x(0) = x_h(0) + x_p(0) = x_0 + 0 = x_0;$$
$$x'(0) = x'_h(0) + x'_p(0) = v_0 + 0 = v_0$$

Therefore, it suffices to find a particular solution that satisfies homogeneous initial conditions. This is achieved by finding appropriate values of t_0 and t_1 in Eq. (3.4.15) that satisfy the homogeneous initial conditions $x_p(0) = 0$ and $x'_p(0) = 0$. We consider first the initial condition $x_p(0) = 0$ so that

$$x_p(0) = x_2(0) \int_{t_1}^{0} \frac{g(u)x_1(u)}{a_0(u)W(u)} \, du - x_1(0) \int_{t_0}^{0} \frac{g(u)x_2(u)}{a_0(u)W(u)} \, du, \tag{3.4.19}$$

implies that

$$x_p(0) = x_2(0) \int_{t_1}^{0} \frac{g(u)x_1(u)}{a_0(u)W(u)} \, du, \tag{3.4.20}$$

provided it is assumed that $x_1 = 0$ and $x_2 \neq 0$. Of course, by setting $t_1 = 0$, we can force $x_p(0) = 0$. Next, differentiating (3.4.15) and subsequently using the initial condition $x'_p(0) = 0$, it follows that

$$x'_p(0) = -x'_1(0) \int_{t_0}^{0} \frac{g(u)x_2(u)}{a_0(u)W(u)} \, du. \tag{3.4.21}$$

Notice that, by taking $x'_1(0) \neq 0$, we may set $t_0 = 0$. Hence, we have

$$x_p(t) = x_2(t) \int_{t_1}^{t} \frac{g(u)x_1(u)}{a_0(u)W(u)} \, du - x_1(t) \int_{t_0}^{t} \frac{g(u)x_2(u)}{a_0(u)W(u)} \, du$$

$$= \int_{0}^{t} \left[\frac{x_1(u)x_2(t) - x_1(t)x_2(u)}{a_0(u)W(u)}\right] g(u) \, du, \tag{3.4.22}$$

which gives the initial values Green's function as

$$G(t, u) = \frac{x_1(u)x_2(t) - x_1(t)x_2(u)}{a_0(u)W(u)}.$$

That is, the solution of the original initial value problem is given by

$$x(t) = x_h(t) + \int_0^t G(t, u) g(u) \, du, \qquad (3.4.23)$$

subject to that the solutions x_1, x_2, and x_h satisfy the following conditions:

$$x_1(0) = 0, \quad x_2(0) \neq 0, \quad x_1'(0) \neq 0, \quad x_2'(0) = 0, \quad x_h(0) = x_0, \quad x_h'(0) = v_0.$$

Example 3.25 For the initial value problem

$$x'' + x = 2 \cos t, \qquad x(0) = 4 \text{ and } x'(0) = 0,$$

the solution of the associated homogeneous equation, with nonhomogeneous initial conditions $x(0) = 4$, $x'(0) = 0$, is given by $x_h(t) = 4 \cos t$. Also, $x_1(t) = \sin t$ and $x_2(t) = \cos t$ are two linearly independent solutions of the homogeneous equation satisfying the homogeneous initial conditions $x(0) = 0$, $x'(0) = 0$, with the Wronskian is $W(t) = -1$. Therefore, in this case, the Green's function can be computed as

$$\begin{aligned} G(t, u) &= \frac{x_1(u)x_2(t) - x_1(t)x_2(u)}{a_0(u)W(u)} \\ &= \sin t \cos u - \sin u \cos t \\ &= \sin(t - u). \end{aligned}$$

It thus follows that

$$\begin{aligned} x_p(t) &= \int_0^t G(t, u) g(u) \, du \\ &= 2 \int_0^t \left[\sin t \cos u - \sin u \cos t \right] \cos u \, du \\ &= 2 \sin t \int_0^t \cos^2 u \, du - 2 \cos t \int_0^t \sin u \cos u \, du \\ &= t \sin t. \end{aligned}$$

Hence, by Eq. (3.4.23), the solution of the problem is obtained as

$$x(t) = 4\cos t + t \sin t.$$

In what follows, we use $y = y(x)$ as unknown function that is assumed to be a twice continuously differentiable function defined over an interval $I = [a, b]$. We describe next how to use the method of Green's function to solve a nonhomogeneous boundary value problem for a differential operator $\mathscr{L}$ of the form

$$\mathscr{L} \equiv \frac{\mathrm{d}}{\mathrm{d}x}\left(p\frac{\mathrm{d}}{\mathrm{d}x}\right) + q, \qquad p, q \in C(I), \qquad (3.4.24)$$

where the boundary conditions are assigned at points $x = a$ and $x = b$. For simplicity, we may take $y(a) = y(b) = 0$. As before, we apply Theorem 3.5 to the nonhomogeneous differential equation $\mathscr{L}[y] = g$. In view of Eq. (3.4.15), we can write the particular solution as

$$y(x) = y_2(x) \int_{x_1}^{x} \frac{g(\xi)y_1(\xi)}{p(\xi)W(\xi)}\,\mathrm{d}\xi - y_1(x) \int_{x_0}^{x} \frac{g(\xi)x_2(\xi)}{p(\xi)W(\xi)}\,\mathrm{d}\xi, \qquad (3.4.25)$$

where the limits x_0 and x_1 are to be found by using the assigned boundary values, and these two values are sought to ensure that the solution of the boundary value problem can be expressed as a single integral involving a suitable Green's function. Notice that the solution $y_h(x)$ of the associated homogeneous problem can be taken to be part of the above solution by making a suitable choice of limits in right-hand side integrals. We use first solutions of the homogeneous equation $\mathscr{L}[y] = 0$ that satisfy the boundary conditions

$$y_1(a) = y_2(b) = 0 \quad \text{and} \quad y_1(b) \neq 0, \quad y_2(a) \neq 0.$$

At $x = 0$, the unknown function y is given by

$$y(a) = y_2(a) \int_{x_1}^{a} \frac{g(\xi)y_1(\xi)}{p(\xi)W(\xi)}\,\mathrm{d}\xi - y_1(a) \int_{x_0}^{a} \frac{g(\xi)x_2(\xi)}{p(\xi)W(\xi)}\,\mathrm{d}\xi$$

$$= y_2(a) \int_{x_1}^{a} \frac{g(\xi)y_1(\xi)}{p(\xi)W(\xi)}\,\mathrm{d}\xi. \qquad (3.4.26)$$

Notice that, by taking $x_1 = a$, the condition is satisfied at the end $x = a$. Similarly, at the end $x = b$, we obtain

$$y(b) = y_2(b) \int_{x_1}^{b} \frac{g(\xi)y_1(\xi)}{p(\xi)W(\xi)} \, d\xi - y_1(b) \int_{x_0}^{b} \frac{g(\xi)x_2(\xi)}{p(\xi)W(\xi)} \, d\xi$$

$$= -y_1(b) \int_{x_0}^{b} \frac{g(\xi)y_2(\xi)}{p(\xi)W(\xi)} \, d\xi. \tag{3.4.27}$$

As before, the above expression gives zero at $x_0 = b$. Therefore, the desired solution can be written as

$$y(x) = y_2(x) \int_{a}^{x} \frac{g(\xi)y_1(\xi)}{p(\xi)W(\xi)} \, d\xi - y_1(x) \int_{b}^{x} \frac{g(\xi)x_2(\xi)}{p(\xi)W(\xi)} \, d\xi$$

$$= \int_{a}^{b} G(x, \xi) \, g(\xi) \, d\xi, \tag{3.4.28}$$

where the Green's function $G(x, \xi)$ in this case is given by

$$G(x, \xi) = \begin{cases} \dfrac{y_1(\xi)y_2(x)}{p(\xi)W(\xi)}, & a \le \xi \le x \\ \dfrac{y_1(x)y_2(\xi)}{p(\xi)W(\xi)}, & x \le \xi \le b \end{cases}, \tag{3.4.29}$$

where the functions y_1, y_2 are solutions of the associated homogeneous boundary value problem, with boundary conditions $y_1(a) = y_2(b) = 0$ and $y_1(b) \ne 0$, $y_2(a) \ne 0$. It should be clear that the above Green's function has many useful properties such as it is symmetric in its arguments and also satisfies the boundary conditions at $x = a$ and $x = b$.

Example 3.26 For the boundary value problem over the interval $[0, 1]$ given by

$$y'' = x^2; \quad \text{with boundary conditions } y(0) = y(1) = 0,$$

the solution of the associated homogeneous equation $y''(x) = 0$ is the function $y(x) = Ax + B$. As we need one solution y_1 satisfying $y_1(0) = 0$, together with the boundary condition $y(0) = 0$, we take $y_1(x) = x$ because the constant A is arbitrary. Further, as the second solution y_2 must satisfy $y_2(1) = 0$, it follows that a natural choice is to take $y_2(x) = 1 - x$. Therefore, in this case, we have

$$p(x)W(x) = y_1(x)y_2'(x) - y_1'(x)y_2(x) = x(-1) - (1 - x) = -1$$

is a constant, as required to be. Hence, by Eq. (3.4.29), we obtain

$$G(x, \xi) = \begin{cases} -\xi(1-x), & 0 \le \xi \le x \\ -x(1-\xi), & x \le \xi \le 1 \end{cases}$$

Finally, the solution of the problem is obtained as

$$y(x) = \int_0^1 G(x, \xi)\, g(\xi)\, d\xi$$

$$= -\int_0^x \xi(1-x)\, \xi^2\, d\xi - \int_x^1 x(1-\xi)\, \xi^2\, d\xi$$

$$= -(1-x)\int_0^x \xi^3\, d\xi - x\int_x^1 x(\xi^2 - \xi^3)\, d\xi$$

$$= \frac{1}{12}(x^4 - x).$$

A theory of Green's function can be developed further to deal with more general types of boundary value problems for the ordinary differential equations of the form

$$\frac{d}{dx}\left(p(x)\frac{dy(x)}{dx}\right) + q(x)y(x) = g(x). \tag{3.4.30}$$

As for Green himself, the central idea relies on the fact that the such a function is actually the system response function to a point source. We may start with the simple observation, namely the boundary value Green's function $G(x, \xi)$ satisfy the homogeneous differential equation $\mathcal{L}[y] = 0$, for $x \ne \xi$. That is,

$$\frac{\partial}{\partial x}\left(p\frac{\partial G(x, \xi)}{\partial x}\right) + q(x)G(x, \xi) = 0, \quad \text{for } x \ne \xi. \tag{3.4.31}$$

We also know that its derivative has a jump discontinuity at $x = \xi$ (see Exercise 3.20). This behavior of the boundary value Green's function $G(x, \xi)$ is similar to the *Heaviside step function* given by

$$H(x) = \begin{cases} 1, & x > 0 \\ 0, & x < 0 \end{cases},$$

with zero everywhere derivative, except at the jump discontinuity, where it has an infinite slope. The situation may be remedied by using the *Dirac delta function* $\delta(x) = H'(x)$ (see Appendix A.3 for further details). More precisely, it can be shown that the Green's function $G(x, \xi)$ satisfy the differential equation

$$\frac{\partial}{\partial x}\left(p\frac{\partial G(x,\xi)}{\partial x}\right) + q(x)G(x,\xi) = \delta(x-\xi),\qquad(3.4.32)$$

and also the homogeneous boundary conditions. We may write Eqs. (3.4.30) and (3.4.32) in compact form as

$$\mathcal{L}[y] = g(x)\quad\text{and}\quad\mathcal{L}[G] = \delta(x-\xi).\qquad(3.4.33)$$

Multiply the first equation by $G(x,\xi)$ and the second equation by $y(x)$, and then subtract to obtain the relation

$$G\,\mathcal{L}[y] - y\,\mathcal{L}[G] = g(x)G(x,\xi) - \delta(x-\xi)y(x).$$

Integrating the above equation over the interval $[a,b]$, the right side gives

$$\int_a^b \left[g(x)G(x,\xi) - \delta(x-\xi)y(x)\right]dx = \int_a^b g(x)G(x,\xi)\,dx - y(\xi).$$

On the other hand, by using the Green's identity, the left side gives

$$\int_a^b \left(G\,\mathcal{L}[y] - y\,\mathcal{L}[G]\right)dx = \left[p(x)\left(G(x,\xi)y'(x) - y(x)\frac{\partial G}{\partial x}(x,\xi)\right)\right]_a^b.$$

Combining the above two equations, and rearranging, we have

$$y(\xi) = \int_a^b g(x)G(x,\xi)\,dx - \left[p(x)\left(G(x,\xi)y'(x) - y(x)\frac{\partial G}{\partial x}(x,\xi)\right)\right]_a^b.\qquad(3.4.34)$$

The above equation is fundamental to solve nonhomogeneous boundary value problems, where the assigned boundary conditions suggest what conditions $G(x,\xi)$ must satisfy. The extra term in Eq. (3.4.34) is called the *Surface term*. We may write

$$S(b,\xi) - S(a,\xi) = \left[p(x)\left(G(x,\xi)y'(x) - y(x)\frac{\partial G}{\partial x}(x,\xi)\right)\right]_a^b\qquad(3.4.35)$$

so that Eq. (3.4.34) can be written as

$$y(\xi) = \int_a^b g(x)G(x,\xi)\,dx - \left[S(b,\xi) - S(a,\xi)\right].\qquad(3.4.36)$$

For example, with $y(a) = y(b) = 0$, the surface term gives

$$y(\xi) = \int_a^b g(x)G(x,\xi)\,dx - \left[p(b)\left(y(b)\frac{\partial G}{\partial x}(b,\xi) - G(b,\xi)y'(b)\right)\right]$$

$$+ \left[p(a)\left(y(a)\frac{\partial G}{\partial x}(a,\xi) - G(a,\xi)y'(a)\right)\right]$$

$$= \int_a^b g(x)G(x,\xi)\,dx + p(b)G(b,\xi)y'(b) - p(a)G(a,\xi)y'(a). \quad (3.4.37)$$

Notice that the right side in above equation vanishes only when $G(x,\xi)$ also satisfies the given homogeneous boundary conditions, which by symmetry property of Green's function leads to previously known solution given by

$$y(x) = \int_a^b g(x)G(x,\xi)\,dx.$$

Further, when $y'(a) = 0$, we can eliminate the term $y(a)\partial_x G(a,\xi)$ by taking $\partial_x G(a,\xi) = 0$. In the same way, it is possible to solve many other types of boundary value problems. More generally, consider the boundary conditions $y(a) = \alpha$ and $y'(b) = \beta$, and suppose we have

$$G(a,\xi) = 0 \quad \text{and} \quad \frac{\partial G}{\partial \xi}(b,\xi) = 0. \quad (3.4.38)$$

By using the symmetry argument, we may write Eq. (3.4.34) as

$$y(x) = \int_a^b G(x,\xi)g(\xi)\,d\xi - \left[p(\xi)\left(y(\xi)\frac{\partial G}{\partial \xi}(x,\xi) - G(x,\xi)y'(\xi)\right)\right]_{\xi=a}^{\xi=b}.$$

$$(3.4.39)$$

In view of (3.4.38), it suffices to analyse the surface term as given in (3.4.35). We have

$$S(b,\xi) - S(a,\xi) = \left[p(b)\left(y(b)\frac{\partial G}{\partial \xi}(x,b) - G(x,b)y'(b)\right)\right]$$

$$- \left[p(a)\left(y(a)\frac{\partial G}{\partial \xi}(x,a) - G(x,a)y'(b)\right)\right]$$

$$= -\beta\, p(b)G(x,b) - \alpha\, p(a)\frac{\partial G}{\partial \xi}(x,a). \quad (3.4.40)$$

Therefore, the general solution of Eq. (3.4.30) satisfying the nonhomogeneous boundary conditions $y(a) = \alpha$ and $y'(b) = \beta$ is given by

$$y(x) = \int_a^b G(x, \xi) g(\xi) \, d\xi + \beta \, p(b) G(x, b) + \alpha \, p(a) \frac{\partial G}{\partial \xi}(x, a). \qquad (3.4.41)$$

3.4.2   Sturm–Liouville Theory

In this part, we discuss an important theory developed separately by the Swiss mathematician *Jacques Sturm* (1803–1855) and the French mathematician *Joseph Liouville* (1809–1882), which provides a *unified scheme* to solve some important *single-parameter* homogeneous BVPs for differential equations of the form (3.4.2). The *Sturm-Liouville theory* is mainly about solving the *eigenvalue problem* $\mathscr{L}\varphi = \lambda \varphi$, subject to assigned (homogeneous) boundary conditions, where $\mathscr{L}$ is some suitably defined *self-adjoint operator* using the linear operator L. The interested reader may refer [13], for further details, and to the advanced text [14], for an in depth study.

Definition 3.15 With L be as in (3.4.2), the operator L^* given by

$$L^* \equiv a_0 D^2 + (2a_0' - a_1) D + (a_0'' - a_1' + a_2), \qquad (3.4.42)$$

is called the *adjoint operator* of the operator L. We say L is **self-adjoint** if $L^* = L$.

In this section, our main focus is to develop the mathematical concepts inasmuch as needed to introduce the concept of *generalised Fourier series* in terms of eigenfunctions of a Sturm–Liouville eigenvalue problem. It follows trivially from the defining condition (3.4.42) that $L^{**} = L$, and that an operator L is *self-adjoint* if and only if $a_0' = a_1$. Therefore, in general, we can write the self-adjoint operator $\mathscr{L}$ associated with the operator L as in (3.4.24).

Example 3.27 In the case of *Legendre equation* given by

$$(1 - x^2) y'' - 2xy' + \lambda y = 0, \quad \text{for } -1 < x < 1, \qquad (3.4.43)$$

the operator L is clearly *self-adjoint*, for any λ, and we have $p(x) = 1 - x^2$ and $q \equiv 0$. To meet certain practical requirements, we prefer to take $\lambda = n(n - 1)$, with $n \neq 0, 1$. A Bessel equation of index λ given in standard form as

$$x^2 y'' + x \, y' + (x^2 - \lambda^2) y = 0, \quad \text{for } x > 0, \qquad (3.4.44)$$

can be written in self-adjoint form as

$$(xy')' + [x - (\lambda^2/x)] \, y = 0. \qquad (3.4.45)$$

In this case, we have $p(x) = q(x) = x$, for all $x > 0$.

Though the operator L in general may not in *self-adjoint* form but, to solve a homogeneous BVP for a differential equation $Ly = 0$, it suffices to solve the equation $\mathscr{L}y = 0$, together with the given homogeneous boundary conditions. The next lemma provides the justification.

Lemma 3.2 *Every operator L as in (3.4.2) can be written in the form $\mathscr{L}$.*

Proof The assertion holds trivially when $a'_0 = a_1$. So, we may assume $a'_0 \neq a_1$. In general, the idea is to find a suitable function $\mu = \mu(x)$ such that the operator μL is self-adjoint. For, it can be shown that we may take

$$\mu(x) := \frac{1}{a_0} \exp\left(\int^x \left(a_1(t)/a_0(t)\right) dt\right), \quad \text{for } x \in I.$$

Notice that $\mu \not\equiv 1$ over I, because it is assumed that $a'_0 \neq a_1$. □

Clearly, multiplication by μ transforms the equation $Ly = f$ to the form $\mathscr{L}y = g$, where the operator $\mathscr{L}$ is as in (3.4.24), with the functions p, q, g are given by the relations

$$p = a_0 \mu, \qquad q = \frac{a_2}{a_0} p, \qquad g = \frac{f}{a_0} p.$$

Definition 3.16 Let $p \in C^1(I)$, and $q, r \in C(I)$. A linear operator of the form

$$\mathscr{S} \equiv \left[\frac{d}{dx}\left(p\frac{d}{dx}\right) + q\right] + \lambda r \tag{3.4.46}$$

is called the *Sturm–Liouville operator* (or simply SLO), and $\mathscr{S}y = 0$ is called a *Sturm-Liouville equation* (or simply SLE), where λ is a scalar (real or complex). We say $\mathscr{S}$ is **regular** if functions p and r are positive over I. Also, an SLE is called **singular** if it is defined over an unbounded interval, or if the function p or r vanishes, or if any one of the functions p, q approaches to ∞, as x approaches to one or both end points of the interval I.

Example 3.28 For $a_0(x) = x^2$, $a_1(x) = x$, $a_2(x) = 2$, and $g \equiv 0$, we have $\mu(x) = 1/x$ so that $p(x) = x$, $q(x) = 2/x$, and $g \equiv 0$, which is same as multiplying the equation $x^2 y'' + xy' + 2y = 0$ by $\mu(x)$. Also, if $a_0 = 1$, $a_1(x) = x$, $a_2 = 0$, then the differential equation $y'' + xy' + 1 = 0$ transforms to the form (3.4.46), with $p(x) = r(x) = e^{x^2/2}$. For Bessel equations in the standard form (3.4.9), we have $\mu(x) = 1/x$, so the Sturm–Liouville equation is obtained by dividing the equation by $1/x$. In this case, as in (3.4.45), we have

$$p(x) = q(x) = x, \quad \text{and} \quad r(x) = -\frac{1}{x}.$$

For the *Chebyshev equation* (3.4.10), the Sturm–Liouville equation is obtained by dividing the equation with $\sqrt{1 - x^2}$ so that we have

$$p(x) = \sqrt{1 - x^2}, \quad q(x) = 0, \quad \text{and} \quad r(x) = -(1 - x^2)^{-1/2}.$$

Notice that, in view of (3.4.24), we can also write an SLE as

$$\mathscr{L} y + \lambda r \, y = 0, \quad \text{for } y \in C^2([a, b]). \tag{3.4.47}$$

By our previous discussion, we know every *regular* SLE has two linearly independent solutions defined over the the interval I. However, values of the parameter λ depends on the assigned boundary conditions.

Remark 3.4 We also come across *singular* Sturm–Liouville equations wherein p or r is a discontinuous function, or the two functions are defined over an unbounded interval such as for a *Bessel equation*. As we shall see in Chap. 8, such types of SLEs arise naturally while applying the method of *separation of variables* to solve a BVP for some linear partial differential equations.

Definition 3.17 A (two-point) *Sturm-Liouville BVP* (or simply SL-BVP) is a Sturm–Liouville equation of the form (3.4.47), where the assigned (separated) boundary conditions are given by (3.4.4). We call it a *regular SL-BVP* if the associated Sturm–Liouville equation is regular.

Example 3.29 The boundary value problems (3.4.5) and (3.4.6) are *regular SL-BVPs*. As we will see in Chaps. 7 and 8, application of the method of *separation of variables* to solve some BVPs for 1-dimensional heat equation gives a regular SL-BVPs involving *Robin type* of boundary conditions such as (3.4.7).

Definition 3.18 A *singular SL-BVP* is a BVP for a singular Sturm–Liouville equation of the form (3.4.47), where linear homogeneous boundary conditions are defined in some appropriate manner.

In this case, the end conditions may not be *separated* and are prescribed to ensure the boundedness of the function $y = y(x)$ at the end points. For an illustration, suppose a Sturm–Liouville equation (3.4.47) has a *singularity* at the end $x = a$. Then, for any $\varepsilon > 0$, and $\varphi, y \in C^2((a, b])$, it follows from the identity

$$\int_{a+\varepsilon}^{b} \left[\varphi \, Ly - y \, L\varphi \right] dx = p(b)\left[y'(b)\varphi(b) - y(b)\varphi'(b) \right]$$

$$- p(a + \varepsilon)\left[y'(a + \varepsilon)\varphi(a + \varepsilon) - y(a + \varepsilon)\varphi'(a + \varepsilon) \right],$$

that if φ, y satisfy the conditions

$$\lim_{x \to a^+} p(x)\left[y'(x)\varphi(x) - y(x)\varphi'(x) \right] = 0; \tag{3.4.48a}$$

$$p(b)\left[y'(b)\varphi(b) - y(b)\varphi'(b) \right] = 0, \tag{3.4.48b}$$

then we have

$$\int\limits_a^b \left[\varphi\, Ly - y\, L\varphi\right] \mathrm{d}x = 0. \tag{3.4.49}$$

Notice that, for the case when $p(a) = 0$, we can replace conditions in (3.4.48) by the two conditions by (1) y and y' remain finite as $x \to a$; and, (2) $d_1 y(b) + d_2 y'(b) = 0$. Therefore, in this case, we say a singular SL-BVP is *self-adjoint* if (3.4.49) holds for any $\varphi,\ y \in C^2(I)$.

Example 3.30 The Sturm–Liouville equations (3.4.8) and (3.4.9) are singular SL-BVPs on the specified interval, and with the given boundary conditions. The latter types of singular SL-BVP arise naturally while applying the method of *separation of variables* to solve a boundary value problem for a linear partial differential equation modeling the *vibrations of a circular membrane*.

Definition 3.19 For functions $p \in C^1([a, b])$, and $q, r \in C[a, b]$, a *periodic SL-BVP* is a BVP for a Sturm–Liouville equation (3.4.47), where both functions p and r are positive periodic functions of period $b - a$, with $p(a) = p(b)$, and the boundary conditions are given by

$$y(a) = y(b) \quad \text{and} \quad y'(a) = y'(b). \tag{3.4.50}$$

For example, (3.4.10) is a singular SL-BVP, with periodic boundary conditions.

We introduce now the concept of eigenvalues and eigenfunctions of a regular or singular SL-BVPs. More generally, for a linear operator $T : V \to V$ of a linear space V over $\mathbb{C}$, a scalar $\lambda \in \mathbb{C}$ is called an *eigenvalue* of T if

$$Tv = \lambda v, \quad \text{for some } 0 \neq v \in V.$$

Each nonzero vector $v \in V$ satisfying the above condition is called an *eigenvector* of T corresponding to the eigenvalue λ. We say the *multiplicity* of an eigenvalue λ is $k \geq 1$ if there exists k linearly independent eigenvectors of T associated with λ. When V consists of functions, we use the terminology *eigenfunction*.

Example 3.31 The linear operator $L = -D^2 : C^2(J) \to C^1(J)$ has the property

$$L\big[\sin(n\,x)\big] = n^2 \sin\big(n\,x\big) \quad \text{and} \quad L\big[\cos(n\,x)\big] = n^2 \cos\big(n\,x\big). \tag{3.4.51}$$

Therefore, for each $n \in \mathbb{N}$, $\lambda_n = n^2$ is an eigenvalue of multiplicity two of the operator $L = -D^2$. In this case, the functions $\cos(n\,x)$ and $\sin(n\,x)$ are two linearly independent eigenfunctions associated with the eigenvalue λ_n. Notice that, by using standard trigonometric identities, it follows that the functions $\varphi_n(x) \in \{\sin nx, \cos nx\}$ satisfy the *orthogonality property* over the interval $[-\pi, \pi]$. More generally, we have

$$\int_{-\pi}^{\pi} \sin{(nx)} \sin{(mx)} \, dx = \begin{cases} 0, & \text{if } n \neq m \\ \pi/2, & \text{if } n = m = 1, 2, \ldots \end{cases};$$

$$\int_{-\pi}^{\pi} \cos{(nx)} \cos{(mx)} \, dx = \begin{cases} 0, & \text{if } n \neq m \\ \pi/2, & \text{if } n = m = 1, 2, \ldots \end{cases}; \qquad (3.4.52)$$

$$\int_{-\pi}^{\pi} \sin{(nx)} \cos{(mx)} \, dx = 0, \quad \text{for all } m, n.$$

Definition 3.20 A scalar λ for which a SL-BVP has a *nontrivial solution* is called an *eigenvalue*,[9] and the corresponding solutions are called the *eigenfunctions*, of the Sturm–Liouville equation. We say λ is *of multiplicity k* if there exists k linearly independent *eigenfunctions* corresponding to λ.

Example 3.32 Consider the SL-BVP (3.4.5) for $\ell = \pi$. That is, we have

$$y'' + \lambda y = 0, \quad \text{for } 0 < x < \pi; \quad \text{with } y(0) = y'(\pi) = 0.$$

For $\lambda = 0$, we obtain $y(x) = cx + d$, so that $y(0) = y'(\pi) = 0$ implies that $y \equiv 0$. Also, for $\lambda < 0$, say $\lambda = -\mu^2$ $(\mu \neq 0)$, we obtain

$$y(x) = c \cosh \mu x + d \sinh \mu x, \quad \text{for some } c, d \in \mathbb{R}.$$

Once again, by using the boundary conditions $y(0) = y'(\pi) = 0$, it follows easily that $y \equiv 0$. Therefore, the above SL-BVP has a *nontrivial solution* only when $\lambda > 0$. In this case, it follows that the SL-BVP has the eigenfunctions given by $\varphi_n(x) = \sin \sqrt{\lambda_n} \, x$, for $x \in [0, \pi]$, corresponding to the eigenvalues

$$\lambda_n = \frac{(2n-1)^2}{4}, \quad \text{for } n \in \mathbb{N},$$

Similarly, it follows that the SL-BVP (3.4.6) has eigenfunctions

$$\varphi_n(x) = \cos\left(\frac{n\pi x}{\ell}\right), \quad \text{for the eigenvalues } \lambda_n = \sqrt{\frac{n\pi}{\ell}}.$$

For the SL-BVP (3.4.6), with boundary conditions $y(0) = y(\ell) = 0$, the eigenfunctions and eigenvalues are respectively given by

$$\varphi_n(x) = \sin\left(\frac{n\pi x}{\ell}\right), \quad \lambda_n = \sqrt{\frac{n^2\pi^2}{\ell^2}}.$$

---

[9] We will see shortly that, in most interesting situations, we have $\lambda \in \mathbb{R}$.

Also, for the SL-BVP (3.4.7), the eigenfunctions corresponding to the eigenvalues $\lambda_n = \mu_n$ are the functions given by $\varphi_n(x) = \sin \mu_n x$. where μ_n is the nth positive root of the transcendental equation

$$\tan(\mu\ell) + \mu = 0.$$

For the singular SL-BVP (3.4.8) related to the Legendre equation, the sequence $\langle P_n \rangle$ of Legendre polynomials are the eigenfunctions corresponding to the eigenvalues $\lambda_n = n(n-1)$, for $n \geq 1$. Notice that $P_n(x) < \infty$ as $x \to \pm 1$. Finally, for the SL-BVP (3.4.9) related to a Bessel equation of index $v > 0$, we have

$$p(x) = x, \quad q(x) = -\frac{v^2}{x}, \quad \text{and} \quad r(x) = x,$$

so that $p(0) = 0$, $q(x) \to \infty$ as $x \to 0^+$, and $r(0) = 0$. For this singular SL-BVP, the eigenfunctions corresponding to the eigenvalues

$$\lambda_n = \mu_n^2 = \frac{\zeta_{vn}^2}{\alpha^2}, \quad \text{for } 0 < x \leq \alpha,$$

are the *Bessel functions* given by $\varphi_n(x) = J_v(\mu_n x)$, where $\zeta_{vn}\alpha$ is the nth zero of the function J_v. Notice that both J_v, $J_v' < \infty$ as $x \to 0^+$.

Example 3.33 Consider a regular SL-BVP, with *Robin type* of boundary conditions, given by

$$y'' + \lambda^2 y = 0, \quad \text{for } 0 \leq x \leq \ell,$$
$$c\, y(0) - y'(0) = 0, \quad d\, y(\ell) - y'(\ell) = 0, \quad \text{with } c, d \in \mathbb{R}^+. \tag{3.4.53}$$

It follows easily that (3.4.53) has a nontrivial solution if

$$\lambda = \frac{1}{\ell} \tan^{-1}\left[\frac{\lambda(c+d)}{\lambda^2 - cd}\right].$$

In the case when both c and d are nonzero, eigenfunctions are given by

$$\varphi_n(x) = \frac{\lambda_n}{c} \cos\left(\lambda_n x\right) + \sin\left(\lambda_n x\right), \quad \text{where } \lambda_n \to \frac{n\pi}{\ell} \text{ as } n \to \infty.$$

In the case when $c \neq 0$ and $d = 0$, eigenfunctions are given by

$$\varphi_n(x) = \frac{\lambda_n}{c} \cos\left(\lambda_n x\right) + \sin\left(\lambda_n x\right)$$
$$= \frac{\cos \lambda_n(\ell - x)}{\sin \lambda_n \ell}, \quad \text{with } \lambda_n \to \frac{n\pi}{\ell} \text{ as } n \to \infty.$$

Finally, in the case when $c \to \infty$ and $d \neq 0$, eigenfunctions are given by

$$\varphi_n(x) = \sin\left(\lambda_n x\right), \quad \text{where } \lambda_n \to \frac{(2n+1)\pi}{2\ell} \quad \text{as } n \to \infty.$$

3.4.3 Eigenfunctions Expansions

As said earlier, the Sturm–Liouville theory studies eigenvalues and eigenfunctions of an SL-BVP of the form (3.4.47). More precisely, we find eigenvalues λ_n of the operator

$$-\frac{1}{r}\mathscr{L}, \quad \text{where } \mathscr{L} \equiv \frac{\mathrm{d}}{\mathrm{d}x}\left(p\frac{\mathrm{d}}{\mathrm{d}x}\right) + q, \tag{3.4.54}$$

such that the corresponding eigenfunctions $\varphi_n \in C^2([a, b])$ are nontrivial solutions of the related SL-BVP. For reasons to be explained later, we need to deal with the cases when SL-BVP is regular, singular, or periodic, separately. Notice that, as illustrated in Example 3.32, corresponding to each eigenvalue of a regular SL-BVP there is exactly one (linearly independent) eigenfunction. However, the same does not hold for a periodic SL-BVP.

Example 3.34 Consider the *periodic* SL-BVP given by

$$\begin{aligned} y'' + \lambda\, y &= 0, \quad \text{for } -\pi < x < \pi; \\ y(-\pi) &= y(\pi), \quad \text{and} \quad y'(-\pi) = y'(\pi). \end{aligned} \tag{3.4.55}$$

When $\lambda > 0$, it follows easily that the eigenfunctions corresponding to the eigenvalue $\lambda_n = n^2$ of the SL-BVP (3.4.55) is given by $\varphi_n(x) \in \left\{\sin nx, \cos nx\right\}$, for $x \in [-\pi, \pi]$. Therefore, for $n \in \mathbb{N}$, the eigenvalue λ_n is of multiplicity two. Also, when $\lambda = 0$, the corresponding eigenfunction is the constant function 1. Indeed, when $\lambda < 0$, no solution of the Sturm–Liouville equation (3.4.55) satisfies the given periodic boundary conditions. Recall that the functions $\varphi_n(x) \in \left\{\sin nx, \cos nx\right\}$ satisfy the *orthogonality property* over the interval $[-\pi, \pi]$ (see (3.4.52)).

The *orthogonality property* for an operator (3.4.54) holds in general, which provides a unified procedure to solve many different types of homogeneous BVPs. In what follows, we study some other interesting properties of the same operator and also prove related theorems. To continue with the general discussion, we introduce some notations. Let I be a subinterval of $\mathbb{R}$, and $\omega \in C(I)$ be a fixed positive *weight function*. For $f, g \in C(I)$, we write

$$\langle f, g \rangle_\omega = \int_a^b [f(x)g(x)]\omega(x)\,dx; \tag{3.4.56a}$$

$$\|f\|_\omega = \sqrt{\langle f, f \rangle_\omega} = \int_a^b (f(x))^2 \omega(x)\,dx, \tag{3.4.56b}$$

It is routine exercise to check that the function $\langle f, g \rangle_\omega$ defines an *inner product* on the space $C(I)$, which is called the *weighted inner product* of f and g with respect to the weight function ω. In particular, we have the concept of *weighted norm* on the space $C(I)$ given by $\|f\|_\omega$. We may write $C_\omega(I)$ for this weighted inner product space.

Definition 3.21 We say $f, g \in C(I)$ are *orthogonal functions* with respect to a weight function $\omega > 0$ if $\langle f, g \rangle_\omega = 0$. Also, $f \in C(I)$ is said to be a *normalised function* with respect to ω if $\|f\|_\omega = 1$. We say $\mathscr{F} \subset C_\omega(I)$ is an *orthonormal set* if every pair of distinct functions in $\mathscr{F}$ are orthogonal, and each $f \in \mathscr{F}$ is normalised.

Theorem 3.12 *If $\varphi_1, \varphi_2 \in C^2([a, b])$ are two eigenfunctions of an operator (3.4.54) corresponding to eigenvalues λ_1 and λ_2, then φ_1 and φ_2 are orthogonal with respect to the weight function $r > 0$.*

Proof By Eq. (3.4.47), we have

$$\frac{d}{dx}\left(p\frac{d\varphi_1}{dx}\right) + (q + \lambda_1 r)\varphi_1 = 0; \tag{3.4.57a}$$

$$\frac{d}{dx}\left(p\frac{d\varphi_2}{dx}\right) + (q + \lambda_2 r)\varphi_2 = 0. \tag{3.4.57b}$$

Multiplying the first equation by φ_2, and the second by φ_1, and then subtracting the resulting equations, we obtain

$$(\lambda_1 - \lambda_2)r\varphi_1\varphi_2 = \varphi_2\frac{d}{dx}\left(p\frac{d\varphi_2}{dx}\right) - \varphi_1\frac{d}{dx}\left(p\frac{d\varphi_2}{dx}\right)$$

$$= \frac{d}{dx}\left[\left(p\frac{d\varphi_1}{dx}\right)\varphi_2 - \left(p\frac{d\varphi_2}{dx}\right)\varphi_1\right].$$

Therefore, by integrating the above equation, we obtain

$$(\lambda_1 - \lambda_2)\int_a^b r\varphi_1\varphi_2 dx = \left[p\left(\frac{d\varphi_1}{dx}\varphi_2 - \frac{d\varphi_2}{dx}\varphi_1\right)\right]_a^b$$

$$= p(b)[\varphi_1'(b)\varphi_2(b) - \varphi_1(b)\varphi_2'(b)]$$
$$- p(a)[\varphi_1'(a)\varphi_2(a) - \varphi_1(a)\varphi_2'(a)]. \tag{3.4.58}$$

By the boundary conditions for φ_1 and φ_2 at the end $x = b$, we have

$$d_1\varphi_1(b) + d_2\varphi_1'(b) = 0 \quad \text{and} \quad d_1\varphi_2(b) + d_2\varphi_2'(b) = 0.$$

In particular, when $d_2 \neq 0$, multiplying the first equation by $\varphi_2(b)$, the second by $\varphi_1(b)$, and subtracting, we have

$$\varphi_1'(b)\varphi_2(b) - \varphi_1(b)\varphi_2'(b) = 0. \tag{3.4.59}$$

When $c_2 \neq 0$ in the boundary conditions for φ_1 and φ_2 at the end $x = a$, by a similar argument, we have

$$\varphi_1'(a)\varphi_2(a) - \varphi_1(a)\varphi_2'(a) = 0. \tag{3.4.60}$$

Hence, by using Eqs. (3.4.59) and (3.4.60), it follows from (3.4.58) that

$$(\lambda_1 - \lambda_2) \int_a^b r\varphi_1\varphi_2 dx = 0. \tag{3.4.61}$$

So, if $\lambda_1 \neq \lambda_2$, then the assertion follows immediately from Definition 3.23.   □

Theorem 3.13 *If $\varphi_1, \varphi_2 \in C^2([a, b])$ are two eigenfunctions of a periodic SL-BVP corresponding to eigenvalues λ_1 and λ_2, then φ_1 and φ_2 are orthogonal with respect to the weight function $r > 0$.*

Proof Since φ_1, φ_2 satisfy the periodic boundary conditions given by

$$\varphi_1(a) = \varphi_1(b) \quad \text{and} \quad \varphi_1'(a) = \varphi_1'(b);$$
$$\varphi_2(a) = \varphi_2(b) \quad \text{and} \quad \varphi_2'(a) = \varphi_2'(b),$$

Equation (3.4.58), in this case, gives

$$(\lambda_1 - \lambda_2) \int_a^b r\varphi_1\varphi_2 dx = [p(b) - p(a)][\varphi_1'(a)\varphi_2(a) - \varphi_1(a)\varphi_2'(a)].$$

But then, since $p(a) = p(b)$, we obtain (3.4.61), and hence the assertion follows immediately from Definition 3.23, provided $\lambda_1 \neq \lambda_2$.   □

Theorem 3.14 *Eigenvalues of a regular SL-BVP are always real numbers.*

Proof Clearly, if $\varphi_1 = u + iv$ is an eigenfunction of a regular SL-BVP corresponding to an eigenvalue $\lambda_1 = \alpha + i\beta$, then $\varphi_2 = \overline{\varphi} = u - iv$ is an eigenfunction corresponding to the eigenvalue $\lambda_2 = \overline{\lambda} = \alpha - i\beta$. Therefore, it follows from Eq. (3.4.61) that

$$2\beta \int_a^b r\left(u^2(x) + v^2(x)\right) dx = 0.$$

Hence, for $r > 0$, we must have $\beta = 0$. □

Notice that the above theorem does not guarantee the existence of an eigenvalue of a regular SL-BVP. However, it can be shown that a self-adjoint regular Sturm–Liouville operator always has infinitely many eigenvalues. The next example illustrates the basic idea.

Example 3.35 Consider the regular SL-BVP with *Robin type* of boundary conditions, as given by (3.4.7). That is, for $\ell = 1$, we have

$$
\begin{aligned}
y'' + \lambda y &= 0, \quad \text{for } 0 \le x \le 1, \\
y(0) &= 0, \quad y(1) + \alpha\, y'(1) = 0, \quad \text{for some } \alpha > 0.
\end{aligned}
\tag{3.4.62}
$$

We see that, with $\mu = \sqrt{\lambda}$, the eigenvalues $\langle \lambda_n \rangle$ are given by

$$
\lambda_n = \mu_n^2, \quad \text{where } \tan \mu_n = -\alpha\, \mu_n, \quad \text{for } n \ge 1,
$$

and the corresponding eigenfunctions are given by $\varphi_n(x) = \sin \sqrt{\lambda_n}x$, for $x \in [0, 1]$. Notice that the sequence $\langle \lambda_n \rangle$ has the properties

$$
\lambda_1 < \lambda_2 < \lambda_3 < \cdots, \quad \text{with } \lim_{n \to \infty} \lambda_n = \infty.
$$

The next two theorems are very important facts of the Sturm–Liouville theory.

Theorem 3.15 *Every self-adjoint regular SL-BVP has infinite number of eigenvalues, say given by $\lambda_0, \lambda_1, \lambda_2, \ldots$, such that*

$$
\lambda_n < \lambda_{n+1}, \quad \text{for all } n \ge 0, \quad \text{with } \lim_{n \to \infty} \lambda_n = \infty.
$$

Moreover, for each $n \ge 0$, the corresponding eigenfunction φ_n is uniquely determined (up to a scalar) and has exactly n zeros in the interval $[a, b]$.

Proof For a proof see [4]. □

Theorem 3.16 *The eigenvalues of a periodic SL-BVP are a sequence of the form*

$$
-\infty < \lambda_0 < \lambda_1 \le \lambda_2 < \lambda_3 \le \lambda_4 < \cdots,
$$

such that the eigenvalue λ_0 has a unique eigenfunction, say φ_0. Also, for $k \ge 0$, eigenfunctions φ_{2k+1} and φ_{2k+2} are unique if $\lambda_{2k+1} < \lambda_{2k+2}$. But, there are two linearly independent eigenfunctions φ_{2k+1} and φ_{2k+2} if $\lambda_{2k+1} = \lambda_{2k+2}$.

Proof For a proof see [4]. □

We have now enough tools to introduce the concept of an *eigenfunctions expansion* of a function φ in some space. More precisely, we need to find out when can we write φ as a series representation of the form

$$\varphi(x) = \alpha_1\,\varphi_1(x) + \alpha_2\,\varphi_2(x) + \alpha_3\,\varphi_3(x) + \cdots, \quad \text{for } x \in I, \tag{3.4.63}$$

where $\{\varphi_n : n \geq 1\}$ is an *orthogonal system*. The relevance of such type of series representation to our present discussion emerges from the fact that, in most interesting cases, a suitable *orthogonal system* is given by the eigenfunctions of some SL-BVP. Recall that the eigenvalues of a regular SL-BVP are always real numbers, with

$$\lambda_1 < \lambda_2 < \lambda_3 < \ldots,$$

such that $\lim_{n\to\infty} \lambda_n = \infty$, and $\lambda_k > 0$ provided $c_1/c_2 < 0$, $d_1/d_2 > 0$, and $q > 0$. Moreover, the associated eigenfunctions φ_n are real-valued *nondegenerate* functions; that is, eigenfunctions are unique up to a scalar multiple, and φ_n have exactly $(n-1)$ zeros in the interval (a, b). We assume eigenfunctions φ_n are normalised with respect to the weight function $r > 0$. The fundamental theorem of Sturm–Liouville theory proves that every function in the space $\mathcal{L}^2(I)$ can be expressed as (3.4.63), where the coefficients functions α_n are determined by applying the *orthogonality property* of the functions φ_n (Theorem 3.17).

Remark 3.5 Unfortunately, no such representation is possible for functions $f \in C_\omega(I)$, mainly due to the fact that the existence of functions φ_n is not always guaranteed. Also, we do not gain anything even when the space $C_\omega(I)$ is enlarged to include the (weighted) Riemann integrable functions. However, in certain situations when I is unbounded, it helps to include the improper Riemann integrable functions.

In what follows, let $I = (a, b)$ be an arbitrary subinterval of $\mathbb{R}$.

Definition 3.22 Let $\omega \in C(I)$ be a fixed positive function, and

$$\mathcal{L}_\omega^2(I) := \left\{\varphi : [a, b] \to \mathbb{C} : \int_a^b |\varphi(x)|^2 \omega(x)\, dx < \infty\right\}. \tag{3.4.64}$$

On this linear space, for $\varphi, \phi \in \mathcal{L}_\omega^2(I)$, we define the *weighted inner product* and *weighted norm*, with respect to a weight function ω, respectively as given by

$$\langle \varphi, \phi \rangle_\omega = \int_a^b \varphi(x)\,\overline{\phi(x)}\,\omega(x)\, dx; \tag{3.4.65a}$$

$$\|\varphi\|_\omega = \sqrt{\langle \varphi, \varphi \rangle_\omega} = \left[\int_a^b |\varphi(x)|^2 \omega(x)\, dx\right]^2. \tag{3.4.65b}$$

It follows from Eq. (3.4.64) that the following *Cauchy-Schwartz inequality* holds in the space $\mathcal{L}_\omega^2(I)$:

$$\left| \int_a^b \varphi(x) \overline{\phi(x)} \, \omega(x) \, dx \right| \leq \left(\int_a^b |\varphi(x)|^2 \omega(x) \, dx \right)^2 \left(\int_a^b |\phi(x)|^2 \omega(x) \, dx \right)^2.$$

(3.4.66)

The standard $\mathscr{L}^2$- space corresponds to the case when $\omega \equiv 1$ over I. It can be shown that $\mathscr{L}_\omega^2(I)$ is a (weighted) *Hilbert space* with respect to the inner product given by (3.4.65a). We may call a function $\varphi \in \mathscr{L}_\omega^2(I)$ as a *weighted square integrable* function defined over the interval I. As usual, the *equality* of two functions in the space $\mathscr{L}_\omega^2(I)$ is defined in *almost everywhere* sense. That is, for $\varphi, \phi \in \mathscr{L}_\omega^2[a, b]$, we have $\varphi = \phi$ if and only if $\|\varphi - \phi\|_\omega = 0$. Equivalently, if the set given by

$$\{x \in [a, b] : \varphi(x) \neq \phi(x)\}$$

has the Lebesgue measure zero.

Definition 3.23 We say $\varphi, \phi \in \mathscr{L}_\omega^2(I)$ are *orthogonal functions* with respect to a weight function $\omega > 0$ if $\langle \varphi, \phi \rangle_\omega = 0$. Also, $\varphi \in \mathscr{L}_\omega^2(I)$ is said to be a *normalised function* with respect to ω if $\|\varphi\|_\omega = 1$. We say $\mathscr{F} \subset \mathscr{L}_\omega^2(I)$ is an *orthogonal system* if every pair of distinct functions in $\mathscr{F}$ are orthogonal, and each function $\varphi \in \mathscr{F}$ is normalised.

Example 3.36 Some useful *orthogonal systems* determined by an SL-BVP are as given below.

1. Since $r(x) = 1$ for the regular SL-BVP (3.4.6), which has the eigenfunctions $\varphi_n(x) = \cos(n\pi x/\ell)$ corresponding to the eigenvalues $\lambda_n = \sqrt{n\pi/\ell}$ (Example 3.32), it follows from Theorem 3.12 that

$$\int_0^\ell \cos\left(\frac{n\pi x}{\ell}\right) \cos\left(\frac{n\pi x}{\ell}\right) dx = 0;$$

$$\int_0^\ell \sin\left(\frac{m\pi x}{\ell}\right) \sin\left(\frac{n\pi x}{\ell}\right) dx = 0, \quad \text{for } m \neq n.$$

2. Since the SL-BVP (3.4.7) has eigenfunctions $\varphi_n(x) = \sin \mu_n x$, we have

$$\int_0^\ell \sin \mu_m x \sin \mu_n x \, dx = 0, \quad \text{for } m \neq n,$$

where μ_n is as defined in Example 3.32.

3. Since $r(x) = 1$ for the periodic SL-BVP (3.4.8), which has the *Legendre polynomials* $P_0, P_1, P_2, \ldots$ given by (3.3.57) as eigenfunctions corresponding to the eigenvalues $\lambda_n = n(n+1)$, it follows that

$$\int_{-1}^{1} P_m(x) P_n(x) dx = 0, \quad \text{for } m \neq n.$$

4. Since $r(x) = x$ for the singular SL-BVP related to Bessel equations (3.4.9) of index v defined over the interval $0 < x < \alpha$, and has the *Bessel functions* $J_v(\lambda_n x)$ for its eigenfunctions corresponding to the eigenvalues $\lambda_n = \zeta_{vn}/\ell$, we obtain

$$\int_{0}^{\alpha} x J_v \left(\frac{\zeta_{vm}^2}{\ell^2} x \right) J_v \left(\frac{\zeta_{vn}^2}{\ell^2} x \right) dx = 0, \quad \text{for } m \neq n,$$

where ζ_{vn} is as defined in Example 3.32.

5. Finally, since $r(x) = (1 - x^2)^{-1/2}$ for the singular SL-BVP (3.4.10), which has the *Chebyshev polynomials*

$$T_n(x) = \cos\left(n \cos^{-1}(x)\right), \quad \text{for } n = 0, 1, 2, \dots,$$

as its eigenfunctions for the eigenvalues $\lambda_n = n^2$, it follows that

$$\int_{-1}^{1} \frac{T_m(x) T_n(x)}{\sqrt{1 - x^2}} dx = 0, \quad \text{for } m \neq n.$$

Definition 3.24 A sequence $\langle \varphi_n \rangle$ in $\mathscr{L}_\omega^2(I)$ is said to *converge* to some $\varphi \in \mathscr{L}_\omega^2[a, b]$ if $\lim\limits_{n \to \infty} \|\varphi_n - \varphi\|_\omega = 0$. As usual, in this case, we write $\varphi_n \to \varphi$ in $\mathscr{L}_\omega^2$-norm.

Example 3.37 Notice that the sequence $\langle \varphi_n \rangle$ in $\mathscr{L}^2(I)$ given by

$$\varphi_n(x) = \begin{cases} 0, & \text{for } x = 0 \\ n, & \text{for } 0 < x \leq 1/n \\ 0, & \text{for } 1/n < x \leq 1 \end{cases}$$

converges pointwise to $\varphi \equiv 0$, but $\|\varphi_n - \varphi\| = \sqrt{n}$. So, *pointwise convergence* may not imply $\mathscr{L}^2$-*convergence*. Also, there exists a sequence in $\mathscr{L}^2(I)$ that converges in $\mathscr{L}^2$-norm, but not pointwise. Further, the inner product space $C(I)$ is contained in the space $\mathscr{L}^2(I)$, as a proper subspace. In fact, it can be shown that $C(I)$ is a *dense subspace* of $\mathscr{L}^2(I)$. The discontinuous function $\varphi \in \mathscr{L}^2[-1, 1]$ given by

$$\varphi(x) = \begin{cases} 0, & \text{for } -1 \leq x < 0 \\ 1, & \text{for } 0 \leq x \leq 1 \end{cases}$$

is the $\mathscr{L}^2$ limit of the sequence $\langle \varphi_n \rangle$ in $C[-1, 1]$ given by

$$\varphi_n(x) = \begin{cases} 0, & \text{for } -1 \leq x \leq -1/n \\ nx + 1, & \text{for } -1/n < x < 0 \\ 1, & \text{for } 0 \leq x \leq 1 \end{cases}$$

Let $\{\varphi_n \in \mathscr{L}_\omega^2(I) \bigcap C(I) : n \geq 1\}$ be an *orthogonal system*, and consider a sequence $\langle s_n \rangle$ in the space $\mathscr{L}_\omega^2(I)$ given by

$$s_n(x) = \sum_{k=1}^{n} \alpha_k \varphi_k(x), \quad \text{for } x \in I.$$

We say the sequence $\langle s_n \rangle$ *converges in mean* to φ on the interval I with respect to the weight function ω if $\varphi \in \mathscr{L}_\omega^2(I)$, and

$$\lim_{n \to \infty} \int_a^b [\varphi(x) - \varphi_n(x)]^2 \omega(x) \, dx = 0. \tag{3.4.67}$$

In this case, we write $\varphi_n \to \varphi$ in $\mathscr{L}_\omega^2$-norm. Therefore, the main problem is to determine appropriate (Fourier) coefficients α_k such that the sequence $\langle s_n \rangle$ gives the best *least square approximation* to the function φ. That is, we seek values of $\alpha_1, \ldots, \alpha_n$ such that the (error) function as given below is minimum

$$E(\alpha_1, \ldots, \alpha_n) = \int_a^b [\varphi(x) - \varphi_n(x)]^2 \omega(x) \, dx$$

$$= \int_a^b \varphi^2 \omega \, dx - 2 \sum_{k=1}^{n} \alpha_k \int_a^b \varphi \varphi_k \omega \, dx + \sum_{k=1}^{n} \alpha_k^2 \int_a^b \varphi_k^2 \omega \, dx \tag{3.4.68}$$

It then follows that E is minimum precisely when, for $1 \leq k \leq n$, we have

$$\alpha_k = \frac{\langle \varphi, \varphi_k \rangle_\omega}{\|\varphi_k\|_\omega^2} = \frac{1}{\|\varphi_k\|_\omega^2} \int_a^b [\varphi \varphi_k] \omega \, dx. \tag{3.4.69}$$

Using the above (Fourier) coefficients α_k in Eq. (3.4.68), we obtain

$$\int_a^b [\varphi - \varphi_n]^2 \omega(x) \, dx = \int_a^b \varphi^2 \omega \, dx - \sum_{k=1}^{n} \alpha_k^2 \int_a^b \varphi_k^2 \omega \, dx. \tag{3.4.70}$$

Therefore, non-negativity of the left-side integral in above equation gives

$$\sum_{k=1}^{\infty} \alpha_k^2 \int_a^b \varphi_k^2 \omega \, dx = \lim_{n \to \infty} \sum_{k=1}^{n} \alpha_k^2 \int_a^b \varphi_k^2 \omega \, dx \le \int_a^b \varphi^2 \omega \, dx, \qquad (3.4.71)$$

which is known as the *Bessel's inequality*. Further, if (3.4.67) holds, then it follows from (3.4.70) that equality holds in Eq. (3.4.71), which gives the *Parseval's identity*. We say an *orthogonal system*

$$\left\{ \varphi_n \in \mathscr{L}_\omega^2(I) \cap C(I) : n \ge 1 \right\}$$

is **complete** with respect to a positive (weight) function $\omega \in C(I)$ if every $\varphi \in \mathscr{L}_\omega^2(I) \cap C(I)$ approximates a series representation of the form (3.4.63). In this case, we write

$$\varphi(x) \sim \sum_{k=1}^{\infty} \alpha_k^2 \varphi_k(x), \quad \text{for } x \in I.$$

Notice that the series on the right side may not even converge pointwise to φ on the interval I. The next fundamental theorem is an important special case of Sturm–Liouville theory.

Theorem 3.17 *The set of eigenfunctions of any regular SL-BVP in the space $PC(I)$ is complete with respect to the weight function r. Moreover, every $\varphi \in PS(I)$ that satisfies the boundary conditions of the regular SL-BVP admits a unique eigenfunctions expansion of the form*

$$\varphi(x) = \sum_{k=1}^{\infty} \alpha_k \varphi_k(x), \quad \text{for each } x \in I, \qquad (3.4.72)$$

where the coefficients α_k are given by the (Fourier) formula (3.4.69).

Proof For a proof of more general statement see [4].   □

As illustrate in Example 3.36, the above theorem also holds for certain periodic/ singular SL-BVPs such as related to the Legendre, Bessel, and Chebyshev differential equations, provided the eigenfunctions belong to $\mathscr{L}_r^2(I)$.

3.5 First Order System of Differential Equations

We start our discussion here with the higher-dimensional analogue of a physical situation as discussed earlier in Example 3.13. Let $x : \mathbb{R} \to \mathbb{R}^n$ be a sufficiently smooth function such that the point $x(t) \in \mathbb{R}^n$ represents the *position* of a moving particle of mass m at time t in a *force field* given by a continuous vector function $f : \Omega \to \mathbb{R}^n$, where $\Omega \subseteq \mathbb{R}^n$ is an open set. We may write $f = (f_1, \ldots, f_n)$, where

$f_i : \Omega \to \mathbb{R}$ is the ith component of the function f, for $i = 1, \ldots, n$. By Newton's *momentum law*, we have

$$m \frac{d^2 x}{dt^2} = f(t; x(t); x'(t); b), \quad \text{for } t \in \mathbb{R}, \tag{3.5.1}$$

where the vector $b = (b_1, \ldots, b_k) \in \mathbb{R}^k$ $(k \geq 1)$ represents the *parameters* influencing the motion of the particle. In general, each b_j may be time-dependent ,i.e. $b_j = b_j(t)$, for $t \in \mathbb{R}$. The *second order system* of differential equations for the unknown function x is a prototypical model that can be used to discuss many different kinds of physical problems.

Example 3.38 With notations as above, if the force field given by the vector f is *conservative*, then there is some function $\varphi \in C^1(\Omega)$ such that $f = -\nabla \varphi$. Recall that we say φ is a *potential function* of the field f, and ∇ is the n-dimensional *del operator* given by

$$\nabla \equiv \left(\frac{\partial}{\partial x_1}, \cdots, \frac{\partial}{\partial x_n} \right). \tag{3.5.2}$$

The *energy function* E_φ of φ at point $x(t)$ is given by

$$E_\varphi(x(t)) := \frac{1}{2} \| x'(t) \|^2 + \varphi(x(t)), \quad \text{for } x(t) \in \Omega,$$

It then follows by using chain rule and Leibniz derivation formula that

$$\frac{d}{dt} E_\varphi(x(t)) = x'' \cdot x' + \nabla \varphi(x) \cdot x' = 0,$$

which is the well known law of *conservation of energy*.

Notice that, by adding the variable $v = x'$, it is possible to write the second order system (3.5.1) as a *first order system* of differential equations given by

$$x'(t) = v(t);$$
$$v'(t) = f(t; x(t); b)$$

For example, if f is (almost constant) *gravitational force* acting on a falling stone of mass m, then the first order system is given by

$$m x_1'' = 0, \quad m x_2'' = 0, \quad \text{and} \quad m x_3'' = -m g,$$

where $g > 0$. In this case, by straightforward integration, we obtain

$$x(t) = x(0) + v(0)t - \frac{g}{2} \begin{pmatrix} 0 \\ 0 \\ 1 \end{pmatrix} t^2.$$

However, with $f(x) = -GmM(x/\|x\|^3)$, it is no longer clear how to solve the associated first order system. Also, any nth order differential equation in normal form (3.3.1) can be expressed as a *first order system* of the form (3.5.4) by taking the *n component functions* $x_i = x_i(t)$ of an unknown function $x = x(t)$ as

$$x_1(t) = x(t), \quad x_2(t) = x'(t), \quad \ldots, \quad x_n(t) = x^{(n-1)}(t),$$

In this latter case, it suffices to assign the initial/boundary values for the n function $x, x', \ldots, x^{(n-1)}$. For example, the *initial values* at $t_0 \in I$ can be written as

$$x(t_0) = \alpha_1, \quad x'(t_0) = \alpha_2, \quad \ldots, \quad x^{(n-1)}(t_0) = \alpha_n, \tag{3.5.3}$$

where $\alpha_i \in \mathbb{R}$. Therefore, at least in theoretical terms, it suffices to develop a theory for a first order system of differential equations.

Definition 3.25 Let $U \subseteq \mathbb{R}^n$ and $V_j \in \mathbb{R}^{k_j}$ be open sets, for $j = 1, \ldots, n$. Suppose each vector function $f_j : I \times U \times V_j \to \mathbb{R}$ is a known continuous function. A **first order system** of differential equations for an unknown differentiable[10] vector function $x : I \to \mathbb{R}^n$ is a collection of n first order differential equations given by

$$\frac{dx_1}{dt} = f_1(t; x_1, \ldots, x_n; b_1)$$

$$\frac{dx_2}{dt} = f_2(t; x_1, \ldots, x_n; b_2)$$

$$\vdots$$

$$\frac{dx_n}{dt} = f_n(t; x_1, \ldots, x_n; b_n) \tag{3.5.4}$$

where it is assumed that $x(t) = (x_1(t), \ldots, x_n(t)) \in U$, for all $t \in I$, and $b_j \in \mathbb{R}^{k_j}$ are called the *parameters* of the system. By taking $f = (f_1, \ldots, f_n)$, we can write the system (3.5.4) more conveniently as

$$\frac{dx(t)}{dt} = f(t, x(t); b), \quad \text{for } t \in I. \tag{3.5.5}$$

The vector function f, for most applications, represents the variations in the *local averages* of some physical quantities measured in some suitable way at points $x = x(t) \in U$. As before, if $x(t_0) = a = (a_1, \ldots, a_n) \in U$, for some $t_0 \in I$, we say the vector a is an *initial value* of the system at t_0.

---

[10] In fact, it can be assumed that the function x is at least continuously differentiable over I.

3.5.1 Existence and Uniqueness Theorem

Let $U \subseteq \mathbb{R}^n$ be an open connected set, $I \subseteq \mathbb{R}$ be an interval, and $f : I \times U \to \mathbb{R}^n$ be at least continuous. We may write $f = (f_1, \ldots, f_n)$. Consider an *initial value problem* for a first order system of differential equations given by

$$\frac{dx}{dt} = f(t; x(t); b), \quad \text{with } x(t_0) = a, \quad \text{for } (t_0, x_0) \in I \times U, \qquad (3.5.6)$$

where $a = (a_1, \ldots, a_n) \in \mathbb{R}^n$ is the vector of *initial values* assigned at $t = t_0 \in I$. As in the case when $n = 1$, in geometrical terms, the graph Γ_x of the function $x = x(t)$ is a curve in $\mathbb{R}^{n+1}$, and f defines a direction field in the domain $I \times U$ such that, if $c = f(s, y)$, then the vector $(1, c) \in \mathbb{R}^{n+1}$ or, equivalently, the line

$$x = y + (t - s)c$$

gives a direction at the point $(s, y) \in I \times U$. The graphs of solutions of the system $x'(t) = f(t, x)$ fits on the direction field. The *existence* and *uniqueness* theorem for the initial value problem (3.5.6), as given below, is very useful to solve many practical problems concerning different types of dynamical systems.

Theorem 3.18 (Existence Theorem) *Let $U \subset \mathbb{R}^n$ and $V \subset \mathbb{R}^k$ be open sets, $c > 0$, and $f_i \in C^1\big[(-c, c) \times V \times U\big]$, for $i = 1, \ldots, n$. Consider a first order system as in (3.5.4), with time-dependent parameters $b = (b_1, \ldots, b_k) \in V$. For any $a = (a_1, \ldots, a_n) \in U$, there exists n smooth functions $x_i = x_i(t; b) : (-\delta, \varepsilon) \times V \to \mathbb{R}$ satisfying the system, and also the initial values given by*

$$x_i(0, b) = a_i, \quad \text{for } i = 1, \ldots, n. \qquad (3.5.7)$$

Proof The statement of the theorem is a straightforward generalisation of Theorem 3.1, and so the proof requires minor modifications. The details are left for the reader as an exercise. $\qquad \square$

Remark 3.6 In general, for an initial value problem of the form (3.5.6), the *smooth dependence* of a solution on initial conditions and parameters hold too. That is, if some other n smooth functions $\widehat{x}_i(t, b) : (-\delta_1, \varepsilon_1) \times V \to \mathbb{R}$ constitute a solution of the initial value problem (3.5.4), then $x_i = \widehat{x}_i$ given on $(-\delta, \varepsilon) \bigcap (-\delta_1, \varepsilon_1)$, for all $i = 1, \ldots, n$. Finally, considering the dependence of these solutions on a, we may write $x_i = x_i(t; a; b)$, and view these as being defined over the open set $(-\delta, \varepsilon) \times V \times U$. Then, there exist neighborhoods $U_a \subset U$ and $V_b \subset V$, and a number $\varepsilon > 0$, such that the solutions $x_i = x_i(t, x, y)$ are defined and smooth on the open set $(-\varepsilon, \varepsilon) \times V_b \times U_a \subset \mathbb{R}^{m+n+1}$.

As said earlier, in general, many sufficient conditions implying uniqueness of a solution of the initial value problem (3.5.6) are known. However, it is highly unlikely to have a necessary and sufficient condition giving the uniqueness of a solution for the problem [10]. Notice that by attaching the n first order equations given by

$$\frac{db_i}{dt} = 0, \quad \text{for } i = 1, \ldots, k,$$

we may assume the linear system (3.5.4) to be without parameters. In all that follows, we may assume $f : \Omega \to \mathbb{R}^n$ is a continuous function satisfying the Lipschitz condition, as given in the next definition.

Definition 3.26 Let $D = I \times U$ be a domain. A function $\mathbf{F} : D \to \mathbb{R}^n$ is said to satisfy the *Lipschitz condition* (or is *Lipschitz continuous*) with respect to x in D if we have

$$\|\mathbf{F}(t, x) - \mathbf{F}(t, y)\| \le L\|x - y\|, \quad \text{for all } t \in I \quad \text{and} \quad x, y \in U, \qquad (3.5.8)$$

for some constant $L > 0$, which is called a Lipschitz constant.

Lemma 3.3 *Let $D = I \times U$ be a convex domain. If $\mathbf{F}(t, x)$ and all the component functions of the Jacobian $\partial \mathbf{F}/\partial x$ are continuous and bounded in D, then $\mathbf{F}$ satisfies the condition (3.5.8), with Lipschitz constant*

$$L = n\,M, \quad \text{with } M := \max_{1 \le i \le n} \sup_{(t,x) \in D} \left|\frac{\partial f_i}{\partial x_j}(t, x)\right|,$$

where the partial derivative $\partial f_i/\partial x_j$ is the (i, j)th entry of the Jacobian $\partial \mathbf{F}/\partial x$.

Proof Notice that, for fixed $t \in I$ and $x, y \in U$, and $0 \le s \le 1$, the point $x + s\,y$ lies on the line segment $[x, x + y]$ joining the points x and $x + y$, and hence in U, by convexity of D. We also have

$$\frac{d}{ds}\left[f_i(t, x + s\,y)\right] = \sum_{j=1}^{n} \frac{\partial f_i}{\partial x_j}(t, x + s\,y)\, y_j.$$

Therefore, by applying the *Mean Value Theorem* to the function g_i given by $g_i(s) = f_i(t, x + s\,y)$, it follows that, for each $1 \le i \le n$, we have

$$f_i(t, x + y) - f_i(t, x) = \sum_{j=1}^{n} \frac{\partial f_i}{\partial x_j}(t, x + s_0\,y)\, y_j,$$

for some $0 < s_0 < 1$. Then, by using Cauchy-Schwartz inequality (2.1.3), we obtain

$$\|f_i(t, x + y) - f_i(t, x)\|^2 \le \left(\sum_{j=1}^{n} \left|\frac{\partial f_i}{\partial x_j}\right|^2\right)\left(\sum_{j=1}^{n} |y_j|\right) \le n^2\, M^2 \|y\|^2.$$

Hence, by summing over all $1 \leq i \leq n$, we have

$$\left\| f_i(t, \mathbf{x} + \mathbf{y}) - f_i(t, \mathbf{x}) \right\|^2 \leq n^2 M^2 \|\mathbf{y}\|^2.$$

Finally, taking the square root, it follows that $\mathbf{F}$ satisfies the condition (3.5.8), with Lipschitz constant $L = n M$. This completes the proof. $\qquad \square$

With f as above, the next theorem proves that the initial value problem (3.5.6) has a unique (local) solution $\varphi \in C^1(J)$, where $J = [t_0, t_0 + c]$, for some $c > 0$. Since analogous assertion holds for the interval $[t_0 - c, t_0]$, we obtain a unique solution over the interval $[t_0 - c, t_0 + c]$, by *pasting together* the two solutions. We may write Ω for the open set $I \times U$ in $\mathbb{R}^{n+1}$ so that $(t_0, \mathbf{a}) \in \Omega$. We shall also use the projection functions $\pi_j : \Omega \to I \times \mathbb{R}$ given by $(t, \mathbf{a}) \mapsto (t, a_j)$, for $j = 1, \ldots, n$.

Theorem 3.19 (Picard-Lindelöf Theorem) *Let $\Omega \subset \mathbb{R} \times \mathbb{R}^n$ be a domain, $(t_0, \mathbf{a}) \in \Omega$ be arbitrary, and $f : \Omega \to \mathbb{R}$ be a continuous function satisfying the Lipschitz condition (3.5.8), with Lipschitz constant $L > 0$. Then there exists a unique solution $\varphi \in C^1(J)$ of the initial value problem (3.5.6), where $J = [t_0 - c, t_0 + c]$, for some $c > 0$.*

Proof The above statement is a straightforward generalisation of Theorem 3.2. As before, let $0 < c < L^{-1}$, and $I = [t_0, t_0 + c]$. Recall that the linear space $X = C(I, \mathbb{R}^n)$ of continuous functions $\mathbf{x} : I \to \mathbb{R}^n$ is a *complete* metric space with respect to the the *supremum metric* d_∞ given by

$$d_\infty(f, g) = \sup \left\{ \|f(t) - g(t)\| : t \in I \right\}, \quad \text{for } f, g \in X.$$

Notice that, for $\mathbf{x} = (x_1, \ldots, x_n) \in X$, the function $x_i = x_i(t)$ is a *local solution* of the *initial value problem*

$$\frac{dx_i}{dt} = f_i(t, x_i(t)), \quad \text{with } x_i(t_0) = a_i, \quad \text{for } (t_0, a_i) \in \pi_i \Omega,$$

if and only if we have

$$x_i(t) = a_i + \int_{t_0}^{t} f_i(s, x_i(s)) \, ds, \quad \text{for } t \in I \quad \text{and} \quad 1 \leq i \leq n.$$

Clearly, any function $\varphi_i = \varphi_i(t)$ satisfying the above integral equation is continuous, with $\varphi_i(t_0) = a_i$. We also have, for each $t \in I$,

$$\left| x_i(t) - a_i \right| = \left| \int_{t_0}^{t} f_i(s, x_i(s)) \, ds \right| \leq \int_{t_0}^{t} |f_i(s, x_i(s))| \, ds \leq c \, B,$$

where B is the bound of f over a compact subdomain of Ω. Let

$$Y := \{y \in X : d_\infty(y, a) < c\,B\},$$

where $a \in X$ is viewed as a constant function. Hence, by above argument, we have $x = (x_1, \ldots, x_n) \in Y$. Consider the linear map $T_a : Y \to Y$ given by

$$T_a(x)(t) := a + \int_{t_0}^{t} f(s, x(s))\,ds, \quad \text{for } t \in I.$$

It thus follows that, for $x_1, x_2 \in Y$, we have

$$d_\infty(T_a(x_1), T_a(x_2)) = \sup_{t \in [0,c]} \left| \int_{t_0}^{t} [f(s, x_1(s)) - f(s, x_2(s))]\,ds \right|$$

$$\leq \int_{0}^{t} |f(s, x_1(s)) - f(s, x_2(s))|\,ds$$

$$\leq L \int_{t_0}^{t} |x_1(s) - x_2(s)|\,ds$$

$$\leq cL\,d_s(x_1, x_2).$$

Therefore, by Theorem A.1, there exists a *unique* $x \in Y$ such that $T(x) = x$, with $x(t_0) = a$. Notice that the above inequality also proves that if f is a C^k function then the mapping $t \mapsto T_a(x)(t)$ is of class C^{k+1}, and hence $t \mapsto x(t)$ is of class C^{k+1}. □

Remark 3.7 The algorithm described in the proof of Theorem 3.19 may help in some concrete situations to obtain an *approximate solution*. However, in most other cases, we apply a more efficient *numerical technique*. The interested reader may refer to text [15] or [16] to find out how numerical methods are applied to obtain an *approximate solution* of some intractable initial-boundary value problems. In general, preference for a particular analytical/numerical method depends entirely on the nature of physical problem under investigation. Notice that a numerical solution of a problem provides only an estimate of the exact solution.

3.5.2 Linear Systems

Notice that, by introducing new dependent variables x_1 and x_2, the second order Eq. (3.3.26) can be written as a first order system given by

$$x_1'(t) = x_2(t);$$

$$x_2'(t) = -\frac{c}{m} x_2(t) - \frac{k}{m} x_1(t) + \frac{a_0}{m} \sin \omega t. \quad (3.5.9)$$

We can also write (3.5.9) in *vector form* as

$$\frac{dx}{dt} = A \cdot x + b, \quad \text{with } x(t) = \left(x_1(t)\ x_2(t)\right)^t = \begin{pmatrix} x_1(t) \\ x_2(t) \end{pmatrix};$$

$$A = \begin{pmatrix} 0 & 1 \\ -k/c & -c/m \end{pmatrix} \quad \text{and} \quad b(t) = \begin{pmatrix} 0 \\ (a_0/m) \sin \omega t \end{pmatrix}. \quad (3.5.10)$$

For many applications related to dynamical system, it is desirable to have a differential equation model given by a *linear system*.

Definition 3.27 For a given $n \times n$ matrix function $A : I \to \mathbb{R}^{n^2}$, with n^2 entries $a_{ij} \in C(I)$, and a function $g : I \to \mathbb{R}^n$, a **linear system** of order n for a differentiable function $x : I \to \mathbb{R}^n$ is a first order system of the form

$$\frac{dx}{dt} = A \cdot x + g, \quad (3.5.11)$$

where both $x = (x_1, \ldots, x_n)$ and $g = (g_1, \ldots, g_n)$ are taken as column vectors. The matrix $A = A(t)$ is called the *coefficients matrix* of the linear system. We say (3.5.11) is a *homogeneous* linear system if $g \equiv 0$. Otherwise, it is called a *nonhomogeneous* linear system.

For example, the initial value problem

$$x'(t) = a\, x(t), \quad \text{with } x(0) = x_0.$$

defines a dynamical system given by the smooth function $\varphi_t(x_0) = x_0 e^{at}$. In general, by taking a square matrix A of order n in place of the scalar a, and a vector function $x = x(t) : \mathbb{R} \to \mathbb{R}^n$ for $x = x(t)$, we obtain a dynamical system given by an initial value problems for a first order linear system of the form (3.5.11) (with $g \equiv 0$). Clearly, by using

$$f(t, x) = A \cdot x + g, \quad \text{for } t \in I,$$

every first order *linear system* is a system of the form (3.5.4).

Remark 3.8 The linear systems of the form (3.5.11) hold special significance in science and engineering, mainly because most dynamical systems are studied in terms of a linear system. In actual applications, the n^2 functions a_{ij} represent the *physical constraints* of a dynamical system under investigation, and the function g is called the *equilibrium function*. In majority of the cases, we come across such types of systems as a linear approximation of a *nonlinear model* of the form (3.5.5) near a *point of equilibria*. Also, by using some efficient numerical tools, it is possible to obtain a *local solution* of a nonlinear problem.

Any initial value problem related to a linear system can be solved explicitly when all a_{ij} are constants. The procedure works on similar lines as given in Theorem 3.6. In this case, a vector function $\mathbf{X} : I \to \mathbb{R}^n$ given by $\mathbf{X}(t) = e^{\lambda t} \mathbf{v}$ is a solution of a homogeneous linear system off the form

$$
\begin{aligned}
\frac{dx_1}{dt} &= a_{11}x_1 + \ldots + a_{1n}x_n \\
\frac{dx_1}{dt} &= a_{21}x_1 + \ldots + a_{2n}x_n \\
&\vdots \\
\frac{dx_n}{dt} &= a_{n1}x_1 + \ldots + a_{nn}x_n
\end{aligned}
\tag{3.5.12}
$$

if and only if $A\mathbf{v} = \lambda\mathbf{v}$. That is, $\mathbf{v}$ is an *eigenvector* associated with the eigenvalue λ_i given by the n roots of the following *characteristic equation* of the matrix A:

$$
|A - \lambda I| = \lambda^n + c_1 \lambda^{n-1} + \cdots + c_{n-1}\lambda + c_n = 0. \tag{3.5.13}
$$

We can assume that $\mathbf{v} \neq \mathbf{0}$. Recall that the eigenvectors of distinct eigenvalues are always linearly independent. Therefore, if there exists n linearly independent vectors $v_1, \ldots, v_n$ such that $A v_i = \lambda_i v_i$, for $i = 1, \ldots, n$, then the n fundamental solutions of the system $\mathbf{X}'(t) = A \cdot \mathbf{X}$ are given by

$$
X_i(t) = e^{\lambda_i t} v_i, \quad \text{for } i = 1, \ldots, n. \tag{3.5.14}
$$

More generally, if the linear system (3.5.13) has k distinct eigenvalues $\lambda_1, \ldots, \lambda_k$ of multiplicities $n_1, \ldots, n_k$, with $n = n_1 + \cdots + n_k$, and also if there are only $m_j < n_j$ linearly independent eigenvectors associated with the eigenvalue λ_j, then the linear equation

$$
\left(A - \lambda_j I\right)^{k+1} v = 0
$$

has at least $m_j + 1$ linearly independent solutions whenever the linear equation

$$
\left(A - \lambda_j I\right) k v = 0
$$

has at least m_j linearly independent solutions. In particular, whenever the coefficients matrix A has $k < n$ linearly independent eigenvectors, we can always find additional solutions of the form $x(t) = e^{\lambda t} v$ by computing vectors v such that, for some $m \geq 1$, we have

$$
(A - \lambda_j I)^{m+1} v = 0, \quad \text{but } (A - \lambda_j I)^m v \neq 0.
$$

The main fact used here (for $m = 1$) is that

$$
e^{At} v = e^{\lambda t} e^{(A - \lambda I)t} v = e^{\lambda t}\left[v + t(A - \lambda I)v\right],
$$

implies that $e^{At}v$ is an additional solution of the system $x'(t) = A \cdot x$, where

$$e^{At} := 1 + At + \frac{A^2 t}{2!} + \cdots + \frac{A^n t}{n!} + \cdots$$

Further, if some $\lambda_i = \alpha_i + i\beta_i$ with an eigenvector $v_i = u_i + iv_i$, then the two fundamental solutions associated with the eigenvalues $\alpha_i \pm i\beta_i$ are given by

$$x_{i1}(t) = e^{\alpha_i t}[u_i \cos \beta_i t - v_i \sin \beta_i t]; \quad \text{and,}$$
$$x_{i2}(t) = e^{\alpha_i t}[u_i \sin \beta_i t + v_i \cos \beta_i t].$$

Therefore, if $X(t)$ is the $n \times n$ *nonsingular* matrix formed having the n fundamental solutions $x_1, \ldots, x_n$ as columns, called the *fundamental matrix*, then the general solution of the system $x'(t) = A \cdot x$ is given by

$$x_c(t) = X(t)c, \quad \text{with } c = (c_1, \ldots, c_n)^t. \tag{3.5.15}$$

Notice that $X'(t) = AX(t)$. The following theorem is a remarkable fact.

Theorem 3.20 *If $X(t)$ is a fundamental matrix of a system $x'(t) = A \cdot x$ then $e^{At} = X(t)X^{-1}(0)$.*

Finally, to solve a system of differential equations of the form (3.2.19) with $A \in M_n(\mathbb{R})$ and initial conditions given by

$$\alpha(t_0) = (\alpha_1, \ldots, \alpha_n), \tag{3.5.16}$$

we use the method of *variation of parameters*. That is, if the particular solution $x_p(t)$ is given by

$$x_p(t) = X(t)u(t), \quad \text{with } u(t) = (u_1(t), \ldots, u_n(t))^t \tag{3.5.17}$$

then by using $x_p(t)$ in Eq. (3.2.19) we have

$$X(t)u'(t) = g(t) \quad \Rightarrow \quad u'(t) = X^{-1}(t)g(t), \tag{3.5.18}$$

and so, by integration, we obtain

$$u(t) = X^{-1}(t_0)\alpha + \int_{t_0}^{t} X^{-1}(s)g(s)ds. \tag{3.5.19}$$

Thus, for the system (3.2.19) having the initial conditions (3.5.16), the *complete solution* is given

$$x(t) = x_c(t) + x_p(t)$$

$$= \mathbf{X}(t)\Big[c + \mathbf{X}^{-1}(t_0)\boldsymbol{\alpha} + \int_{t_0}^{t} \mathbf{X}^{-1}(s)g(s)ds\Big], \qquad (3.5.20)$$

where c is computed using (3.5.16). Notice that, using the observation $\mathbf{X}(t) = e^{At} \Rightarrow \mathbf{X}^{-1}(s) = e^{-As}$, we have

$$x(t) = e^{At}c + e^{A(t-t_0)}\boldsymbol{\alpha} + \int_{t_0}^{t} e^{A(t-s)}g(s)ds. \qquad (3.5.21)$$

Autonomous Systems

In many practical situations, we need consider only Cauchy problems for a first order *autonomous system* of the form

$$\frac{dx}{dt} = f(x), \quad \text{with } x(t_0) = x_0, \qquad (3.5.22)$$

where f is some locally Lipschitz continuous function defined over a domain $U \subseteq \mathbb{R}^n$. Recall that a function $f : U \to \mathbb{R}$ is said to be *locally Lipschitz* with respect to $x \in U$ if f is Lipschitz over $U \cap N$, for some neighborhood $N = N(x)$, where Lipschitz constant depends on N. As before, we can be show that f is locally Lipschitz if both f and $\partial f / \partial x$ are continuous in U. Notice that, by attaching $(m + 1)$ equations given by

$$\frac{dt}{dt} = 1, \quad \frac{db_i}{dt} = 0, \quad \text{for } i = 1, \dots, m,$$

any first order system (3.5.4) can be viewed as an autonomous system, without parameters.

As said before, by the fundamental theorem of calculus, the initial value problem (3.5.22) for the case when $n = 1$ has a unique solution for any point $(t_0, x_0) \in I \times J$ if and only if the function $f : J \to \mathbb{R}$ is continuous, for some open interval $J \subseteq I$. In general, a curve $\varphi : (t_0 - c, t_0 + c) \to \mathbb{R}^n$ is an *integral curve* of the field f if and only if

$$\varphi(t) = \varphi(t_0) + \int_{t_0}^{t} f(\varphi(s)) \, ds, \quad \text{for } t \in (t_0 - c, t_0 + c).$$

By Theorem 3.19, we know that the initial value problem (3.5.22) has a unique *local solution* if $f \in C^1(I \times U)$, i.e., if each component $f_i : I \times U \to \mathbb{R}$ of f is continuously differentiable function, for $i = 1, \dots, n$.

Example 3.39 Let $x_0 = (1, 1)$ and suppose a field $f = (f_1, f_2)$ is given by

$$f_1(x, y) = x(t) \quad \text{and} \quad f_2(x, y) = y(t).$$

By Definition 2.29, a curve C_γ passing through the point x_0 is integral to f if and only if $\gamma(t) = (x(t), y(t))$, where (x, y) is a unique solution of the initial value problem given by

$$\frac{dx}{dt} = x(t) \quad \text{and} \quad \frac{dy}{dt} = y(t), \quad \text{with } x_0 = (1, 1). \tag{3.5.23}$$

Clearly, the integral curve C_γ is given by the parametrisation $\gamma(t) = (e^t, e^t)$, for $t \in \mathbb{R}$, which follow the radial trajectories (as *orbits*) starting from the point $(1, 1)$, and having the field f as the associated *velocity field* given by (x', y'). On the other hand, if $f_1 = (f_1, f_2)$ is given by

$$f_1(x, y) = x(t) \quad \text{and} \quad f_2(x, y) = -y(t),$$

then the trajectories of the integral curve C_ϕ given by the parametrisation $\phi(t) = (e^t, e^{-t})$, for $t \in \mathbb{R}$, stays on the on x-axis moving towards the point $(1, 1)$ (without ever meeting the origin); stays on the y-axis heading away from the point $(0, 0)$; and, the remaining trajectories follow the hyperbolic paths asymptotic to the two axes.

As a further illustration of the ideas introduced above, we consider now a general *autonomous system* (3.5.22) for the case when $n = 2$, and $f = (f_1, f_2)$. That, let

$$\frac{dx}{dt} = f_1(x, y);$$
$$\frac{dy}{dt} = f_2(x, y). \tag{3.5.24}$$

The central idea is to study the *orbits* of the points (x, y) satisfying the above equations. A unique solution $\varphi(t) = (\varphi_1(t), \varphi_2(t))$ of the above system can be interpreted as the parametric representation of the associated integral curve C_φ in xy-plane, which is called the *phase plane* of the system (3.5.24), and the *trajectories* (or *orbits*) given by the mutually disjoint integral curves define a *phase diagram* (or a *phase portrait*). It is common to add arrows to orient trajectories in order of increasing t. The phase diagram of a system (3.5.24) helps in studying quantitative behavior of a solution, without even having an explicit expression for the same.

Definition 3.28 A solution $\varphi = \varphi(t)$ of (3.5.24) is said to be *periodic* of period ℓ if $\varphi(t_1) = \varphi(t_2)$, for some $t_1 < t_2$, with $\ell = t_2 - t_1$. Also, a point $a \in U$ is called a *critical point* (or an *equilibrium point*) if $f(a) = 0$. In latter case, $\varphi \equiv a$ is a solution with the orbit given by $\{a\}$.

Notice that, if $\varphi(t)$ exists for $t \geq t_1$ such that $\lim_{t \to \infty} \varphi(t) = a$, then $f(a) = 0$. In general, we seek a function $F = F(x, y)$ that is constant along trajectories $F(x, y) = c$, each called the *first integral* of the system (3.5.24).

Remark 3.9 It is interesting fact that an autonomous system (3.5.24) is related to a first order differential equation of the form

$$- f_2(x, y) \frac{dx}{ds} + f_1(x, y) \frac{dy}{ds} = 0, \quad \text{for } x = x(s) \text{ and } y = y(s). \quad (3.5.25)$$

As said above, a solution of the above equation given in parametric form as $\gamma(s) = (x(s), y(s))$ is a C^2-function such that the condition

$$\left(\frac{dx}{ds}\right)^2 + \left(\frac{dy}{ds}\right)^2 > 0,$$

ensures that the integral curve C_γ is a regular curve, and also that it can be written locally as $x = \varphi(t)$ and $y = \phi(t)$, for some C^1-functions φ and ϕ. Clearly, a solution of system (3.5.24) (with parameter s) is a solution of Eq. (3.5.25), and also, more generally, of the equation

$$- \mu(x, y) f_2(x, y) \frac{dx}{ds} + \mu(x, y) f_1(x, y) \frac{dy}{ds} = 0. \quad (3.5.26)$$

Notice that, if $\mu \neq 0$ is an *integrating factor* of Eq. (3.5.26), then we can find a *potential function* $F = F(x, y)$ that is constant along trajectories. More precisely, if $\nabla F = \left(- \mu f_2, \mu f_1 \right)$, then F is constant along solutions of the system (3.5.24) (with parameter s). The trajectories of the system (3.5.24) are the level sets given by

$$L_c = F^{-1}(c) = \{ (x, y) \in D : F(x, y) = c \}.$$

For example, with

$$\frac{dx}{ds} = -y(s) \quad \text{and} \quad \frac{dy}{ds} = x(s),$$

the corresponding first order differential equation is $x \, dx + y \, dy = 0$, and so if the function $F(x, y) = x^2 + y^2$ is constant along each solution then the trajectories are circles centered at the origin. We can have more general formulations in the case when the level sets L_c are curves

Definition 3.29 A (smooth) *dynamical system* defined over a domain U is a C^1 function $\varphi : \mathbb{R}^+ \times U \to \mathbb{R}^n$ such that for $\varphi_t(x) := \varphi(t, x)$ we have

1. $\varphi_0(a) = a$, for any $a \in U$;
2. $\varphi_{t+s} = \varphi_t \circ \varphi_s$, for all $t, s \in \mathbb{R}$.

In actual practice, the function φ_t for any fixed $t \in \mathbb{R}^+$ represents the *contour* associated with a *flow* modelled by a first order autonomous system as in (3.5.22). Therefore, to analyse the time evolution of the underlying phenomena, and hence determine the *state*[11] of the dynamical system, we seek a unique solution of the initial value problems (3.5.22).

Example 3.40 The *state* of a mass–spring–dashpot system is provided by the (column) vector $x = (x_1\ x_2)^t$, where $x_1(t) = x(t)$ and $x_2(t) = x'(t)$. Notice that, in case of the system of equations such as (3.3.3), it possible to determine the exact *trajectory* $x = x(t)$ of the particle provided the *initial position* $x(t_0) = x_0 \in \mathbb{R}^n$ and the *initial velocity* $x'(t_0) = v_0$ are known.

A prototypical example of a first order autonomous linear system is the *ecological model*, as given in the next illustration, was developed by the American biophysicist *Alfred Lotka* (1880–1949) and the Italian mathematician *Vito Volterra* (1860–1940).

Example 3.41 (*Prey-Predator Model*) We consider here a linear autonomous dynamical system consisting of two species, a prey species and a predator species. Let $x = x(t)$ and $y = y(t)$ respectively represent the population sizes of prey and predator at a time t. It is assumed that, in the absence of predator ($y \equiv 0$), prey population $x(t)$ has enough resources to grow at a rate

$$a = \text{birth rate} - \text{death rate} > 0,$$

so that it grows exponentially as $x(t) = x(0)e^{at}$. However, if the prey population is hunted down by predator at a rate $b\,y(t)$, with $b > 0$, then the growth rate decreases from a to $a - b\,y(t)$. On the other hand, suppose predator growth is proportional to $x(t)$, with proportionality constant $d > 0$ and, in the absence of prey population ($x \equiv 0$), the predator population decay exponentially according to the equation $dy/dt = -c\,y(t)$, with $c > 0$. Of course, in the presence of prey population, the growth rate improves to $-c + d\,y(t)$. Therefore, the autonomous linear system given by

$$\frac{dx}{dt} = x(t)\big(a - b\,y(t)\big);$$
$$\frac{dy}{dt} = y(t)\big(-c + d\,x(t)\big), \tag{3.5.27}$$

describes the dynamics within the total population. By Theorem 3.19, for every pair of initial values (x_0, y_0) at $t = 0$, the system (3.5.27) has a unique solution. It can be shown that the *trajectories* of the system (3.5.27) are periodic and they remain positive when both x_0, y_0 are positive. Also, at any time $t > 0$, the averages $\bar{x}$ and $\bar{y}$ respectively of functions x and y in a period are given by

$$\bar{x}(t) = \frac{c}{d} \quad \text{and} \quad \bar{y}(t) = \frac{a}{b}.$$

---

[11] The **state** of a dynamical system refers to the *minimal data* needed to decode the complete information about the system at any time, past or future, provided the same is known at some initial time.

Notice that, by taking $X(t) = \ln x(t)$ and $Y(t) = \ln y(t)$, we can write the system (3.5.27) as

$$\frac{dX}{dt} = a - b\,e^{Y(t)} = -\frac{\partial \mathcal{H}}{\partial Y};$$
$$\frac{dY}{dt} = -c + d\,e^{X(t)} = \frac{\partial \mathcal{H}}{\partial X},$$

(3.5.28)

where the Hamiltonian $\mathcal{H}$ is given by

$$\mathcal{H}(X, Y) := -a\,Y + b^Y + d^X - c\,X,$$

which proves $\mathcal{H}(X(t), Y(t)) = \mathcal{H}(X(0), Y(0))$.

Exercises 3

3.1 Show that the first order equation $|x'(t)| + |x(t)| = -1$ has no real solution.

3.2 Solve the equation $x^2 dt - t(t + x)dx = 0$. What will happen to $x(t)$ as $t \to \infty$?

3.3 Solve the initial value problem $x' = 1 + x^2$, with $x(0) = 1$. Specify the interval on which the solution is defined.

3.4 Show that a solution $\varphi = \varphi(t)$ of the initial value problem (3.2.2) is given by

$$\varphi(t) = \frac{a p_0}{b p_0 + (a - b p_0)e^{-a(t - t_0)}}.$$

Hence deduce that, as $t \to \infty$, we have $\varphi(t) \to a/b$, for $p_0 > 0$; and, 0, for $p_0 = 0$. Also, study the monotone property of the growth function $p(t)$. What is the shape of the curve of $p(t)$?

3.5 Suppose $\varphi = \varphi(t)$ is a solution of a *Riccati equation* given by (3.2.28). Show that, by using the substitution $x(t) = \varphi(t) + (1/y)$, the above equation transforms to a first order linear equation, and hence write a general solution. Also, show that the substitution $x(t) = \varphi(t) + y$ transforms a Riccati equation to a Bernoulli equation of the form (3.2.27), hence solve the differential equation

$$x' = 1 + t^2 - 2t\,x + x^2, \quad \text{with } \varphi(t) = t.$$

3.6 Show that, by using the substitution $y = \exp\left(-\int q\,x\,dt\right)$, a *Riccati equation* of the form (3.2.28), with $0 \neq q \in C^1(I)$, transforms to the linear differential equation

$$y'' + [p + (q'/q)]y' - rq\,y = 0.$$

Conversely, if $\varphi > 0$ is solution of the above equation, then $\phi = -\left(\log \varphi\right)'/q$ is a solution of the Riccati equation. Hence, solve the initial value problem

$$x' - x + e^t x^2 + 5e^{-t} = 0, \quad \text{with } x(0) = \alpha.$$

3.7 Prove uniqueness part of Theorem 3.2 without using the Banach contraction principle.

3.8 (Continuous dependence) Let the function $f = f(t, x)$ as in Eq. (3.2.1) be continuous, and satisfies the Lipschitz condition, over a rectangular region $\Omega' = [t_0 - a, t_0 + a] \times [x_0 - b, x_0 + b]$. Suppose $|x_0 - x_0^*| < \varepsilon$, for some $\varepsilon > 0$. Show that if φ is a solution then we have $|\varphi(t_0) - \varphi(t_0^*)| < \varepsilon$.

3.9 Show that, by using the transformation given by

$$x(t) = y \exp\left(-\frac{1}{2} \int^t \left(a_1(s)/a_0(s)\right) ds\right), \quad \text{for } t \in I, \tag{3.5.29}$$

a homogeneous equation (3.3.10) changes to the form

$$y''(t) + q(t)y(t) = 0, \quad \text{with } q(t) = \frac{a_2}{a_0} - \frac{a_1^2}{4a_0^2} - \frac{a_1'a_0 - a_1a_0'}{2a_0^2}. \tag{3.5.30}$$

3.10 Use the method of undetermined coefficients to find a particular solution of the differential equation given by $x'' - 2x' + x = 2e^t + 2x$.

3.11 Solve the BVP for the Cauchy–Euler equation $x^2y'' + x y' + \lambda y = 0$, with $y'(1) = 0$ and $y(\ell) = 0$.

3.12 Solve the BVP for the Bessel equation $x^2y'' + x y' + \lambda^2x^2 y = 0$, with $|y(0)| < \infty$ and $y'(\ell) = 0$.

3.13 Verify the Rodrigues' formula (3.3.57).

3.14 Expand the (generating) function $G(x, t) = \left(1 - 2xt + x^2\right)^{-1/2}$ in a power series, and hence deduce that

$$G(x, t) = \sum_{n=0}^{\infty} P_n(t)x^n,$$

where $P_n(t)$ is the Legendre polynomial of order n.

3.15 For $v > 0$, let J_v denote the Bessel function of the first kind of order v. Show that

a. J_v and J_{-v} are linearly dependent if $v \in \mathbb{Z}$;

b. for $t > 0$, we have

$$J_{1/2}(t) = \sqrt{\frac{2}{\pi t}} \sin t \quad \text{and} \quad J_{-1/2}(t) = \sqrt{\frac{2}{\pi t}} \cos t.$$

3.16 Show that the transformation given by $x = y/\sqrt{t}$ changes a Bessel equation of index $v \in \mathbb{R}^+$, as given by (3.4.45), to the equation given by

$$y''(t) + q(t)y(t) = 0, \quad \text{with } q(t) = \left(1 + \frac{1 - 4v^2}{4t^2}\right). \tag{3.5.31}$$

3.17 Let $y(x)$ be the position of a perfectly elastic hanging cable such as a suspension bridge, with fixed end points at $x = 0$ and $x = \ell$. Suppose the distributed load is given by the function $g(x)$. Show that the function $y = y(x)$ is given by the BVP:

$$y''(x) - c\sqrt{1 + \left[y'(x)\right]^2} = 0, \quad \text{with } y(0) = a \quad \text{and} \quad y(\ell) = b,$$

where the constant c is the ratio of the weight of the cable to the tension. Further, show that the general solution of the above differential equation is a *catenary*.

3.18 Suppose $y(x)$ represents the temperature of a sufficiently thin rod dipped in nuclear fuel. Show that the function y is given by the BVP:

$$y''(x) + (g/c) = (hC/kA)\left[y(x) - T\right], \quad \text{with } y(0) = y(\ell) = T,$$

where g is the rate at which heat is generated, the terms on the right side represent the heat transfer to surroundings by convection. Also, find $y = y(x)$.

3.19 Find values of λ for which the BVPs given below have nontrivial solution.

   a. $y''(x) + \lambda^2 y = 0$, with $y(0) = 0$ and $y'(\ell) = 0$.
   b. $y''(x) + \lambda^2 y = 0$, with $y'(0) = 0$ and $y(\ell) = 0$.
   c. $y''(x) + \lambda^2 y = 0$, with $y'(0) = 0$ and $y'(\ell) = 0$.

3.20 Let $G(x, \xi)$ be the boundary value Green's function as defined in Sect. 3.4.1. Show that the Green's function

   a. satisfy the differential equation

$$\frac{\partial}{\partial x}\left(p\frac{\partial G(x, \xi)}{\partial x}\right) + q(x)G(x, \xi) = 0, \quad \text{for } x \neq \xi;$$

   b. is continuous at $x = \xi$. That is, $G(\xi^+, \xi) = G(\xi^-, \xi)$;
   c. has the property that the derivative $\partial_x G(x, \xi)$ has a jump discontinuity at $x = \xi$. More precisely, we have

$$\frac{\partial G(\xi^+, \xi)}{\partial x} - \frac{\partial G(\xi^-, \xi)}{\partial x} = \frac{1}{p(\xi)}.$$

3.21 Construct the Green's function for the boundary value problem given by

$$y'' + \lambda^2 y = g(x); \quad \text{with boundary conditions } y(0) = y(1) = 0.$$

3.22 Solve the boundary value problem given by

$$y'' = x^2; \quad \text{with boundary conditions } y(0) = 1 \quad \text{and} \quad y(1) = 2,$$

by using the boundary value Green's function.

3.23 Let $G(x, \xi)$ be the boundary value Green's function as defined in Sect. 3.4.1. Prove the jump condition given by

$$\lim_{\varepsilon \to 0} \left[p(x) \frac{\partial G(x, \xi)}{\partial x} \right]_{\xi-\varepsilon}^{\xi+\varepsilon} = 1.$$

3.24 Let the operator L be as in (3.4.2), and its *adjoint operator* L^* be as given in (3.4.42). Show that, for any $\varphi \in C^2(I)$, we have

$$\int_a^x \left(\varphi \, Ly - y \, L^* \varphi \right) dx = \left[a_0 \left(y' \varphi - y \varphi' \right) + \left(a_1 - a_0' \right) y \varphi \right]_a^x. \tag{3.5.32}$$

Hence, deduce the *Lagrange's identity* and *Green's identity* for L respectively given by

$$\varphi \, Ly - y \, L^* \varphi = \frac{d}{dx} \left[a_0 \left(y' \varphi - y \varphi' \right) + \left(a_1 - a_0' \right) y \varphi \right]; \tag{3.5.33}$$

$$\int_a^b \left(\varphi \, Ly - y \, L^* \varphi \right) dx = \left[a_0 \left(y' \varphi - y \varphi' \right) + \left(a_1 - a_0' \right) y \varphi \right]_a^b. \tag{3.5.34}$$

What if $L = L^*$, i.e. if L is a self-adjoint operator.

3.25 (*Abel's formula*) Suppose φ_1 and φ_2 are two solutions of Eq. (3.4.47). Show that the function $p \, W(\varphi_1, \varphi_2)$ is identically constant over $[a, b]$, where W is the Wronskian. Hence deduce that each eigenfunction of a regular SL-BVP is unique, up to a scalar multiple.

3.26 (*Sturm Separation Theorem*) Let the two solutions φ_1 and φ_2 of a homogeneous equation of the form (3.3.10) be linearly independent over the interval $J \subseteq I$. Show that between any two consecutive zeros of φ_1 there exists exactly one zero of φ_2.

3.27 (*Sturm Comparison Theorem*) Let φ and ϕ be respectively *nontrivial* solutions of the homogeneous equations of the form

$$x''(t) + p(t)x(t) = 0 \quad \text{and} \quad x''(t) + q(t)x(t) = 0, \quad \text{for } t \in J.$$

such that $p - q > 0$ over I. Show that between any two consecutive zeros of ϕ there is at least one zero of φ. Hence, deduce that every solution of the equation $x''(t) + t^2 x(t) = 0$ has infinitely many zeros in $[1, \infty)$.

3.28 Suppose q as in Exercise 3.19 satisfies the condition that $q \leq 0$ in some interval (t_1, t_2). Show that any *nontrivial* solution of Eq. (3.5.31) has at most one zero in (t_1, t_2).

3.29 Considering the two cases when $\nu \in [0, 1/2)$ and $\nu > 1/2$, show that the distance between consecutive zeros of any nontrivial solution of the Bessel equation approach to π as $t \to \infty$. What could be said about the case when $v = 1/2$.

3.30 Consider a BVP for the Euler equation given by

$$x^2 y'' + x y + \lambda y = 0, \quad \text{for } 1 < x < e;$$
$$y(1) = 0 \quad \text{and} \quad y(e) = 0,$$

Find its Sturm–Liouville form, and hence the eigenvalues and eigenfunction.

3.31 Find the self-adjoint form of the *Hermite equation* given by

$$y'' - 2x y' + \lambda y = 0, \quad \text{for } x \in \mathbb{R},$$

and hence the eigenvalues and eigenfunctions. Show that, with the end conditions $y \to 0$ as $x \to \pm\infty$, the eigenfunctions $\varphi_n(x) = e^{-x^2/2} H_n(x)$ are orthogonal over $\mathbb{R}$, where H_n are Hermite polynomials.

3.32 Find a series expansion of $\varphi(x) = \sin x$ in terms of eigenfunctions of SL-BVP given by

$$y'' + \lambda y = 0, \quad \text{for } x \in [0, \pi];$$
$$y(0) = 0 \quad \text{and} \quad y(\pi) + y'(\pi) = 0.$$

3.33 Suppose a function $f = f(t, x) : I \times U \to \mathbb{R}^n$ is *locally Lipschitz* with respect to x in some domain $U \subseteq \mathbb{R}^n$. Show that f satisfies the Lipschitz condition in x on every compact set $C \subset U$.

References

1. Arnold, V. I. (1992). *Ordinary differential equations* (3rd ed.). Springer.
2. Birkhoff, G., & Rota, G.-C. (1969). *Ordinary differential equations* (2nd ed.). Wiley.
3. Braun, M. (1993). *Differential equations and their applications* (4th ed.). Springer.
4. Coddington, E. A., & Levinson, N. (1987). *Theory of ordinary differential equations*. Tata McGraw-Hill Publishing Co. Ltd.
5. Hartman, P. (1964). *Ordinary differential equations*. Wiley.
6. Myint-U, T. (1978). *Ordinary differential equations*. Elsevier.
7. Xie, W.-C. (2010). *Differential equations for engineers*. Cambridge University Press.
8. Rudin, W. (1976). *Principles of mathematical analysis* (3rd ed.). McGraw-Hill Book Co.
9. Arnold, V. I. (1988). *Geometric methods in the theory of ordinary differential equations* (2nd ed.). Springer.
10. Kharazishvili, A. B. (2000). *Strange functions in real analysis*. Deker.
11. Agarwal, R. P., & Lakshmikantham, V. (1993). *Uniqueness and nonuniqueness criteria for ordinary differential equations*. World Scientific.
12. Keller, H. (1968). *Numerical methods for two point boundary value problems*. Waltham.
13. Al-Gwaiz, M. A. (2008). *Sturm-Liouville theory and its applications*. Springer.
14. Zettl, A. (2005). *Sturm-Liouville theory* (Vol. 121). AMS.

15. Acheson, D. (1997). *From calculus to chaos: An introduction to dynamics*. Oxford University Press.
16. Hirsch, M. W., Smale, S., Devaney, R. L. (2004). *Differential equations, dynamical systems, and introduction to chaos* (2nd ed.). Elsevier.

Chapter 4
Partial Differential Equation Models

> *Although to penetrate into the intimate mysteries of nature and thence to learn the true causes of phenomena is not allowed to us, nevertheless it can happen that a certain fictive hypothesis may suffice for explaining many phenomena.*
>
> *Leonhard Euler (1707–1783)*

A partial differential equation is an implicit or explicit equation involving derivatives of an unknown functions of more than one independent variables. A *differential equation model* of a physical phenomenon is a reasoned and controlled approximation of involved natural processes that considers all relevant scientific facts and also laws governing the phenomenon. In this chapter, we obtain mathematical description of some practical problems focusing mainly on significant aspects of processes under study. As inclusion of any superficial variables or parameters increases the *complexity* of the associated mathematical problem, an appropriate approach is to start with a simpler model by making several simplifying assumptions.

In Sect. 4.1, we discuss briefly the procedure adopted to formulate a model of a given practical problem. Section 4.2 is mainly about the derivation of the three prototypical partial differential equations concerning phenomena such as *wave motion* related to vibrating strings and drumheads, *heat conduction* in solids, and the *Laplace equation* modeling problems related to potential theory. More generally, in Sect. 4.3, we give derivation of three fundamental partial differential equations governing phenomena such as related to the transfer of mass through a moving fluid (*continuity equation*), the transfer of momentum as in bulk flows (*equations of motion*), and the transfer of energy (*diffusion equation*). Such types of differential equations play an important role in analysing diverse types of multivariate phenomena related to continuum mechanics.

© The Author(s), under exclusive license to Springer Nature Singapore Pte Ltd. 2022   175
A. K. Razdan and V. Ravichandran, *Fundamentals of Partial Differential Equations*,
https://doi.org/10.1007/978-981-16-9865-1_4

In all that follows, unless otherwise specified, we may call a partial differential equation (or a system of such types of equations) simply as a differential equation involving certain number of independent variables. Mathematical prerequisites for the topics discussed in this chapter are covered in Chap. 2.

4.1 Mathematical Modelling

The human body is an important *large physical system* consisting of many different types of *subsystems* down to the cellular levels. Each subsystem involves complex natural processes governed by certain universal laws.[1] It is a perfect example of an *efficient system* that has inbuilt optimisation schema for each of its component subsystems mostly controlled by the brain, which uses (chemical) controls all the time to perform a multilayer nonlinear analysis at a very high speed to ensure enhanced functioning of each vital organ for better survival chances against all odds. Some of the differential equations derived in this chapter also find applications in studies related to certain important problems in medical sciences. For example, according to recent studies, developmental biology uses differential equation models to study molecular mechanisms responsible for cell signalling and aggregation when nutrients are scarce. The motivated reader may refer the text [1] to know more.

In general, application of mathematical tools in analysing a practical situation requires a proper articulation of scientific facts so that the *solution* of the associated problem admits a convenient physical interpretations. Further, an approximation technique used in a particular situation depends on how appropriately one may choose the applicable postulates or laws such as Newton's for classical mechanics, Schrödinger's for quantum mechanics, Navier–Stokes' for fluid flow, and Maxwell's for electrodynamics.

In most cases, a mathematical description of processes involved with a physical system leads to problem involving an unknown function of certain number of independent variables, and also some known functions representing the *system parameters*. It is common to make some *simplifying assumptions* and reduce the number of parameters involved, subject to that the state of the system under study do not change drastically. As a rule of thumb, we first formulate an easy-to-deal-with model and upgrade to a more realistic model as we go along. The equation (or a system of equations) so obtained is used to analyse the associated practical problems. In most applications, the ultimate aim is to facilitate the analysis and design of *controls* for the related dynamical systems.

---

[1] It is in general true that each physical system is governed by some universal laws.

A mathematical study of a physical system involves any of the three *modelling considerations* as given below.

1. Use postulates and/or governing physical laws such as Newton's for continuum mechanics, Schrödinger's for quantum mechanics to obtain a *differential equation model* involving derivatives of an known function [2]. In this case, *model equation* invariably involve *parameters* that have a direct influence on the system performance under investigation.
2. Use mathematical reasoning and basic scientific information to approximate the associated natural processes, considering how do they work and also physical constraints influencing the system [3, 4]. In this case, we also use some *constitutive relations* such as Fourier's for heat flux, Hooke's for stress and strain, and Fick's for diffusion to remove the excess variables from the model equations. The constitutive relations, also known as the equations of state, arise from the physical assumptions typically imposed on the nature or type of the medium.
3. Use relevant data for involved *system variables* and *system parameters*, preferably from diverse sources, to *simulate processes* by using a reasoned and controlled statistical/numerical approach. In this case, the enhanced computational capabilities achieved due to technological innovations in past few decades have helped in improving simulation outcomes in terms of better estimations, and hence more precise predictions. In recent years, the related developments in several critical scientific areas led to the discovery of certain *powerful techniques*, which provided new more efficient numerical and stochastic models to deal with large scale multi-parameter data.

The *state* of an evolutionary physical system under study refers to the *data* that provides complete information about the system at any time, subject to that the same is known at some initial time. Further, it is important to understand the effects of the *environment* of the system on the functioning of the involved natural processes. In most applications, the complexities of natural processes of a phenomenon are analysed in terms of analytical and geometrical properties of solutions of the associated initial-boundary value problems. Therefore, a significant role is played by certain fundamental theorems related to multivariable scalar or vector fields and hence are prerequisites for the topics presented in this book.

In applications dealing with problems studied at micro-level, like particle interactions at molecular level, mathematical formulation of a phenomenon leads to discrete relations like *difference equation* or a *matrix equation* involving an unknown function, where involved independent variables take only discrete values. Such type of mathematical descriptions is called *discrete models*. For example, an important class of discrete models arise while studying the *linear time invariant* systems, as discussed later in Chap. 8.

On the other hand, most physical laws idealising natural processes of a dynamic system are invariably expressed mathematically as equation(s) involving time rate of change and *gradient* of some physical quantity represented by a function of certain number of independent variables specifying *position*, and a time variable. So, we end up having for the system a *continuous model* given by one or more differential equations. A system of differential equations is said to be *ordinary* if all space variables are functions of time. It helps to have a working knowledge about some simple dynamical systems modelled by ordinary differential equations. For their use in the sequel, we provide a brief discussion about such type of equations in the next section. And, a partial differential equation involves derivatives of a scalar function of both space and time variables.

Model Formulation

At the first stage of *model formulation* of a phenomenon, we make some simplifying assumptions considering all related scientific facts about the involved natural processes. Each such assumption has to have some relevance to the purpose of the study, and their precision and correctness help in choosing an appropriate set of postulates and laws governing the involved processes. For example, mass is constant for models based on Newtonian physics, but it must be assumed a variable while using Einstein's special theory of relativity. Next, we identify important *system parameters* concerning all relevant processes occurring in the system. Accordingly, dependent and independent variables are identified. For example, while modelling for weather forecast, the *position* and *time* are the independent variables, and dependent variables specify temperature and humidity. Notice that, in this case, earth's gravity field and rotational speed act as *nonadjustable* parameters.

In actual practice, modelling of a dynamic system takes as input large volume of *real data* for the involved *variables* and *parameters*. Also, a good assessment of *system environment* and *natural constraints* is needed to undertake a meaningful study of the phenomena. So, at the final stage, we collect data for chosen variables from diverse sources, if possible, and subsequently use postulates and/or laws governing the involved processes to derive model equation(s). In particular, to formulate a statistical model of *best fit*, suitable interpolation techniques are applied to the source data collected earlier for variables and system parameters. In this latter situation, type of data used classifies a model as *deterministic* or *probabilistic*. If the independent variables are random, and probabilistic data is known, then the associated distributions related equation(s) gives a *stochastic model*.

All through the process, in general, a model needs to be *calibrated* several times using the collected real data. Such an exercise is important because it ensures that the difference between the model outputs and the real values largely remains below the acceptable *error tolerance*, and hence *validates* the model.

A fruitful way of formulating a model involving only dimensionless quantities is known as *nondimensionalising a model*, which provides an insight into how to scale relations of the system so that the total number of variables and/or parameters are *minimal*. Since variables and parameters in general have physical dimensions, the techniques of *nondimensionalising* and *scaling* in terms of their transformations are useful tools to simplify and analyse a mathematical model (mainly due to Buckingham π-theorem). As the present book is not about modelling techniques *per se*, we will not discuss dimension analysis and scaling aspects any further. The interested reader may like to read more from the book [5] or [6].

Solution and Analysis

Time evolution of processes associated with a physical system is analysed using analytical properties of a solution of the equation modelling the system. Notice that a mathematical model is only an approximation of reality. And, since working of related phenomena is subject to certain *natural constraints* determining the *system parameters*, so we need to find only a solution of the associated equation(s) that depends on the *initial data*, and estimated values of the system *parameters*. Accordingly, we choose an appropriate analytical or numerical method to find a unique *solution*, if the one exists, subject to that it also satisfies the initial data specified by some *auxiliary conditions*.

Every system is taken to be surrounded by an *environment* that defines the *boundary conditions* for the related differential equation model. On the other hand, the *initial conditions* are specified by the **state** of the system, which refers to the characteristic data providing information about the involved processes as they evolve over time, subject to that the same is known at some initial time.

In mathematical terms, the conditions satisfied by the unknown function and/or some of its derivatives on the boundary of its *domain of definition* are known as the *boundary conditions* of the differential equation. And, the condition satisfied by the function and/or some of its derivatives at an initial time $t = 0$ are called the *initial conditions*. An *initial-boundary value problem* (or *IBVP*, in short) related to a dynamic system is a problem that needs to solve the differential equation subject to both these types of conditions. Therefore, in actual practice, we seek a *unique solution* of an IBVP for a differential equation that is compatible with specified behaviour of the phenomenon.

Interpretations

Finally, based on our analysis of a unique solution, we infer *physical interpretations* using functional analytical techniques and/or geometric methods facilitated by packages like MATLAB or Mathematica. This way, modelling helps to understand natural processes in terms of system responses to a given set of input values and hence to suggest a scheme for evaluation of the phenomenon under study. However, in this text, we are not discussing this part of the theory.

To conclude our discussion here, we remark that the whole exercise of model formulation of a physical system is undertaken to assess the *current state* of the system and to design, subsequently, *controls* for the same. An efficiently designed numerical/statistical *control* is applied whenever some component of a system fails to operate as expected.

4.2 Three Prototypical Equations

It is largely believed that use of partial differential equations in solving problems concerning some physical phenomena started with Jean d' Alembert's work on vibrations of a string. Most concepts we shall discuss in rest of the book arose from the related *vibrating string controversy*[2] that started in 1747 soon after *Jean d' Alembert* (1717–1783) published his *travelling wave solution* of the one dimensional wave equation. The main issue was how to explain the connection between a physical problem and the proposed mathematical descriptions. The controversy spanned over the eighteenth century, and it got involved many eminent scientist with diverse backgrounds such as *Daniel Bernoulli* (1700–1782), *Leonhard Euler* (1707–1783), and *Joseph-Louis Lagrange* (1736–1813). As discussed later in Chap. 8, Lagrange resolved the main issue partially in 1759.

More generally, let $\Omega \subseteq \mathbb{R}^n$ be a *nice domain*.[3] Suppose $u = u(x, t) : \Omega \times \mathbb{R}^+ \to \mathbb{R}$ is a sufficiently smooth function representing a physical quantity such as the density, velocity, pressure, viscosity, and temperature. A typical *partial differential equation* for a (unknown) function $u = u(x, t)$ is an equation that gives a relation between the time rate of change $u_t = \partial u / \partial t$ of a physical quantity $u(x, t)$, its *flux* across the boundary surface S, and a source or sink function $g(x, t)$ representing the amount of quantity being created or destroyed within the region Ω. Therefore, a mathematical formulation of various types of problems concerning continuum mechanics leads to partial differential equations. We will write the partial derivatives of a function $u = u(x, t)$ as given below:

$$q = u_t = \partial_t u = \frac{\partial u}{\partial t}; \qquad p_i = u_{x_i} = \partial_{x_i} u = \frac{\partial u}{\partial x_i};$$

$$r_{ij} = u_{x_i x_j} = \partial^2_{x_i x_j} u = \frac{\partial^2 u}{\partial x_i \partial x_j} = \frac{\partial^2 u}{\partial x_j \partial x_i}; \quad \text{etc.} \tag{4.2.1}$$

Also, the *del operator* ∇ in n variables is written as

$$\nabla \equiv \left(\frac{\partial}{\partial x_1}, \cdots, \frac{\partial}{\partial x_n} \right).$$

In this part of the chapter, we use typical engineering approach to model phenomena concerning vibrations of strings and drums, heat conduction in solids, and potential theory. The fundamental models derived here are second order differential equations in two and three variables that are used to solve many important practical problems in science and engineering. We shall use the three archetypal equations throughout the book to discuss concepts related to specific types of differential equations. In all that follows, it is assumed that the reader is familiar with basic concepts of continuum mechanics. Appendix A.2 provides details related to electromagnetic theory. We recommend the standard texts [7], for any further details.

As said earlier, we usually start with a simple model by focusing only on major factors affecting the *state* of a system under study. This forces us to make several

---

[2] We will give a detailed account of the controversy in Sect. 8.1.

[3] It is assumed that Ω is *simply connected* open set with a piecewise smooth closed boundary $S = \partial \Omega$. In most situations of practical importance, as discussed in this chapter, it suffices to assume that $u \in C^2(\Omega \times \mathbb{R}^+)$.

simplifying assumptions about the involved processes and the environment of the system. Of course, after having a good practice with the *first model*, we relax one or more simplifying assumptions at a time to reformulate a more realistic model of the system.

Wave Equations

On historical note, these types of equations were first studied in eighteenth century by Daniel Bernoulli, D'Alembert, L. Euler, and J. Lagrange. Collectively, the related issues are known as the *vibrating strings problems*, as discussed later in Sect. 8.1. First, we formulate a 1-dimensional model to study *transverse motions* of a vibrating string of a finite length, say ℓ, with ends fastened along the x-axis at $x = 0$ and $x = \ell$. Suppose the linear density of the string is given by $\rho = \rho(x)$. When *struck* by a force or set in motion by being *plucked* at a point, we write $y = y(x, t)$ for the *transverse deflection* of the string at a point $0 < x < \ell$ and time $t \geq 0$. The model is formulated considering the following reasonable simplifying assumptions:

1. The string is perfectly elastic so that it offers no resistance to bending.
2. The weight of the string is negligible so that the effect of the gravity on vibrations is zero.
3. The vertical deflections $y = y(x, t)$ are small in comparison with the length ℓ of the string.
4. The string is sufficiently stretched before fastening its ends so that the effect of gravity on the *tensile force* $\mathbf{T}$ at every point is negligible.
5. At any fixed time t, the *gradient* $y_x(x, t)$ equals the slope $\frac{d}{dx}(y(x, t))$ of the string at the position x.
6. For any fixed point $x \in (0, \ell)$, the time derivative $y_t(x, t)$ equals the velocity $\frac{d}{dt}(y(x, t))$ of the string at any time t.

Example 4.1 (**1-*dimensional Wave Equation***) By assumption (3), the tension in the string is always directed along the tangents to its instantaneous position. Let T_1 and T_2 be the components of the *tensile force* $\mathbf{T}$ exerted on an infinitesimal arc ds of the string, say between points $P(x, y)$ and $Q(x + \delta x, y + \delta y)$, so that by (2.2.7)

$$
ds = \int\limits_{x}^{x+\delta x} \sqrt{1 + y_x^2}\, dx \approx \delta x.
$$

We may thus assume that there is no elongation effect on string due to vibrations. Then, by Hooke's law, the tension $\mathbf{T}$ at any point is independent of time. This justifies the assumption (5). Also, for each point $x \in (0, \ell)$, the horizontal component of the force $\mathbf{T}$ is constant. For, with $|\mathbf{T}| = T$, if T_x, T_y are the projections of tension $\mathbf{T}$, respectively, on x- and y-axes, then

$$T_x(x) = T(x) \cos \theta = \frac{T}{\sqrt{1 + y_x^2}} \approx T(x);$$

$$T_y(x) = T(x) \sin \theta = T(x) \tan \alpha \approx T(x) y_x,$$

where θ is the angle between the tangent to curve $y(x, t)$ and the x-axis. Since only transverse motion are allowed, the sum total of projections of tensile force on the x-axis must be zero. Also, since forces of inertia and external forces are taken to be directed along y-axis, it follows that

$$T(x) = T(x + \delta x), \quad \text{for all } x \in (0, \ell).$$

Therefore, the tension $\mathbf{T}$ is independent of x. This justifies the assumption (6). So, we may write $\mathbf{T}(x) = T_0$.

Now, consideration of the fact that there are two vertical forces acting on the arc $ds \approx \delta x$ between the points $P(x, y)$ and $Q(x + \delta x, y + \delta y)$ implies that the y-component of the momentum of this arc is given by

$$\int_x^{x+\delta x} y_t(\xi, t) \rho(\xi) d\xi.$$

So, during the time $[t, t + \delta t]$, the change in momentum is given by

$$\int_x^{x+\delta x} \rho(\xi) [y_t(\xi, t + \delta t) - y_t(\xi, t)] d\xi.$$

Equating this to the sum of the *impulse force* $T_0 y_x(x, t)$ acting on the arc and a continuously distributed external force given by a (load) function $f(x, t)$, we obtain the following integral form of equation modelling the transverse vibrations of the string:

$$\int_x^{x+\delta x} \rho(\xi) [y_t(\xi, t + \delta t) - y_t(\xi, t)] d\xi$$

$$= \int_t^{t+\delta t} T_0(\xi) [y_x(x + \delta x, \tau) - y_x(x, \tau)] d\tau + \int_x^{x+\delta x} \int_t^{t+\delta t} f(\xi, \tau) d\xi d\tau. \quad (4.2.2)$$

Thus, assuming $y \in C^2([0, \ell] \times \mathbb{R}^+)$ and applying mean value theorem twice, it follows that

$$y_{tt}(\xi_1, t_t) \rho(\xi^*) \delta t \delta x = [T_0 y_{xx}(\xi_2, t_2) + f(\xi_3, t_3)] \delta t \delta x,$$

where $\xi_1, \xi_2, \xi_3 \in (x, x + \delta x)$ and $t_1, t_2, t_3 \in (t, t + \delta t)$. Dividing by $\delta x \delta t$, and taking limits $\delta x \to 0$, $\delta t \to 0$, we conclude the function $y = y(x, t)$ satisfies the following 1-dimensional nonhomogeneous *wave equation*

$$\rho\, y_{tt} = T_0 y_{xx} + f(x, t). \tag{4.2.3}$$

Notice that ρ is constant if the string is *homogeneous*. In this case, the constant $c = \sqrt{T_0/\rho}$ determines the *wave speed*, and wave equation (4.2.3) takes the form

$$y_{tt} = c^2 y_{xx} + F(x, t), \quad \text{where } F(x, t) = \frac{1}{\rho} f(x, t), \tag{4.2.4}$$

is the (concentrated) force function referred to as *unit mass*. We also call it *d'Alembert's equation*.[4] As a model for a communications system, Eq. (4.2.4) has a nice physical interpretation: *If the medium is stimulated by a source function $f(x, t)$, say a noise, then the same propagates in all directions with the velocity c.*

Alternatively, for the vertical components $-T_1 \sin \alpha$ at P and $T_2 \sin \beta$ at the point Q, Newton's momentum law gives

$$\rho\, \delta x\, y_{tt} = -T_1 \sin \alpha + T_2 \sin \beta.$$

Dividing by $T_2 \cos \beta = T_1 \cos \alpha = T$, we obtain

$$\tan \beta - \tan \alpha = \frac{\rho\, \Delta x}{T} y_{tt}.$$

Also, using the assumption (5), it follows that

$$\tan \alpha = y_x \big|_x \quad \text{and} \quad \tan \beta = y_x \big|_{x+\delta x}.$$

Dividing the last equation by δx, we obtain

$$\frac{1}{\delta x}\left[\frac{\partial y}{\partial x}\Big|_x - \frac{\partial y}{\partial x}\Big|_{x+\delta x} \right] = \frac{\rho}{T} y_{tt}.$$

Therefore, by letting $\delta x \to 0$, it follows that the function $y = y(x, t)$ satisfies the following 1-dimensional homogeneous *wave equation*

$$y_{tt} = c^2\, y_{xx}, \quad \text{with } c^2 = \frac{T}{\rho}, \tag{4.2.5}$$

Notice that if a concentrated force $f_0(t)$ is applied at a point $x_0 \in [x, x + \delta x]$ (creating a *kink*) Eq. (4.2.2) can be expressed as

---

[4] D'Alembert was a French music theorist and a scientist who first derived the equation.

$$\int\limits_{x}^{x+\delta x} \rho(\xi)\big[y_t(\xi, t+\delta t) - y_t(\xi, t)\big]d\xi - \int\limits_{x}^{x+\delta x}\int\limits_{t}^{t+\delta t} f(\xi, \tau)d\xi\, d\tau.$$

$$= \int\limits_{t}^{t+\delta t} T_0\big[y_x(x+\delta x, \tau) - y_x(x, \tau)\big]d\tau + \int\limits_{t}^{t+\delta t} f_0(\tau)d\tau.$$

Now, using the assumption that velocities at every point on the string is finite, and allowing $x, x + \delta x \to x_0$, the left side of the last equation is zero. So, we obtain

$$\int\limits_{t}^{t+\delta t} T_0\big[y_x(x_0^+, \tau) - y_x(x_0^-, \tau)\big]d\tau = -\int\limits_{t}^{t+\delta t} f_0(\tau)d\tau$$

Once again, using mean value theorem and passing to limit $\delta t \to 0$, we have

$$y_x(x, t)\Big|_{x_0^-}^{x_0^+} = -\frac{1}{T_0} f_0(t),$$

proving that $y_x(x, t)$ are discontinuous at the point of application of concentrated force. So, for a valid wave equation, we must have

$$y(x_0^+, t) = y(x_0^-, t);$$
$$y_x(x_0^+, t) - y_x(x_0^-, t) = -\frac{1}{T_0} f_0(t).$$

By assuming more realistic physical conditions, we can derive the following wave equations using similar argument:

1. Taking into consideration the *time-dependent* air resistance $r = r(t) > 0$ present in the system, the wave equation gets an additional term *proportional* to the speed y_t so that we obtain

$$y_{tt} - c^2 y_{xx} + r(t)y_t = 0. \tag{4.2.6}$$

2. Likewise, taking into consideration the presence of a transversal *elastic force*, the wave equation gets an additional term *proportional* to the displacement $y = y(x, t)$ so that we obtain

$$y_{tt} - c^2 y_{xx} + k\, y = 0, \quad \text{where } k > 0. \tag{4.2.7}$$

Next, we model 2-dimensional transverse motions of a perfectly flexible vibrating membrane like a drumhead, which we may view as a part of the xy-plane, say $\Omega \subset \mathbb{R}^2$. The membrane is assumed to be *homogeneous*, i.e. the density $\rho = \rho(x, y)$ is constant, for all $(x, y) \in \Omega$. In this case, to model *transverse deflections* $u =$

$u(x, y, t)$ of the vibrating membrane, we derive the equation using the following simplifying assumptions:

1. The membrane is sufficiently thin so that it cannot resist bending moments.
2. The weight of the membrane is negligible so that the total force on the boundary is comparatively large.
3. Before fastening its ends in xy-plane, the membrane is sufficiently stretched so that the *tensile force* **T** per unit length is constant and everywhere normal to the boundary of the membrane.
4. In comparison with verticle deflections, the lateral displacement are negligible.
5. In comparison with the lateral dimensions of the membrane, the *deflections* $u = u(x, y, t)$ are small.
6. At any fixed time t, the *gradients* $u_x(x, y, t)$ and $u_y(x, y, t)$, respectively, equals the slopes $\frac{d}{dx}(u(x, y, t))$ and $\frac{d}{dy}(u(x, y, t))$ of the membrane, for each point $(x, y) \in \Omega$.
7. For any fixed point $(x, y) \in \Omega$, the time derivative $u_t(x, y, t)$ equals the velocity $\frac{d}{dt}(u(x, y, t))$ of the string at any time t.

Example 4.2 (*2-dimensional Wave Equation*) Consider an *infinitesimal curvilinear* element $ABCD$ of the vibrating membrane with curvilinear sides δx and δy. By assumption (5), at every point $(x, y) \in \Omega$ and at any time $t \geq 0$, the deflection $u = u(x, y, t)$ of the membrane is very small in comparison with the size of the membrane. Thus, the inclination angles are small, and so the rectangular projection onto the xy plane of $ABCD$ could be assumed to have the edges approximated by lengths δx and δy. Whence, the *tangential forces* acting on the infinitesimal curvilinear sides can be approximated by $T \delta x$ and $T \delta y$ so that if T_{hx} denotes the horizontal component of $T \delta y$ in the xu-plane, then

$$T \delta y(\cos \beta - \cos \alpha) = 0 \quad \Rightarrow \quad T_{hx} = T \delta y \cos \beta = T \delta y \cos \alpha.$$

Similarly, if T_{hy} denotes the horizontal component of $T \delta x$ in the yu-plane, then

$$T \delta x(\cos \beta_1 - \cos \alpha_1) = 0 \quad \Rightarrow \quad T_{hy} = T \delta y \cos \beta_1 = T \delta y \cos \alpha_1.$$

In view the previous two equations, we have

$$T \delta y = \frac{T_{hx}}{\cos \beta} = \frac{T_{hx}}{\cos \alpha} \quad \text{and} \quad T \delta x = \frac{T_{hy}}{\cos \beta_1} = \frac{T_{hy}}{\cos \alpha_1}. \tag{4.2.8}$$

Now, since the sum of forces in the vertical direction is

$$T \delta y(\sin \beta - \sin \alpha) + T \delta x(\sin \beta_1 - \sin \alpha_1),$$

Newton's second law of motion, together with Eq. (4.2.8) gives

$$T_{hx}(\tan \beta - \tan \alpha) + T_{hy}(\tan \beta_1 - \tan \alpha_1) = \rho \delta x \delta y \, u_{tt}. \tag{4.2.9}$$

Also, taking (x, y) for the projection of the point D, we have

$$\tan \beta = u_x(x + \delta x, y, t) \quad \text{and} \quad \tan \alpha = u_x;$$
$$\tan \beta_1 = u_y(x, y + \delta y, t) \quad \text{and} \quad \tan \alpha_1 = u_y.$$

Using these in Eq. (4.2.9), we obtain

$$T_{hx}(u_x(x + \delta x, y, t) - u_x) + T_{hy}(u_y(x, y + \delta y, t) - u_y) = \rho \delta x \delta y \, u_{tt}. \quad (4.2.10)$$

Approximating all cosines of inclinations to 1, Eq. (4.2.10) yields

$$T\delta y(u_x(x + \delta x, y, t) - u_x) + T\delta x(u_y(x, y + \delta y, t) - u_y) = \rho \delta x \delta y \, u_{tt}.$$

Divide the last equation throughout by $\rho \delta x \delta y$, and taking limit $(\delta x, \delta y) \to (0, 0)$, we finally obtain

$$c^2[u_{xx} + u_{yy}] = u_{tt}, \quad \text{where } c^2 = \frac{T}{\rho}, \quad (4.2.11)$$

which is the 2-dimensional wave equation.

With little more effort, these arguments can be extended to derive a 3-dimensional wave equation modelling vibrations of an elastic body in $\mathbb{R}^3$.

Heat Equations

Here, we model *heat flow* through a solid body having the *mass density* given by the function $\rho = \rho(x)$. We may view the body as a compact domain $\Omega_n \subset \mathbb{R}^n$, with $1 \leq n \leq 3$. The partial differential equations derived here are called *heat equations*. In historical terms, these type of equations were first derived by Joseph Fourier in early nineteenth century (see Sect. 8.2 for details). Recall that the *thermal energy* moves by *conduction* and *convection*. The conduction refers to transfer of thermal energy caused due to random collisions of molecules like in a diffusion process. In the case, heat energy is concentrated initially at a position $x \in \Omega_n$, and *convection* takes place if the agitated molecules move from one region to other. However, for simplicity, it is assumed that contribution of conduction in heat flow is far more significant than due to convection. That is, field velocity of the heat flow is negligible.

Since heat distribution inside Ω_n is not uniform, the *thermal energy density* e is a function of both position x and the time $t \geq 0$. We may thus write $e = e(x, t)$. For any infinitesimal control volume dV of the region Ω_n, with a smooth boundary surface dS, the fixed *heat energy* stored within is given by $h(x, t) = e(x, t)dV$. So, the time rate of change of heat energy at a position x is given by

$$h_t(x, t) = \partial_t[e(x, t)dV]. \quad (4.2.12)$$

Further, allowing for internal source or sink of heat energy, let $f = f(x, t)$ be the heat energy *generated* or *consumed*[5] per unit volume and per unit time. Notice that $f(x, t)$ in dV is approximately a constant, and so the total thermal energy generated or consumed inside dV per unit time is given by $f(x, t)dV$. Now, if $q = q(x, t)$ is time rate flow (i.e. the *flux*) of thermal energy along the the (outward) normal n to the surface element dS, then by *conservation of heat energy* applied to dV we have

$$h_t(x, t) = e_t(x, t)dV \approx -q \cdot ndS + f(x, t)dV. \qquad (4.2.13)$$

In totality, over the region Ω_n, we obtain

$$\int_{\Omega_n} e_t(x, t)dV = -\int_{\partial\Omega_n} q \cdot ndS + \int_{\Omega_n} f(x, t)dV$$

$$= \int_{\Omega_n} \nabla q dV + \int_{\Omega_n} f(x, t)dV,$$

where the second equality follows from divergence theorem. Therefore,

$$e_t(x, t) = -\nabla q + f(x, t). \qquad (4.2.14)$$

This equation is of little use because physical measurement related to heat flow is not given in terms of $e = e(x, t)$. It is described using *temperature* $u = u(x, t)$ at a position $x \in \Omega_n$, and at time $t \geq 0$. Recall that, in general, the *specific heat* $c = c(x)$ of a material is the amount of heat energy required to raise the temperature of a unit mass of the material by one unit. So, the heat energy per unit mass is given by $c(x)u(x, t)$. Therefore, the total energy inside dV is given by the relation

$$e(x, t)dV = c(x)u(x, t)\rho(x)dV.$$

It thus follows by routine argument that

$$e(x, t) = c(x)u(x, t)\rho(x), \qquad (4.2.15)$$

which is an important relation connecting the thermal energy and the temperature. Substituting in (4.2.14), we obtain

$$c(x)\rho(x) \frac{\partial u(x, t)}{\partial t} = -\nabla q + f(x, t). \qquad (4.2.16)$$

Notice that this equation involves two unknowns, viz. temperature u and flux q. However, we are interested in an equation governing heat flow in terms of temperature $u = u(x, t)$ only. We are thus led to consider the following *constitutive relations* related to heat conduction:

---

[5] For example, as in case of a chemical reaction or electrical heating.

1. The Fourier's *first principle* states that every infinitesimal change δQ in heat flow $Q = Q(t)$ of an infinitesimal element of Ω_n in time δt due to heat loss or gain is proportional to the mass δm of the material and the change in the temperature, say δu. The proportionality constant is given by the *specific heat c*. So, we write $dQ = c\,\delta m\,\delta u$.

2. According to *Fourier's law* of heat conduction, the time rate of heat flow $Q'(t)$ across an infinitesimal surface element bounding a volume $dV \subset \Omega_n$ is proportional to cross-sectional surface area of the element $d\mathbf{S} = \mathbf{n}dS$, and to the temperature gradient normal to this area given by $\partial_n u$. The *positional* proportionality constant in this case gives the *thermal conductivity* of the material, denoted by $k = k(\mathbf{x})$, which is usually assumed to be independent of the point $\mathbf{x} \in \Omega$. So, we have

$$Q'(t) = -k\,\partial_n u\,dS. \qquad (4.2.17)$$

Notice that higher the value of k more would be the amount of heat flow, even when the temperature difference between two positions is same. The *minus* indicates that the heat flows from hotter to colder region.

For a body of *homogeneous material*, the constant $\alpha^2 = k/(c\rho)$ governs the rate and direction of heat flow: *Higher the thermal diffusivity of a material the faster would be the heat conduction*. It is called the *thermal diffusivity* of the material. We consider first the case of heat conduction in a body $\Omega_3 \subset \mathbb{R}^3$.

Example 4.3 (*3-dimensional Heat Equation*) Suppose $\delta P \subset \Omega_3$ is an infinitesimal element of mass δm having boundaries approximated by lengths δx, δy and δz. Then, the weight of δP is given by

$$\delta\omega = \rho\delta x\delta y\delta z = g\delta m.$$

At the point $\mathbf{x} = (x, y, z) \in \Omega$ and at time $t \geq 0$, the thermal *energy density* in terms of temperature $u = u(\mathbf{x}, t)$ is given by (4.2.15). By constitutive relation (1), if δu is the change in temperature in time δt, then we have

$$\delta Q = c\,\delta m\,\delta u = \frac{c\rho}{g}\,\delta x\delta y\delta z\,\delta u$$

gives the quantity of heat stored in δP. So, the rate of temperature change in this element is given by

$$\frac{\delta Q}{\delta t} = \frac{c\rho}{g}\,\delta x\delta y\delta z\,\frac{\delta u}{\delta t}. \qquad (4.2.18)$$

Clearly, the amount of heat that causes a change in temperature Δu in time δt is due to heat transfer either from within δP or through its faces.

Assign labels A, B, C, D to the front face and E, F, G, H to the rare face of δP, written anti-clockwise, with A and E being on the line parallel to x-axis. Notice that if δA is the area of a face, then $\delta A = \delta y\delta z$ for both front and rare faces of δP.

Similarly, $\delta A = \delta x \delta z$ for both left and right faces, and $\delta A = \delta x \delta y$ for both top and bottom faces.

By constitutive relation (2), the rate of heat flow out of the front face of the element and into the rare face of the element are, respectively, given by

$$k \, \delta A \, u_x \quad \text{and} \quad - k \, \delta A \, u_x. \tag{4.2.19}$$

So, the net rate of change of heat entering the rare face and leaving the front face of the element is given by

$$k \, \delta y \delta z \left[u_x \Big|_{x+\delta x} - u_x \Big|_x \right],$$

where the partial derivative $u_x \big|_{x+\delta x}$ is evaluated at the centroid of the face. Equating the rate in Eq. (4.2.18) with the total δA in (4.2.19), and considering the contribution from all the six faces of δP, we obtain

$$\frac{c\rho}{g} \delta x \delta y \delta z \, \frac{\delta u}{\delta t} = k \left[\delta y \delta z (u_x|_{x+\delta x} - u_x|_x) + \delta x \delta z (u_y|_{y+\delta y} - u_y|_y) \right.$$
$$\left. + \delta x \delta y (u_z|_{z+\delta z} - u_z|_z) \right] + \delta x \delta y \delta z \, f(x, t),$$

where $f(x, t)$ is the internal heat change per unit volume. Dividing the last equation throughout by $(c\rho/g)\delta x \delta y \delta z$, and letting δx, δy, δz, and δt approach zero, we obtain

$$u_t = a^2 (u_{xx} + u_{yy} + u_{zz}) + \frac{g}{c\rho} f(\mathbf{x}, t), \quad \text{with } a^2 = g\,\alpha^2, \tag{4.2.20}$$

which is called the 3-dimensional *heat equation*. In practice, we usually work with the assumption that $f(x, t) = 0$ when heat is neither generated nor lost within the solid. Notice that (4.2.20) is same as the *diffusion equation* (4.3.23) for $n = 3$, where $k = a^2$ represents the coefficient of diffusion.

We model next the problem of heat conduction in a thin bar or a wire of finite length, say ℓ, oriented along the x-axis with ends at points $x = 0$ and $x = \ell$. So, here, we have $\Omega_1 = [0, \ell]$.

Example 4.4 (1-*dimensional Heat Equation*) It may be assumed that the lateral surface of the bar is perfectly insulated to ensure 1-dimensional heat flows along the x-axis. Then, at a point $0 \le x \le \ell$ and at a time $t \ge 0$, the function $u(x, t)$ representing the temperature distribution of the bar satisfies the *heat equation*

$$u_t = a^2 \, u_{xx}, \tag{4.2.21}$$

which is also known as the *Fourier heat equation*. A complete derivation of Eq. (4.2.21) can be obtained by adapting procedure used in previous example. In fact, details are simpler in this case because heat flow is 1 dimensional. Further, as in previous case, we can derive wide range of 1-dimensional heat equation by considering more realistic physical conditions. For example,

$$u_t = \frac{a^2}{c(x)} \, \partial_x \Big(k(x) \, u_x \Big), \tag{4.2.22}$$

is the *generalised heat equation* modelling heat flow in a nonuniform bar. Also, the *damped heat equation*

$$u_t = a^2 \big[u_{xx} - \lambda u \big], \quad \lambda > 0, \tag{4.2.23}$$

models temperature distribution of the bar that is losing some heat due to radiation.

A model of heat conduction in a thin flat plate of uniform thickness, as given below, takes the plate to be the rectangle in xy-plane having the corners at points

$$(0,0), \quad (0,b), \quad (a,0), \quad \text{and} \quad (a,b).$$

Therefore, in this case, we have $\Omega_2 = [0, a] \times [0, b]$. Take $b \to \infty$ for an infinite plate.

Example 4.5 (*2-dimensional Heat Equation*) As before, to ensure 2-dimensional flow, it is assumed that both faces of the plate are properly insulated. Adapting procedure used in Example 4.3, it follows that the temperature distribution $u = u(x, y, t)$ in the plate at a point $(x, y) \in \Omega_2$ and at a time $t \geq 0$ satisfies the equation

$$u_t = a^2 (u_{xx} + u_{yy}), \quad (x, y) \in \Omega_2, \tag{4.2.24}$$

which is called the 2-dimensional heat equation.

Notice that, as before, the steady-state equation satisfied by the function $u = u(x, y)$ is given by

$$\nabla^2 u = u_{xx} + u_{yy} = 0, \tag{4.2.25}$$

which is called the Laplace equation in two variables.

Laplace and Poisson Equations

Let $\Omega \subset \mathbb{R}^3$ be a compact region, and $x = (x, y, z)$. Recall that the differential operator

$$\nabla^2 := \partial_{xx} + \partial_{yy} + \partial_{zz}$$

is a 3-dimensional *Laplacian* in variables x, y, z. In view of Eq. (4.2.26), when $u_t = 0$, the equation modelling *steady state* of a process in a system without source or sink is given by

$$\nabla^2 u = u_{xx} + u_{yy} + u_{zz} = 0. \tag{4.2.26}$$

As said before, this is known as the *Laplace equation*, say for a function $u = u(x)$ representing the steady-state temperature distribution at the position x. In addition to such type of steady-state cases of *conservation equations* modelling transport phenomena, we also come across Laplace equations in applications concerning gravitational potential, electrostatic, and electrodynamic potential.

Example 4.6 Suppose a point mass m is at point $x = x(t)$ at time $t \geq 0$. It is assumed that every point mass around the (source) mass m experience the force of attraction governed by Newton's gravitational law. At any fixed time t, the one-parameter *gravitational field* $\mathbf{g}(x, t)$ defined by the point mass m is given by

$$\mathbf{g}(x, t) = G \, m \, \frac{x - x(t)}{\|x - x(t)\|^3}, \quad \mathbf{x} \in \Omega \subset \mathbb{R}^3, \tag{4.2.27}$$

where G is the (relative) gravitational constant. So, the force this field exerts on any other point mass M located at a point $x \in \Omega$ is given by

$$f(x, t) = M\mathbf{g}(x, t) = G \, mM \, \frac{x - x(t)}{\|x - x(t)\|^3}. \tag{4.2.28}$$

Writing $r(t) = \|x - x(t)\|$ for the distance between the point masses m and M, the force function

$$u = u(x, t) = \frac{c}{r}, \quad \text{with } c = G \, mM,$$

represents the *gravitational potential* at point x and at time $t \geq 0$, because the (spatial) *gradient* of the function $u = u(x, t)$ defines the field $\mathbf{g}(x, t)$. More generally, if m is unit mass, and $\rho = \rho(x)$ is the mass density of an infinitesimal volume dV around the point $x = (X, Y, Z)$ in a continuous distribution of masses within the volume V of the domain Ω, then the function

$$u(x) = G \iiint_V \frac{\rho(x)}{r} \, dV.$$

defines the *potential* at the point x. Notice that

$$u_x = -Gk \iiint_V \rho \, \frac{(x - X)}{r^3} \, dV;$$

$$u_{xx} = -Gk \iiint_V \rho \left[\frac{1}{r^3} - 3 \frac{(x - X)^2}{r^5} \right] dV;$$

$$u_{yy} = -G \iiint_V \rho \left[\frac{1}{r^3} - 3 \frac{(y - Y)^2}{r^5} \right] dV;$$

$$u_{zz} = -G \iiint_V \rho \left[\frac{1}{r^3} - 3 \frac{(z - Z)^2}{r^5} \right] dV.$$

Adding the last three equations, we obtain the Laplace equation

$$\nabla^2 u = u_{xx} + u_{yy} + u_{zz} = 0,$$

for the function $u = u(x)$. Therefore, the gravitational potential in $\mathbb{R}^3$ is modelled by a Laplace equation.

The *electrostatic potential* also satisfy the Laplace equation in three variables because Coulomb's law for electrostatics force of attraction between charged bodies is same as Newton's gravitational law. Further, for the same reason, the *magnetic potential* satisfies a Laplace equation in three variables.

Definition 4.1 A partial differential equations of form

$$\nabla^2 u = g(x), \quad u = u(x) \in \Omega \subseteq \mathbb{R}^2, \tag{4.2.29}$$

is known as a *Poisson equation*, where $g(x)$ is called the *source term*.

Poisson equations arise naturally in many applications of science and engineering. Some such types of equations have already been derived in Sect. 4.3, while writing model equations for diffusion process in solid, liquid or gases. For example, that is precisely the case when we model problems related to equilibrium temperature distribution $u = u(x, y)$ in a material with a time-independent source $g(x, y)$ of the thermal energy. As said before, a Laplace equation is obtained when $g \equiv 0$. Further, it follows from the Maxwell equations that

$$\nabla \times \mathbf{E} = \mathbf{0} \quad \nabla \cdot \mathbf{E} = \frac{\rho}{\varepsilon_0};$$
$$\nabla \times \mathbf{B} = \mu_0 \mathbf{J} \quad \nabla \cdot \mathbf{B} = 0,$$

where $\mathbf{E}$ and $\mathbf{B}$ are, respectively, the electrostatic and magnetostatics fields, ρ is the charge density, and $\mathbf{J}$ is the field of current flux (see Appendix A.2 for details). The relation $\nabla \times \mathbf{E} = \mathbf{0}$ implies that $\mathbf{E} = -\nabla \varphi$, where φ is an *electrostatic potential*. So,

$$\nabla \cdot \mathbf{E} = \frac{\rho}{\varepsilon} \quad \Rightarrow \quad \nabla^2 \varphi = -\frac{\rho}{\varepsilon_0},$$

which is a Poisson equation of *electrostatics*. Clearly, this reduces to a Laplace equation in a region where there is no charge or current. Also, in this latter situation, the relation $\nabla \times \mathbf{B} = \mathbf{0}$ implies that $\mathbf{B} = -\mu_0 \nabla \psi$, where ψ is a *magnetostatics potential*. So, as said before, we obtain the Laplace equation $\nabla^2 \psi = 0$ for magnetostatics potential.

More generally, let us derive Poisson equations modelling potential of a time-varying electric field $\mathbf{E} = \mathbf{E}(t)$ in a charged medium with associated time varying magnetic field given by $\mathbf{B} = \mathbf{B}(t)$. By Theorem 2.28, if $\mathbf{V} = V(\mathbf{r})$ is a vector field like $\mathbf{E}$ is specified over a region Ω, then we can write

$$\mathbf{V} = -\nabla u(\mathbf{r}) + \nabla U(\mathbf{r}), \tag{4.2.30}$$

provided $V(\mathbf{r}) \to 0$, and $\nabla \times \mathbf{V}$, $\nabla \cdot \mathbf{V}$ approach to zero faster than the function $1/r^2$, as $r = \|\mathbf{r}\| \to \infty$. Recall that $u(\mathbf{r})$ and $\mathbf{U} = U(\mathbf{r})$ are, respectively, the *scalar* and *vector potential* of the field $V(\mathbf{r})$. Therefore, for

$$\nabla \cdot \mathbf{V} = f \quad \text{and} \quad \nabla \times \mathbf{V} = \mathbf{F},$$

Equation (4.2.30) implies that

$$f = \nabla \cdot \mathbf{V} = -\nabla^2 u(\mathbf{r}) \quad \Rightarrow \quad \nabla^2 u = -f;$$
$$\mathbf{F} = \nabla \times \mathbf{V} = \nabla^2 \mathbf{U} \quad \Rightarrow \quad \nabla^2 \mathbf{U} = \mathbf{F}.$$

That is, the scalar potential function $u = u(\mathbf{r})$ and also the component functions $u_i = u_i(\mathbf{r})$ of the vector potential $\mathbf{U} = U(\mathbf{r})$, satisfy the *Poisson equation* of the form (4.2.29).

4.3 Models for Transport Phenomena

Fundamental to the continuum mechanics is the natural processes governed by *diffusion* and *convection* in continuous materials such as liquids, gases, and solids. Recall that a *cross process* is the one wherein the gradient of one type of physical entity causes the transport of some other types. In mathematical terms, the balance laws related to *transport phenomena* are usually stated as the theorems involving the *line*, *surface*, and/or the *volume integrals* of scalar and/or vector fields, as discussed in Chap. 2.

As we shall see in the subsequent chapters, many practical problems in science and technology concerning physical processes such as vibrations, wave motions, heat conduction, fluid flow, and electrodynamics can be solved as *initial-boundary value problems* for some linear partial differential equations. In this section, we focus mainly on phenomena governed by the transfer of the mass, momentum, and energy. For example, the molecular diffusion in a gaseous medium is a case of the *mass transfer*, the viscosity of a moving fluid is a case of the *momentum transfer*, and the heat conduction in solids is a case of *energy transfer*. In general, the related governing *balance laws* are the conservation principle expressed in terms of the partial derivatives of a function $u = u(\boldsymbol{x}, t) \in C^2(\Omega \times \mathbb{R}^+)$.

In practical terms, observations at the macroscopic scale taken with reference to a *control volume* dV of the region Ω are actually averages taken over small packet of particles in the control volume. That is, for example, computations related to a motion of a material in a medium concerns movements of small packets represented by the spatial variable $\boldsymbol{x} \in \Omega$. It is assumed that the amount of material in a control volume in comparison with the whole is large enough to contain a considerable amount of *physical quantity* under investigation. Observations about a physical quantity taken at a fixed time are *Eulerian* and those taken along the trajectory of a flow are *Lagrangian*.

Mass Transfer: Continuity Equations

The mass transfer refers to a noninteracting chemical in a fluid freely flowing through a physical volume $\Omega \subseteq \mathbb{R}^n$, with velocity represented by a function $v = v(x, t) \in C^2(\Omega_1)$, where $\Omega_1 = \Omega \times \mathbb{R}^+$. Let $\rho = \rho(x, t) \in C^1(\Omega_1)$ represent the mass density of the material in a control volume dV at point $x \in \Omega$ and time $t \geq 0$. In this part of the section, we derive *continuity equation* modelling the time rate of change of ρ. Recall that *linear convection* (or an *advection*) is a process wherein particles of the material carry their energy as they move in the direction of the flow. In general, the *flux density vector* (or simply *flux vector*) given by

$$\mathbf{Q} = \mathbf{Q}(x, t) = \rho(x, t)\, v(x, t), \quad \text{for } x \in \Omega \quad \text{and} \quad t \geq 0,$$

represents the *flux density*. This vector is used to model the mass transfer of a material in an ideal fluid. The associated *conservation law* states that the sum of the time rate of increase in the mass density (ρ_t) and the divergence of the flux vector $\mathbf{Q}$ ($\nabla \cdot \mathbf{Q}$) equals the volume at the source.

More generally, if the flux across a *surface element* dS is given by $\mathbf{V} \cdot \mathbf{n}\, dS$ then the vector $\mathbf{V}$ represents the net rate of the mass flux per unit volume. Let a function $\rho = \rho(x, t)$ represent the mass density at the position x and at time $t \geq 0$ of a *noninteracting* material flowing freely in a fluid with a velocity $v = v(x, t)$ across the *closed* surface S that bounds a volume $V = V(\Omega)$. At any time $t \geq 0$, the total mass of the material inside the volume of Ω is given by

$$M(t) = \int_\Omega \rho(x, t)\, dV. \tag{4.3.1}$$

Since an infinitesimal surface element of the smooth boundary surface S is *almost flat*, say with a surface area dS, the quantities $\mathbf{n}, v, \rho$ can be assumed to remain constant during an infinitesimal time interval $[t, t + dt]$. So, if θ is the angle between the velocity vector v and the unit outward normal $\mathbf{n} = \mathbf{n}(x)$, then the net mass of the material flowing through dS during this infinitesimal time period is given by

$$dM = \rho|v|dt \cos\theta\, dS = \rho v \cdot \mathbf{n} dS\, dt.$$

Therefore, the *flux* across the surface element dS is given by

$$M'(t) = \rho v \cdot \mathbf{n} dS = \mathbf{Q} \cdot \mathbf{n} dS.$$

Hence, the rate at which the mass enters Ω equals the surface integral of the mass flux $\mathbf{Q} = \rho v$. That is,

$$\text{rate of mass in-flow} = -\int_S \rho v \cdot \mathbf{n} dS, \tag{4.3.2}$$

where the *minus sign* indicates the fact that $v \cdot \mathbf{n} < 0$. Notice that, if a material of density ρ is added to Ω (or consumed inside) according to formula $g(x, t)$, then the related *balance law* can be expressed in integral form as

$$\frac{\mathrm{d}}{\mathrm{d}t} \int_\Omega \rho \, \mathrm{d}V + \int_S \rho v \cdot \mathbf{n} \mathrm{d}S = \int_\Omega g(x, t) \mathrm{d}V. \qquad (4.3.3)$$

Now, using the Leibniz rule and applying Theorem 2.25, we obtain

$$\int_\Omega \rho_t \, \mathrm{d}V + \int_\Omega \nabla \cdot \rho v \, \mathrm{d}V = \int_\Omega g(x, t) \mathrm{d}V. \qquad (4.3.4)$$

Since this holds for every control volume $\mathrm{d}V$ of Ω, and assuming that the *integrand* has desired smoothness properties, the balance law in differential form is given by

$$\rho_t + \nabla \cdot \rho v = g(x, t). \qquad (4.3.5)$$

In the above equation, only the function $\rho = \rho(x, t)$ is unknown, whereas the functions $\mathbf{Q} = \rho v$ and $g(x, t)$ are usually determined by using some *constitutive relations*. In actual applications, the flux vector $\mathbf{Q} = \mathbf{Q}(x, t)$ depends on ρ, ρ', and/or some *parameters*. Notice that if no material is being created or lost so that $g \equiv 0$ then, by *law of conservation* for mass transfer, we must have

$$\rho_t + \nabla \cdot (\rho v) = \rho_t + \nabla \rho \cdot v + \rho(\nabla v) = 0, \qquad (4.3.6)$$

which is known as the *continuity equation*. As said earlier, Eq. (4.3.6) implies that the sum total of the time-rate increase of the density and the divergence of the flux vector is zero. Notice that, by the total derivative rule, we can also write

$$\frac{\mathrm{d}\rho}{\mathrm{d}t} = \rho_t + \nabla \rho \cdot x'(t).$$

Therefore, assuming that each control volume $\mathrm{d}V$ moves with the velocity v so that $x'(t) = v$, we obtain

$$\frac{D}{Dt} \rho = \rho_t + \nabla \rho \cdot v, \qquad (4.3.7)$$

which is called the *material derivative* of the density function $\rho = \rho(x, t)$. Hence, it follows from Eq. (4.3.6) that we have

$$\frac{D}{Dt} \rho = -\rho(\nabla \cdot v). \qquad (4.3.8)$$

That is, the time rate of change in the density leads to a compression or expansion of the mass in the control volume $\mathrm{d}V$. Therefore, if the fluid is *incompressible* (equivalently, if ρ is constant), we must have $\nabla \cdot v = 0$. That is, for an incompressible flow, the divergence of the flow velocity must vanish. Further, it follows from (4.3.6) that

$$\rho_t + \nabla \rho \cdot v = 0, \qquad (4.3.9)$$

This is differential form of the law of conservation for the mass transfer in an *incompressible fluid*. It is possible to derive similar partial differential equations of laws of conservation for other type of transport phenomena. In certain situations, we can view some spatial variables can be treated as *parameters* of the equation (Exercise 4.9).

Momentum Transfer: Equations of Motion

We derive here *system* of partial differential equations as a model to study practical problems related to the dynamics of *real fluids*, which are called *equations of motion*. The related initial-boundary value problems help study physical processes dealing with momentum transfer.

In this case, in addition to three components of velocity vector $\mathbf{u} = \mathbf{u}(x, t)$, the pressure $p = p(x, t) > 0$ (as global interaction among all particles), and the mass density $\rho = \rho(x, t)$, we also take into consideration amount of the energy dissipation in terms of viscosity and heat exchange among different parts of it. Therefore, the proposed model must also include internal energy density function $e = e(x, t)$ and the heat flux density function $q = q(x, t)$. Notice that the frictional forces inside the fluid enhance the local coherence of the flow leading to a lower velocity if the fluid particles move faster than the averages of their neighbourhoods. We consider next two forces acting on a control volume:

1. The *body forces* due to gravity $\mathbf{g}$ given by

$$\mathbf{F}_v := \int_V \rho \mathbf{g} \, dV. \qquad (4.3.10)$$

2. The *surface forces* given by

$$\mathbf{F}_S := \int_S (-p) \mathbf{n} \, dS. \qquad (4.3.11)$$

Such types of forces are caused due to collisions between fluid molecules on either side of the surface S, which produce a flux of momentum across the boundary in the direction of the normal $\mathbf{n}$. We have considered only the effect of the pressure $p > 0$ on the volume V bounded by the surface S.

By Newton's momentum law, the sum total of forces acting on a control volume dV must equals the rate of change of momentum. Therefore, by using the divergence theorem, we have

$$\int_V \rho \frac{D\mathbf{u}}{Dt} \, dV = \int_V [-\nabla p + \rho \mathbf{g}] dV. \qquad (4.3.12)$$

Hence, by routine argument and Eq. (4.3.7), we obtain

$$\rho \frac{D\mathbf{u}}{Dt} = \rho\left(u_t + (\mathbf{u} \cdot \nabla)\mathbf{u}\right) = -\nabla p + \rho\mathbf{g}, \qquad (4.3.13)$$

which is the *Euler equation*. In particular, when the fluid is at rest, $\mathbf{u} \equiv 0$ gives the *equation of hydrostatic balance* as given by

$$\nabla p = \rho\mathbf{g} \quad \Leftrightarrow \quad p(\mathbf{x}) = \rho\mathbf{g} \cdot \mathbf{x} + c, \qquad (4.3.14)$$

where c is a constant. Further, considering tangential surface forces due to velocity gradients (viscosity ν), an extra term $\nu\nabla^2\mathbf{u}$ is introduced so that we have

$$\rho \frac{D\mathbf{u}}{Dt} = -\nabla p + \rho\mathbf{g} + \nu\nabla^2\mathbf{u}. \qquad (4.3.15)$$

These are *Navier-Stokes equations*. For the nonlinear term in acceleration

$$\frac{D\mathbf{u}}{Dt} = u_t + (\mathbf{u} \cdot \nabla)\mathbf{u}, \qquad (4.3.16)$$

we can write

$$(\mathbf{u} \cdot \nabla)\mathbf{u} = \nabla\left(\frac{\|\mathbf{u}\|^2}{2}\right) - \mathbf{u} \times \boldsymbol{\omega}, \quad \text{with } \boldsymbol{\omega} = \nabla \times \mathbf{u}. \qquad (4.3.17)$$

Therefore, the Euler equations (4.3.13) take the form

$$\mathbf{u}_t + \nabla\left(\frac{\|\mathbf{u}\|^2}{2}\right) - \mathbf{u} \times \boldsymbol{\omega} = -\nabla(p/\rho) + \mathbf{g}, \qquad (4.3.18)$$

where ρ is constant. Hence, taking the curl of both sides, we obtain

$$\frac{\partial\boldsymbol{\omega}}{\partial t} - \nabla \times (\mathbf{u} \times \boldsymbol{\omega}) = 0. \qquad (4.3.19)$$

In particular, using the vector identity (2.3.11b),

$$\nabla \times (\mathbf{u} \times \boldsymbol{\omega}) = (\boldsymbol{\omega} \cdot \nabla)\mathbf{u} - (\mathbf{u} \cdot \nabla)\boldsymbol{\omega} + (\nabla \cdot \boldsymbol{\omega})\mathbf{u} - (\nabla \cdot \mathbf{u})\boldsymbol{\omega},$$
$$= (\boldsymbol{\omega} \cdot \nabla)\mathbf{u} - (\mathbf{u} \cdot \nabla)\boldsymbol{\omega} - (\nabla \cdot \mathbf{u})\boldsymbol{\omega}, \quad (\text{because } \nabla \cdot \boldsymbol{\omega} = 0)$$

it follows that for *incompressible flows* ($\nabla \cdot \mathbf{u} = 0$), we have

$$\frac{D\boldsymbol{\omega}}{Dt} = \frac{\partial\boldsymbol{\omega}}{\partial t} + (\mathbf{u} \cdot \nabla)\boldsymbol{\omega}) = (\boldsymbol{\omega} \cdot \nabla)\mathbf{u}. \qquad (4.3.20)$$

This is known as the *vorticity equation*, which says that the change in vorticity of a fluid particle occurs due to the gradient of the velocity vector **u** in the direction of the vector $\omega = \nabla \times \mathbf{u}$. In particular, we conclude that a flow remains *rotation-free* if it starts in that state. Also, in the case of a 2-dimensional planar flow, we have

$$\omega = \left(0, 0, \partial_x u_2 - \partial_y u_1\right) \quad \Rightarrow \quad (\omega \cdot \nabla)\mathbf{u} = \omega \frac{d}{dz}\mathbf{u}(x, y) = 0;$$

so Eq. (4.3.20) reduces to

$$\frac{D\omega}{Dt} = \frac{\partial \omega}{\partial t} + (\mathbf{u} \cdot \nabla)\omega) = 0, \qquad (4.3.21)$$

which shows that the vorticity of a fluid particle remains constant. In addition, if the flow is steady, i.e. $\omega_t = 0$, then the vorticity along streamlines is constant.

Energy Transfer: Diffusion Equations

Finally, as an important case of energy transfer, we consider diffusion of a substance like a chemical through a homogeneous liquid (or a gas in air) contained in a bounded volume. Recall that diffusion process takes place due to the collisions of neighbouring molecules by which the kinetic energy of molecules is transferred from one to its nearest neighbour. It is assumed that the fluid is nearly motionless to avoid convection. Over a period of time, the substance diffuses throughout the fluid randomly moving from regions of higher concentration to the lower concentration. Under these conditions, we derive partial differential equations that are used to study time rate change in concentration levels of the diffusing substance at a position. These are called *diffusion equations*.

Let the mass density of the diffusing substance at position $x \in \Omega \subset \mathbb{R}^n$ and at time $t \geq 0$ be given by a sufficiently smooth function $u = u(x, t)$. The *initial concentration* may be assumed to be given by a function $f(x) = u(x, 0)$. Then, the total mass $M = M(t)$ inside the volume V of the region Ω at time t is given by

$$M(t) = \int_V u(x, t)\, dV, \quad \text{for } t \geq 0. \qquad (4.3.22)$$

If the flux vector $\mathbf{Q} = \mathbf{Q}(x)$ changes only when it goes through the surface element dS of a control volume dV, then we have

$$\frac{d}{dt} \int_V u(x, t)\, dV = - \int_S \mathbf{Q} \cdot \mathbf{n}\, dS,$$

where **n** is the (outward) normal to the surface S. Also, by Theorem 2.25, we have

$$\int_S \mathbf{Q} \cdot \mathbf{n} \, dS = \int_V \nabla \cdot \mathbf{Q} \, dV.$$

Putting together the previous two equations, we obtain

$$\int_V \left[u_t(\mathbf{x}, t) + \nabla \cdot \mathbf{Q} \right] dV = 0.$$

Assuming the necessary smoothness of the vector $\mathbf{Q}$, and using the fact that control volume dV is arbitrary, we obtain

$$u_t + \nabla \cdot \mathbf{Q} = 0, \tag{4.3.23}$$

which is the differential form of an n-dimensional *balance equation*. Many physical phenomena in science and engineering such as *heat conduction, Brownian motion, population dynamics* are best described by Eq. (4.3.23).

In particular, we obtain usual conservation law when the flux vector $\mathbf{Q}$ is a function of u, i.e. $\mathbf{Q} = \mathbf{Q}(u)$. For example, by Fick's law, we know that the diffusion flux is given by the relation

$$\mathbf{Q} = -k \, \nabla u(\mathbf{x}, t), \quad \mathbf{x} \in \Omega,$$

where k is the diffusion coefficient, which may be assumed to be constant. Therefore, we obtain

$$u_t - \nabla(k \, \nabla u) = 0 \quad \Rightarrow \quad u_t = k \, \nabla^2 u, \tag{4.3.24}$$

which is the *diffusion equation* that describes the concentration fluctuations of a diffusing substance in the region Ω. More complex models are obtained by allowing diffusion coefficient k to be a function of u, $\mathbf{x}$ or t. For example, the first equation in (4.3.24) with $k = k(u)$ gives

$$u_t - \nabla(k(u) \, \nabla u) = 0 \quad \Rightarrow \quad u_t = \nabla k(u) \cdot \nabla u + k(u)\nabla^2 u = 0, \tag{4.3.25}$$

which is a *nonlinear diffusion equation*.

Furthermore, if $g(\mathbf{x}, t)$ is the net rate at which the diffusing substance is being *sourced* into the fluid (per unit volume) at the position $\mathbf{x}$ or it is getting dissolved (say, due to a *sink* created by a chemical agent), then we obtain the following *non-homogeneous* diffusion equation:

$$u_t = k \, \nabla^2 u + g(\mathbf{x}, t). \tag{4.3.26}$$

Another interesting particular case is when ρ represents the mass density of a gas spreading by diffusion process only. In this case, as $-\nabla \rho$ gives the direction of steepest decrease, so it is plausible to assume that for some $d > 0$ the flux vector $\mathbf{Q} = -d \, \nabla \rho$, where d is called the *gas diffusion constant*. Notice that it is yet another case of a *constitutive relation*. Thus, by (4.3.23), we obtain

$$\rho_t - \nabla(\mathrm{d}\,\nabla\rho) = 0 \quad \Rightarrow \quad \rho_t = \mathrm{d}\,\nabla^2\rho, \tag{4.3.27}$$

which is also known as Fourier's *heat equation*. More precisely, if $T(\boldsymbol{x}, t)$ is a measure of temperature then according to *Fourier's law* $\mathbf{Q} = -c\,\nabla\,T(\boldsymbol{x}, t)$, where c is the thermal conductivity of the material. A detailed account of heat equations in dimension ≤ 3 is given in the next section.

Remark 4.1 In 1905, Albert Einstein used probabilistic arguments to derive equation (4.3.27) to model Brownian motions of microscopic particles in many diverse situations in physics, chemistry, biology, medicine, finance, etc., that provided the strong evidence of the fact that these type of motions are caused due to collisions with a large number of particles having randomly distributed velocities.

Exercises 4

4.1 Give full details about derivation of wave equation (4.2.6).

4.2 Give full details about derivation of wave equation (4.2.7).

4.3 Modify suitably the details given in Example 4.2 to derive a 3-dimensional wave equation modelling vibrations of an elastic body in $\mathbb{R}^3$.

4.4 Give a complete derivation of Eq. (4.2.21) by adapting procedure described Example 4.3.

4.5 Derive the *generalised heat equation* (4.2.22) that models heat flow in a nonuniform bar.

4.6 Derive the heat equation (4.2.23) that models the temperature distribution in a bar, which loses some heat due to radiation.

4.7 Modify the procedure described in Example 4.3 to give a full derivation of the 2-dimensional heat equation (4.2.25).

4.8 Consider distribution of a mass in a region $\Omega \subset \mathbb{R}^3$, with density function given by $\rho = \rho(\boldsymbol{x})$. Let $f(\boldsymbol{x})$ be the associated gravitational field, with a *gravitational potential* $u = u(\boldsymbol{x})$. Derive Eq. (4.2.29), with $g(\boldsymbol{x}) = 4\pi\,G\rho(\boldsymbol{x})$.

4.9 Consider flow of a *noninteracting* contaminant in an *incompressible* fluid flowing through a 2-dimensional region given by

$$\Omega = \{(x, y) \mid -\infty < x < \infty, \ |y| < 1\},$$

with time-dependent velocity field directed only towards x-axis. Let $\rho(x, y, t)$ denotes the mass density of the contaminant in the fluid at a position $(x, y) \in \Omega$ and at the time $t \geq 0$. Show that $\rho = \rho(x, y, t)$ satisfies the equation

$$\rho_t + (1 - y^2)\rho_x = 0, \quad -\infty < x < \infty, \quad t \geq 0. \tag{4.3.28}$$

4.10 (*Kelvin circulation theorem*) Show that, in an inviscid fluid of uniform density, the *circulation around a closed material curve remains constant*.

References

1. Murray, J. D. (2002). *Mathematical biology—An introduction*. Springer.
2. Arnold, V. I., Kozlov, V. V., & Neishtadt, A. I. (2006). *Mathematical aspects of classical and celestial mechanics* (3rd ed.). Springer.
3. Hirsch, M. W., Smale, S., & Devaney, R. L. (2004). *Differential equations, dynamical systems, and introduction to chaos* (2nd ed.). Elsevier.
4. Stakgold, I., & Holst, M. J. (2011). *Green's functions and boundary value problems*. Wiley.
5. Howison, S. (2005). *Practical applied mathematics: modelling, analysis, approximation*. Cambridge University Press.
6. Lin, C. C., & Segel, L. A. (1988). *Mathematical applied to deterministic problems in natural sciences*. Classics in applied mathematics. SIAM.
7. Griffiths, D. J. (2013). *Introduction to electrodynamics* (5nd ed.). Pearson.

Chapter 5
Partial Differential Equations

> One would have to have completely forgotten the history of
> science so as to not remember that the desire to know nature has
> had the most constant and happiest influence on the
> development of mathematics.
>
> Henri Poincáre (1854–1912)

This chapter provides necessary background to discuss more formally the main topics on the subject. Let Ω be an open set that, in practical terms, represents a *physical volume* in the space $\mathbb{R}^n$. We may write the elements of the set $\Omega_1 = \Omega \times \mathbb{R}^+$ as (x, t), where $x_1, \ldots, x_n$ give the *spatial coordinates* of a point $x \in \Omega$ and t as a time variable. In Sect. 5.1, we introduce some basic concepts such as *order* and *linearity type* of a general partial differential equation for a sufficiently smooth function $u = u(x, t) : \Omega_1 \to \mathbb{R}$ representing some scalar quantity at a point $x \in \Omega$ and at time $t \geq 0$. In fact, for most cases we are dealing with in this book, it suffices to assume that $u \in C^2(\Omega_1)$. Unless specified otherwise, by a differential equation we always mean a partial differential equation.

An interesting classification of second order linear differential equations is about the *geometry type* of their respective *solution spaces*. In Sect. 5.2, we show that each second order linear differential equation in two variables can be transformed to one of the three *normal forms*, by using a suitable change of coordinates: A wave equation of *hyperbolic type*; a heat equation of *parabolic type*; or, a Laplace equation of *elliptical type*. In particular, the first order linear differential equations are the simplest types of *hyperbolic equations*. By what is seen in Chap. 4, we understand why such types of differential equations are of so much importance to problems in science and engineering.

By a *classical solution* of a kth order differential equation in $(n + 1)$-variables we mean a C^k-function φ that satisfies the equation identically over a domain

$\Omega_1 = \Omega \times \mathbb{R}^+$. In Sect. 5.3, we discuss the related concepts. Unlike an ordinary differential equation, the *general solution* of a differential equation contains functions as *parameters*. Therefore, in the present case, the *solution space* is an infinite dimensional function space. Further, the topology of the domain Ω plays a significant role. In the next chapter, we will discuss some standard analytical methods due to Lagrange and Charpit of finding the general solution or the complete integral of a first order differential equation in two variables.

A *differential equation model* of a multivariate phenomenon involving an unknown function $u = u(x, t)$ is of practical use only when the governing equation admits a *unique* solution subject to it meeting certain *auxiliary conditions* that are usually given as functional specifications for $u = u(x, t)$. In Sect. 5.4, we discuss most common types of auxiliary conditions, namely, an *initial condition* for $u = u(x, t)$ specified at some $t = t_0 \geq 0$, and a *boundary condition* for $u = u(x, t)$ specified over the boundary $\Gamma = \partial\Omega$ of the domain Ω. In the process, we formulate some important *initial-boundary value problems* for phenomena such as dealing with vibrations, heat conduction, fluid dynamics, and electrodynamics. We also introduce Hadamard's concept of *well-posed problem*.

In the concluding Sect. 5.5, we consider simple situations to introduce some standard sufficient conditions that help prove uniqueness of a solution of an initial-boundary value problem, if it exists. We illustrate use of Green's function based *energy method* for initial-boundary value problems related to a wave equation. We also see an application of Lemma 2.3 in proving uniqueness of a solution of some interesting initial-boundary value problems related to Laplace and Poisson equations. We will also discuss the concept of *stability of the solution*, which is important mainly due to the fact that, in some cases, there may be a possibility that even a small error in boundary data may lead to some drastic changes in the solution of an initial-boundary value problem. The reader may refer [1, 6–10] for more details.

5.1 Preliminaries

It helps in some situations to assume that the open set $\Omega \subseteq \mathbb{R}^n$ is *simply connected*,[1] and has a *piecewise smooth* closed boundary $\Gamma = \partial\Omega$. That is, when $\Omega \subseteq \mathbb{R}^n$ is seen as an immersed submanifold of dimension $n - 1$, the boundary Γ is assumed to be a finite union of smooth surfaces of dimension $n - 1$. For example, the interior of a rectangular region $\overline{\Omega} = [a, b] \times [c, d]$ fits the above description. Further, while modelling the equilibrium state of a phenomenon such as dealing with heat conduction or potential of a conservative force field, we may also take Ω to be bounded. In most practical situations discussed earlier, the restriction of the function $u = u(x, t) : \Omega_1 \to \mathbb{R}$ to the boundary Γ is assumed to be continuous. Each *differential equation model* derived in Chapter 4 can be expressed implicitly as

---

[1] A path connected open set $D \subseteq \mathbb{R}^n$ is said to be *simply connected* if every simple closed curve in D can be *continuously shrunk* to a point. That is, the region D has no *holes*, in topological sense.

$$F\left(x;\ t;\ u;\ p_i;\ q;\ r_{ij};\dots\right) = 0, \qquad \text{for } (x,t) \in \Omega_1, \qquad (5.1.1)$$

where F is some *nice function*, and the symbols p_i, q, r_{ij}, etc., are the partial derivatives of the function u given by

$$p_i = u_{x_i} = \partial_{x_i} u = \frac{\partial u}{\partial x_i}; \qquad q = u_t = \partial_t u = \frac{\partial u}{\partial t};$$

$$r_{ij} = u_{x_i x_j} = \partial^2_{x_i x_j} u = \frac{\partial^2 u}{\partial x_i \partial x_j} = \frac{\partial^2 u}{\partial x_j \partial x_i}; \qquad \text{etc.}$$

Also, the *del operator* ∇ in n variables $x_1, \dots, x_n$ is written as

$$\nabla \equiv \left(\frac{\partial}{\partial x_1}, \cdots, \frac{\partial}{\partial x_n} \right).$$

Definition 5.1 For a function F that is sufficiently smooth with respect to variables x, t, u, p_i, q, r_{ij}, $\dots$ such that at least one of the derivatives F_{p_i}, F_q, $F_{r_{ij}}, \dots$, is nonzero over a suitable domain, an equation of the form (5.1.1) is called a general *partial differential equation* for a function $u = u(x, t)$, with $(x, t) \in \Omega_1$. The **order** of (5.1.1) is the order of the highest derivative of u appearing in the equation. The sum of highest order terms is called the **principle part** of a differential equation.

Clearly, we must have $u = u(x, t) \in C^k(\Omega_1)$ when Eq. (5.1.1) is of order $k \geq 1$. Further, as said earlier, a *differential equation model* of a phenomenon involving a vector function $u = u(x, t) \in \mathbb{R}^n$ is given by a *system of differential equations* of the form (5.1.1). Whenever it is possible to express Eq. (5.1.1) as

$$G\left(x;\ t;\ u;\ p_i;\ q;\ r_{ij};\dots\right) = g(x, t), \qquad \text{for } (x,t) \in \Omega_1, \qquad (5.1.2)$$

we say (5.1.2) is a *nonhomogeneous equation*, where $g : \Omega_1 \to \mathbb{R}$ is some nice function that represents a *source* or a *sink* of the physical system under investigation.

As said earlier, in this text, the discussion on the subject is restricted to equations involving two or three independent variables. That is, we are mainly dealing with differential equations having x, y as the spatial variables. Accordingly, a differential equation for a function $u = u(x, t)$ is a 1-dimensional equation defined over and open set $\Omega = (a, b) \times [0, \infty)] \subseteq \mathbb{R}^2$; and, for a function $u = u(x, y, t)$, it is a 2-dimensional equation defined over and open set $\Omega = U \times [0, \infty)] \subseteq \mathbb{R}^3$, where $U \subseteq \mathbb{R}^2$ is an open set. In certain simpler situations, it is possible to extend the argument to the case involving n spatial variables. However, for differential equations of order ≥ 2, it may not be feasible to deal with the involved geometric complexities [2].

In most interesting situations, Eq. (5.1.1) also involves functions $a_1, a_2, \dots$ of variables x, t, $p_i, \dots$, which appear as *coefficients* in various terms containing derivatives of the function u. As in the case of ordinary differential equations, such *coefficient functions* in actual applications represent adjustable physical parameters such as *speed, decay rate, specific heat, wave number, viscosity, elasticity coefficients*, etc. To avoid the geometric complexities of the involved argument, it is assumed that all functions a_i's defined over some suitable domain are sufficiently smooth. A more critical role is played by coefficients appearing in the *principle part* of a differential equation.

Linearity Type of a Differential Equation

Other than the *order* of a differential equation, as defined above, the *linearity type* of the function F with respect to the variables u, p_i, q, r_{ij}, ... provides a first stage classification of equations of the form (5.1.1), determines the complexity level of the geometry involved with any related problem, and hence also of the *solution method* to be applied in a particular situation.

Definition 5.2 A differential equation (5.1.1) is said to be a

1. *Linear equation* if the function F is linear with respect to variables u, p_i, q, r_{ij}, ..., whichever is present in the equation, and the coefficient functions depend on the independent variables $x_1, \ldots, x_n$ and t only;
2. *Semilinear equation* (or *almost linear*) if the principle part of the equation is linear, and the coefficients appearing in the principle part are functions of $(n + 1)$ independent variables $x_1, \ldots, x_n$, t only;
3. *Quasilinear equation* if the principle part of the equation is linear, the coefficients appearing in the principle part are functions of $(n + 1)$ independent variables and also of the dependent variable $u(x, t)$, and at least one coefficient function appearing in lower order terms depends on $(n + 1)$ independent variables and also on some lower order derivative of the function u.

A differential equation for a function $u = u(x, t)$ is called *fully nonlinear* if it is not a quasilinear equation.

As in the case of an ordinary differential equation, the linear differential equations are simplest to deal with. Notice that, however, though each linear ordinary differential equation admits a global solutions, but the same is not true for latter types of differential equations. A general kth order linear differential equation for a function $u = u(x, t)$ can be written in operator notation as given below. For $x = (x_1, \ldots, x_n)$ and an n-tuple $a = (a_1, \ldots, a_n)$ of non-negative integers, we write

$$x^a = x_1^{a_1} \cdots x_n^{a_n} \quad \text{and} \quad D^a = D_1^{a_1} \cdots D_n^{a_n},$$

where D_j denote the partial differential operator $\partial/\partial x_j$, for $j = 1, \ldots, n$. We write $|a| = a_1 + \cdots + a_n$. Then the symbol

$$D^a = \frac{\partial^{|a|}}{\partial x_1^{a_1} \cdots \partial x_n^{a_n}}$$

is a *partial differential operator* of order $|a|$. Therefore, a general linear differential equation of order $k \geq 1$ can be written as

$$\sum_{|a| \leq k} \alpha^a(x) D^a u = f(x), \quad \text{for } x \in \Omega, \tag{5.1.3}$$

where the summation on the left is taken over all possible values of *index vector* a with $|a| \leq k$. By Theorem 3.3, linearity of $\mathscr{P}_1(D)$ is equivalent to *principle of superposition*. When $f \equiv 0$ over the domain Ω, we say (5.1.3) is a *homogeneous equation*. Otherwise, it is called a *nonhomogeneous equation*. Let $\mathscr{P}(x, D)$ denote the *linear partial differential operator* on the left of Eq. (5.1.3), and we shall reserve the notation $\mathscr{P}(D)$ for case when all the coefficients $\alpha^a(x)$ are constants.

Some important partial differential operators such as the *Laplacian, wave*, and the *heat operators* in three variables are respectively given by

$$D_1^2 + D_2^2 + D_3^2, \quad D_1^2 + D_2^2 - D_3^2, \quad \text{and} \quad D_1^2 + D_2^2 - D_3,$$

which are examples of linear differential equations with *constant coefficients*, where $x_1 = x$, $x_2 = y$ are spatial variables, and $x_3 = t$ is a time variable. Further, the *biharmonic operator* given by

$$\mathscr{P}(D) = D_1^4 + 2D_1^2 D_2^2 + D_2^4 = \left(D_1^2 + D_2^2\right)^2,$$

is used to study practical problems related to *elasticity*.

It is an important fact that, for a general kth order linear differential equation, the geometry of a solution of each related problem depends on the *principle part* of the equation only, which we may write as $P_k(x, D)$. For example, as shown in the next section, the canonical forms such as *hyperbolic, parabolic* or *elliptical type* of a second order linear differential equations in two variables depend on the coefficients appearing in $P_2(x, D)$ only.

Example 5.1 The three differential equations for a function $u = u(x, t)$ given by

$$x u_x - t u_t - u = 0; \tag{5.1.4}$$

$$\sqrt{1 - x^2}\, u_x + t u_t - 5u = 0, \quad \text{with } x^2 < 1; \tag{5.1.5}$$

$$u u_x + u_t - 2 = 0, \quad \text{with } u(x, x) = x, \, x \neq 1, \tag{5.1.6}$$

may be written in operator form as

$$\mathscr{P}_1(D_x, D_t)u = u, \quad \mathscr{P}_1(D_x, D_t)u = 5u, \quad \text{and} \quad \mathscr{P}_1(D_x, D_t)u = 2,$$

where the differential operator $\mathscr{P}_1(D_x, D_t)$ are respectively given by

$$\mathscr{P}_1(D_x, D_t) \equiv x \, D_x - t \, D_t;$$
$$\mathscr{P}_1(D_x, D_t) \equiv \sqrt{1 - x^2} \, D_x + t \, D_t;$$
$$\mathscr{P}_1(D_x, D_t) \equiv u \, D_x + D_t.$$

In Chap. 6, we shall discuss a method of finding the general solution of linear differential equations in two variables, with constant coefficients. Notice that, for an kth order nonhomogeneous linear equation of the form

$$\mathscr{P}_k(x, D) \, u = g(x), \quad \text{with} \quad x \in \Omega \subset \mathbb{R}^n,$$

if the function $g = g(x)$ is given by

$$g(x) = c_1 g_1(x) + \cdots + c_m g_m(x),$$

where $g_1, \ldots, g_m \in C(\Omega)$ and $c_1, \ldots, c_m \in \mathbb{R}$ are arbitrary, the modified *principle of superposition* as stated in the next theorem says that it suffices to solve the associated simpler problems. A proof follows directly from Theorem 3.3. It also proves to be an effective tool while solving some nonlinear problems, by using an appropriate *linearisation technique*.

Theorem 5.1 *Suppose the n functions* $u_1, \ldots, u_m \in C^k(\Omega)$ *satisfy an kth order linear differential equation* $\mathscr{P}_k(x, D) \, u_i = g_i(x)$ *such that* $F(u_i) = f_i$ *over the boundary* $\Gamma = \partial \Omega$, *for* $i = 1, \ldots, m$. *Then, for* $u = c_1 u_1 + \cdots + c_m u_m$, *we have*

$$\mathscr{P}_k(x, D) \, u = c_1 g_1 + \cdots + c_m g_m,$$

and $F(u) = c_1 f_1 + \cdots + c_m f_m$.

As said earlier, the discussion in this book is restricted to differential equations of order $k \leq 2$ in one or two spatial variables. The *order* and *linearity type* of a differential equation are two important notions, providing an elementary level classification of such types of equations into families with similar properties. A large number of practical problems related to completely different phenomena are modelled by the first and second order differential equations of simpler linearity type. The theory of first order differential equations in n variables is largely well established. We also have some advanced mathematical tools to deal with certain important problems related to second order differential equations.

First Order Differential Equations

Let $\Omega \subseteq \mathbb{R}^n$ be an open set. By Definition 5.1, a first order differential equation for a function $u = u(x) \in C^1(\Omega)$ is implicitly defined equation given by

$$F(x; u; p_1, \ldots, p_n) = 0, \quad \text{with} \quad p_i = \frac{\partial u}{\partial x_i}, \tag{5.1.7}$$

where F is a sufficiently smooth function of variables x, u, $p_1, \ldots, p_n$, with

$$\left(\frac{\partial F}{\partial p_1}\right)^2 + \cdots + \left(\frac{\partial F}{\partial p_n}\right)^2 \not\equiv 0, \qquad \text{over} \quad \Omega \times \mathbb{R}^{n+1}.$$

For $n = 2$, we have $\Omega \subseteq \mathbb{R}^2$ and $u = u(x, y)$. Let F be a sufficiently smooth function of variables x, y, u, p, q, with $F_p^2 + F_q^2 \not\equiv 0$ over Ω, such that we have

$$F(x, y; u; p, q) = 0, \qquad \text{for} \quad p = u_x \text{ and } q = u_y. \tag{5.1.8}$$

In this case, we say Eq. (5.1.8) is a first order differential equation in two variables. As defined earlier, a first order differential equation for a C^1-function $u = u(x, y)$ given by

$$a(x, y)\, p + b(x, y)\, q + c(x, y)\, u + d(x, y) = 0, \tag{5.1.9}$$

is called a *linear equation*, where the coefficients a, b, c some C^1-function defined over Ω. When $d \equiv 0$ over Ω, it is called a *homogeneous equation*. Otherwise, we say (5.1.9) is a *nonhomogeneous equation*. Next, a first order *semilinear equation* for function u is given by

$$a(x, y)\, p + b(x, y)\, q = c(x, y; z), \tag{5.1.10}$$

and a *quasilinear equation* for function u is given by

$$a(x, y, z)\, u_x + b(x, y, z)\, u_y = c(x, y, z; u), \tag{5.1.11}$$

which is also known as a *Lagrange equation*. For example, the three first order differential equations given by (5.1.4)–(5.1.6) are respectively linear, semilinear, and a quasilinear equation.

First order differential equations of simpler *linearity types* arise naturally both in mathematics and in various practical applications. For example, as seen earlier in Chap. 3, *integrating factor* $\mu = \mu(x, y)$ of an ordinary differential equation of the form

$$M(x, y)\, dx + N(x, y)\, dy = 0,$$

is a solution of a first order linear differential equation given by (3.2.12). In practical terms, the *linear equations* derived below governs 1-dimension transport phenomena.

Example 5.2 (*Transport Equation*) Consider a noninteractive contaminant flowing in a fluid through a thin cylindrical pipe of variable cross-sectional area given by $a = a(x)$, for $x \in \mathbb{R}$. Suppose the *concentration* (mass density per unit length) of the contaminant be given by a C^1-function written as $\rho = \rho(x, t)$, and the continuous function $v = v(x, t)$ represents the speed of the flow, at point $x \in \mathbb{R}$ and time $t \geq 0$. Therefore, for $x_1, x_2 \in \mathbb{R}$ and $t_1, t_2 \in [0, \infty)$, we have

$$\int_{x_1}^{x_2} \rho(x, t_2) \, A \, dx = \text{contaminant mass in } [x_1, x_2] \text{ at time } t_2;$$

$$\int_{x_1}^{x_2} \rho(x, t_1) \, A \, dx = \text{contaminant mass in } [x_1, x_2] \text{ at the time } t_1;$$

$$\int_{t_1}^{t_2} \rho(x_1, t) \, v \, A \, dt = \text{contaminant mass passing the point } x_1 \text{ during } [t_1, t_2];$$

$$\int_{t_1}^{t_2} \rho(x_2, t) \, v \, A \, dt = \text{contaminant mass passing the point } x_2 \text{ during } [t_1, t_2].$$

Hence, in the absence of a *source* or a *sink*, the the function $\rho = \rho(x, t)$ satisfies the *balance equation* given by

$$\int_{x_1}^{x_2} \rho(x, t_2) \, A \, dx = \int_{x_1}^{x_2} \rho(x, t_1) \, A \, dx + \int_{t_1}^{t_2} \rho(x_1, t) \, v \, A \, dt - \int_{t_1}^{t_2} \rho(x_2, t) \, v \, A \, dt.$$

By the Fundamental Theorem of Calculus, we also have

$$\int_{x_1}^{x_2} \rho(x, t_2) \, A \, dx - \int_{x_1}^{x_2} \rho(x, t_1) \, A \, dx = \int_{x_1}^{x_2} \int_{t_1}^{t_2} \partial_t (\rho \, A) \, dt \, dx$$

$$\int_{t_1}^{t_2} \rho(x_1, t) \, v \, A \, dt - \int_{t_1}^{t_2} \rho(x_2, t) \, v \, A \, dt = - \int_{t_1}^{t_2} \int_{x_1}^{x_2} \partial_x (\rho \, v \, A) \, dx \, dt.$$

It thus follows from the above balance equation that

$$\int_{x_1}^{x_2} \int_{t_1}^{t_2} \partial_t (\rho(x, t) A) \, dt \, dx + \int_{t_1}^{t} \int_{x_1}^{x_2} \partial_x (\rho(x, t) \, v \, A) \, dx \, dt = 0.$$

Now, since x_1, $x_2 \in \mathbb{R}$ and t_1, $t_2 \in [0, \infty)$ are arbitrary, we find that the concentration function $\rho = \rho(x, t)$ satisfies the differential equation given by

$$A \, \rho_t + \partial_x (\rho \, v \, A) = 0, \qquad \text{for } (x, t) \in \Omega \subset \mathbb{R} \times \mathbb{R}^+, \tag{5.1.12}$$

which is a 1-dimensional version of *continuity equation* (4.3.6). In particular, when $a(x)$ is constant for all x, we obtain the equation

$$\rho_t + \partial_x\big(\rho(x,t)v(x)\big) = 0, \qquad \text{for} \ \ (x,t) \in \Omega \subset \mathbb{R} \times \mathbb{R}^+. \tag{5.1.13}$$

Moreover, when the speed $v(x,t) = c$ is a constant for all (x,t), we obtain

$$\rho_t + c\,\rho_x = 0, \qquad \text{for} \ \ -\infty < x < \infty, \ \ \text{and} \ \ t \ge 0, \tag{5.1.14}$$

which is known as the 1-dimensional *transport equation* for constant speed. It is used to study the *concentration* of a material in a fluid flowing with a constant velocity. The transport equation (5.1.13) is homogeneous because fluid is incompressible, and no *diffusion term* appears as the contaminant is assumed to be noninteractive.

In abstract terms, a first order linear equation in n-variables is studied in terms of the linear operator $\mathscr{P}_1(D)$ it defines on some Banach space V of functions, where the norm of a function in V is defined as a (modified) *supremum norm*.

Definition 5.3 With terminology as above, a differential operator $\mathscr{P} = \mathscr{P}(D) : V \to V$ is said to be *linear* if, for any $u_1, u_2 \in V$ and $c_1, c_2 \in \mathbb{R}$, we have

$$\mathscr{P}(c_1 u_1 + c_2 u_2) = c_1 \mathscr{P}(u_1) + c_2 \mathscr{P}(u_2).$$

For example, on the open set $U \subset \mathbb{R}^2 \setminus \{x = 0\}$, the operator $\mathscr{P}_1(D_x, D_t)$ defined on the Banach space $V = C^1(U)$ by

$$\mathscr{P}_1(D_x, D_t) \equiv D_x^2 - 4x^2\,D_t^2 - \frac{1}{x}\,D_x,$$

is a linear differential operator.

Next, for $z = u(x,y) \in C^1(\Omega)$, consider an implicitly defined surface given by

$$G\big(\varphi(x,y,z), \phi(x,y,z)\big) = 0, \qquad \text{for an arbitrary} \ \ G \in C^1(\mathbb{R}^2), \tag{5.1.15}$$

where $\varphi, \phi \in C^1(\Omega \times \mathbb{R})$ are two *functionally independent* functions. Differentiating with respect to x and y, and using the relation $dG = G_x\,dx + G_y\,dy = 0$, we obtain

$$G_x = G_\varphi[\varphi_x + p\,\varphi_z] + G_\phi[\phi_x + p\,\phi_z] = 0;$$
$$G_y = G_\varphi[\varphi_y + q\,\varphi_z] + G_\phi[\phi_y + q\,\phi_z] = 0.$$

The above system of equations in unknowns G_φ, G_ϕ has a nontrivial solution if and only if we have

$$\det \begin{pmatrix} \varphi_x + p\,\varphi_z & \phi_x + p\,\phi_z \\ \varphi_y + q\,\varphi_z & \phi_y + q\,\phi_z \end{pmatrix} = 0. \tag{5.1.16}$$

Expanding the determinant in (5.1.16), we obtain the *quasilinear equation*

$$\frac{\partial(\varphi, \phi)}{\partial(y, z)} \, p + \frac{\partial(\varphi, \phi)}{\partial(z, x)} \, q = \frac{\partial(\varphi, \phi)}{\partial(x, y)}, \tag{5.1.17}$$

where the three *Jacobians* are respectively given by

$$\frac{\partial(\varphi, \phi)}{\partial(y, z)} = \det \begin{pmatrix} \varphi_y & \varphi_z \\ \phi_y & \phi_z \end{pmatrix}, \qquad \frac{\partial(\varphi, \phi)}{\partial(z, x)} = \det \begin{pmatrix} \varphi_z & \varphi_x \\ \phi_z & \phi_x \end{pmatrix}, \qquad \text{etc.}$$

On the practical side, the mass density of vehicles along a sufficiently long stretch of a road is a typical real-life situation that is modelled by a quasilinear equation.

Example 5.3 (*Traffic Equations*) It is assumed that vehicles are of the same type, say all are cars, and that the road is a single-lane free way, without a *source* (entry point) or a *sink* (exit point). We observe the traffic movement from a distance so that the *flow* may be seen as a continuous fluid. Therefore, the law of mas conservation applies in the present situation. For a fixed but arbitrary point x_0, let $N(x, t)$ denote the number of cars at time $t \geq 0$ between points $x_0 < x$, with $x - x_0 = \ell$. Suppose the *mass density* of vehicles over a length ℓ along the road is given by a function $\rho = \rho(x, t) \in C^1(\mathbb{R} \times \mathbb{R}^+)$. The function N is a continuous by our assumption so that the mass density $\rho(x, t)$ is obtained from $N(x, t)$ by *averaging*. That is, we have

$$N(x, t) = \int_{x_0}^{x} \rho(u, t) \, du.$$

Therefore, the following implication holds:

$$\overline{\rho}(x, t) = \frac{1}{\ell} \, N(x, t) \quad \Rightarrow \quad \rho(x, t) = \frac{1}{T} \int_{t-(T/2)}^{t+(T/2)} \overline{\rho}(x, \tau) \, d\tau,$$

where the road length is assumed to be large enough so that each interval of length ℓ contains sufficiently many cars, and T is small. If the velocity of cars at a point $(x, t) \in \mathbb{R} \times \mathbb{R}^+$ is written as $v(x, t)$, then we have

$$\int_{x_0}^{x} \partial_t[\rho(u, t)] \, du = \frac{d}{dt} \int_{x_0}^{x} \rho(u, t) \, du = \frac{dN}{dt}$$

$$= \text{flux at } x_0 - \text{flux at } x$$

$$= \rho(x_0, t) \, v(x_0, t) - \rho(x, t) \, v(x, t)$$

$$= -\int_{x_0}^{x} \partial_u[\rho(u, t)v(u, t)] \, du.$$

Since above equation holds for all $x \geq x_0$, it follows by *continuity argument* that

$$\rho_t(x, t) + \partial_x[\rho(x, t)v(x, t)] = 0, \qquad (5.1.18)$$

which is *Lighthill–Whitham model* for road traffic or simply *traffic equations*. As in previous chapter, we may write $q(x, t) = \rho v$ for the *flux density* at point $x \in \mathbb{R}$ and time $t \geq 0$. The two special cases of practical importance are as given below:

1. Suppose an individual driver changes speed according to the local density, i.e. by keeping a watch on the intensity of traffic behind and in front of him. For simplicity, we may take

$$v(\rho) = a(\rho_{max} - \rho),$$

where the parameters a and ρ_{max} are determined from the data fitting the above equation, and also considering the fact that $v_{max} = a\rho_{max}$ gives the *maximum speed* a car can attain (as per the permissible speed limit) when the road is open, i.e. when $\rho = 0$. Also, notice that ρ_{max} indicates the traffic jam so that $v = 0$. Therefore, by the substituting relation given by

$$q(\rho) = \rho v = a\rho(\rho_{max} - \rho)$$

into Eq. (5.1.18), we obtain

$$\rho_t + v_{max}\left(1 - \frac{2\rho}{\rho_{max}}\right)\rho_x = 0. \qquad (5.1.19)$$

The derivative of flux density q with respect to ρ is called the *density velocity*, denoted by $V(q)$, and is given by the relation

$$V(q) := \frac{dq}{d\rho} = a(\rho_{max} - 2\rho). \qquad (5.1.20)$$

The density velocity is smaller than the (average) car velocity, which is always positive. However, we have

$$V < 0 \quad \Leftrightarrow \quad \rho > \frac{\rho_{max}}{2} \quad \Rightarrow \quad \text{heavy traffic;}$$
$$V > 0 \quad \Leftrightarrow \quad \rho < \frac{\rho_{max}}{2} \quad \Rightarrow \quad \text{light traffic.}$$

Finally, using the *scaled variables* given by

$$u(x, t) = \frac{\rho}{\rho_{max}} \quad \text{and} \quad t' = v_{max}\, t,$$

we obtain a simpler model as given below to study road traffic:

$$u_t + (1 - 2u)\, u_x = 0, \qquad\qquad (5.1.21)$$

which is a first order quasilinear equation. For $(x, t) \in \mathbb{R} \times \mathbb{R}^+$, if $g(x, t)$ denotes the flux density per unit length and unit time at a source (or a sink) then a general *differential equation model* to study traffic movement on a long stretch of road is given by

$$u_t + (1 - 2u)u_x = g(x, t), \qquad \text{for } x \in \mathbb{R}, \text{ and } t \geq 0. \qquad (5.1.22)$$

2. If most drivers are assumed to change speed considering the *local density*, and also according to variations in spatial density ρ_x of cars ahead of him, we need to modify the previous model by using the following *constitutive relation* for the velocity $v(x, t)$:

$$v(x, t) = a(\rho_{max} - \rho) - b\, \rho_x, \quad \text{with } b > 0.$$

where the parameters a, b and ρ_{max} are determined from the data fitting the above equation. Using Eq. (5.1.18), it is easy to verify that the traffic equation in this case is given by

$$\rho_t + V(q)\rho_x = \frac{a}{2}\,\partial_{xx}\rho^2, \qquad\qquad (5.1.23)$$

where $V(q)$ is as in Eq. (5.1.20).

We list below some other important first order quasilinear equations that apply to study phenomena such as mentioned alongside.

1. The nonhomogeneous quasilinear equation given by (5.1.22) is known as the *traffic model*, where $u = u(x, t)$ is the mass density of the traffic at a point $x \in \mathbb{R}$ and time $t \geq 0$. The function $g = g(x, t) \in C(\mathbb{R} \times [0, \infty))$ represents the *flux density* at an entry point or at an exit point along the road (Example 5.3).
2. If the speed of a moving nonviscous fluid in a thin pipe of uniform dimension is written as $u = u(x, t)$, then the homogeneous quasilinear equation given by

$$u_t + u\, u_x = 0, \quad -\infty < x < \infty, \ t \geq 0, \qquad\qquad (5.1.24)$$

is called the *inviscid Burger equation*, which was introduced by Harry Bateman in 1915. Equation (5.1.24) can also be seen as the *transport equation* with variable speed given by a C^1-function $u = u(x, t)$.

In general, the geometric complexity of a *solution method* applied in a particular situation depends on the nature of the coefficient functions appearing in the *principle part* of an equation of the form (5.1.7). It is convenient to deal with the first order linear or a semilinear equations.

Remark 5.1 In Chap. 6, our main focus is to discuss methods of finding the general solution of a first order differential equation in two variables due to Lagrange, Charpit, and Jacobi. In Chap. 7, we use the method of *characteristic curves* due to Cauchy, Monge, Lagrange, and Hamilton to solve some Cauchy problems for first order differential equation in two variables. We also discuss some applications of such types of equations in these two chapters.

Second Order Differential Equations

A second order differential equation in two variables x and y is given by

$$F(x, y; u; p, q; u_{xx}, u_{xy}, u_{yy}) = 0, \quad \text{for} \quad z = u(x, y) \in C^2(\Omega), \quad (5.1.25)$$

where function F is sufficiently smooth with respect to all involved variables, and $F_p^2 + F_q^2 \neq 0$ over Ω. In particular, a second order *quasilinear equation* is given by

$$a u_{xx} + b u_{xy} + c u_{yy} + F_1(x, y; u; u_x, u_y) = 0, \quad (5.1.26)$$

where the coefficients a, b, c are functions of the independent variables x, y, and also of the dependent variable $z = u(x, y)$. As said earlier, (5.1.26) is a *semilinear equation* when functions a, b, c depend on variables x and y only. Also, a general second order *linear equation* for a function $u \in C^2(\Omega)$ is given by

$$a u_{xx} + b u_{xy} + c u_{yy} + d u_x + e u_y + f u + g = 0, \quad (5.1.27)$$

where the coefficients $a, \ldots, g$ are functions of the independent variables x and y only. As in the case of a first order differential equation in two variables, we say (5.1.27) is a *homogeneous equation* if the $g \equiv 0$. Otherwise, it is called a *nonhomogeneous equation*.

The next two examples illustrate that the second order differential equations of simpler *linearity types* arise naturally in mathematics, and also in practical situations. The main idea is to eliminate all parameters from the given functional relation. For convenience, we may write the second order partial derivatives of a C^2-function $u = u(x, y)$ as

$$r = u_{xx} = \frac{\partial^2 u}{\partial x^2}, \quad s = u_{xy} = \frac{\partial^2 u}{\partial x \partial y}, \quad t = u_{yy} = \frac{\partial^2 u}{\partial y^2}.$$

Example 5.4 For $a, b, c \in \mathbb{R}$, consider the function given by

$$u(x, y) = a x^2 + 2b x y + c y^2.$$

Differentiating u twice with respect to x and y, we obtain

$$p = 2ax + 2by, \quad q = 2bx + 2cy$$
$$r = 2a, \quad s = 2b, \quad \text{and} \quad t = 2c.$$

Eliminating a, b, c from the above equations, we obtain the second order differential equation given by

$$x^2 u_{xx} + 2xy\, u_{xy} + y^2 u_{yy} = 2\,u. \tag{5.1.28}$$

Next, for any two C^2-functions f, g, consider the function given by

$$u(x, y) = f(x + ay) + g(x - ay).$$

Differentiating the function u twice with respect to x and y, we obtain

$$p = f'(x + ay) + g'(x - ay), \qquad q = af'(x + ay) - ag'(x - ay),$$
$$r = f''(x + ay) + g''(x - ay), \quad t = a^2 f'(x + ay) + a^2 g'(x - ay).$$

In this case, we obtain the second order differential equation given by

$$a^2 u_{xx} - u_{yy} = 0. \tag{5.1.29}$$

Further, for arbitrary C^2-functions f, g, consider the function given by

$$u(x, y) = f(x^2 + y) + g(x^2 - y).$$

Differentiating the function u twice with respect to x and y, we obtain

$$p = 2x[f'(x^2 + y) + g'(x^2 - y)], \qquad q = f'(x^2 + y) - g'(x^2 - y),$$
$$r = 4x^2[f''(x^2 + y) + g''(x^2 - y)] + 2[f'(x^2 + y) + g'(x^2 - y)],$$
$$t = f''(x^2 + y) + g''(x^2 - y).$$

In this case, we obtain the second order differential equation given by

$$x\, u_{xx} - 4x^3 u_{yy} - u_x = 0. \tag{5.1.30}$$

A significant number of problems concerning different types of phenomena in physics and engineering involve second order differential equations. For example, a C^2-function $u = u(x, y)$ defines a *minimal surface* in Monge's sense if and only if u is given by the second order quasilinear equation

$$\left[1 + u_y^2\right] u_{xx} - 2 u_x\, u_y u_{xy} + \left[1 + u_x^2\right] u_{yy} = 0.$$

In practical terms, we derive here 1-dimensional *diffusion equation*.

Example 5.5 (*Diffusion Equation*) Consider a closed cylindrical tube, say of fixed length ℓ, that is filled with a *noninteractive* fluid, and suppose we inject into the fluid a chemical substance, say an ink. The tube is taken to be sufficiently thin so that it can be assumed that the diffusion takes place along the x-axis. Suppose a C^2 function

$u = u(x, t)$ represents the *concentration* of ink at the point $x \in [0, \ell]$ and at time $t \geq 0$. At any fixed time $t \geq 0$, as the mass $M(t)$ of the ink in the part $[x_0, x] \subseteq [0, \ell]$ does not change without flowing-in or flowing-out of the ends of the tube, we have

$$M(t) = \int_{x_0}^{x} u(\xi, t) \, d\xi \quad \Rightarrow \quad M'(t) = \int_{x_0}^{x} u_t(\xi, t) \, d\xi.$$

By Fick's law, we know that the diffusion flux is proportional to the *concentration gradient* $u_x = \partial_x u$ so that we also have

$$M'(t) = k \left[u_x(x, t) - u_x(x_0, t) \right],$$

where $k = k(u, x)$ is the *diffusion coefficient*. Indeed, we can take k to be constant when it is independent of u. It follows from previous two equations that

$$\int_{x_0}^{x} u_t(\xi, t) \, d\xi = k \left[u_x(x, t) - u_x(x_0, t) \right].$$

Differentiating the above equation with respect to x, we obtain

$$u_t(x, t) = k \, u_{xx}(x, t), \quad \text{for } 0 \leq x \leq \ell \text{ and } t \geq 0, \tag{5.1.31}$$

which is the 1-dimensional *diffusion equation*. More generally, let a C^2-function $u = u(x, t)$ represents the mass density of a diffusing substance such as the heat in a thin bar, and suppose the flux $f = f(x, t)$ is a smooth function. As shown in Sect. 4.3, the function u satisfies the nonlinear diffusion equation given by

$$u_t + \frac{\partial}{\partial x} [f(u)] = 0. \tag{5.1.32}$$

The above equation also models heat transfer in a medium where the thermal conductivity depends on the temperature distribution. In particular, when $f'(u) = \alpha(u)$, we obtain a homogeneous quasilinear equation of the form

$$u_t + \alpha(u) \, u_x = 0, \tag{5.1.33}$$

which is known as the (inviscid) *Burgers' equation*, named after *Johannes Burgers* (1895–1981). The above model was first introduced in 1915 by *Harry Bateman* (1882–1946). Equation (5.1.33) also models several other phenomena such as related to the fluid dynamics, population genetics, neurology, etc.

We list below some other important second order differential equations[2] that model various types of phenomena such as mentioned alongside.

1. The *Ronald Fisher equation* given by

$$u_t - u_{xx} = r\,u\,(1 - u) \tag{5.1.34}$$

models *reaction-diffusion process*, which was initially formulated to study the spatial distribution of an advantageous genes represented by a C^2-function u, and explore the related travelling wave solutions. The above quasilinear equation also models phenomena related to wave propagation, chemical kinetics, nuclear reactors, combustion, etc. Notice the resemblance to the 1-dimensional *logistic equation* discussed earlier in previous chapter. Unlike the transport equation, with constant speed, the determination of speeds of *travelling waves* is part of the problem for Eq. (5.1.34).

2. In 1948, Burgers introduced the quasilinear equation given by

$$u_t + u\,u_x = \nu\,u_{xx}, \quad \nu \text{ being the kinematic viscosity,} \tag{5.1.35}$$

which is known as *viscous Burgers' equation*, and is a prototype conservation equation to model many different types of nonlinear phenomena involving *shock waves*. The above equation is also seen as the heat equation with a nonlinear term uu_x, which occur due to *convection*. It also models several other natural processes such as related to *turbulence, nonlinear acoustic, traffic flow*, etc. By using *Cole-Hopf transformation* given by

$$u(x, t) = -2\nu\,(v_x/v), \tag{5.1.36}$$

Equation (5.1.35) reduces to standard nonhomogeneous heat equation.

3. A model for a piece of telegraph wire as an electric circuit that has a register of resistance $R\,dx$ and a coil of inductance $L\,dx$ is linear equation given by

$$u_{tt} + (a + b)\,u_t + ab\,u = c^2 u_{xx}, \tag{5.1.37}$$

where $u(x, t)$ represents the voltage at position x and time t. It is known as the *Telegraph equation*, initially formulated in 1876 by *Oliver Heaviside* (1850–1925) as part of the *transmission line model*, where constant coefficients are given by

$$a = \frac{G}{C}, \quad b = \frac{R}{L}, \quad \text{and} \quad c^2 = \frac{1}{LC}.$$

It is assumed that the current $i(x, t)$ through the wire can escape to ground through a register of conductance $G\,dx$ or through a capacitor of capacitance $C\,dx$. The

---

[2] A large number of practical problems in physics and engineering can be solved as *initial-boundary value problems* related to such types of differential equations.

term $(a + b)\, u_t$ is called the *dissipation term* and $ab\, u$ the *dispersion term*. Notice that (5.1.37) reduces to standard wave equation (5.1.38) when both these terms are zero.

4. If a C^2-function $y = y(x, t)$ represents the transverse deflection of a vibrating string at a position $x \in [0, \ell]$ and at a time $t > 0$, then we know that y satisfies the 1-dimensional *wave equation* given by

$$y_{tt} = c^2 y_{xx}. \tag{5.1.38}$$

5. If a C^2-function $u = u(x, y)$ is a time-harmonic wave function in a free space with a localised source $f(x, y)$, then $u = u(x, y)$ satisfies the *Helmholtz equation* following 2-dimensional:

$$u_{xx} + u_{yy} + k^2 u = -f(x, y), \tag{5.1.39}$$

where $k > 0$ is called the *wave number*.

6. If a C^2-function $u = u(x, y, t)$ represents the temperature distribution of an infinite laterally insulated rectangular plate, then we know that u satisfies the 2-dimensional *heat equation* given by

$$u_t = c^2 [u_{xx} + u_{yy}]. \tag{5.1.40}$$

7. If a C^2-function $u = u(x, t)$ represents the temperature distribution of a bar that loses some heat due to radiation, then u satisfies the 1-dimensional linear *damped heat equation* given by

$$u_t = a^2 [u_{xx} - \lambda u], \quad \text{for } \lambda > 0. \tag{5.1.41}$$

8. If a C^2-function $u = u(x, y)$ represents the equilibrium temperature distribution of a laterally insulated rectangular plate at a position $(x, y) \in \mathbb{R}^2$, then we know that u satisfies the 2-dimensional homogeneous linear *Laplace equation* given by

$$u_{xx} + u_{yy} = 0. \tag{5.1.42}$$

9. If a C^2-function $u = u(x, y)$ represents the equilibrium temperature distribution of a body receiving thermal energy as given by $f(x, y)$, then we know that u satisfies the 2-dimensional nonhomogeneous linear *Poisson equation* given by

$$u_{xx} + u_{yy} = f(x, y). \tag{5.1.43}$$

5.2 Classification and Canonical Forms

Let $\Omega \subseteq \mathbb{R}^2$ be an open set, and consider the general second order linear differential equation for a function $u \in C^2(\Omega)$ given by

$$a\,u_{xx} + 2b\,u_{xy} + c\,u_{yy} + F_1(x, y; u; p, q) = 0, \qquad \text{for } z = u(x, y), \quad (5.2.1)$$

where the coefficients $a, b, c \in C^2(\Omega)$ are such that the condition $a^2 + b^2 + c^2 \neq 0$ holds over Ω. In this section, our main concern is the *principle part* given by

$$a\,u_{xx} + 2b\,u_{xy} + c\,u_{yy}, \qquad\qquad\qquad\qquad (5.2.2)$$

because only it participates in the classification procedure described below. When the coefficients a, b, c are constants, the geometry type of Eq. (5.2.1) remains uniform over a domain Ω. However, in the general case, the equation may be of different types across various regions of Ω. We will study Eq. (5.2.1) over a domain $\Omega_1 \subseteq \Omega$ such that the *discriminant* given by

$$D := b^2 - a\,c \qquad\qquad\qquad\qquad (5.2.3)$$

has the same sign at each point of Ω_1. We show that, for $(x_0, y_0) \in \Omega_1$, there exists a neighbourhood U_0 of the point (x_0, y_0) and sufficiently smooth functions φ, ϕ such that the transformation $(x, y) \mapsto (\xi, \eta)$ given by

$$\xi = \varphi(x, y) \qquad \text{and} \qquad \eta = \phi(x, y),$$

changes Eq. (5.2.1) to a differential equation that has one of the three geometry types[3] such as given below:

1. A *hyperbolic type* such as the wave equation (5.1.38).
2. A *parabolic type* such as the heat equation (5.1.40).
3. An *elliptic type* such as the Laplace equation (5.1.42).

The terminology used above are adopted from the theory of second degree curves in two variables. Notice that, since a transformation of coordinates is meaningful only when it is sufficiently smooth and nonsingular, we may assume $\varphi, \phi \in C^2(\Omega_0)$, and also that the Jacobian given by

$$J[\varphi, \phi; x, y] = \frac{\partial(\varphi, \phi)}{\partial(x, y)} = \det \begin{pmatrix} \varphi_x & \varphi_y \\ \phi_x & \phi_y \end{pmatrix} \qquad (5.2.4)$$

---

[3] The argument given in this section can be modified suitably using tools from linear algebra to extended the procedure for a second order linear differential equations in n variables. However, canonical forms obtained in the general case have more than three types such as *ultrahyperbolic, elliptically parabolic, hyperbolically parabolic,* etc. [3].

is nonzero over a neighbourhood V_0 of point (x_0, y_0) contained in U_0. In this case, we also say that the two functions φ, ϕ are *functionally independent* over V_0.

The transformed differential equation in variables ξ and η is called the *canonical form* (or the *normal form*) of the differential equation (5.2.1). Each type of equation has some specific geometric properties, and hence requires a separate analytical treatment while solving a related initial-boundary value problem. Now, to find the canonical form of a differential equation (5.2.1), differentiate twice the function

$$u(x, y) = v(\xi, \eta) = v(\varphi(x, y), \phi(x, y))$$

so that by the chain rule we obtain the five relations given by

$$u_x = v_\xi \, \xi_x + v_\eta \, \eta_x,$$
$$u_y = v_\xi \, \xi_y + v_\eta \, \eta_y;$$
$$u_{xx} = v_{\xi\xi}\xi_x^2 + 2v_{\xi\eta}\xi_x\eta_x + v_\xi \, \xi_{xx} + u_{\eta\eta}\eta_x^2 + v_\eta \, \eta_{xx},$$
$$u_{xy} = v_\xi \, \xi_{xy} + v_{\xi\xi}\xi_x\xi_y + [\xi_x\eta_y + \eta_x\xi_y]v_{\xi\eta} + v_{\eta\eta}\eta_x\eta_y + v_\eta \, \eta_{xy},$$
$$u_{yy} = v_\xi \, \xi_{yy} + v_{\xi\xi}\xi_y^2 + 2v_{\xi\eta}\xi_y\eta_y + v_{\eta\eta}\eta_y^2 + v_\eta \, \eta_{yy}.$$

Substituting the above relations in Eq. (5.2.1), and rearranging the terms, the new differential equation in variables ξ and η is given by

$$A\,v_{\xi\xi} + 2B\,v_{\xi\eta} + C\,v_{\eta\eta} + G(\xi, \eta; v; v_\xi, v_\eta) = 0, \tag{5.2.5}$$

where G is a first order differential equations in variables ξ and η, and the coefficients A, B, C are given by the relations

$$A = a\,\xi_x^2 + 2b\,\xi_x\xi_y + c\,\xi_y^2; \tag{5.2.6}$$
$$B = a\,\xi_x\eta_x + b[\xi_x\eta_y + \xi_y\eta_x] + c\xi_y\eta_y; \tag{5.2.7}$$
$$C = a\,\eta_x^2 + 2b\,\eta_x\eta_y + c\,\eta_y^2, \tag{5.2.8}$$

which may be written in matrix form as

$$\begin{pmatrix} A \\ B \\ C \end{pmatrix} = \begin{pmatrix} \xi_x & \eta_x \\ \xi_y & \eta_y \end{pmatrix} \begin{pmatrix} a & b \\ b & c \end{pmatrix} \begin{pmatrix} \xi_x & \xi_y \\ \eta_x & \eta_y \end{pmatrix} \tag{5.2.9}$$

Comparing the principle parts of the differential equations (5.2.1) and (5.2.5), it follows that

$$(B^2 - AC) = (b^2 - a\,c)\,J^2,$$

so that the sign of the respective *discriminants* $D := b^2 - ac$ and $D' := B^2 - AC$ remains the same, because $J \neq 0$. Said differently, the differential equation (5.2.1) is invariant with respect to any smooth coordinates transformation satisfying the *nondegeneracy* condition (5.2.4). Therefore, the value of the discriminant $D := b^2 - ac$ is an *intrinsic property* of Eq. (5.2.1). More precisely, we have

$$D > 0 \text{ over } U_0 \implies \text{Eq. (5.2.1) is of hyperbolic type;}$$
$$D = 0 \text{ over } U_0 \implies \text{Eq. (5.2.1) is of parabolic type;}$$
$$D < 0 \text{ over } U_0 \implies \text{Eq. (5.2.1) is of elliptic type.}$$

Remark 5.2 As a special case, it follows from the above discussion that if we start with the first order linear differential equation for a function $u \in C^1(\Omega)$ given by

$$a(x, y) u_x + b(x, y) u_y + c(x, y) u + d(x, y) = 0, \quad \text{for } z = u(x, y), \tag{5.2.10}$$

where the coefficients a, b, c, d are defined over Ω, with $a, b \in C^1(\Omega)$ such that $a^2 + b^2 \not\equiv 0$ over Ω, then in some neighbourhood Ω_0 of (x_0, y_0) Eq. (5.2.10) can be transformed to the differential equation given by

$$v_\xi + \alpha(\xi, \eta) v + \beta(\xi, \eta) = 0, \tag{5.2.11}$$

where coefficients α, $\beta \in C^1(\Omega_0)$, and η is a *parameter*. That is, we obtain a 1-parameter family of ordinary differential equations in the independent variable ξ. We call Eq. (5.2.11) the *canonical form* of Eq. (5.2.10). In Sect. 6.1, we use the above argument to find the general solution of a semilinear equation.

We consider now Eq. (5.2.5) to conclude our discussion. The main idea is to find functionally independent functions φ, $\phi \in C^2(U_0)$ such that at least one coefficient among A, B, C is zero. Notice that, for any solution $\varphi \in C^1(\Omega)$ of the equation

$$a \varphi_x^2 + 2b \varphi_x \varphi_y + c \varphi_y^2 = 0, \quad \text{for } z = \varphi(x, y), \tag{5.2.12}$$

we have $A = 0$ if $\xi = \varphi(x, y)$. Therefore, the problem of finding suitable variables ξ and η boils down to solving the above equation.

Lemma 5.1 *A function $\varphi \in C^1(\Omega)$ is a solution of Eq. (5.2.12) if and only if $\varphi(x, y) = c_1$ is the integral curve of the ordinary differential equation given by*

$$a \left(\frac{dy}{dx}\right)^2 - 2b \left(\frac{dy}{dx}\right) + c = 0. \tag{5.2.13}$$

Further, if a second integral curve $\phi(x, y) = c_2$ of Eq. (5.2.13) is independent of $\varphi(x, y) = c_1$, then it follows by taking $\eta = \phi(x, y)$ that the coefficient $C = 0$.

Proof Left for the reader as a simple exercise.   □

Lemma 5.1 can also be proved by resolving the left side of (5.2.12) as a product of two linear differential operators. We can also write Eq. (5.2.13) as two first order ordinary differential equations given by

$$\frac{dy}{dx} = \frac{b + \sqrt{b^2 - ac}}{a};$$ (5.2.14a)

$$\frac{dy}{dx} = \frac{b - \sqrt{b^2 - ac}}{a},$$ (5.2.14b)

which are called the *characteristic equation* of Eq. (5.2.1), and the associated solutions as the *characteristic curves*. Therefore, we may identify the principle part (5.2.2) of Eq. (5.2.1) with the quadratic equation given by

$$a\lambda^2 + 2b\lambda + c = 0,$$ (5.2.15)

which has the *discriminant* given by $D := 4(b^2 - ac)$. Hence, an appropriate transformation $(x, y) \mapsto (\xi, \eta)$ is obtained as solutions of the first order ordinary differential equations given by

$$\frac{\varphi_x}{\varphi_y} = \frac{-b - \sqrt{b^2 - ac}}{a};$$ (5.2.16a)

$$\frac{\phi_x}{\phi_y} = \frac{-b + \sqrt{b^2 - ac}}{a}.$$ (5.2.16b)

In the three separate cases that follow, we show that the characteristic curves for a hyperbolic type equation are real and distinct; for a elliptic type equation they are complex conjugate; and, for a parabolic type equation, the characteristic curves are real and coincident. As said earlier, the conclusion in each case is based on the value of the *discernment* $D = b^2 - ac$ at a point $(x, y) \in \Omega_1$:

1. **Case-I**: Suppose $D = b^2 - ac > 0$. Notice that, by changing variables (x, y) to new variable (x', y') such that $x' = x + y$ and $y' = x - y$, it can be assumed that a, c do not vanish simultaneously at (x_0, y_0). We may also assume that the coefficient $a \neq 0$ at (x_0, y_0). Then the quadratic equation (5.2.15) has two real distinct roots, say λ_1 and λ_2. We may thus choose new variables[4] $\xi = \varphi(x, y)$ and $\eta = \phi(x, y)$ such that λ_1 and λ_2 are respectively the characteristic curves of ordinary differential equations (5.2.16). It then follow from equations in (5.2.6) and (5.2.16) that the coefficient $A = 0$ (Lemma 5.1). Further, by putting together equations in (5.2.8) and (5.2.16), we obtain $C = 0$. Therefore, by using the relations given by

$$\lambda_1 \lambda_2 = \frac{c}{a} \quad \text{and} \quad \lambda_1 + \lambda_2 = -2\frac{b}{a},$$

it follows from Eq. (5.2.7) that

$$B = \frac{2}{a}(ac - b^2)\varphi_y\phi_y \quad \Rightarrow \quad B \neq 0.$$

---

[4] Recall that a sufficient condition is given by the condition (5.2.4).

From the equations $\varphi_x = \lambda_1 \varphi_y$ and $\phi_x = \lambda_2 \phi_y$, we also have that the Jacobian $J = \varphi_x \phi_y - \varphi_y \phi_x \neq 0$, provided both φ_y and ϕ_y are nonzero in some neighbourhood of point (x_0, y_0). Therefore, we need integral curves of the ordinary differential equations given by

$$\frac{dy}{dx} + \lambda_1(x, y) = 0 \quad \text{and} \quad \frac{dy}{dx} + \lambda_2(x, y) = 0, \qquad (5.2.17)$$

which we can find by assigning suitable initial conditions. Finally, dividing Eq. (5.2.5) by B, and using the 2-parameter family of characteristic curves obtained as above, it follows that the canonical form of Eq. (5.2.1) in this case is given by

$$v_{\xi\eta} + \text{ lower order terms} = 0.$$

That is, Eq. (5.2.1) is of *hyperbolic type*. We may write characteristic curves respectively given by the first order equations (5.2.17) as

$$f_1(x, y) = c_1 \quad \text{and} \quad f_2(x, y) = c_2, \qquad (5.2.18)$$

where c_1, c_2 are some constants.

Example 5.6 For the differential equation given by $u_{xx} - x^2 u_{yy} = 0$, as the roots of the associated quadratic equation $\lambda^2 - x^2 = 0$ are given by $\lambda_1(x, y) = x$ and $\lambda_2(x, y) = -x$, we may thus obtain the *characteristic curves* from the two ordinary differential equations given by

$$\frac{dy}{dx} = x \quad \text{and} \quad \frac{dy}{dx} = -x.$$

Therefore, taking new variables ξ and η as given by

$$\xi = \varphi(x, y) = y + \frac{1}{2}x^2;$$

$$\eta = \phi(x, y) = y - \frac{1}{2}x^2,$$

it follows that the given equation has the canonical form given by

$$v_{\xi\eta} = \frac{1}{4(\xi - \eta)}(v_\xi - v_\eta),$$

which is indeed a *hyperbolic type* equation.

2. **Case-II**: If $D = b^2 - ac = 0$, then the quadratic equation (5.2.15) has a double real root given by

$$\lambda(x, y) = \frac{\varphi_x}{\varphi_y} = -\frac{b}{a}.$$

It thus follows from Eq. (5.2.6) that the coefficient $A = 0$. Further, to have ϕ independent of φ, we may take $u(x, y) = \phi_x/\phi_y = \lambda(x, y)$ so that $C \neq 0$. Also, by using Eq. (5.2.7), we have

$$B = \left(-\frac{b^2}{a} + c\right) \varphi_y \phi_y \quad \Rightarrow \quad B = 0.$$

Therefore, by using the characteristic curves given by

$$\varphi(x, y) = \lambda(x, y) = \text{constant}$$

and any ϕ, independent of φ, the canonical form of Eq. (5.2.1) is given by

$$v_{\eta\eta} + \text{lower order terms} = 0.$$

Hence, in this case, the given differential equation is of *parabolic type*. As in the previous case, the characteristic curves of Eq. (5.2.1) are given by the integrals of the differential equation given by

$$\frac{dy}{dx} = -\lambda(x, y) = \frac{b}{a}.$$

In particular, it follows that a second order linear parabolic equation cannot be solved by the *method of characteristics*.

Example 5.7 For the equation $u_{xx} + 2u_{xy} + u_{yy} = 0$, since $\lambda = -1$ is a double root of the quadratic equation $\lambda^2 + 2\lambda + 1 = 0$ given by (5.2.15), it follows that a 1-parameter family of characteristic curves is given by the equation $dy/dx = 1$. Therefore, in this case, we may take $\xi = \varphi(x, y) = x - y$. Accordingly, we take $\eta = \phi(x, y) = x + y$, which is independent of φ. Hence, the canonical form of the given differential equation is given by the equation $v_{\eta\eta} = 0$.

3. **Case-III**: When $D = b^2 - ac < 0$, the two roots of the quadratic equation given by (5.2.15) are *complex conjugates* so that the associated functions are also complex conjugate, i.e. if $\xi = \varphi(x, y)$ then $\eta = \overline{\varphi}$. Therefore, to keep the transformation real, we may use the transformation of variables $(\xi, \eta) \to (\alpha, \beta)$ given by

$$\alpha = \frac{1}{2}[\varphi + \overline{\varphi}] \quad \text{and} \quad \beta = \frac{1}{2i}(\varphi - \overline{\varphi}).$$

Hence, in this case, the above transformation takes $u_{\xi\eta}$ to $w_{\alpha\alpha} + w_{\beta\beta}$, where we may write $u(\xi, \eta) = w(\alpha, \beta)$. It then follows that Eq. (5.2.1) reduces to the canonical form as given by

$$w_{\alpha\alpha} + w_{\beta\beta} + \text{lower order terms} = 0,$$

which is an equation of *elliptic type*. It also implies that the elliptic type equation has *no real characteristics*. In particular, it follows that second order linear elliptic equations cannot be solved by the *method of characteristics*.

Example 5.8 For the equation $u_{xx} + x^2 u_{yy} = 0$, complex conjugate roots of the quadratic equation $\lambda^2 + x^2 = 0$ given by (5.2.15) are $\lambda_1 = ix$ and $\lambda_2 = -ix$. Therefore, in this case, by taking

$$\xi = \varphi(x, y) = iy + \frac{1}{2}x^2 \quad \text{and} \quad \eta = \phi(x, y) = -iy + \frac{1}{2}x^2,$$

it follows that for the transformations given by $\alpha = x^2/2$ and $\beta = y$ the canonical form is given by

$$w_{\alpha\alpha} + w_{\beta\beta} - \frac{1}{2\alpha} u_\alpha = 0,$$

which is an elliptical type equation.

The next example shows that the three prototypical second order linear differential equations with constant coefficients are indeed of the geometry types as mentioned in the beginning of the section.

Example 5.9 The Laplace equation $u_{xx} + u_{yy} = 0$ is elliptic because $a = c = 1$ and $b = 0$ implies $b^2 - ac = -1 < 0$. The Poisson equation $\nabla^2 u = u_{xx} + u_{yy} = f(x, y)$ is elliptic for the same reason. The heat equation $u_{xx} = u_t$ is parabolic because $a = 1, b = c = 0$ implies that $b^2 - ac = 0$. Finally, the wave equation $\alpha^2 u_{xx} = u_{tt}$, with $\alpha > 0$, is hyperbolic because $a = \alpha^2, b = 0, c = -1$ implies that $b^2 - ac = \alpha^2 > 0$.

The next example is about second order differential equations with variable coefficients that have different geometry types in different regions of their respective domains of definition.

Example 5.10 (a) For the differential equation given by

$$u_{xx} - 4x^2 u_{yy} = \frac{1}{x} u_x,$$

we have $a = 1, b = 0, c = -4x^2$ so that $b^2 - ac = -16x^4 \leq 0$. Therefore, the equation is parabolic along the y-axis and it is elliptic at all other points of $\mathbb{R}^2$.
(b) For the *Tricomi's equation* given by $y u_{xx} + u_{yy} = 0$, we have $b^2 - ac = -y$ so that the equation is elliptic type over the upper-half plane $y > 0$; parabolic type along the x-axis; and, it is hyperbolic type over the lower-half plane $y < 0$.

(c) The differential equation given by

$$\frac{1}{1 - m^2} u_{xx} + u_{yy} = 0, \quad \text{for} \quad m^2 < 1,$$

is a model to study problems related to *disturbances in incompressible flow*. In this case, the discriminant is given by

$$b^2 - ac = -\frac{1}{1 - m^2}.$$

Therefore, the equation is elliptic type, for $m < 1$; and, hyperbolic type, for $m > 1$.

Remark 5.3 As we will see in Chap. 6, though the above technique extends to quasi-linear equation, but the conclusion about the three canonical forms do not hold. In fact, the characteristic equations of a quasilinear equation also involve the dependent variable $z = u(x, y)$. Fortunately, by applying Legendre's *Hodograph transformation*, we can transform a quasilinear equation to a linear equation by reversing the roles of the dependent and independent variables (see Exercise 5.16).

5.3 Classical Solution

Recall that $\Omega_1 = \Omega \times \mathbb{R}^+$, where $\Omega \subseteq \mathbb{R}^n$ is an open set. Suppose a differential equation given by (5.1.1) has order $k \geq 1$. In general, a function $u = u(x, t) \in C^k(\Omega_1)$ is called a *classical solution* (or simple a *solution*) if it satisfies the differential equation identically over Ω_1. An implicit solution of the form

$$\Psi(x; t; u; f_1, \ldots, f_k) = 0, \qquad (5.3.1)$$

is called the *general solution* of the differential equation, where $f_1, \ldots, f_k$ are arbitrary univariate functions. Also, a solution of the form

$$\Psi(x; t; u; a_1, \ldots, a_k) = 0, \qquad (5.3.2)$$

is called a *complete integral* of the differential equation, where $a_1, \ldots, a_k$ are arbitrary *independent* scalars. It is difficult to solve a general differential equation of the form (5.1.1). However, it is easy to verify whether a given explicit or implicit function is indeed a solution of a differential equation.

In particular, the *general solution* of Eq. (5.1.8) is given implicitly as

$$G = G(x, y; \varphi, \phi), \qquad \text{with } \varphi, \phi \in C^1(\mathbb{R}). \qquad (5.3.3)$$

For example, integration of equation $u_x = 0$ gives that $u(x, y) = f(y)$ as the general solution, where f is an arbitrary univariate C^1-function. Also, as we shall find shortly, implicitly defined function given by

$$u(x, y) = g(x^2 - y^2), \qquad \text{for some } g \in C^1(\mathbb{R}),$$

is the general solution of the equation $y\,u_x + x\,u_y = 0$. Also, the *complete integral* is given by a function of the form

$$G = G(x, y; a, b), \qquad \text{with} \quad a, b \in \mathbb{R}, \tag{5.3.4}$$

where the function G in each case is assumed to be sufficiently smooth, and the matrix given by

$$\begin{pmatrix} \varphi_a & \varphi_{xa} & \varphi_{ya} \\ \varphi_b & \varphi_{xb} & \varphi_{yb} \end{pmatrix} \tag{5.3.5}$$

has the rank two. Notice that the above rank condition ensures that scalars a and b are *functionally independent*. It is always possible to derive the general solution from a complete integral. For example, by using the relation $b = \phi(a)$ in (5.3.4), we can write the general solution as

$$G(x, y; u; a, \phi(a)) = 0,$$

which is a 1-parameter family of *integral surfaces* of Eq. (5.1.8). The explicit expression is obtained by eliminating the parameter a from the two associated equations. Therefore, in most cases, it suffices to focus only on finding a complete integral. The next example illustrates the procedure.

Example 5.11 Consider the linear equation given by $u_x + u_y = 1$, where $u = u(x, y) \in C^1(\Omega)$ for some open set $\Omega \subseteq \mathbb{R}^2$. As we shall see in the next section, the function $u(x, y) = ax + (1 - a)y + b$ is a complete integral of the equation. Therefore, by taking $b = \phi(a)$, we can write

$$u(x, y) = ax + (1 - a)y + \phi(a).$$

Differentiating the above equation with respect to a, we obtain

$$0 = x - y + \phi'(a).$$

Hence, by solving for a, we have $a = h(x - y)$, for some function h. Substituting this expression back into in the last equation, we obtain

$$u - y = a(x - y) + \phi(h(x - y)) = f(x - y),$$

which gives the general solution of the equation $u_x + u_y = 1$. Notice that we can also write the above solution implicitly as $F(u - y, x - y) = 0$, for some C^1-function F.

Example 5.12 Consider the first order linear differential equation $p = 2xq$. As we shall see in the next section, the function $u(x, y) = a(x^2 + y) + b$ is a complete integral of the equation. Taking $b = \phi(a)$, we have

$$u(x, y) = a(x^2 + y) + \phi(a).$$

Differentiating this equation with respect to a, we get $0 = x^2 + y + \phi'(a)$. So, solving for a, we get $a = h(x^2 + y)$. Substituting this in the last equation, we get $u(x, y) = a(x^2 + y) + \phi(h(x^2 + y)) = f(x^2 + y)$. This is the general integral of the equation $p = 2xq$. We can also write the above solution in implicit form as $F(x^2 + y, u) = 0$.

In general, when the function F has a simpler form, it is possible to find the general solution of Eq. (5.1.8) more convenient by using some integration methods.

1. Suppose $F(p, q) = 0$ is a linear equation. In this case, we may take $u(x, y) = ax + by + c$ as a trial solution so that

$$p = u_x = a \quad \text{and} \quad q = u_y = b$$

implies that the function $u(x, y) = ax + by + c$ is a complete integral, provided a and b are such that $F(a, b) = 0$. Further, by applying the *implicit function theorem*, a locally defined smooth function g can be found such that $F(a, b) = 0 \Rightarrow b = g(a)$. Therefore, the general solution of the equation $F(p, q) = 0$ is given by $u(x, y) = ax + g(a)y + c$. The next example illustrates the procedure.

Example 5.13 (i) Let $u(x, y) = ax + by + c$ be a complete integral of the nonlinear equation $p^2 + q = p$. Then $a^2 + b = a$ implies that $b = a - a^2$, and so $\varphi(x, y) = ax + a(1 - a)y + c$ is the complete integral. (ii) Let $u(x, y) = ax + by + c$ be the complete integral of the nonlinear equation $p - q = pq$. Then $a - b = ab$ implies that $b = a/(1 + a)$, and so $\varphi(x, y) = ax + (a/(1 + a))y + c$ is the complete integral.

2. Suppose $F(x, y; p, q) = px + qy + f(p, q)$, where $f \in C(\mathbb{R}^2)$. Such type of first order differential equation is also known as the *Clairaut's form* of linear equation. In this case, we may take $p = a$ and $q = b$ so that $u(x, y) = ax + by + f(a, b)$ can be written as $u(x, y) = px + qy + f(p, q)$. Therefore, the general solution of a Clairaut's equation is given by

$$u(x, y) = ax + by + f(a, b), \quad \text{with} \quad f \in C(\mathbb{R}^2).$$

For example, the function $u(x, y) = ax + by + a^2 - b^2$ is the general solution of the *Clairaut equation* given by $u(x, y) = px + qy + p^2 - q^2$.

3. An equation of the form (5.1.8) is called *variable separable* if, for some functions $\lambda, \mu \in C^1(\Omega)$, we have

$$F \equiv \lambda(x, p) - \mu(y, q),$$

In this case, we may take $\lambda(x, p) = a$ and $\mu(y, q) = a$. Solving the above equations for p and q, we obtain $p = \phi(x, a)$ and $q = \psi(y, a)$. Therefore, we have

$$du = p\,dx + q\,dy = \phi(x, a)dx + \psi(y, a)dy.$$

Hence, by integrating the above equation, the general solution is obtained as

$$u(x, y) = \int [\phi(x, a)\,dx + \psi(y, a)\,dy] + b,$$

The next example illustrates the procedure.

Example 5.14 (i) Clearly, the equation $p - x^2 = q + y^2$ is a variable separable form. Taking $p - x^2 = a$ and $q + y^2 = a$, it follows that a complete integral is given by

$$u(x, y) = \int p\,dx + q\,dy$$

$$= \int (a + x^2)dx + (a - y^2)dy$$

$$= a(x + y) + \frac{1}{3}(x^3 - y^3) + b.$$

(ii) The nonlinear equation $pq = xy$ can be changed to variable separable form by expressing it as $p/x = y/q$. So, we may take $p/x = a$ and $y/q = a$. Then,

$$u(x, y) = \int p\,dx + q\,dy$$

$$= \int ax\,dx + y/a\,dy$$

$$= ax^2/2 + y^2/(2a) + b/2.$$

Thus, a complete integral is given by $2z = ax^2 + y^2/a + b$.

Notice that the two equations of the form

$$f(x, p, q) = 0 \quad \text{and} \quad g(y, p, q) = 0$$

are special cases of the *variable separable* form. Solving for q, the equation $f(x, p, q) = 0$ can be written as $g(x, p) = q$. Therefore, by taking $q = a$ and $g(x, p) = a$, we can solve as

$$u(x, y) = \int \phi(x, a)dx + ady + b = \int \phi(x, y)dx + ay + b,$$

where $p = \phi(x, a)$ is obtained from the relation $g(x, p) = a$ by solving it for p. The next example illustrates the procedure.

Example 5.15 The equation $p = q \cos x$ is of the form $f(x, p, q) = 0$. Writing $q = a$ and solving for p, we get $p = a \cos x$ and hence a complete integral is given by

$$u(x, y) = \int p dx + q dy$$

$$= \int a \cos x dx + a dy = a \sin x + ay + b.$$

5. Suppose $F(u, p, q) = 0$. In this case, by using a relation of the form $u = u(\xi)$ with $\xi(x, y) = x + ay$, we can find the general solution by solving an associated ordinary differential equation. We obtain $p = u_x = u'(\xi)$ and $q = u_y = au'(\xi)$ so that the equation transforms to

$$F\left(u, \frac{du}{d\xi}, a\frac{du}{d\xi}\right) = 0.$$

Solving the above differential equation, we obtain the general solution as $u = u(\xi) = u(x + ay)$. The next example illustrates the procedure.

Example 5.16 (i) For the *nonlinear* equation $u = p^2 + q^2$, let $\xi(x, y) = x + ay$ and $u = u(\xi)$. Then, the given equation becomes

$$u = \left[\frac{du}{d\xi}\right]^2 (1 + a^2), \quad \text{i.e.,} \quad \sqrt{1 + a^2}\frac{du}{d\xi} = \sqrt{u}.$$

Solving this differential equation, we obtain a complete integral as

$$2\sqrt{1 + a^2}\sqrt{u} = \xi + b \quad \text{or} \quad 4(1 + a^2)u = (x + ay + b)^2.$$

(ii) For the equation $p(1 + q) = q u$, let $\xi(x, y) = x + ay$ and $u = u(\xi)$. Then, the given equation becomes

$$\frac{du}{d\xi}\left(1 + a\frac{du}{d\xi}\right) = az\frac{du}{d\xi}.$$

This equation is satisfied if $du/d\xi = 0$ or $a(du/d\xi) - az = -1$. The first equation gives $z = a$ and the second $\log(u - 1) = \xi + b = x + ay + b$. Solving the latter equation, it follows that the complete integral of the given equation is

$$2\sqrt{1 + a^2}\sqrt{u} = \xi + b \quad \text{i.e.,} \quad 4(1 + a^2)u = (x + ay + b)^2.$$

A powerful solution technique applicable to homogeneous linear differential equations, say in two independent variables x and y, works on the assumption that the equation has a *product solution* of the form $u(x, y) = f(x) g(y)$, where functions f and g satisfy appropriate smoothness conditions. It is known as the method of *sepa-*

ration of variables, which d'Alembert introduced in 1773 to give a series solution of the wave equation. In next few examples, we illustrate the procedure involved with the method.

Example 5.17 Suppose the first order homogeneous equation given by

$$u_x - u_y = 0, \quad \text{for} \quad u = u(x, y) \in C^1(\Omega),$$

has a solution of the form $u(x, y) = f(x)g(y)$, where f and g are some differentiable functions. By using the relations $u_x = f' g$ and $u_y = f g'$, it follows from the above equation that

$$f' g - f g' = 0 \quad \Rightarrow \quad f'(x) = k f(x) \text{ and } g'(y) = k g(y),$$

for some scalar k, which is usually called the *separation constant* of the equation. It thus follows that the given differential equation decomposes into two ordinary differential equations in unknown functions f and g. Solving the two ordinary differential equations, we obtain

$$u(x, y) = f(x)g(y) = \left(ae^{kx}\right)\left(be^{ky}\right) = ab\, e^{k(x+y)},$$

which is the general solution of the given homogeneous equation. The parameters a and b are determined by using auxiliary conditions, if any assigned.

The same method also applies to some equations with variable coefficients.

Example 5.18 Consider the differential equation for function $u \in C^1$ given by

$$y\, u_x + x\, u_y = 0, \quad \text{for} \quad z = u(x, y).$$

Suppose $u(x, y) = f(x)g(y)$ is a solution of the above equation. Substituting the relations $u_x = f' g$ and $u_y = f g'$ back into the equation, it follows that

$$y f' g + x f g' = 0 \quad \text{i.e.} \quad \frac{f'}{x f} = -\frac{g'}{y g}.$$

Therefore, for separation constant $k = 2b$, we may write $g'(y) = -2b\, g(y)$ so that $f'(x) = 2b\, f(x)$. Solving the ordinary differential equations given by

$$f' - 2bx\, f = 0 \quad \text{and} \quad g' + 2by\, g = 0,$$

it follows that

$$f(x) = A\, e^{bx^2} \quad \text{and} \quad g(y) = B\, e^{-by^2}.$$

Hence, in this case, the general solution of the given equation is obtained as

$$u(x, y) = f(x)g(y) = AB\, e^{b(x^2 - y^2)} = ae^{b(x^2 - y^2)}.$$

As said earlier, the parameters a and b are determined by using auxiliary conditions, if any assigned.

Fourier's work related to heat conduction in solids, leading to the fundamental concept of *Fourier series*, rendered *separation of variables* to be one of the most widely used solution methods. We will use the method in Chaps. 8 and 9 to solve some important problems related to certain fundamental second order linear differential equations, as described in the next section.

5.4 Initial-Boundary Value Problems

As said earlier in Chap. 3, a differential equation model of a univariate phenomenon involving a function $\varphi : I \to \mathbb{R}^n$ serves its purpose only when an *initial value problem* or a *boundary value problem* for associated dynamical system has a unique solution over some interval $J \subseteq I$. Recall that the values assigned to functions $\varphi, \varphi', \ldots$ at a single point $t_0 \in I$ are known as the initial values, whereas boundary conditions arise as various types of sum of values of $\varphi, \varphi', \ldots$, at the two end points of the interval I. In any particular situation, specification of above two types of *initial data*[5] for the involved unknown function is a part of the modelling process of any univariate phenomenon.

Let Γ denote the boundary of an open set $\Omega \subseteq \mathbb{R}^n$, and write $\Omega_1 = \Omega \times \mathbb{R}^+$. In the present case, if a multivariate phenomenon is modelled by a second order differential equation for a function $u = u(x, t) \in C^2(\Omega_1)$, a functional relation defined for $x \in \Omega$ of the form

$$u(x, 0) = \alpha(x), \quad u_t(x, 0) = \beta(x), \quad \text{and/or} \quad u_{x_i}(x, 0) = \alpha_i(x), \qquad (5.4.1)$$

is called an *initial condition*, where α, β, α_i are some nice function. On the other hand, a functional relation defined for $x \in \Gamma$ (and for all t) of the form

$$u(x, t) = \alpha(x), \quad \text{or} \quad u_{x_i}(x, t) = \alpha_i(x), \quad \text{or} \quad \partial_n u(x, t) = \beta(x), \qquad (5.4.2)$$

is called a *boundary condition*, where $\partial_n u$ denotes the *normal derivative*[6] of u over Ω. In the same way, we may define the other terminologies such as *homogeneous boundary condition, linear boundary condition*, etc. Therefore, once again, the main problem is to find a solution φ of a differential equation over a domain $D \subseteq \Omega_1$ that is *unique* subject to that it also satisfies assigned auxiliary data of the type as mentioned above.

---

[5] In practical terms, specification of *initial values* is related to the *state* of the system, and *boundary conditions* takes into consideration the physical environment of the problem.

[6] As said earlier in Chap. 2, $\partial_n u$ is defined at every point of the boundary Γ, except along infinitesimal common faces of constituent surfaces.

Definition 5.4 An *initial value problem* (or simply IVP) is about finding a function $\varphi = \varphi(x, t) \in C^2(D)$ that satisfies the differential equation, and also specified initial conditions such as (5.4.1). A *boundary value problem* (or simply a BVP) is about finding a function $\varphi = \varphi(x, t) \in C^2(D)$ that satisfies the differential equation, and also specified boundary conditions such as (5.4.2).

Notice that the above definition can be modified in straightforward way for differential equations of any order $k \geq 1$. As said earlier, in most situations, we deal with the differential equations of simpler *linearity types* only. *Cauchy problems* constitute an important class of initial value problems. For example, consider the quasilinear equation for the function $z = u(x, y)$ given by

$$a(x, y, u)u_x + b(x, y, u)u_y = c(x, y, u),$$

and let Γ be the regular curve given parametrically by

$$\Gamma: \quad x = x_0(s), \quad y = y_0(s), \quad z = z_0(s), \quad \text{for} \quad -\infty < s_1 \leq s \leq s_2 < \infty.$$

Suppose C is the orthogonal projection of Γ onto the xy-plane. The *Cauchy problem* amounts to finding a C^1-solution $u = u(x, y)$ such that $u(x_0(s), y_0(s)) = z_0(s)$, i.e. we need to find the surface S defined by $z = u(x, y)$ which contains the curve Γ. In Chaps. 7 and 11, we discuss some interesting solution method to deal with such types of problems for hyperbolic type of differential equations in two variables.

Recall that a scalar function $u \in C^2(\Omega)$ is said to be *harmonic over* Ω if it satisfies the *Laplace equation*, for all $x = (x_1, \ldots, x_n) \in \Omega$. That is,

$$\frac{\partial^2 u}{\partial x_1^2} + \cdots + \frac{\partial^2 u}{\partial x_n^2} = 0.$$

As seen earlier in Chap. 4, in practical terms, the above differential equation governs the equilibrium state of a phenomenon such as related to a study of temperature distribution given by the function $u(x, t)$. Therefore, a typical BVP for a Laplace equation is about finding a *harmonic function* $u \in C^2(\Omega)$ that equals some function specified across the boundary Γ of the open set Ω. The two types of boundary conditions given below arise more frequently, which are respectively known as the *Dirichlet condition* and *Neumann condition*:

$$\nabla^2 u = 0, \quad \text{with} \quad u = \varphi \quad \text{on} \quad \Gamma;$$
$$\nabla^2 u = 0, \quad \text{with} \quad \partial_n u = \phi \quad \text{on} \quad \Gamma,$$

where φ, ϕ are some nice functions defined on Γ, and $\partial_n u = \partial u / \partial n$ is the *normal derivative* to the graph surface of the function $u = u(x)$, which represents the direction of the *outward flux* across Γ. For $\Omega = [0, \ell]$, we have $\partial_n u(0) = -u_x(0)$ and $\partial_n u(\ell) = u_x(\ell)$. Similarly, for $\Omega = [0, a] \times [0, b]$, we have the following four cases

$$\partial_n u(0, y) = -u_x(0, y), \qquad \partial_n u(a, y) = u_x(a, y)$$
$$\partial_n u(x, 0) = -u_y(x, 0), \qquad \partial_n u(x, b) = u_y(x, b).$$

Notice that if more than one conditions hold, say at $(0, 0)$, then $\partial_n u(x, y)$ is not defined at the corner point. We usually assume that the system under investigation is without any source or sink.

Definition 5.5 A *Dirichlet Problem* is a BVP for a function $u = u(x) \in C^2(\Omega)$ that satisfies a Laplace equation, and also specified functional relations for u given on the boundary Γ. Similarly, a *Neumann Problem* is a BVP for a function $u = u(x) \in C^2(\Omega)$ that satisfies a Laplace equation, and also specified functional relations for the normal derivative $\partial_n u$ defined on the boundary Γ.

Example 5.19 Let a function $u = u(x, y) \in C^2(\Omega)$ represents the equilibrium temperature at point (x, y) on a rectangular plate $\Omega = [0, a] \times [0, b]$, with insulated lateral surfaces. We may taking $u(x, y)$ to be identically zero on the three edges of the plate, say along the lines $x = 0$, $x = a$, and $y = b$, so that we obtain three *homogeneous* boundary conditions. Further, suppose the function $u(x, y)$ on the fourth edge along the line $x = a$ is given by a sufficiently smooth function $f(x)$. Therefore, in the present situation, Dirichlet problem is about finding a *harmonic function* $u = u(x, y) \in C^2(\Omega)$ such that

$$u(0, y) = 0, \quad \& \quad u(a, y) = 0, \quad \text{for } 0 \le y \le b; \text{ and}$$
$$u(x, 0) = f(x), \quad \& \quad u(x, b) = 0, \quad \text{for } 0 \le x \le a.$$

Similar types of Dirichlet problems arise when a harmonic function $u = u(x, y, z)$ represents an *electrostatic potential*, with boundaries maintained at different fixed voltages, in many other situations such as concerning vibrating strings or a drumhead, and also while dealing with mathematical problems related to finding the *minimal energy* of a surface such as a soap film. Many such types of Dirichlet problems can be solved by using method of separation of variables and Fourier series (see Chaps. 8 and 9).

Example 5.20 Consider the case of diffusion of a chemical in a fluid contained in a closed container Ω, with the *mass density* of diffusing chemical being given a function $u \in C^2(\Omega)$ so that $\nabla u(x)$ is the flux of the chemical at point $x \in \Omega$. Then, a typical 2-dimensional Neumann problem related to this situation is to find the function $u = u(x)$ that satisfies the Laplace equation $\nabla^2 u = 0$ or a Poisson equation $\nabla^2 u = f(x)$, and also the homogeneous Neumann condition $\partial_n u = 0$. On the other hand, if a function $u \in C^2(\Omega)$ represents a temperature distribution across a metal container Ω so that $\nabla u(x)$ is the flux of the heat at point x, then for $x \in \Gamma$, $\partial_n u(x) = f(x)$ is the flux of heat *pumping* across the boundary Γ at the point x. Similar Neumann problems arise when $u \in C^2(\Omega)$ represents an *electric potential* so that $\nabla u(x)$ is the electric field at point x. Notice that, in this case, homogeneous Neumann condition amounts to saying that no field lines penetrate the boundary Γ.

Most such types of Neumann problems are solved by using the method of separation of variables and Fourier series (see Chaps. 8 and 9).

In some applications, boundary conditions for a function $u = u(x, t)$ are specified in terms of functions defined differently on various parts of the boundary $\Gamma = \partial\Omega$. For example, in the case of a 1-dimensional differential equation defined over an interval $\Omega = [a, b]$, separated boundary conditions at the two end points may be of the Sturm–Liouville type as given below:

$$\alpha_1 u(a, t) + \beta_1 u_x(a, t) = g(t); \quad \text{and,}$$
$$\alpha_2 u(b, t) + \beta_2 u_x(b, t) = h(t), \ \forall \ t > 0.$$

We may also have a *mixed type* of boundary conditions of the form

$$u + \partial_n u = f,$$

which is known as a *Robin conditions* (or a *Churchill condition*), where $\partial_n u = \partial u / \partial \mathbf{n}$ is the *normal derivative* of $u = u(x)$ that represents the direction of the *outward flux* across the boundary Γ. More generally, a mixed type of nonhomogeneous boundary condition for functions $\alpha, \ \beta, \ f : \Gamma \to \mathbb{R}$ is given by

$$\alpha(x)u(x) + \beta(x)\partial_n u = f(x), \quad \text{for} \ \ x \in \Gamma.$$

For example, if an imperfectly insulated body of thermal conductivity c is dipped in a hot liquid with the ambient temperature T, then according to Newton's *law of cooling* the heat flux across the boundary of the body is given by the relation

$$\partial_n u = c(u - T) \quad \Leftrightarrow \quad c\, u - \partial_n u = f, \ \text{with} \ \ f = cT.$$

Therefore, Newton's law of cooling is a nonhomogeneous boundary conditions of mixed type. Also, recall that Sturm–Liouville theory introduce briefly in Section 3.4.2 provides some powerful tools to solve important boundary value problems, with *periodic boundary conditions* such as given by

$$u(-x) = u(x), \quad \partial_n u(-x) = \partial_n u(x), \quad x \in \Gamma,$$

where the domain Ω is assumed to have a *symmetric structure*.

Definition 5.6 An *initial-boundary value problem* (or simply IBVP) related to a second order differential equation is about finding a function $\varphi = \varphi(x, t) \in C^2(\Omega_1)$ defined on a domain $D \subseteq \Omega_1$ such that $\varphi = \varphi(x, t)$ satisfies the differential equation, and also specified initial and boundary conditions.

Clearly, we can formulate the above definition for differential equation of any order $k \geq 1$. For many applications, it suffices to use some analytical method to find a unique solution of an initial-boundary value problems for a first or a second order differential equation of simpler *linearity type*. In what follows, we formulate initial-boundary value problems concerning some important phenomena in physics and engineering. Let us start with the next definition.

Example 5.21 The problems formulated here are related to the second order linear differential equation that models the transverse vibrations of a perfectly elastic string of finite length, say ℓ, with two ends fastened at points $x = 0$ and $x = \ell$ along the x-axis. We know that the *displacement function* $y = y(x, t)$ satisfies the homogeneous wave equation given by

$$y_{tt} = c^2\, y_{xx}, \qquad \text{for } 0 \le x \le \ell, \text{ and } t \ge 0. \tag{5.4.3}$$

Suppose the *initial deflection* of the string due to *plucking* is given by $y(x, 0) = f(x)$, $x \in (0, \ell)$. Also, we may write the *initial velocity* as $y_t(x, 0) = g(x)$, $x \in (0, \ell)$ when the string in equilibrium state is *struck* with a force. In this case, we say that the function $y = y(x, t)$ satisfies the *nonhomogeneous* initial conditions given by

$$y(x, 0) = f(x), \quad \text{and} \quad y_t(x, 0) = g(x), \text{ for } 0 < x < \ell; \tag{5.4.4}$$

and also the following homogeneous boundary conditions

$$y(0, t) = 0, \quad \text{and} \quad y(\ell, t) = 0, \qquad \text{for } t \ge 0. \tag{5.4.5}$$

Notice that the conditions (5.4.5) hold because the ends of the string are fixed. A typical wave equation-based initial-boundary value problem needs us to find a solution of the equation (5.4.3) that is unique subject to that it satisfies conditions (5.4.4) and (5.4.5) for some functions $f(x)$ and/or $g(x)$.

Example 5.22 With notations as in Example 4.2 related to a vibrating membrane, let $(x, y) \in [a, b] \times [c, d]$. Since boundary edges are fixed, the *deflection function* $u = u(x, y, t)$ satisfies the following homogeneous boundary conditions:

$$u(a, y, t) = u(b, y, t) = 0, \quad \text{and} \quad u(x, c, t) = u(x, d, t) = 0. \tag{5.4.6}$$

Also, if the *initial deflection* of the membrane is given by the function $f(x, y)$ and the *initial velocity* is given by a function $g(x, y)$, then $u = u(x, y, t)$ satisfies the nonhomogeneous initial conditions given by

$$u(x, y, 0) = f(x, y), \quad \text{and} \quad u_t(x, y, 0) = g(x, y). \tag{5.4.7}$$

A typical 2-dimensional wave equation-based initial-boundary value problem needs us to find a unique function $u(x, y, t)$ that satisfies the equation

$$u_{tt} = c^2\big[u_{xx} + u_{yy}\big], \tag{5.4.8}$$

and also the assigned initial and boundary conditions for various choices of functions $f(x, y)$ and $g(x, y)$.

Example 5.23 We consider here initial-boundary value problems related to heat flow through a uniform cylindrical rod of length ℓ. As said earlier, it may be assumed that the rod is placed along the x-axis with ends at points $x = 0$ and $x = \ell$. We also assume that the lateral surface of the rod is properly insulated, and that there is no internal source of heat. Suppose the initial temperature distribution of the rod is given by a function $f(x)$, $0 < x < \ell$. Recall that, in this case, the unknown temperature distribution function $u = u(x, t)$ satisfies the heat equation given by

$$u_t = a^2 u_{xx}, \quad \text{for } 0 \leq x \leq \ell \text{ and } t \geq 0. \tag{5.4.9}$$

We consider now different types of boundary conditions that one may come across in actual applications. To start with, we first suppose that the both ends of the rod are insulated so that we have the *Neumann boundary conditions* such as given by

$$u_x(0, t) = 0, \quad \text{and} \quad u_x(\ell, t) = 0, \quad \text{for } t > 0; \tag{5.4.10}$$

if one end is maintained at zero temperature, and the other is insulated, then we have

$$u(0, t) = 0, \quad \text{and} \quad u_x(\ell, t) = 0, \quad \text{for } t > 0; \tag{5.4.11}$$

and, when both ends are maintained at zero temperature, we have the *Dirichlet boundary conditions* such as given by

$$u(0, t) = 0, \quad \text{and} \quad u(\ell, t) = 0, \quad \text{for } t > 0. \tag{5.4.12}$$

The types of boundary conditions as given below arise while studying heat flow in a circular ring, say of parameter ℓ:

$$u(0, t) = u(\ell, t), \quad \text{and} \quad u_x(0, t) = u_x(\ell, t), \quad \text{for } t > 0. \tag{5.4.13}$$

Recall that the above types of boundary conditions are also known as *periodic boundary conditions*. In a more general situation, when heat is allowed to flow across ends to the environment assumed to be at zero temperature, Newton's *law of cooling* leads to boundary conditions such as given by

$$u_x(0, t) = \alpha\, u(0, t), \quad \text{and} \quad u_x(\ell, t) = -\alpha\, u(\ell, t), \quad \text{for } t > 0, \tag{5.4.14}$$

where $\alpha > 0$ is the ratio of the surface conductance to the thermal conductivity of the bar. In each one of such situations, a typical initial-boundary value problem needs us to find a sufficiently smooth function that satisfies the equation (5.4.9), given boundary conditions, and also an initial condition of the form

$$u(x, 0) = f(x), \qquad \text{for } 0 < x < \ell. \tag{5.4.15}$$

Example 5.24 We consider here the case of heat flow in a long thin rectangular plate that is infinite in y-direction. With notations as in Example 4.5, let $(x, y) \in [0, \pi] \times [0, \infty)$. At time $t = 0$, the lower edge $[0, \pi] \times \{0\}$ of the plate is supplied heat given by a function $f \in C([0, \pi])$, and the vertical edges $\{0\} \times [0, \infty)$ and $\{\pi\} \times [0, \infty)$ are held constant at the zero temperature, which may not be the *absolute zero*. We may write $u = u(x, y, t)$ for the temperature distribution of the plate at any point $(x, y) \in (0, \pi] \times (0, \infty)$ and at time $t > 0$. We may further assume that the faces of the plate are properly insulated so that no heat flows across the plate. Therefore, the function u satisfies the boundary conditions given by

$$u(0, y, t) = 0, \qquad u(\pi, y, t) = 0, \qquad \text{for } y > 0; \tag{5.4.16a}$$

$$u(x, 0, 0) = f(x) \qquad \text{for } 0 < x < \pi. \tag{5.4.16b}$$

As the temperature in the plate drop to zero when (x, y) is far off from the source, we may take the boundary condition on the *edge at infinity* given by

$$\lim_{y \to \infty} u(x, y, t) = 0, \qquad \text{for } 0 < x < \pi \text{ and for large } t > 0. \tag{5.4.17}$$

More generally, if the temperatures of the vertical edges $\{0\} \times [0, \infty)$ and $\{\pi\} \times [0, \infty)$ at $t = 0$ are respectively given by continuous functions g and h, then we write the initial conditions at vertical edges as

$$u(0, y, 0) = g(y), \qquad \text{for } y > 0; \tag{5.4.18a}$$

$$u(\pi, y, 0) = h(y), \qquad \text{for } y > 0, \tag{5.4.18b}$$

In this case, a typical initial-boundary value problem needs us to find a unique function that satisfies the 2-dimensional heat equation given by

$$u_{tt} = c^2 [u_{xx} + u_{yy}], \qquad \text{for } (x, y) \in [0, \pi] \times [0, \infty) \text{ and } t \ge 0, \tag{5.4.19}$$

and also the assigned boundary and initial conditions for various choices of functions f, g and h. In particular, for the *steady state situations*, the temperature distribution function $u = u(x, y)$ satisfies *Laplace equation*

$$u_{xx} + u_{yy} = 0, \qquad \text{for } 0 < x < \pi \text{ and } y > 0. \tag{5.4.20}$$

Notice that a situation such as above arises when $f \not\equiv 0$ for $t > 0$, and the boundary conditions mentioned above are of *Dirichlet type*. However, we may also stipulate boundary conditions of *Neumann type* for a Laplace equation.

Well-posed Problems

Our choice for a particular *solution method*[7] always depends on the type of phenomenon we are studying in a given situation. For example, to solve initial-boundary value problems related to wave propagation, the best strategy is to apply *oscillatory approximations* offered by Fourier series and integrals, as discussed briefly in Chaps. 8 and 10.

Definition 5.7 A *classical solution* (or simply a *solution*) of an initial-boundary value problem is a function $\varphi = \varphi(x, t)$ in some suitable space $C^n(\Omega \times \mathbb{R}^+)$ that satisfies the differential equation, and also the assigned initial/boundary conditions.

Usually, the existence and uniqueness of a solution of an initial-boundary value problem over a suitable domain is an abstract statement specifying analytical properties that a solution must satisfy. Sometimes, the statement also specifies some topological conditions on the domain of existence. According to *Jacques Hadamard* (1865–1963) [4], we say an initial-boundary value problem is *well posed* if the following three conditions hold:

1. *Existence:* There exists a solution that satisfies the equation and also the specified initial-boundary conditions. In abstract terms, this part of the problem is usually settled by using some advance analysis tools. The topological structure of the domain also plays a significant role.
2. *Uniqueness:* It is *unique* subject to these initial and boundary conditions. It is comparatively easier part of the problem, because in most cases uniqueness of a solution follows directly from some known theorem in analysis.
3. *Continuous Dependence:* The solution is *stable* in the sense that a small perturbation in the specified initial/boundary conditions, and also values of associated *parameters*, leads to a small change in the solution.

In some situations, we may also have to take care of the *regularity property* of the solution. An initial-boundary value problem is said to be *ill-posed* if any one of the above three conditions is violated. Some *ill-posed* problems arise naturally in applications such as control theory, inverse scattering, etc. It is mainly due to factors such as the nonhomogeneity of the involved equations or the type of initial data given that may render an initial-boundary value problem *ill-posed*.

In any particular situation, when a physical phenomenon is analysed in terms of some analytical properties of a solution of the associated initial-boundary value problem, it makes sense to seek a mathematical argument to prove the *existence* and *uniqueness* of a solution of the governing differential equation, which also satisfies the assigned auxiliary conditions. Therefore, it is necessary to verify the compatible of a solution of the differential equation with initial and boundary conditions specified for the involved unknown function. That is precisely what we will do for certain well known differential equations in later chapters. More precisely, in Chaps. 7 and 8, we apply the method *separation of variables* and *Fourier series* to solve some simple

---

[7] In some situations, instead of finding an explicit expression for a possible solution, we are more keen to study the solution for some analytical or geometric properties.

types of initial-boundary value problems over *bounded domains*. In Chaps. 9 and 10, we apply *transform techniques* due to Fourier and Laplace to solve some similar types of initial-boundary value problems given over an *unbounded domain*.

Many new theories such as related to mathematical analysis and differential geometry were developed during nineteenth century, while establishing the existence and uniqueness of a solution of some important initial-boundary value problems. More recently, mainly due to improved numerical methods and availability of enhanced computational facilities, it is now possible to find an *approximate solution* of certain intractable practical problems in physics and engineering.

While dealing with a simpler situation, it is more convenient to apply some analytical method or a geometrical argument to find an explicit solution of an initial-boundary value problem. In most such cases, it is possible to reduce the given problem to a corresponding initial-boundary value problem for a first order system of ordinary differential equations. Therefore, we can apply some known method to obtain an explicit solution of the main problem. For example, the *method of characteristics* uses geometric approach to solve a Cauchy problem for *hyperbolic type* differential equation (Chap. 6). The same technique also applies while approximating a solution by using some *discretisation* procedure. For more complex situations, some appropriate numerical technique is used to obtain a converging sequence of approximate solutions.

As seen earlier in Sect. 4.2, a boundary value problem may arise for a nonhomogeneous differential equation or for a homogeneous differential equation with nonhomogeneous boundary conditions. So, as long as G is a finite sum of functions, we apply *separation of variables* and *Fourier series* to solve the given IBVP over a bounded domain. Chap. 8 provides the further details.

However, the separation of variables technique fails when the equation and/or some boundary condition is nonhomogeneous. It is though possible in some cases to reduce a general BVP for a nonhomogeneous equation like a Poisson or a Laplace equation, with nonhomogeneous boundary conditions, to a nonhomogeneous BVP with homogeneous boundary conditions. In this case, G is expressed as an infinite series in the sense of *Strum-Liouville theory*, and then we apply *generalised Fourier series* to compute the coefficient functions c_i. Chapter 7 provides the related details about the method of *eigenfunctions expansion*. Theorem 5.1 also applies while using *transform methods* of Fourier and Laplace to solve some interesting BVPs defined over an unbounded domain. Chapters 8 and 9 provide the related details.

In general case, existence of a solution is established by applying some analytical and/or geometric reasoning that is usually nonconstructive. An algorithmic approach to establish the existence of a solution of a differential equation takes into consideration the *type* of the equation, *nature* of assigned initial and boundary conditions, and also *choice* for a particular coordinates system of the domain of definition of the differential equation.

Remark 5.4 *Differential equation models* we come across in most applications are quasilinear or fully nonlinear differential equations. In such cases, certain concepts from advanced analysis and differential geometry come handy while establishing the existence and uniqueness of a solution of a related *initial-boundary value problem*, and also to analyse the related physical interpretations. Numerical techniques help find *approximate solutions* while dealing with such types of problems for higher order fully nonlinear differential equations. The reader may refer [5], for further details.

5.5   Uniqueness Theorems and Stability Issues

In general, there are some standard theorems giving sufficient conditions to prove uniqueness of a solution. For wave equations related IBVPs, we may use Green's function based *energy method*. The reader is referred to ([1], Sect. 5D), for a general discussion. We will consider here some simple situations. For example, Lemma 2.3 applies to prove uniqueness of a solution of Laplace and Poisson equations.

Theorem 5.2 (Uniqueness of Solution for Laplace Equation) *Suppose u_1 and u_2 are two solutions of Laplace equation*

$$\nabla^2 u = u_{xx} + u_{yy} = g(x, y)$$

in a domain Ω, with $u|_{\partial\Omega} = 0$, then $u_1 = u_2$ over Ω.

Proof For $w = u_1 - u_2$, $\nabla^2 w = 0$ implies

$$\int_{\partial\Omega} w\, \nabla w \cdot \mathbf{n}\, dS = \int_{\Omega} \nabla \cdot (w\, \nabla w) dV = \int_{\Omega} (w\nabla^2 w + (\nabla w)^2) dV,$$

using divergence theorem, where $\mathbf{n}$ is the normal to Ω, pointing outwards. Again, by divergence theorem,

$$\int_{\Omega} (\nabla w)^2 dV = \int_{\partial\Omega} w\, \frac{\partial w}{n} dS = 0.$$

Now the integrand $(\nabla w)^2$ is non-negative in Ω and hence for the equality to hold we must have $\nabla w = 0$ i.e., w is constant in Ω. Since $w = 0$ on $\partial\Omega$ and the solution is smooth, we must have $w \equiv 0$ on Ω, by *maximum principle*. That is, $u_1 = u_2$ over Ω.   □

The same proof works when the *normal derivative*[8] $\partial_n u = \partial u/\partial n$ is given on the boundary $\Gamma = \partial\Omega$ or for mixed conditions, as we shall state shortly.

---

[8] Recall that a surface in $\mathbb{R}^3$ is *piecewise smooth* if *normal vector field* is defined at every point except at points on the common edges between constituent faces of its boundary.

Theorem 5.3 (Uniqueness of Solution for Wave Equation) *Let $\Omega := \mathbb{R} \times \mathbb{R}^+$. Suppose $y_1, y_2 \in C^2(\Omega)$ be such that both satisfy the Cauchy problem:*

$$y_{tt} = c^2 \, y_{xx} + f(x, t), \quad \text{for } x \in \mathbb{R} \text{ and } t > 0;$$
$$y(x, 0) = f(x), \quad \text{and } y_t(x, 0) = g(x), \quad \text{for } x \in \mathbb{R}, \tag{5.5.1}$$

where $f(x, t)$ represents time-dependent distribution of force, say due to gravity. Then $y_1 \equiv y_2$ over Ω.

Proof Clearly, the function $w = y_1 - y_2$ satisfies the *homogeneous* Cauchy problem:

$$w_{tt} = c^2 \, w_{xx}, \quad \text{for } x \in \mathbb{R} \text{ and } t > 0;$$
$$w(x, 0) = 0, \quad \text{and } w_t(x, 0) = 0, \quad \text{for } x \in \mathbb{R}. \tag{5.5.2}$$

Now, multiplying the equation by w_t and integrating by parts, we obtain

$$\int_{-\infty}^{\infty} \partial_t [w_t^2 / 2] dx = -c^2 \int_{-\infty}^{\infty} \partial_t [w_x^2 / 2] dx \tag{5.5.3}$$

Consider the continuous function

$$E(t) := \int_{-\infty}^{\infty} \left[\frac{1}{2} w_t^2 + \frac{c^2}{2} w_x^2 \right] dx, \quad t \geq 0. \tag{5.5.4}$$

It then follows from Eq. (5.5.11) that $E'(t) = 0$ or, equivalently, $E(t) = constant$ for all $t \geq 0$. However, by homogeneity of initial conditions in (5.5.10), we have $E(0) = 0$. Therefore, we must have

$$E(t) = E(0) = 0, \quad t > 0.$$

Hence, by definition of $E(t)$, we conclude that $w_t = w_x = 0$ for all $(x, t) \in \Omega$. That is, the function $w = w(x, t)$ is identically constant. But then, $w(x, 0) = 0$ concludes the proof. $\square$

In physical terms, the function

$$E(t) = \int_{-\infty}^{\infty} \left[\frac{1}{2} y_t^2 + \frac{c^2}{2} y_x^2 \right] dx,$$

at any moment $t \geq 0$ represents the sum of the *kinetic energy* and *potential energy*, where the latter is due to tension in the string. For this reason, the procedure used above is called the *energy method*. Notice that

$$E(0) = \frac{1}{2} \int\limits_{-\infty}^{\infty} \left[f(x)^2 + c^2 g(x)^2 \right] \mathrm{d}x.$$

A same method works to prove uniqueness of solution $u = u(x, t)$ of a heat equation (5.4.9). But, in this case, procedure is more straightforward due to the fact the *heat function*

$$H(t) := \int u(x, t)\mathrm{d}x, \quad t > 0, \tag{5.5.5}$$

decays in time. If $x \in [\alpha, \beta]$ then

$$H'(t) := \int\limits_{\alpha}^{\beta} u_t(x, t)\mathrm{d}x \tag{5.5.6}$$

$$= \int\limits_{\alpha}^{\beta} a^2 u_{xx}(x, t)\mathrm{d}x \tag{5.5.7}$$

$$= a^2 \left[u_x(b, t) - u_x(a, t) \right]. \tag{5.5.8}$$

Therefore, the *heat flux through ends* controls the heat energy. In particular, if the ends are insulated so that we have *Neumann type* of boundary conditions (5.4.10), then $H \equiv constant$ over the interval $[\alpha, \beta]$. The same conclusion holds over $\mathbb{R}$ subject to that integral on the right of (5.5.5) makes sense,[9] and there is no loss of heat at infinity. Of course, then $H'(t) = 0$ for $t > 0$.

Theorem 5.4 (Uniqueness of Solution for Heat Equation) *Let $\Omega := [0, \ell] \times \mathbb{R}^+$. Suppose $u_1, u_2 \in C^2(\mathbb{R} \times \mathbb{R}^+)$ be such that both satisfy the IBVP:*

$$u_t = a^2 u_{xx} + f(x, t), \quad for \ 0 < x < \ell \ and \ t > 0;$$
$$u(0, t) = f(t), \quad and \ \ u(\ell, t) = g(t), \quad for \ t > 0; \tag{5.5.9}$$
$$u(x, 0) = \phi(x), \quad for \ 0 < x < \ell.$$

where the function $f(x, t)$ represents time-dependent heat source. Then $u_1 \equiv u_2$ over Ω.

Proof Clearly, the function $w = u_1 - u_2$ satisfies the *homogeneous* Cauchy problem:

$$w_t = a^2 w_{xx}, \quad for \ 0 < x < \ell \ and \ t > 0;$$
$$w(0, t) = 0, \quad and \ \ w(\ell, t) = 0, \quad for \ t > 0; \tag{5.5.10}$$
$$w(x, 0)0, \quad for \ 0 < x < \ell.$$

---

[9] That is, $u \in L(\mathbb{R})$, $u(x, t) \to 0$ as $x \to \pm\infty$, and u_x is bounded over $\mathbb{R}$.

Now, multiplying the equation by w_t and integrating over $[0, \ell]$, we obtain

$$\frac{d}{dt} \int_0^\ell \frac{1}{2} w^2 dx = \left[a^2 w w_x \right]_0^\ell - c^2 \int_0^\ell w_x^2 dx \qquad (5.5.11)$$

Thus the *heat energy* function

$$E(t) := \int_0^\ell \frac{1}{2} w^2 dx, \quad t \ge 0 \qquad (5.5.12)$$

is a decreasing function whenever $w w_x$ is nonpositive over the interval $[0, \ell]$. For example, this is case for boundary conditions (5.4.11) and (5.4.12). In this case, $E(t) \le E(0)$ for $t > 0$, implies that

$$\int_0^\ell \frac{1}{2} w^2(x, t) dx \le \int_0^\ell \frac{1}{2} w^2(x, 0) dx = 0, \quad \text{for } t > 0.$$

Hence, by definition of $E(t)$, we conclude that $w(x, t) = 0$ for all $(x, t) \in \Omega$. This concludes the proof. □

  Stability of the solution is important mainly due to the fact that, in some cases, there may be a possibility that even a small error in boundary data could result into some drastic changes in the solution of a problem.

Exercises 5

5.1 Derive telegraph equation (5.1.37) as a model for an infinitesimal telegraph wire to study it as an electric circuit that consists of a register of residence $R\, dx$ and a coil of induction $L\, dx$. It may be assumed that the current $i(x, t)$ through the wire can escape to ground through a register of conductance $G\, dx$ or through a capacitor of capacitance $C\, dx$.

5.2 (One-dimensional *Euler Equation*) For a sufficiently smooth function f, and

$$V = \{\{v \in C^2[a, b] \mid v(a) = u_a, \ v(b) = u_b, \ \text{for some } u_a, u_b \in \mathbb{R}\}.$$

Let $u \in V$ be a solution of the *variational problem* of finding

$$\min_{v \in V} \int_a^b f(x, v(x), v'(x)) \, dx.$$

Use Exercise 2 to show that

$$\frac{d}{dx}\big[f_{u'}\big(x, u(x), u'(x)\big)\big] = f_u\big(x, u(x), u'(x)\big), \quad \text{for } x \in (a, b).$$

5.3 (n-dimensional *Euler Equation*) Suppose $\Omega \subset \mathbb{R}^n$ is a domain, $x = (x_1, \ldots, x_n)$, and F be a sufficiently smooth scalar function of its arguments. For an arbitrary scalar function h defined on the boundary $\partial\Omega$, take

$$V = \{\{f \in C^2(\overline{\Omega}) \mid f = h \text{ on } \partial\Omega\}.$$

Let $g \in V$ be a solution of the *variational problem* of finding

$$\min_{f\in V} \int_{\Omega} F\big(x, f(x), f'(x)\big) \, dx.$$

Prove that

$$\sum_{i=1}^{n} \frac{\partial}{\partial x_i} F_{g_{x_i}} - F_g = 0, \quad \text{for } x \in \Omega.$$

Further, explain how boundary value problem:

$$u_{xx} + u_{yy} = 0, \quad \text{with } u = h \text{ on } \partial\Omega \subset \mathbb{R}^2$$

is related to the variational problem

$$\min_{v\in V} \int_{\Omega} \big(v_x^2 + v_y^2\big) \, dx dy.$$

5.4 If u_0 satisfy a nonhomogeneous linear differential equation $L(u) = f$, and $u_1, u_2, \ldots u_n$ satisfy the corresponding homogeneous linear differential equation $L(u) = 0$, then show that $u_0 + c_1 u_1 + c_2 u_2 + \cdots + c_n u_n$ also satisfy the equation $L(u) = f$.

5.5 Suppose a C^1-function $u(x, y) = \varphi(x, y; a, b)$ defines a 2-parameter family of surfaces such that the rank of the matrix as given below is two:

$$\begin{pmatrix} \varphi_a & \varphi_{xa} & \varphi_{ya} \\ \varphi_b & \varphi_{xb} & \varphi_{yb} \end{pmatrix}$$

Show that, by eliminating parameters a and b from the three relations given by

$$u = \varphi(x, y; a, b), \quad u_x = \varphi_x(x, y; a, b), \quad u_y = \varphi_y(x, y; a, b),$$

we obtain a first order differential equation of the form $F\big(x, y; u; u_x, u_y\big) = 0$.

5.6 Find nontrivial solutions $u = u(x, y)$ of the differential equation $y u_x - x u_y = 0$.

5.7 Prove that the Laplace equation $u_{xx} + u_{yy} = 0$ has infinitely many linearly independent solutions in the space $C^2(\mathbb{R}^2)$.

5.8 Recall that a function $u = u(x_1, \ldots, x_n)$ is said to be *radially symmetric* if $u(x_1, \ldots, x_n) = f(r)$, for some function f, where $r = \|x_1, \ldots, x_n\|$. Find all radially symmetric functions satisfying the n-variable Laplace equation over the domain $\mathbb{R}^n \setminus \{0\}$, for $n \geq 2$.

5.9 Show that implicit relation $\varphi(x + y + z, x^2 + y^2 + z^2) = 0$ is the general solution of the the differential equation $(y - z)u_x + (z - x)u_y = x - y$, where $z = u(x, y)$ is a C^1-function.

5.10 Show that implicitly relation $\varphi(xy, y/z) = 0$ is the general solution of the differential equation $y u_y - x u_x = z$, where $z = u(x, y)$ is a C^1-function.

5.11 Show that the function $z = u(x, y) = (x + a)(y + b)$ is the complete integral of the the differential equation $pq = z$, where $p = u_x$ and $q = u_y$.

5.12 Show that the function $z = u(x, y) = ax^2 + by^2 + u^2 - 1$ is the complete integral of the differential equation $x u\, u_x + y z\, u_y = z^2 - 1$.

5.13 Show that if $u = u(x, y)$ is a solution of the equation given by

$$a(u_x, u_y)u_{xx} + 2b(u_x, u_y)u_{xy} + c(u_x, u_y)u_{yy} = 0,$$

and $v = v(\xi(x, y), \eta(x, y)) = x u_x + y u_y - u$, then $v_\xi = x$, v_η, and $v = v(\xi, \eta)$ satisfies the linear differential equation

$$a(\xi, \eta)v_{\eta\eta} - 2b(\xi, \eta)v_{\xi\eta} + c(\xi, \eta)v_{\xi\xi} = 0.$$

5.14 Show that in a neighbourhood of point (x_0, y_0) there exists C^1 functions φ, ϕ satisfying Eq. (5.2.16) such that $\varphi_y \neq 0$ and $\phi_y \neq 0$.

5.15 Let λ_1 and λ_2 be the roots of Eq. (5.2.15), and C^1 functions φ, ϕ respectively be solutions of Eq. (5.2.16). Show that the level curves $\varphi(x, y) = c_1$ and $\phi(x, y) = c_2$ are characteristic curves of Eq. (5.2.1).

5.16 Show that if $u = u(x, y)$ is a solution of a differential equation given by

$$a(u_x, u_y)u_{xx} + 2b(u_x, u_y)u_{xy} + c(u_x, u_y)u_{yy} = 0,$$

and $v = v(\xi(x, y), \eta(x, y)) = x u_x + y u_y - u$, then $v_\xi = x$, v_η, and $v = v(\xi, \eta)$ satisfies the linear differential equation

$$a(\xi, \eta)v_{\eta\eta} - 2b(\xi, \eta)v_{\xi\eta} + c(\xi, \eta)v_{\xi\xi} = 0.$$

5.17 Classify the following differential equations as elliptic, parabolic, or hyperbolic:

(a) $u_{xx} + a^2 u_{yy} = 0$, $a > 0$;   (b) $4u_{xx} + 5u_{xy} + u_{yy} + u_x + u_y = 2$;

(c) $y u_{xx} + x u_{yy} = 0$;   (d) $x^2 u_{xx} + 2xy u_{xy} + y^2 u_{yy} + u_x + u_y = 0$.

Moreover, obtain their normal forms, clearly specifying the characteristics coordinates used.

5.18 Find a general solution of the linear differential equation given by

$$x^2 u_{xx} + 2x u_{xy} + y^2 u_{yy} = 0,$$

by first reducing it to the canonical form.

5.19 Find a general solution of the linear differential equation given by

$$x u_{xx} + (x - y) u_{xy} - y u_{yy} = 0, \quad x > 0, \quad \text{for } y > 0,$$

by first reducing it to the canonical form.

5.20 Find a general solution of the linear differential equation given by

$$y^2 u_{xx} - 2y u_{xy} + u_{yy} = u_x + 6.$$

by first reducing it to the canonical form.

5.21 Find a general solution of the linear differential equation given by

$$x^2 u_{xx} + 2x u_{xy} + y^2 u_{yy} = 0,$$

by first reducing it to the canonical form.

5.22 Show that the differential equation $3u_{xx} + 7u_{xy} + 2u_{yy} = 0$ is hyperbolic for all x, y, and hence find the general solution.

5.23 For the differential equation

$$y u_{xx} - 2 u_{xy} + e^x u_{yy} + x^2 u_x - u = 0$$

specify the regions in $\mathbb{R}^2$ where it is hyperbolic, parabolic, or elliptical.

5.24 Write the quasilinear equation for a function $u = u(x, y)$ given by

$$P(x, y, u) u_x + Q(x, y, u) u_y = R(x, y, u)$$

as a homogeneous equation for a function of three variables.

5.25 Consider the Hamilton–Jacobi equation in polar coordinates given by

$$\varphi_t + \frac{1}{2}\left(\varphi_r^2 + \frac{1}{r^2}\varphi_\theta^2\right) = \frac{k^2}{r},$$

associated with the Hamiltonian

$$H = \frac{1}{2}(p^2 + q^2) - U(x, y), \quad \text{with } p = x' \text{ and } q = y',$$

where $x''(t) = U_x$, $y''(t) = U_y$, and $U(x, y) = k^2(x^2 + y^2)^{-1/2}$. Show that a two-parameter solution of the Hamilton–Jacobi equation given by

$$\varphi = \varphi(\alpha, \beta; \theta, r, t)$$

is a complete integral.

5.26 For $x \in \mathbb{R}$ and $t \geq 0$, let a C^1-function $\rho = \rho(x, t)$ represents the concentration (mass density per unit length) of a noninteractive contaminant moving along a fluid flowing at a constant speed $c = 4$ through a thin tube of uniform cross-sectional area. Suppose the initial contaminant increases at a rate three times the instantaneous concentration, and $\rho(0, t) = e^{-5t}$, for all $t \geq 0$. Find an explicit expression for the function ρ.

References

1. Pivato, M. (2010). *Linear partial differential equations and Fourier theory*. Cambridge University Press.
2. Arnold, V. I. (2004). *Lecturers on partial differential equations*. Springer.
3. Zachmanoglou, E. C., & Thoe, D. W. (2012). *Introduction to partial differential equations with applications*. Dover.
4. Hadamard, J. (1952). *Lectures on Cauchy's problem in linear partial differential equations*. Yale University Press.
5. Olver, P. J. (2014). *Introduction to partial differential equations*. UTM Springer.
6. Folland, G. B. (1976). *Introduction to partial differential equations*. Princeton University Press.
7. Logan, J. D. (2004). *Applied partial differential equations* (Vol. 2). Springer.
8. Renardy, M., & Rogers, R. C. (2004). *An introduction to partial differential equations*. TAM-13. Springer.
9. Sauderer, E. (2006). *Partial differential equations of applied mathematics*. Wiley.
10. Weinberger, H. F. (1965). *A first course in partial differential equations*. Wiley.

Chapter 6
General Solution and Complete Integral

As long as algebra and geometry have been separated, their progress has been slow and their uses limited; but when these two sciences have been united, they have lent each mutual force and have marched together towards perfection.

Josep-Louis Lagrange (1736–1813)

This chapter is about some standard methods that are useful to find the general solution or the complete integral of a first order differential equation in two variables. In Sect. 6.1, we use the concept of characteristics coordinates to find the general solution of semilinear equations. In Sect. 6.2, we apply Lagrange's method to find the general solution of quasilinear equations. In Sect. 6.3, we use Lagrange's method to solve some higher order linear equations with constant coefficients. In Sect. 6.4, we discuss Lagrange–Charpit method of finding the complete integral of a fully nonlinear equations in two variables. In each case, the main idea is to decompose the original differential equation into one or more first order ordinary differential equations, which are solved by applying some appropriate method discussed earlier in Chap. 3.

The concepts introduced in this chapter are used in the next chapter to discuss applications of the *method of characteristics* in solving some important Cauchy problems related to first order differential equations.

6.1 Characteristics Coordinates

For an open set $\Omega \subseteq \mathbb{R}^n$, we write $\Omega_1 = \Omega \times \mathbb{R}^+$. A *general solution* of a kth order differential equation of the form (5.1.1) is an implicit equation given by

$$\Psi\left(x;\ t;\ u;\ f_1, \ldots, f_n\right) = 0, \tag{6.1.1}$$

© The Author(s), under exclusive license to Springer Nature Singapore Pte Ltd. 2022
A. K. Razdan and V. Ravichandran, *Fundamentals of Partial Differential Equations*,
https://doi.org/10.1007/978-981-16-9865-1_6

where $u = u(x, t) \in C^k(\Omega_1)$ and $f_1, \ldots, f_n$ are arbitrary univariate functions. The method of *coordinates transformation*, introduced earlier in Chap. 5, is a convenient way to find the general solution of a first order semilinear equations for a function $u = u(x, y) \in C^1(\Omega)$. We first generalise the assertions made in Remark 5.2 to semilinear equation. That is, we discuss the procedure of obtaining a *nonsingular* transformation of coordinates such as

$$\xi = \varphi(x, y) \quad \text{and} \quad \tau = \phi(x, y), \tag{6.1.2}$$

which helps reduce Eq. (5.1.10) to a family of first order ordinary differential equations in the independent variable ξ, parametrised by η. Recall that a transformation (φ, ϕ) defined over a neighbourhood Ω_0 of a point $(x_0, y_0) \in \Omega$ is nonsingular (*nondegenerate*) if and only if the Jacobian of transformation (6.1.2) given by

$$J(\varphi, \phi; x, y) = \frac{\partial(\varphi, \phi)}{\partial(x, y)} = \begin{vmatrix} \varphi_x & \varphi_y \\ \phi_x & \phi_y \end{vmatrix}$$

is nonzero over the neighbourhood Ω_0. More precisely, whenever $J(\varphi, \phi; x, y) \neq 0$ at a point (x_0, y_0), the *inverse function theorem* ensures the existence of a transformation $(\xi, \eta) \mapsto (x, y)$ over the neighbourhood Ω_0.

Let us start the discussion with an illustration concerning an interesting practical situation, as given in the next example. The argument given below may be seen as an introduction to the *method of characteristics*, as described later in this chapter.

Example 6.1 Let a C^1-function $u = u(x, t)$ describe the motion of a substance in a fluid moving with a speed $c \in \mathbb{R}$. Then, as we know, u satisfies the transport equation

$$u_t + c\, u_x = 0, \quad \text{for} \; -\infty < x < \infty, \; \text{and} \; t \geq 0. \tag{6.1.3}$$

To find the general solution of the above equation, we apply a coordinates transformation such as (6.1.2) to reduce it a 1-parameter family of ordinary differential equations. We may take

$$\xi = \varphi(x, t) = x + ct \quad \text{and} \quad \tau = \phi(x, t) = x - ct.$$

Then, for $u(x, t) = v(\xi, \tau)$, it follows by chain rule that we can write

$$u_x = v_\xi \, \xi_x + v_\tau \, \tau_x = v_\xi + v_\tau;$$
$$u_t = v_\xi \, \xi_t + v_\tau \, \tau_t = cv_\xi - cv_\tau.$$

The above two relations imply that we have

$$u_t + cu_x = (1 + c^2)\, v_\xi.$$

Therefore, as $1 + c^2 \neq 0$, a general solution of Eq. (6.1.3) is produced by integrating the ordinary differential equation $v_\xi = 0$. Therefore, over the xt-plane,

$$v(\xi, \tau) = g(\tau) \quad \Rightarrow \quad u(x, t) = g(x - ct),$$

where $g \in C^1(\mathbb{R})$ is an arbitrary function. Therefore, the *integral surface* of the transport equation (6.1.3) given by

$$\Gamma_u = \{(x, t, z) \in \Omega \times \mathbb{R}^+ : z = u(x, t)\}$$

consists of the *characteristic curves* $g(x - ct)$, which are lines shaped by the function g. For each fixed $t = t_0$, the function $u = u(x, t_0)$ defines a *wave profile* so that the integral surface Γ_u is a continuous sequence of evolving waveforms over time $t > t_0$. Therefore, we can view characteristics as a part of *travelling wave*, moving to right (for $c > 0$) or to the left (for $c < 0$). Further, since we have

$$c\, u_x + u_t = c\, g'(x - ct) - c\, g'(x - ct) = 0,$$

the function $u = u(x, t)$ is constant along the curves $x = x_0 + ct$, for some fixed $x_0 \in \mathbb{R}$. Such types of lines in the xt-plane are known as the (projected) *characteristics* of the transport equation (Definition 7.1).

As said earlier, in the general case, the main issue is to find a nonsingular transformation (6.1.2) such that a first order semilinear equation for a function $z = u(x, y) \in C^1(\Omega)$ reduces to a 1-parameter ordinary differential equation. The integral surface of the equation is then found by *putting together* the characteristics of 1-parameter family of ordinary differential equations. We start with the case of a homogeneous linear equation given by

$$a(x, y)\, p + b(x, y)\, q - c(x, y)\, u = 0, \quad \text{for } (x, y) \in \Omega, \tag{6.1.4}$$

where coefficients a, b, c are constants. As before, we may assume $a^2 + b^2 \neq 0$ and write the integral surface of Eq. (6.1.4) as

$$\Gamma_u := \{(x, y, z) \in \Omega \times \mathbb{R} : z = u(x, y)\}.$$

At each point of the surface $\Gamma_u : u(x, y) - z = 0$, the normal is given by the vector $\mathbf{n} = (p, q, -1)$. It thus follows from Eq. (6.1.4) that, at each point of the surface Γ_u, the normal vector $\mathbf{n}$ satisfies the condition

$$\mathbf{n} \cdot \mathbf{v} = ap + bq - cu = 0, \quad \text{where } \mathbf{v} = (a, b, c).$$

Said differently, at each point of the surface Γ_u, the tangent to the level curves $u(x, y) = constant$ is given by the vector $\mathbf{v} = (a, b, c)$. Therefore, to eliminate the derivative $q = u_y$ from Eq. (6.1.4), we choose transformation (6.1.2) in such a manner so that the ξ-axis in new coordinates is parallel to the vector v. Or, equivalently, the *characteristics* are given by the integral curves of the differential equation $y'(x) = b/a$, assuming $a \neq 0$. Hence, we may take

$$bx - ay = \tau, \quad \text{with } \tau \text{ being a constant.}$$

We are thus led to consider a transformation given by

$$\xi = \varphi(x, y) = ax + by \quad \text{and} \quad \tau = \phi(x, y) = bx - ay. \tag{6.1.5}$$

Applying chain rule to the function given by

$$u(x, y) = v(\xi, \tau) = v(ax + by, bx - ay),$$

we obtain the relations given by

$$p = v_x(\xi, \tau) = v_\xi \, \varphi_x + v_\tau \, \phi_x = a \, v_\xi + b \, v_\tau;$$
$$q = v_y(\xi, \tau) = v_\xi \, \varphi_y + v_\tau \, \phi_y = b \, v_\xi - a \, v_\tau.$$

Substituting back into Eq. (6.1.4), we obtain

$$(a^2 + b^2) \frac{\partial v(\xi, \tau)}{\partial \xi} = c \, v(\xi, \tau).$$

Treating the coordinate τ as parameter, we can write the above equation as 1-parameter family of ordinary differential equations given by

$$\frac{dv}{d\xi} = \frac{c}{a^2 + b^2} \, v,$$

which has a unique solution given by

$$v(\xi, \tau) = g(\tau)e^{\alpha \xi}, \quad \text{for some nice function } g \in C^1(\mathbb{R}).$$

As the Jacobian of the transformation (6.1.5) is nonzero, the inverse transformation $(\xi, \tau) \mapsto (x, y)$ can be found so that the general solution of Eq. (6.1.4) is given by

$$u(x, y) = g(bx - ay)e^{\alpha(ax + by)}.$$

As long as the coefficients of a semilinear equation are constants, the characteristics are straight lines. As we shall see shortly, the situation is not the same for a semilinear equation when coefficients in the principle part are functions of variables x and y.

Example 6.2 For a point $x \in \mathbb{R}$ and time $t > 0$, let $u(x, t)$ represent the *mass density* of a (radioactive) chemical in a fluid flowing freely at a constant speed c through a thin tube, say of uniform cross-sectional area. Suppose the decay rate of u is $b > 0$. Then, the function $u = u(x, t)$ satisfies the homogeneous *transport equation* given by

$$u_t + cu_x = -bu, \quad \text{for } x \in \mathbb{R} \text{ and } t \geq 0, \tag{6.1.6}$$

which is known as the 1-dimensional *advection-decay equation*. In this case, by the above discussion, using the coordinates transformation given by

$$\xi = \varphi(x, t) = ax + t \quad \text{and} \quad \tau = \phi(x, t) = x - at,$$

it follows that the 1-parameter ordinary differential equation is given by

$$v_\xi - cv = 0, \quad \text{where } c = \frac{-b}{1 + a^2}.$$

Therefore, it follows that

$$v(\xi, \eta) = g(\tau)e^{c\xi} \quad \Rightarrow \quad u(x, t) = g(x - at)e^{ac\,x + ct},$$

where g is some function. Equation (6.1.6) has a unique solution, for any specified initial condition. Suppose the function $u(x, 0) = e^{-5x}$ represents the initial concentration of the contaminant, for $x \in \mathbb{R}$. Then, the complete profile of the concentration of the contaminant is given by that the function

$$u(x, t) = e^{-(ac+5)x + a(ac+5)t}\,e^{ac\,x + ct} = e^{(a^2 c + 5a + c)t - 5x}.$$

The same procedure applies to a *nonhomogeneous* transport equation given by

$$au_x + u_t + bu = f(x, t), \quad \text{for } -\infty < x < \infty \text{ and } t \geq 0,$$

In this case, the 1-parameter ordinary differential equation is given by

$$v_\xi - cv = \frac{c}{b} f(\xi, \tau), \quad \text{where } c = \frac{-b}{1 + a^2}.$$

Therefore, by a standard argument, we obtain

$$v(\xi, \tau) = e^{c\xi} \left[-\frac{c}{b} \int e^{-c\xi} f(\xi, \tau)\,d\xi + g(\tau) \right].$$

The general solution of the equation is obtained by inverting the variables.

For an analytical treatment, consider the first order homogeneous linear equation given by

$$a(x, y)\,p + b(x, y)\,q = 0, \quad \text{for some } z = u(x, y) \in C^1(\Omega). \tag{6.1.7}$$

Suppose there is some nonsingular transformation given by (6.1.2) that helps reduce
Eq. (6.1.7) to a 1-parameter homogeneous ordinary differential equation of the form

$$h(\varphi, \phi)\, v_\xi = 0, \tag{6.1.8}$$

Applying chain rule to the function $u(x, y) = v(\varphi(x, y), \phi(x, y))$, we have

$$u_x = v_\xi\, \varphi_x + v_\eta\, \phi_x \quad \text{and} \quad u_y = v_\xi\, \varphi_y + v_\eta\, \phi_y.$$

Substituting back into Eq. (6.1.7), and rearranging the terms, we obtain

$$\left(a\, \xi_x + b\, \xi_y\right) v_\xi + \left(a\, \eta_x + b\, \eta_y\right) v_\eta = 0.$$

Hence, to get an ordinary differential equation (6.1.8), we may take $a\, \eta_x + b\, \eta_y = 0$
so that, for $a \neq 0$, the transformation $\eta = \phi(x, y)$ must satisfy the condition $\phi_x / \phi_y = -(b/a)$. Also, for $\phi(x, y) = constant$, we have

$$0 = d\phi = \phi_x\, dx + \phi_y\, dy \quad \Rightarrow \quad \frac{dy}{dx} = -\frac{\phi_x}{\phi_y} = \frac{b}{a},$$

which proves that $\phi(x, y) = constant$ is a solution of the *characteristic equation*

$$\frac{dy}{dx} = \frac{b(x, y)}{a(x, y)}, \tag{6.1.9}$$

that defines *characteristics* of Eq. (6.1.7). The infinitely many solutions may thus be
written as

$$G(x, y, \eta) = 0, \quad \text{with parameter } \eta \text{ being a constant.}$$

In general, 1-parameter family of level curves $\eta(x, y) = constant$ of the integral
surface of a homogeneous semilinear equation given by

$$a(x, y)\, u_x + b(x, y)\, u_y = c(x, y;\, z), \quad \text{for } z = u(x, y); \tag{6.1.10}$$

is called **characteristic curves** of the equation, which is a *unique* solution of the
ordinary differential equation given by

$$a(x, y)\, \eta_x + b(x, y)\, \eta_y = c(\xi, \eta;\, v). \tag{6.1.11}$$

Example 6.3 Let $\Omega \subseteq \mathbb{R}^2$ be an open set. Consider the first order differential equation for a function $u \in C^1(\Omega)$ given by

$$x\, p + 2y\, q = x - 3\, u, \quad \text{for } z = u(x, y).$$

In this case, we have $a = x$ and $b = 2y$. Therefore, we may obtain the *characteristics* of the above equation from the ordinary differential equation given by

$$\frac{dy}{dx} = \frac{b}{a} = \frac{2y}{x},$$

which has the general solution given by

$$y(x) = A e^{x^2}, \quad \text{for} \ -\infty < x < \infty, \ \text{and} \ y > 0.$$

Notice that $y(0) = A$. Now, as $y = y(x)$ represents curves of constant η, we may choose

$$\eta =: \phi(x, y) = \ln y - x^2, \quad \text{and} \quad \xi = \varphi(x, y) = x,$$

so that $J = \phi_y = y^{-1} \neq 0$. In view of the last equation, with $A = 1$, we have $y(\xi) = e^{\xi^2 + \eta}$, where η is a parameter. Thus, with $u(x, y) = v(\xi, \eta)$, the transformations yield

$$p = v_\xi \varphi_x + v_\eta \phi_x = v_\xi - 2\xi \, v_\eta;$$

$$q = v_\xi \varphi_y + v_\eta \phi_y = \frac{1}{y} v_\eta = \frac{1}{e^{\xi^2 + \eta}} v_\eta.$$

Hence, the given differential equation becomes

$$\xi \left(v_\xi - 2\xi \, v_\eta \right) + 2 \, e^{\xi^2 + \eta} \frac{1}{e^{\xi^2 + \eta}} v_\eta + 3 v = \xi,$$

which simplifies to the ordinary differential equation given by

$$v_\xi + \frac{3}{\xi} v = 1, \quad \text{with} \ \xi \neq 0. \tag{6.1.12}$$

In general, one family of new coordinates is provided by the characteristics of Eq. (6.1.10), whereas we may take the second set of coordinate curves $\xi(x, y) = constant$ to be any parametrised family of smooth curves that are *nowhere tangent* to the characteristics. Notice that, in previous discussion, we chose to work with lines parallel to y-axis as the second types of coordinate curves.

Example 6.4 Consider the semilinear equation for $u \in C^1(\Omega)$ given by

$$x^2 p + y q + x u = x^2, \quad \text{for} \ z = u(x, y).$$

Since we have $a(x, y) = x^2$ and $b(x, y) = y$, the integral of characteristic equation

$$\frac{dy}{dx} = \frac{b}{a} = \frac{y}{x^2},$$

is given by the functions $y(x) = Ae^{-1/x}$, with $x \neq 0$ and $y > 0$. Now, for $y = y(x)$ to represent the characteristics of constant ϕ, we may take

$$\eta := \phi(x, y) = \ln y + \frac{1}{x} \quad \text{and} \quad \xi = \varphi(x, y) = x.$$

Clearly, $J = \eta_y = y^{-1} \neq 0$. Therefore, we can write $y = e^{\eta - (1/\xi)}$ so that, with $v(\xi, \eta) = u(x, y)$, the above transformation gives

$$p = u_x = v_\xi \varphi_x + v_\eta \phi_x = v_\xi - \frac{1}{\xi^2} v_\eta;$$

$$q = u_y = v_\xi \varphi_y + v_\eta \phi_y = \frac{1}{y} v_\eta = \frac{1}{e^{\eta - (1/\xi)}} v_\eta.$$

Hence, the given semilinear equation reduces to the differential equation

$$\xi^2 \left(v_\xi - \xi^{-2} v_\eta\right) + e^{\eta - (1/\xi)} \frac{1}{e^{\eta - (1/\xi)}} v_\eta + \xi \, v = \xi^2,$$

which simplifies to give

$$v_\xi + \frac{1}{\varphi} v = 1. \tag{6.1.13}$$

The above equation is valid for any (ξ, η), with $\xi \neq 0$, and the integration factor is given by

$$e^{\int (1/\xi) d\xi} = e^{\ln \xi} = \xi.$$

It thus follows that we have

$$u(\xi, \eta) = \xi^{-1} \int \xi \, d\xi + \xi^{-1} g(\eta) = \frac{\xi}{2} + \xi^{-1} g(\eta),$$

where g is some function. Finally, the general solution of the given semilinear equation is obtained as the function given by

$$u(x, y) = \frac{x}{2} + \frac{1}{x} g\big(\ln y + (1/x)\big).$$

More generally, a non-degenerate coordinates transformation reduces a semilinear equation to an ordinary differential equations of the form

$$h(\xi, \eta) v_\xi + k(\xi, \eta) v = c(\xi, \eta, v), \tag{6.1.14}$$

which we may solve by any standard solution method, where η is taken as a parameter. Notice that, as long as the characteristic equation is linear, the above procedure gives a simple solution of the semilinear equation. More precisely, in a suitable region such that $h(\xi, \eta) \neq 0$, we find the integrating factor by using

$$\alpha(\xi, \eta) = \frac{k(\xi, \eta)}{h(\xi, \eta)} \quad \text{and} \quad \beta(\xi, \eta) = \frac{c(\xi, \eta, v)}{h(\xi, \eta)},$$

so that we can write

$$e^{\int \alpha(\xi, \eta)d\xi} v_\xi + e^{\int k(\xi, \eta)d\xi} a(\xi, \eta)v = \beta(\xi, \eta)e^{\int \alpha(\xi, \eta)d\xi}$$

$$\Rightarrow \quad \frac{\partial}{\partial \xi}\left(e^{\int \alpha(\xi, \eta)d\xi} v\right) = \beta(\xi, \eta)e^{\int \alpha(\xi, \eta)d\xi}$$

$$\Rightarrow \quad e^{\int \alpha(\xi, \eta)d\xi} v = \int \beta(\xi, \eta)e^{\int \alpha(\xi, \eta)d\xi} d\xi + g(\eta),$$

where g is some function. Hence, the solution of a linear equation of the form (6.1.14) is given by

$$v(\xi, \eta) = e^{-\int a(\xi, \eta)d\xi} \left[\int b(\xi, \eta)e^{\int a(\xi, \eta)d\xi} d\xi + g(\eta)\right].$$

Finally, a general solution $u = u(x, y)$ of a semilinear equation is obtained by substituting back $\xi(x, y)$ and $\eta(x, y)$ into the last equation. However, when the characteristic equation is a nonlinear ordinary differential equation, the best we can hope for is to obtain a *local solution* uniquely, by applying Picard–Lindelöf theorem.

Example 6.5 Let $\Omega \subseteq \mathbb{R}^2$ be an open set. Consider the first order differential equation for a function $u \in C^1(\Omega)$ given by

$$\sqrt{1 - x^2}\, p + y q = 0, \quad \text{with } u(0, y) = y.$$

In this case, we have

$$a(x, y) = \sqrt{1 - x^2} \quad \text{and} \quad b(x, y) = y.$$

Therefore, the characteristics of the above equation are the integral curves of the ordinary differential equation given by

$$\frac{dy}{dx} = \frac{b(x, y)}{a(x, y)} = \frac{y}{\sqrt{1 - x^2}},$$

which has the general solution given by

$$y(x) = A\, e^{\sin^{-1} x}, \quad \text{for } -1 < x < 1 \text{ and } -2 < y < 2.$$

Notice that $y(0) = A$. Since $y = y(x)$ represent curves of constant η, we may take

$$\eta =: \phi(x, y) = \ln y - \sin^{-1} x, \quad \text{and} \quad \xi = \varphi(x, y) = x.$$

Clearly, $J = \phi_y = y^{-1} \neq 0$. By above equation, we have $y(\xi) = e^{\sin^{-1} \xi + \eta}$, where η is a parameter. Therefore, with $u(x, y) = v(\xi, \eta)$, the above transformation gives

$$p = v_\xi \varphi_x + v_\eta \phi_x = v_\xi - \frac{1}{\sqrt{1 - \xi^2}} v_\eta;$$

$$q = v_\xi \varphi_y + v_\eta \phi_y = \frac{1}{y} v_\eta = \frac{1}{e^{\sin^{-1} \xi + \eta}} v_\eta.$$

Hence, the given equation reduces to the ordinary differential equation

$$\sqrt{1 - \xi^2}\, v_\xi - v_\eta + v_\eta = 0, \quad \text{i.e.,} \quad v_\xi = 0,$$

which is valid for any (ξ, η), with $|\xi| < 1$. It thus follows that

$$u(x, y) = g(\ln y - \sin^{-1} x), \quad \text{for } |x| < 1 \text{ and } |y| < 2,$$

where g is an arbitrary univariate C^1-function. Finally, since the condition $u(0, y) = y$ implies that $y = g(\ln y)$, i.e. $g(x) = e^x$, a unique solution of the given Cauchy problem is obtained as

$$u(x, y) = y\, e^{-\sin^{-1} x}, \quad \text{with } |x| < 1.$$

As illustrated in Example 6.5, it is in general possible to find a unique solution of Eq. (6.1.11) by assigning an *initial data* on a *noncharacteristic* initial curve. With $a \neq 0$ over a neighbourhood Ω_0 of point $(x_0, y_0) \in \Omega$, since the *initial curve* $x = x_0$ is not characteristic with respect to Eq. (6.1.7) at point (x_0, y_0), a unique solution $\eta(x, y)$ satisfying the condition $\eta(x_0, y) = y$ exists in a neighbourhood $U \subseteq \Omega_0$ of (x_0, y_0). A requisite transformation is nonsingular if we choose $\xi = \varphi(x, y)$ arbitrarily subject to the Jacobian of the transformation which is nonzero at the point $(x_0, y_0) \in U$ (see (5.2.4)). Notice that we also have that $a \neq 0$ over U. In most situations, since $\eta_y(x_0, y_0) = 1$, a natural choice is to take $\xi = \varphi(x, y) = x$ so that we automatically have that $J = \phi_y \neq 0$ over Ω_0 (by usual continuity argument). By reversing the roles of variables x and y, a similar approach applies when $b \neq 0$ at a point $(x_0, y_0) \in \Omega$. The next examples illustrate the above argument.

Example 6.6 Let $\Omega \subseteq \mathbb{R}^2$ be an open set, and $u \in C^1(\Omega)$. Consider the Cauchy problem given by

$$u_x + x u_y = y, \quad \text{with } u(0, y) = y.$$

Take $(x_0, y_0) = (0, 0)$. Since η is given by a unique solution of the characteristic equation

$$\eta_x + x \eta_y = 0, \quad \text{with } \eta(0, y) = y,$$

we obtain $y - (x^2/2) = constant$. Therefore, the general solution of above ordinary differential equation is given by $\eta = \phi(y - (x^2/2))$. In particular, for it to satisfy the initial condition $\eta(0, y) = y$, we must have $\phi(y) = y$, and hence it follows that

$$\eta = \phi(x, y) = y - \frac{x^2}{2}, \quad \text{for } (x, y) \in \mathbb{R}^2.$$

Taking $\xi = \varphi(x, y) = x$, the Jacobian of the transformation $J = \varphi_x \eta_y - \varphi_y \eta_x = 1$, and the inverse relations are given by $x = \xi$, $y = \eta + (\xi^2/2)$. In new coordinates, the given problem reduces to the initial value problem given by

$$v_\xi = \eta + \frac{\xi^2}{2}, \quad \text{with } v(0, \eta) = \eta + (\xi^2/2).$$

The general solution of the above ordinary differential equation is given by

$$v(\xi, \eta) = \eta\xi + \frac{\xi^3}{6} + f(\eta), \quad \text{for an arbitrary univariate function } f.$$

Hence, the general solution of the given semilinear equation is obtained as

$$u(x, y) = xy - \frac{x^3}{3} + f(y - (x^2/2)), \quad \text{for } (x, y) \in \mathbb{R}^2.$$

Second Order Linear Hyperbolic Equation

The procedure described above also applies to a second order linear hyperbolic equation in two variables, mainly due to the fact that such a type of equation always has two distinct real characteristics. The next example illustrates the procedure.

Example 6.7 For $x \in [0, 1]$ and $y > 0$, consider the initial value problem given by

$$u_{xx} + u_{xy} - 2u_{yy} = -1, \quad \text{with } u(x, 0) = u_y(x, 0) = x.$$

As $\lambda_1 = -2$ and $\lambda_2 = 1$ are the two real distinct roots of the quadratic equation $\lambda^2 + \lambda - 2 = 0$, the two characteristics are given by the ordinary differential equations

$$\frac{dy}{dx} = 1 \quad \text{and} \quad \frac{dy}{dx} = -2,$$

which suggests that we may take

$$\xi = \varphi(x, y) = x - \frac{y}{2} \quad \text{and} \quad \eta = \phi(x, y) = x + y.$$

Therefore, the canonical form of the given equation is given by the differential equation $v_{\xi\eta} + (2/9) = 0$, which has the solution given by

$$v(\xi, \eta) = -\frac{2}{9}\xi\eta + f(\xi) + g(\eta),$$

where f, g are some functions. Notice that the initial condition $u(x, 0) = x$ implies

$$u(\xi = x, \eta = x) = -\frac{2}{9} \quad \text{and} \quad x^2 + f(x) + g(x) = x.$$

Hence, we obtain $f(x) + g(x) = x + (2/9)x^2$. Again, since

$$u_y = -(1/2)v_\xi + v_\eta = (1/9)\eta - (1/2)f'(\xi) + g'(\eta),$$

it follows from the initial condition $u_y(x, 0) = x$ that

$$u_y(\xi = x, \eta = x) = \frac{1}{9}x - \frac{1}{2}f'(x) + g'(x) = x,$$

and hence by applying integration we obtain

$$g(x) - (1/2)f(x) = (5/9)x^2 + k.$$

Solving for f and g, we obtain

$$f(x) = \frac{2}{3}x - \frac{2}{9}x^2 - \frac{2}{3}k \quad \text{and} \quad g(x) = \frac{1}{3}x + \frac{4}{9}x^2 + \frac{2}{3}k.$$

We thus conclude that

$$v(\xi, \eta) = -\frac{2}{9}\xi\eta + \frac{2}{3}\xi - \frac{2}{9}\xi^2 + \frac{1}{3}\eta + \frac{4}{9}\eta^2,$$

which shows that a unique solution of the given problem is obtained as

$$u(x, y) = x + xy + \frac{y^2}{2},$$

by using the *inverse transformation* given by

$$x(\xi, \eta) = \frac{2\xi + \eta}{2} \quad \text{and} \quad y(\xi, \eta) = \frac{2(\eta - \xi)}{3}.$$

We consider now a Cauchy problem related to the *transport equation* (6.1.3) for the function $u = u(x, t) \in C^1(\mathbb{R} \times [0, \infty))$ that represents the concentration of a diffusing substance in some medium, along the x-axis and at time $t > 0$. Suppose the *initial concentration* of the substance is given by

$$f(x) = u(x, 0) = \begin{cases} x, & 0 < x < 1 \\ 0, & \text{otherwise} \end{cases}, \quad \text{for } x \in \mathbb{R}.$$

It is assumed that the initial curve specified by f satisfies the (compatibility) condition as stated in Definition 7.2. Then, the unique solution of this Cauchy problem is given by

$$u(x, t) = f(x - ct)$$
$$= \begin{cases} x - ct, & ct < x < ct + 1 \\ 0, & \text{otherwise} \end{cases}, \quad x \in \mathbb{R}, \quad t \geq 0,$$

Therefore, for $c > 0$, the contaminant $u(x, t)$ moves from left to right after every fixed point in time $0 = t_0 < t_1 < t_2 \cdots$, shifted every time to the right by 0, ct_1, ct_2, ... units. And, for $c < 0$, it travels with the speed $|c|$ from right to left in the same way. Hence, the integral surface of the constant speed transport equation (6.1.3) describes the movement of the contaminant $u = u(x, t)$ from the point $(0, \xi)$ to some other point (t, x) along the *lines* $x = ct + \xi$. More generally, the *method of characteristics*, as discussed later in Chap. 7, provides a geometric procedure to construct the integral surface of a Cauchy problem for a first order differential equation, along a specified (*compatible*) initial curve.

In what follows, we apply the above procedure to find a solution of a Cauchy problem for an important second order hyperbolic equations in two variables, namely the wave equation given by

$$y_{tt} - c^2 y_{xx} = 0, \quad \text{for } -\infty < x < \infty \text{ and } t \geq 0, \tag{6.1.15}$$

where $y(x, t)$ represents the transversal displacement[1] of a vibrating string at point $(x, t) \in [0, \ell] \times [0, \infty)$, and the scalar $c > 0$ represents the *wave speed*. We may view the function $y : \mathbb{R} \times [0, \infty) \to \mathbb{R}$ which is an odd 2ℓ-*periodic extension* of the *displacement function* $y(x, t)$ that agrees with the initial curve defined over the interval $0 < x < \ell$. More precisely, we prove *d'Alembert's formula* of finding a unique solution of the wave equation (6.1.15).

Theorem 6.1 (d'Alembert's Formula) *The Cauchy problem for the 1-dimensional wave equation given by*

$$y_{tt} - c^2 y_{xx} = 0, \quad \text{for } (x, t) \in \Omega = \mathbb{R} \times \mathbb{R}^+;$$
$$y(x, 0) = d(x) \quad \text{and} \quad y_t(x, 0) = v(x), \quad \text{with } d, v \in C^2(\Omega), \tag{6.1.16}$$

---

[1] It is also possible to take the function $y(x, t)$ to represent the lateral displacement of a cylindrical elastic rod of small cross-sectional area.

has a unique solution given by

$$y(x, t) = \frac{1}{2}[d(x - ct) + d(x + ct)] + \frac{1}{2c} \int\limits_{x-ct}^{x+ct} v(u)du. \qquad (6.1.17)$$

Proof We know that $y(x, t)$ satisfies the wave equation (6.1.15). And, since $\lambda_1 = -1/c$ and $\lambda_2 = 1/c$ are the two real distinct roots of the quadratic equation $c^2\lambda^2 - 1 = 0$, it follows the 2-parameter family of curves of the characteristics system given by

$$\frac{dt}{dx} = \frac{1}{c} \quad \text{and} \quad \frac{dt}{dx} = -\frac{1}{c}, \qquad (6.1.18)$$

which suggests to consider the transformation of the form

$$\xi(x, t) = x - ct \quad \text{and} \quad \eta(x, t) = x + ct.$$

In practical terms, we say that *information* along these characteristics is propagated, respectively, with speeds c and $-c$. As shown earlier, the equation $z_{\xi\eta} = 0$ is the canonical form of the hyperbolic wave equation. Integrating twice, the solution is thus given by

$$z(\xi, \eta) = F(\xi) + G(\eta) = F(x - ct) + G(x + ct),$$

where $F, G \in C^2(\mathbb{R})$ are arbitrary. Notice that

$$y(x, 0) = d(x)$$
$$\Rightarrow \quad z(\xi = x, \eta = x) = F(x) + G(x) = d(x);$$

Also, since $y_t = -cz_\xi + cz_\eta = c(z_\eta - z_\xi)$, we have

$$y_t(x, 0) = v(x)$$
$$\Rightarrow \quad y_t(\xi = x, \eta = x) = c(G'(x) - F'(x)) = v(x).$$

Integrating, we get

$$-cF(x) + cG(x) = \int\limits_0^x v(u)du + A.$$

Solving the last two equations for F and G, we obtain

$$F(x) = \frac{1}{2c}\left[cd(x) - \int_0^x v(u)du - A \right] ; \text{ and}$$

$$G(x) = \frac{1}{2c}\left[cd(x) + \int_0^x v(u)du + A \right].$$

Hence, it follows that

$$z(\xi, \eta) =$$

$$\frac{1}{2c}\left[cd(x - ct) - \int_0^{x-ct} v(u)du \right] + \frac{1}{2c}\left[cd(x + ct) + \int_0^{x+ct} v(u)du \right],$$

which proves that

$$y(x, t) = \frac{1}{2}\left[d(x - ct) + d(x + ct) \right] + \frac{1}{2c}\int_{x-ct}^{x+ct} v(u)du, \tag{6.1.19}$$

for every $-\infty < x < \infty$ and $t \geq 0$. This is called the *d'Alembert formula*. The uniqueness follows by applying *energy method*, as discussed earlier. ☐

In particular, with $c = 1$, $d(x) = \cos x$, and $v(x) = x$, we have

$$y(x, t) = \cos x \cos t + xt, \quad \text{for all } (x, t) \in \Omega.$$

The *d'Alembert's formula* in general provides a nice physical interpretation of the Cauchy problem in terms of *domain of dependence* of $y(x, t)$ at a point $(x_0, t_0) \in \Omega$ and the *region of influence* over an interval $I = [a, b]$. For any $x_0 \in I$, the projected characteristics in this case are given by the lines $x - ct = x_0$ and $x + ct = x_0$, starting from the point $(x_0, 0)$. To analyse the geometry over xt-plane, we may write

$$x - ct = x_0 \implies t = \frac{1}{c}x - \frac{1}{c}x_0;$$
$$x + ct = x_0 \implies t = -\frac{1}{c}x + \frac{1}{c}x_0. \tag{6.1.20}$$

It follows from Eq. (6.1.19) that, for any $(x_0, t_0) \in \Omega$, the solution depends on the values of the *displacement function* $d(x)$ at points $x_0 + ct_0$ and $x_0 - ct_0$, and also on the values of the function $v(x)$ in the interval $[x_0 - ct_0, x_0 + ct_0]$. Said differently, $y(x_0, t_0)$ depends only the *initial data* assigned on the interval $[x_0 - ct_0, x_0 + ct_0]$.

Definition 6.1 The region of the xt-plane bounded by the lines $x = x_0 - ct_0$ and $x = x_0 + ct_0$ passing through the point (x_0, t_0) is called the *characteristic triangle*, and the base interval $[x_0 - ct_0, x_0 + ct_0]$ is called the *domain of dependence* of the solution $y = y(x, t)$ at point (x_0, t_0).

Clearly, we have $y(x_0, t_0) = 0$ if both $d(x)$ and $v(x)$ are zero over this domain. Conversely, given any interval $[a, b] \subset \Omega$, we ask: *which points $(x, t) \in \Omega$ are influenced by the initial conditions on the $[a, b]$*? The *region of influence* of the interval $[a, b]$ consists precisely of such points. Notice that these are the points that have the *domain of dependence* overlapping the interval $[a, b]$. Therefore, for any (x_0, t_0) *outside the region of influence* of $[a, b]$, no initial data assigned over the interval can determine the values $y(x_0, t_0)$. Hence, the initial conditions assigned over $[a, b]$ influence those points (x, t) in xt-plane that satisfy the conditions as given below:

$$x - ct \leq b \quad \text{and} \quad x + ct \geq a.$$

Clearly, the region of influence is the truncated *characteristic cone* bounded by the interval $[a, b]$ and also by the lines given by

$$x + ct = a \quad \text{and} \quad x - ct = b, \quad \text{for } t > 0,$$

so that $y \equiv 0$ over $\mathbb{R} \setminus [x - ct, x + ct]$ if both $d(x)$ and $v(x)$ are zero outside $[a, b]$.

The above considerations lead directly to the conclusion that the procedure, known as the *graphical method*, applies while solving Cauchy problems for wave equation (6.1.15). Notice that, for any $x_0 \in \Omega$, we have

$$y(x, t) = F(x - ct) + G(x + ct), \quad (x, t) \in \Omega \times \mathbb{R}^+, \tag{6.1.21}$$

where the functions F and G over Ω are given by

$$F(x) = \frac{1}{2}\left[d(x) - \frac{1}{c}\int_{x_0}^{x} v(u)du\right];$$

$$\tag{6.1.22}$$

$$G(x) = \frac{1}{2}\left[d(x) + \frac{1}{c}\int_{x_0}^{x} v(u)du\right].$$

The above two functions help identify the *right propagation* and *left propagation* along the characteristics of the solution (6.1.19): the part $F(x - ct)$ represents the *forward wave*, and $G(x + ct)$ represents the *backward wave*. Using the following procedure, and assuming that $F, G \in C^2(\Omega)$, the *graphical method* provides a *classic solution* at time t:

1. Graph $F(x - ct)$ and $G(x + ct)$ at $t = 0$.
2. Translate the forward wave to right by ct units, and the backward wave to left by ct units.
3. Add together the resulting forward and backward waves.

Notice that, however, if the two functions F and G are only piecewise continuous over Ω, then $y = y(x, t)$ in (6.1.21) is also piecewise continuous, and so we are led to the notion of *generalised solution* of the wave equation obtained by using sequences of smooth functions $\langle F_n \rangle$ and $\langle G_n \rangle$ converging, respectively, to F and G such that

$$y_n(x, t) = F_n(x - ct) + G_n(x, t), \quad \text{for all } n$$
$$\Rightarrow \quad y(x, t) = F(x - ct) + G(x + ct).$$

Further, unlike parabolic equations, for a hyperbolic equation like (6.1.15) a typical situation arises at a point (x_0, t_0) where the solution $y = y(x, t)$ is not smooth. In this case, since either F is not smooth at $x_0 - ct_0$ or G is not smooth at $x_0 + ct_0$, the *singularities* would travel along the *forward wave* or along the *backward wave*.

Example 6.8 With notations as above, and $x \in \Omega$, suppose

$$u(x, 0) = f(x) = \begin{cases} 2, & |x| \le a \\ 0, & \text{otherwise} \end{cases};$$

$$u_t(x, 0) = g(x) = 0.$$

Then, the *forward wave* is obtained as

$$F(x - ct) = \begin{cases} 1, & |x - ct| \le a \\ 0, & \text{otherwise} \end{cases}, \quad \text{for } t \ge 0,$$

and the *backward wave* is obtained as

$$G(x + ct) = \begin{cases} 1, & |x + ct| \le a \\ 0, & \text{otherwise} \end{cases}, \quad \text{for } t \ge 0.$$

Example 6.9 With notations as above, and $x \in \Omega$, suppose

$$u(x, 0) = f(x) = 0;$$

$$u_t(x, 0) = g(x) = \begin{cases} 0, & x < 0 \\ 1, & x \ge 0 \end{cases}.$$

We may take $c = 1$. Here, the *forward wave* is obtained as

$$F(x - ct) = -\frac{1}{2} \int_0^{x-t} g(u)\mathrm{d}u = -\frac{\max\{0, x - t\}}{2},$$

and the *backward* wave is obtained as

$$G(x + ct) = \frac{1}{2} \int\limits_0^{x+t} g(u)du = \frac{\max\{0, x + t\}}{2}.$$

In this case, we can use characteristic lines $x = -t$ and $x = t$ to graph the translated *backward* and *forward* waves, where the latter is zero for $x > t$. Therefore, the solution is given by

$$y(x, t) = \begin{cases} 0, & x < -t \\ \frac{1}{2}(x + t), & -t \leq x \leq t, \\ t, & x > t \end{cases} \quad (x, t) \in \Omega \times \mathbb{R}^+.$$

Unlike a boundary value problem, the domain where a Cauchy problem admits a unique solution is not known in advance. Nevertheless, we can formulate a Cauchy problem for a differential equation on part of the boundary of the *domain of dependence* as the limiting case of the desired solution.

We discuss next an application of *Duhamel principle* to the nonhomogeneous wave equation, with homogeneous initial conditions. It uses Dirichlet formula for the solution of the associated homogeneous problem and also the fact as given below:

$$\frac{d}{dt} \int\limits_0^t f(t, \tau)d\tau = f(t, t) + \int\limits_0^t f_t(t, \tau)d\tau, \quad \text{for all } f \in C^1([0, t]). \qquad (6.1.23)$$

Theorem 6.2 (Duhamel Principle) *A unique solution of the following Cauchy problem related to nonhomogeneous wave equation*

$$y_{tt} = c^2 y_{xx} + f(x, t), \quad \text{for } x \in \mathbb{R} \text{ and } t > 0;$$
$$y(x, 0) = 0 \quad \text{and} \quad y_t(x, 0) = 0, \quad \text{for all } x \in \mathbb{R}, \qquad (6.1.24)$$

is given by the function

$$y(x, t) = \frac{1}{2c} \int\limits_0^t \int\limits_{x-c(t-s)}^{x+c(t-s)t} f(\xi, \tau)d\xi d\tau. \qquad (6.1.25)$$

Proof Let $y(x, t; \tau)$ be a one-parameter solution of the *homogeneous* wave equation

$$y_{tt}(x, t; \tau) = c^2 y_{xx}(x, t; \tau), \quad \text{for } x \in \mathbb{R} \text{ and } t > \tau;$$
$$y(x, \tau; \tau) = 0 \quad \text{and} \quad y_t(x, \tau; \tau) = f(x, \tau), \quad \text{for } x \in \mathbb{R}. \qquad (6.1.26)$$

Notice that this is a Cauchy problem with initial conditions specified at a *delayed* time $t = \tau$. The main argument of *Duhamel principle* is that the function

$$y(x, t) = \int_0^t y(x, t; \tau)d\tau, \quad (x, t) \in \mathbb{R} \times \mathbb{R}^+, \tag{6.1.27}$$

is a solution of the problem (6.1.24). By (6.1.23), we have

$$y_t(x, t) = y(x, t; t) + \int_0^t y_t(x, t; \tau)d\tau$$

$$= \int_0^t y_t(x, t; \tau)d\tau, \quad \text{for all } x \in \mathbb{R}. \text{ (by first IC in (6.1.26))}$$

Applying Eq. (6.1.23) again, we obtain

$$y_{tt}(x, t) = y_t(x, t; t) + \int_0^t y_{tt}(x, t; \tau)d\tau$$

$$= f(x, t) + \int_0^t y_{tt}(x, t; \tau)d\tau, \quad \text{for all } x \in \mathbb{R}. \text{ (by second IC in (6.1.26))}$$

It thus follows from the *parametric* wave equation in (6.1.26) that

$$y_{tt}(x, t) = f(x, t) + \int_0^t y_{tt}(x, t; \tau)d\tau$$

$$= f(x, t) + c^2 \int_0^t y_{xx}(x, t; \tau)d\tau$$

$$= f(x, t) + c^2 y_{xx}(x, t),$$

where the last equality holds by (6.1.27). Therefore, to complete the proof, we have to find the solution $y(x, t; \tau)$ of the *homogeneous* problem (6.1.26). However, it follows directly from *Dirichlet formula* after rescaling. That is, we have

$$y(x, t; \tau) = \frac{1}{2c} \int_{x-c(t-\tau)}^{x+c(t-\tau)} f(\xi, \tau)d\xi, \tag{6.1.28}$$

for every $-\infty < x < \infty$ and $t > 0$. Hence,

$$y(x, t) := \frac{1}{2c} \int\limits_{0}^{t} \int\limits_{x-c(t-\tau)}^{x+c(t-\tau)} f(\xi, \tau) d\xi d\tau,$$

over the (triangular) *domain of dependence* with respect to point (x, t). □

Notice that the method extends easily to situation when both the *displacement function* $d(x) = y(x, 0) = f(x)$ and the *initial velocity* $v(x) = y_t(x, 0) = g(x)$ are *nonzero* for all $x \in \mathbb{R}$. In this case, solution is sum of the *Dirichlet formula* and the solution obtained by the previous theorem (Exercise 6.4).

Remark 6.1 Unlike a hyperbolic differential equation such as wave equation, a parabolic differential equation like heat equation $u_t = k \, \nabla^2 u \, (k = a^2)$ has the *regularity property*; i.e. the solution is immediately smoothed even when the initial data is rough. However, in this case, the *method of characteristics* is not applicable.

In Chapter 11, we obtain the fundamental solution of 1-dimensional heat equation over $\mathbb{R}^+$ of the form

$$u(x, t) = \frac{1}{2\sqrt{\pi kt}} e^{-x^2/4kt},$$

by using Laplace transform technique. The case of heat equation defined over bounded domain, and with homogeneous boundary conditions, is dealt with in Chap. 8, using method of *separation of variables* and Fourier series. And, for nonhomogeneous boundary conditions, we will use in Chap. 9 the method of *eigenfunction expansion* and generalised Fourier series.

We conclude our discussion here by solving the Cauchy problem for homogeneous heat equation as given below:

$$\begin{aligned} u_t &= k \, u_{xx} + f(x, t), \quad \text{for } x \in \mathbb{R} \text{ and } t > 0; \\ u(x, 0) &= \phi(x) \quad x \in \mathbb{R}. \end{aligned} \tag{6.1.29}$$

6.2 Lagrange's Method

Let $\Omega \subseteq \mathbb{R}^2$ be an open set and $u \in C^1(\Omega)$. We may write $D = \Omega \times u(\Omega)$. Consider a quasilinear equation for function $z = u(x, y)$ given by

$$a(x, y, z) \, p + b(x, y, z) \, q = c(x, y, z), \quad \text{with } p = u_x \text{ and } q = u_y, \tag{6.2.1}$$

where $a, b, c \in C^1(D)$ are such that not both a and b vanish simultaneously over D. The method of characteristic coordinates, as discussed in the previous section, cannot be applied in this case. However, we can find the general solution of

Eq. (6.2.1) by applying a method due to *Joseph-Louis Lagrange* (1736–1813). The main argument uses the *integral curves* of the vector field

$$f(x) = (a(x), b(x), c(x)), \quad \text{for } x = (x, y, z) \in D. \tag{6.2.2}$$

We will need some of the terminology introduced earlier in Sect. 2.2. More generally, suppose D is an open set in $\mathbb{R}^3$, and let the graph surface of a function $F \in C^1(D)$ be written as Γ_F. For an arbitrary (regular) point $x_0 = (x_0, y_0, z_0) \in D$,

$$F(x, y, z) = F(x_0, y_0, z_0),$$

is a level surface of Γ_F, which we may write as Γ_{x_0}. Therefore, by Theorem 2.20, the normal $\mathbf{n}(x_0)$ to the surface Γ_{x_0} at point x_0 is given by

$$\mathbf{n}(x_0) := \nabla F\Big|_{x_0} = (F_x(x_0), F_y(x_0), F_z(x_0)).$$

Hence, the tangent plane T_{x_0} to the surface Γ_{x_0} at point x_0 is given by

$$T_{x_0} : \quad (x - x_0) F_x + (y - y_0) F_y + (z - z_0) F_z = 0.$$

In particular, if $u \in C^1(\Omega)$ is a solution of Eq. (6.2.1), and we write the corresponding graph surface Γ_u as

$$\Gamma_u : \quad F(x, y, z) = u(x, y) - z, \quad \text{with } z = u(x, y),$$

it follows that the normal to the surface Γ_u is given by

$$\mathbf{n}(x, y, z) = \nabla F = (p, q, -1), \quad \text{for } (x, y) \in \Omega.$$

As Eq. (6.2.1) in vector notation is given by

$$(a, b, c) \cdot (p, q, -1) = 0, \tag{6.2.3}$$

we conclude that Γ_u is an integral surface of Eq. (6.2.1) if and only if the tangent plane T_{x_0} at each point $x_0 = (x_0, y_0, z_0) \in \Gamma_u$ contains the vector $f(x_0)$ given by (6.2.2). The central idea of *Lagrange's method* is to find the general solution of Eq. (6.2.1) by using a 2-parameter family of curves obtained from the intersection of two (*nondegenerate*) integral surfaces of the equation. Suppose $\varphi, \psi \in C^1(D)$ are *functionally independent* functions such that the level surfaces given by

$$\varphi(x, y, z) = c_1 \quad \text{and} \quad \psi(x, y, z) = c_2, \tag{6.2.4}$$

are integral surfaces of Eq.(6.2.1). Since the condition of *nondegeneracy* of two surfaces is equivalent to the property that

$$\nabla \varphi \times \nabla \psi \neq 0, \quad \text{at each point } x_0 = (x_0, y_0, z_0) \in D,$$

it follows from Eq. (6.2.3) that the intersection of the two tangent planes $\Pi_\varphi(x_0)$ and $\Pi_\psi(x_0)$, respectively, of the surfaces Γ_φ and Γ_ψ contains the direction $f(x_0)$. Said differently, the equation of the tangential curve to surfaces Γ_φ and Γ_ψ at each point x_0 is given by the equation

$$\frac{x - x_0}{a(x_0)} = \frac{y - y_0}{b(x_0)} = \frac{z - z_0}{c(x_0)}. \tag{6.2.5}$$

That is, an integral surface of Eq. (6.2.1) satisfies the property that the tangent plane $\Pi_\phi(x_0)$ at each point $x_0 = (x_0, y_0, z_0)$ contains the line (6.2.5). Therefore, at each point of the curve C, *infinitesimal* line element $ds = (dx, dy, dz)$ satisfies the condition $ds \times f = 0$. Hence, through each point on an integral surface of Eq. (6.2.1) there passes an integral curve of the *autonomous system* given by

$$\frac{dx}{ds} = a(x, y, z)$$

$$\frac{dy}{ds} = b(x, y, z) \tag{6.2.6}$$

$$\frac{dz}{ds} = c(x, y, z)$$

For a subinterval $I \subseteq \mathbb{R}$, suppose $\gamma(s) = (x(s), y(s), z(s))$ is a *solution curve* of the system (6.2.6). By eliminating parameter s from the system (6.2.6), we may also write any two of the three variables x, y, z as a function of the third variable. For example, taking $a \neq 0$, variables y and z may be treated as functions of variable x so that it is possible to write the above system as

$$\frac{dy}{dx} = \frac{b(x, y, z)}{a(x, y, z)} \quad \text{and} \quad \frac{dz}{dx} = \frac{c(x, y, z)}{a(x, y, z)}, \tag{6.2.7}$$

With $a \neq 0$ over the domain D, the above pair of ordinary differential equations is called the *characteristic system* of a quasilinear equation (6.2.1), and the associated integral curves are called the *characteristic curves*. A similar remark applies for the case when $b \neq 0$ over the domain D. In view of relation $ds \times f = 0$, we can also write the system (6.2.6) as

$$\frac{dx}{a(x, y, z)} = \frac{dy}{b(x, y, z)} = \frac{dz}{c(x, y, z)}. \tag{6.2.8}$$

Suppose C is an integral curve of the vector field $f = (a, b, c)$ represented by the relations given in (6.2.4). Sometimes, we also say that C is a *first integral* of the autonomous system (6.2.6). As said earlier, since $\nabla \varphi$ and $\nabla \psi$ are normal to the curve C, the vector $f = (a, b, c)$ on the curve C at each point is tangential. Therefore, we also have

$$f \cdot \nabla \varphi = 0 \quad \text{and} \quad f \cdot \nabla \psi = 0, \tag{6.2.9}$$

In the present situation, we can reformulate Definition 2.25 as given below.

Definition 6.2 Let $D \subseteq \mathbb{R}^3$ be a domain. A function $\varphi = \varphi(x, y, z) \in C^1(D)$ is called a *first integral* of the characteristic system (6.2.8) if $f \cdot \nabla \varphi = 0$ over the domain D.

It follows from Lemma 2.1 that if $\varphi, \psi \in C^1(\Omega \times \mathbb{R})$ are two functionally independent first integrals of the vector field $f = (a, b, c)$, then the two equations in (6.2.4) describe the collection of all integral curves of f over the domain D. However, as said earlier, not necessarily uniquely. Therefore, the collection of first integrals of a vector field f is a 2-parameter family of curves.

Further, by Lemma 2.2, an infinite number of first integrals of $f = (a, b, c)$ can be formed from any given first integral of a vector field f. That is, if $\varphi \in C^1(D)$ is a first integral of the vector field $f = (a, b, c)$, then $\psi = G(\varphi)$ is also a first integral of f, for any single-variable C^1-function defined on a suitable subinterval of $\mathbb{R}$. Moreover, if $\varphi, \psi \in C^1(\Omega \times \mathbb{R})$ are first integrals of f, then $G(\varphi, \psi) = 0$ is also a first integral of f, for any two-variable C^1-function defined on a suitable domain in $\mathbb{R}^2$.

The next theorem proves that a general solution of Eq. (6.2.1) can be written in terms of 2-parameter family of characteristic curves, obtained from the intersections of nondegenerate integral surfaces of the quasilinear equation. Recall that two C^1-functions φ and ψ defined over an open set $D \subset \mathbb{R}^3$ are *functional independent* if $\nabla \varphi \times \nabla \psi \neq 0$ over D.

Theorem 6.3 (Lagrange Theorem) *Suppose $\varphi, \psi \in C^1(\Omega \times \mathbb{R})$ are functionally independent functions such that the integral curves of the vector field $f = (a, b, c)$ given by Eq. (6.2.4) are the characteristic curves of the autonomous system (6.2.8). Then, the implicit equation*

$$G(\varphi, \psi) = 0, \quad \text{for any } G \in C^1(\Omega \times \mathbb{R}), \tag{6.2.10}$$

is a general solution of the quasilinear equation (6.2.1).

Proof As $G(\varphi, \psi) = 0$, it follows from the equation given by

$$dG = G_x dx + G_y dy = 0,$$

that we have

$$G_x = G_\varphi [\varphi_x + \varphi_z p] + G_\psi [\psi_x + \psi_z p] = 0;$$
$$G_y = G_\varphi [\varphi_y + \varphi_z q] + G_\psi [\psi_y + \psi_z q] = 0$$

The above system of equations in unknowns G_φ and G_ψ has a nontrivial solution if and only if we have

$$\det \begin{pmatrix} \varphi_x + \varphi_z p & \psi_x + \psi_z p \\ \varphi_y + \varphi_z q & \psi_y + \psi_z q \end{pmatrix} = 0. \tag{6.2.11}$$

Expanding the determinant in (6.2.11), it follows that Eq. (6.2.1) holds, with functions a, b, c being replaced, respectively, by the relations given by

$$J[\varphi, \psi; y, z] = \det \begin{pmatrix} \varphi_y & \varphi_z \\ \psi_y & \psi_z \end{pmatrix}; \tag{6.2.12}$$

$$J[\varphi, \psi; z, x] = \det \begin{pmatrix} \varphi_z & \varphi_x \\ \psi_z & \psi_x \end{pmatrix}; \tag{6.2.13}$$

$$J[\varphi, \psi; x, y] = \det \begin{pmatrix} \varphi_x & \varphi_y \\ \psi_x & \psi_y \end{pmatrix}, \tag{6.2.14}$$

where J is the Jacobian as defined in (2.2.11). Next, taking differentials of functions φ and ψ, it follows from (6.2.4) that

$$d\varphi = \varphi_x \, dx + \varphi_y \, dy + \varphi_z \, dz = 0;$$
$$d\psi = \psi_x \, dx + \psi_y \, dy + \psi_z \, dz = 0.$$

Therefore, both vectors $\nabla\varphi$ and $\nabla\psi$ are orthogonal to the *differential vector* (dx, dy, dz), and hence the (nonzero) vector

$$\nabla\varphi \times \nabla\psi = \left(J[\varphi, \psi; y, z], J[\varphi, \psi; z, x], J[\varphi, \psi; , y, x] \right). \tag{6.2.15}$$

is parallel to (dx, dy, dz). Equivalently, we can write

$$\frac{dx}{J[\varphi, \psi; y, z]} = \frac{dy}{J[\varphi, \psi; z, x]} = \frac{dz}{J[\varphi, \psi; , y, x]}, \tag{6.2.16}$$

On the other hand, by Eq. (6.2.3), we also have

$$f \cdot \nabla\varphi = 0 \quad \text{and} \quad f \cdot \nabla\psi = 0, \tag{6.2.17}$$

where $f = (a, b, c)$. That is, at each point $(x, y, z) \in \Omega \times \mathbb{R}$, the vector f is parallel to the (nonzero) vector $\nabla\varphi \times \nabla\psi$. A simple comparison of the above observation with (6.2.16) implies that

$$\frac{J[\varphi, \psi; y, z]}{b} = \frac{J[\varphi, \psi; z, x]}{b} = \frac{J[\varphi, \psi; , x, y]}{c}. \tag{6.2.18}$$

It thus follows from (6.2.11) that $G(\varphi, \psi) = 0$ is a general solution of quasilinear equation (6.2.1). □

Notice that, in Theorem 6.3, we are not asserting that every general solution of Eq. (6.2.1) is of the form (6.2.10). It simply provides a procedure to find a general solution of a quasilinear equation (6.2.1) in terms of *any two* functionally independent C^1-functions φ and ψ, as obtained from the associated characteristic system (6.2.8). For the latter part, we usually apply methods such as given below:

1. (*Grouping Method*) Consider any two of these three ratios in (6.2.8), and solve the resulting first order differential equations.
2. (*Multiplier Method*) Use constants or functions as multipliers, while applying *componendo–dividendo* to any two or all the three ratios in (6.2.8), and then solve the resulting first order differential equations.

The next three examples illustrate the above two methods.

Example 6.10 Consider a semilinear equation $ap + bq + dz = 0$ for a C^1-function $z = u(x, y)$, where a, b, d are constants, with $a^2 + b^2 \neq 0$, and $c = -Cz$. Recall that the integral curves of the vector field $f = (a, b, -dz)$, with $(x, y) \neq (0, 0)$, are represented by Eq. (6.2.8), where φ and ψ are any two functionally independent C^1-functions obtained from the characteristic system given by

$$\frac{dx}{a} = \frac{dy}{b} = \frac{dz}{-Cz}.$$

From the first equality, we have $b\,dx - a\,dy = 0$, and so we may take $\varphi(x, y, z) = bx - ay = c_1$. On the other hand, by taking the first and last ratios, we have

$$\frac{dz}{dx} + \frac{C}{a}z = 0 \quad \Rightarrow \quad \psi(x, y, z) = e^{(C/a)x}z = c_2.$$

Notice that we have

$$\nabla\varphi = (b, -a, 0) \quad \text{and} \quad \nabla\psi = \left((C/a)e^{(C/a)x}z, 0, e^{(C/a)x}\right)$$

so that we obtain

$$\nabla\varphi \times \nabla\psi = \left(-ae^{(C/a)x}, -be^{(C/a)x}, Ce^{(C/a)x}z\right),$$

which implies φ and ψ are functionally independent over $\Omega = \mathbb{R}^2$. Therefore, a general solution of the Lagrange equation $ap + bq + Cz = 0$ is obtained as

$$u(x, y) = e^{-(C/a)x}g(bx - ay), \quad \text{for some } g \in C^1(\mathbb{R}^2).$$

As seen earlier in previous section, one of the integral curves of a semilinear equation is lines whenever the coefficients appearing in the principle part are constants. However, the same may not be true in general.

Example 6.11 Consider the homogeneous semilinear equation $yp + xq = 0$. In this case, the integral curves of the vector field $f = (y, x, 0)$, with $(x, y) \neq (0, 0)$, are represented by Eq. (6.2.8), where φ and ψ are any two functionally independent C^1-functions obtained from the characteristic system given by

$$\frac{dx}{y} = \frac{dy}{x} = \frac{dz}{0}.$$

By using the last ratio, it follows that we may take $\varphi(x, y, z) = z = c_1$, i.e. planes parallel to the xy-plane. Further, it follows from the first equality that $xdx - ydy = 0$ so that we may take $\psi(x, y, z) = x^2 - y^2 = c_2$, which is a family of hyperbolas in the xy-plane. For $(x, y) \neq (0, 0)$, we have $\nabla\varphi \times \nabla\psi \neq 0$ over the domain $\Omega = \mathbb{R}^2 \setminus \{(0, 0)\}$. Therefore, a general solution of the given equation may be written as

$$z = u(x, y) = g(x^2 - y^2), \quad \text{for some } g \in C^1(\Omega).$$

Example 6.12 Consider the quasilinear equation $xz\,p - yz\,q = y^2 - x^2$. In this case, the integral curves of the vector field $f = (xz, yz, y^2 - x^2)$, with $(x, y) \neq (0, 0)$, are represented by Eq. (6.2.8), where φ and ψ are any two functionally independent C^1-functions obtained from the characteristic system given by

$$\frac{dx}{xz} = \frac{dy}{-yz} = \frac{dz}{x^2 - y^2}.$$

By the first equality, we have $ydx + xdy = 0$ so that $\log x + \log y = \log a$. Therefore, we may take $\varphi(x, y, z) = xy = c_1$, which is a family of *rectangular hyperbolas*. Further, by using x and y as multipliers in the first equality, and also the third ratio, we have

$$\frac{xdx + ydy}{z(x^2 - y^2)} = \frac{dz}{x^2 - y^2} \quad \Rightarrow \quad x\,dx + y\,dy - z\,dz = 0.$$

We may thus take $\psi(x, y, z) = x^2 + y^2 - z^2 = c_2$, which is a family of *single-sheet hyperboloids*. The nondegeneracy condition can be verified as in the above examples. Hence, a general solution of the given quasilinear equation is given by $g(xy, x^2 + y^2 - u^2) = 0$, where g is a C^1-function defined on $\Omega = \mathbb{R}^2 \setminus \{(0, 0)\}$.

In general, unlike a semilinear equation, the main difficulty arises when each ordinary differential in system (6.2.8) involves all the three variables x, y, z. However, in case one of the three variables is missing from a characteristic equation, we can use some other method.[2] For example, when a and b are functions of variables x and y only so that the general solution of $dy/dx = b/a$ is given by $\varphi(x, y) = c_1$, it follows easily that φ is a first integral of system (6.2.8). Therefore, in addition to *grouping* and *multiplier* methods, we can also use *substitution* in the way as described below:

---

[2] One such method is known as *separation of variables*, which we shall discuss in Chap. 7.

3. (*Substitution Method*) Find the second functionally independent function ψ by eliminating one of the variables from one of the ordinary differential equations in (6.2.8) so that the reduced differential equations involve two variables only.

Example 6.13 Consider the semilinear equation $x\,p + y\,q = xy(z^2 + 1)$. In this case, the integral curves of the vector field $f = (xz, yz, y^2 - x^2)$, with $(x, y) \neq (0, 0)$, are represented by Eq. (6.2.8), where φ and ψ are any two functionally independent C^1-functions obtained from the characteristic system given by

$$\frac{dx}{x} = \frac{dy}{y} = \frac{dz}{xy(z^2 + 1)}.$$

It follows from the first equality that we may take $\varphi(x, y, z) = y/x = c_1$, with $x \neq 0$, which is a family of *planes*. Therefore, by using the substitution $y = c_1 x$, it follows from the first and last ratios in the above characteristic system that we may take

$$\psi(x, y, z) = c_1 \frac{x^2}{2} - \tan^{-1} z = \frac{xy}{2} - \tan^{-1} z = c_2.$$

It is easy to verify that φ and ψ are functionally independent at every point $(x, y, z) \in \mathbb{R}^3$ such that $x \neq 0$. Hence, a general solution of the given quasilinear equation is given by

$$g\left(y/x, (xy/2) - \tan^{-1} z\right) = 0,$$

where g is a C^1-function defined on $\Omega = \mathbb{R}^2 \setminus \{x = 0\}$.

Notice that, to do away with implicit relation $G(\varphi, \psi) = 0$, we construct the integral surface Γ_u by using characteristic curves passing through the distinct points of a given (initial) curve Γ_0 on the surface Γ_u. Let Γ_0 be the curve of intersection of the surfaces

$$\phi_1(x, y, u(x, y)) = 0 \quad \text{and} \quad \phi_1(x, y, u(x, y)) = 0;$$

and Γ be a characteristic curve defined by the level surfaces

$$\varphi(x, y, z) = c_1 \quad \text{and} \quad \psi(x, y, z) = c_2.$$

If Γ_0 is not a characteristic curve, then we can obtain an implicit relation $f(c_1, c_2) = 0$ by eliminating x, y, z from the last two equations. However, if Γ_0 is a characteristic curve, then the integral surface is a union of one-parameter family of curves of which Γ_0 is a member. Notice that, if

$$\phi(r, s) = (f_1(r, s), f_2(r, s), f_3(r, s)), \quad \text{with } J[f_1, f_2; r, s] \neq 0,$$

is a parametrisation of 1-parameter surface $\varphi = c$, where each $f_i : I \times K \to \mathbb{R}$ is a smooth function, then by the Jacobian condition as given in (2.2.11) it is always possible to find functions $\alpha, \beta : \Omega \to \mathbb{R}$ such that

$$r = \alpha(x, y) \quad \text{and} \quad s = \beta(x, y).$$

So, we may write $u(x, y) = f_3\big(\alpha(x, y), \beta(x, y)\big)$.

Remark 6.2 Let $\Omega \subset \mathbb{R}^n$ be a domain, and $x = (x_1, \ldots, x_n) \in \Omega$. A general first order quasilinear equation in n variables $x_1, \ldots, x_n$ is given by

$$a_1\big(x; u(x)\big)p_1 + \cdots + a_1\big(x; u(x)\big)p_n = b\big(x; u(x)\big), \qquad (6.2.19)$$

where $u = u(x) \in C^1(\Omega)$ is an unknown function, with $p_i = \partial_{x_i} u$. Partial differential equations of the form (6.2.19) have applications in many areas of scientific investigations such as quantum physics, gas dynamics, and geometric optics. The assertions we proved above, for the case when $n = 2$, remain valid in general. More precisely, suppose an integral curve C of a C^1 nonvanishing vector field given by

$$f\big(x; u(x)\big) = \big(a_1\big(x; u(x)\big), \ldots, a_n\big(x; u(x)\big)\big), \qquad (6.2.20)$$

is represented by the equations

$$\varphi_1\big(x, u(x)\big) = c_1, \quad \ldots, \varphi_n\big(x, u(x)\big) = c_n, \qquad (6.2.21)$$

where $\varphi_1, \ldots, \varphi_n$ are any n functionally independent C^1-functions obtained from the characteristic system given by

$$\frac{dx_1}{a_1} = \cdots = \frac{dx_n}{a_n} = \frac{du}{b}. \qquad (6.2.22)$$

As before, we say a function $u \in C^1(\Omega)$ is a *first integral* of f (or, equivalently, of the system (6.2.22)) if u satisfies the quasilinear equation (6.2.19). Then, for any C^1-function g, the implicit relation

$$g\big(\varphi_1, \ldots, \varphi_n\big) = 0 \qquad (6.2.23)$$

is a general solution of the quasilinear equation (6.2.19). The first order autonomous system of $(n + 1)$ ordinary differential equations given by

$$\frac{dx_i}{ds} = a_i\big(x; u\big), \quad \text{for} \ \ 1 \leq i \leq n; \ \ \text{and}, \ \ \frac{du}{ds} = b\big(x; u\big),$$

defines the characteristic curves of Eq. (6.2.19). The vector field f is defined in a region of the Euclidean space $\mathbb{R}^{n+1}$, and in terms of *Lie derivatives*, the left side of the quasilinear equation is of the form $L_f(u)$, where

$$L_f := a_1 \frac{\partial}{\partial x_1} + \cdots + a_n \frac{\partial}{\partial x_n}.$$

Therefore, a function $u = u(x)$ is a solution of the equation $L_f(u) = 0$ if and only if it is constant along the integral curves of the vector field f. However, this result no longer holds for higher order partial differential equations. An interesting situation arises when $a_i(x; u(x)) = x_i$, for $i = 1, \ldots, n$.

Recall that a Cauchy problem for a first order quasilinear equation (7.1.8), with initial condition $u(x, 0) = \phi(x)$, is about the existence of an integral surface that contains the *initial curve* $\phi : I \to D$ given by

$$\Gamma_\phi = \{(s, 0, \phi(s)) : s \in I\},$$

so that $u(x(s), 0) = \phi(s)$, for all $s \in I$. The curve $\phi = \phi(s)$ acts like a spine about which the whole integral surface is *knitted* as the union of characteristic curves. To begin with the construction of the integral surface along the curve Γ_ϕ, take a point $(x_0, 0, \phi(x_0)) \in \Gamma_\phi$. The first step is to find a characteristic curve from the system (7.1.10) for the initial condition as specified by the point $(x_0, 0, \phi(x_0))$. We obtain the desired integral surface as the union of all characteristics by repeating the same procedure for each point on Γ_ϕ. Example 6.14 illustrates the fact that, unlike semilinear equations, all the three characteristic equations are required to define the characteristics.

Example 6.14 Consider the Cauchy problem

$$(y + u)u_x + yu_y = x - y, \quad \text{with } u(x, 1) = 1 + x \text{ for } y > 0.$$

In this case, we need to solve the characteristic equations

$$\frac{dx}{ds} = y + u, \quad \frac{dy}{ds} = y, \quad \text{and} \quad \frac{du}{ds} = x - y.$$

Putting the first and third equations together, we obtain the differential equation $(x + u)' = x + u$. Also, as we can write $dy/ds = y$ as $y^{-1}dy/ds = 1$, it follows that

$$\frac{d}{ds}\left(\frac{x+u}{y}\right) = \frac{1}{y}\frac{d}{ds}(x+u) - \frac{x+u}{y^2}\frac{dy}{ds} = 0.$$

Therefore, we have $x + u/y = c_1$, where c_1 is a constant. Next, putting the first and second equation together, we obtain $d/ds(x - y) = u$, so that the third equation can also be written as

$$(x - y)\frac{d}{ds}(x - y) = u\frac{du}{ds}.$$

Hence, it follows from the equation

$$\frac{d}{ds}\left[(x - y)^2 - u^2\right] = 2(x - y)\frac{d}{ds}(x - y) - 2u\frac{du}{ds} = 0$$

that $(x - y)^2 - u^2 = c_2$, where c_2 is a constant. The general solution of the quasilinear equation is thus obtained as $(x - y)^2 - u^2 = g(x + u/y)$, where g is an arbitrary single-variable function. Finally, applying the initial condition $u(x, 1) = 1 + x$ and making the substitution $t = 2x + 1$, we obtain

$$(x - 1)^2 - (x + 1)^2 = g(2x + 1) \quad \Rightarrow \quad g(t) = 2(1 - t).$$

It thus follows that we have

$$(x - y)^2 - u^2 = 2(1 - (x + u)/y) = \frac{2}{y}(y - x - u).$$

Solving the above equation as quadratic in u, we have

$$u_{\pm}(x, y) = \frac{1}{y} \pm (x - y + (1/y)),$$

which together with the assigned initial condition $u(x, 1) = 1 + x$ gives $u_{\pm}(x, 1) = 1 \pm x$. This last relation suggests to take the positive root. Hence, a unique solution of the given Cauchy problem is obtained as

$$u(x, y) = x - y + \frac{2}{y}.$$

The next example illustrates the fact that, in general, a Cauchy problem for a linear differential equation may have a unique solution or no solution at all or an infinite number of solutions.

Example 6.15 As seen in Example 6.10, the general solution of the semilinear equation

$$2u_x + 3u_y + 8u = 0$$

is given by $u(x, y) = e^{-4x} g(3x - 2y)$, where g is some smooth function. Notice that since specifying $u = u(x, y)$ at a point would not determine the function g uniquely, so we need to consider a characteristic curve Γ satisfying a given initial conditions. First, let Γ_0 be the initial curve defined by the condition $u(x, 0) = \sin x$. Then, by putting $t = 3x$,

$$\sin x = u(x, 0) = e^{-4x} g(3x) \quad \Rightarrow \quad g(t) = \sin(t/3) e^{4t/3}.$$

Therefore, a unique solution in this case is given by

$$u(x, y) = \sin(x - 2y/3) e^{-8y/3}.$$

Also, using Γ_0 defined by the initial condition $u(x, x) = x^4$, we obtain a unique solution $u(x, y) = (3x - 2y)^4 e^{8(x-y)}$ along this initial curve. Next, suppose Γ_0 is defined by the initial condition $u(x, y) = x^2$ along the line $3x - 2y = 1$. It then turns out that we need to have $g(1) = x^2 e^{4x}$, $\forall x$. Since this is impossible, there is no solution of the associated Cauchy problem. Notice that $u = u(x, y)$ takes the value x^2 at points (x, y) on the line $3x - 2y = 1$. Finally, using the initial condition $u(x, y) = u(x, (3x - 1)/2) = e^{-4x}$ along the line $3x - 2y = 1$, we have $g(1) = 1$, a condition satisfied by infinite number of functions, and hence in this case there are infinitely many solutions of the associated Cauchy problem.

An interesting deduction that can be drawn from Example 6.15 is that among the three initial conditions considered above, only the line $3x - 2y = 1$ corresponds to a characteristic curve of the equation. Therefore, it is plausible to expect that the *existence* of a solution of a Cauchy problem depends on whether or not the initial curve is a characteristic curve.

6.3 Linear Equations with Constant Coefficients

As a simple application of Lagrange's method, we describe here a procedure to find the general solution of a linear partial differential equation in two variables, say x and y, with *constant coefficients*. Suppose the involved unknown function $z = u(x, y)$ is sufficiently smooth over a domain $\Omega \subseteq \mathbb{R}^2$. In this case, it is convenient to write the partial derivatives of u in *operator notations* as given by

$$Du = \frac{\partial u}{\partial x} = \partial_x u; \quad D'u = \frac{\partial u}{\partial y} = \partial_y u;$$

$$D^2 u = \frac{\partial^2 u}{vx^2} = \partial_{xx} u; \quad D'^2 u = \frac{\partial^2 u}{\partial y^2} = \partial_{yy} u;$$

$$D'Du = \frac{\partial u}{\partial x \partial y} = \frac{\partial u}{\partial y \partial x} = DD'u = \partial_{xy} u, \quad \text{etc.}$$

For example, if $u(x, y) = x^2 \cos y$, then

$$D^2 D'(x^2 \cos y) = D^2(-2x^2 \sin y) = -4D(x \sin y) = -4 \sin y.$$

Therefore, as a simple modification in notations introduced earlier in Sect. 5.1, we write $\mathscr{P}(D, D')$ for the nth order linear *partial differential operator* in two variables x and y, which can be expressed as a sum of (homogeneous) partial differential operator $\mathscr{P}_n(D, D')$ of the degree n as given by

$$\mathscr{P}_n(D, D') = \left[\sum_{k=0}^{n} a_k^{(n)} D^{n-k} D'^k \right], \quad \text{for } a_k^{(n)} \in \mathbb{R}. \tag{6.3.1}$$

Therefore, we can write a general nth order linear partial differential equation with constant coefficients as

$$\mathscr{P}(D, D')\, u = g(x, y), \quad \text{for } z = u(x, y) \in C^n(\Omega), \tag{6.3.2}$$

where the operator $\mathscr{P}(D, D')$ is given by

$$\mathscr{P}(D, D') \equiv \mathscr{P}_n(D, D') + \mathscr{P}_{n-1}(D, D') + \cdots + \mathscr{P}_1(D, D') + \mathscr{P}_0(D, D'),$$

with $\mathscr{P}_0(D, D') = a \in \mathbb{R}$. Recall that Eq. (6.3.2) is linear because each differential operator of degree n given by (6.3.1) is linear with respect to the two basic partial differential operators $D = \partial/\partial x$ and $D' = \partial/\partial y$. When $g \equiv 0$ over the domain Ω, we say (6.3.2) is a *homogeneous equation*. Otherwise, it is a *nonhomogeneous equation*.

The basic idea of the *solution method* as given below is similar to the procedure adopted earlier to find the general solution of a linear ordinary differential equation with constant coefficients. Therefore, we write the *complete solution* of Eq. (6.3.2) as

$$u(x, y) = u_h(x, y) + u_p(x, y), \quad \text{for } (x, y) \in \Omega, \tag{6.3.3}$$

where $u_h(x, y)$ is the *general solution* of the homogeneous equation

$$\mathscr{P}(D, D')u = 0, \tag{6.3.4}$$

and $u_p(x, y)$ is a *particular integral* Eq. (6.3.2) given by

$$u_p(x, y) := \left[\mathscr{P}(D, D')\right]^{-1} g(x, y). \tag{6.3.5}$$

We use Lagrange's method specifically to find the function u_h. Notice that, as for polynomials in two variables, the operator $\mathscr{P}(D, D')$ may have two types of *factors*:

1. A *reducible* homogeneous operator $\mathscr{P}_n(D, D')$, having factor such as a first order linear differential operators of the form $(aD + bD' + c)$, with $(a, b) \neq (0, 0)$ may be complex numbers. For example, the two homogeneous operators given by

$$\mathscr{P}_2(D, D') = D^2 - D'^2;$$
$$\mathscr{P}_3(D, D') = D^3 - 6D^2D' + 11DD'^2 - 6D'^3,$$

   are reducible such that we have

$$\mathscr{P}_2(D, D') = (D + D')(D - D');$$
$$\mathscr{P}_3(D, D') = (D - D')(D - 2D')(D - 3D').$$

2. An *irreducible inhomogeneous factor* of the form

$$2D^2 - D' \quad \text{or} \quad D^3 + DD' + D'^3.$$

Clearly, in general, an operator $\mathscr{P}(D, D')$ may contain terms that have both types of factors. For example, the operator

$$\mathscr{P}(D, D') = D^3 - D^2 D' + DD'^2 - D'^3$$

has a *reducible homogeneous* factor given by $(D - D')$ and *irreducible homogeneous operator* $D^2 + D'^2 - DD'$ as the other factor.

In any of the above situations, if $\mathscr{P}(D, D')$ has a factor of the form $aD + bD' + c$, then we need to solve a linear equation of the form

$$\big[aD + bD'\big]u = -cz, \quad \text{for } z = u(x, y), \tag{6.3.6}$$

with the associated characteristic system given by

$$\frac{dx}{a} = \frac{dy}{b} = \frac{dz}{-cz}.$$

Therefore, by Example 6.10, the general solution is given by

$$z = u(x, y) = e^{(-c/a)x} g(bx - ay), \tag{6.3.7}$$

where g is an arbitrary single-variable function. The general solution (6.3.7) simplifies to $u(x, y) = g(bx - ay)$ when $c = 0$ and to $u(x, y) = e^{(-c/b)y} g(x)$ when $a = 0$. In general, if an nth order differential operator $\mathscr{P}(D, D')$ has factors such as given by

$$\mathscr{P}(D, D') = \prod_{i=1}^{n} (a_i D + b_i D' + c_i), \quad \text{where } a_i \neq 0, \tag{6.3.8}$$

then we can write the general solution $z = u(x, y)$ as

$$u(x, y) = \sum_{i=1}^{n} e^{(-c_i/a_i)x} g_i(b_i x - a_i y), \tag{6.3.9}$$

provided all the factors on the right-hand side of (6.3.8) are *distinct*. The next two examples illustrate the procedure.

Example 6.16 For the third order homogeneous linear differential equation

$$(D^3 - 6D^2 D' + DD'^2 - 6D'^3) z = 0,$$

the associated differential operator $\mathscr{P}(D, D')$ factorises as

$$\mathscr{P}(D, D') = (D - 6D')(D^2 + D'^2) = (D - 6D')(D + iD')(D - iD').$$

So, the general function is given by

$$u(x, y) = g_1(y + 6x) + g_2(y - ix) + g_3(y + ix),$$

where g_1, g_2, g_3 are arbitrary single-variable function.

Example 6.17 For the second order homogeneous linear differential equation

$$(D^2 - 3DD' - 2D'^2 + 4D + 7D' - 6)z = 0,$$

the associated differential operator $\mathscr{P}(D, D')$ factorises as

$$\mathscr{L}(D, D') = (D - 2D' + 3)(2D + D' - 2).$$

So, the general function is given by

$$u(x, y) = e^{-3x}g_1(y + 2x) + e^x g_2(2y - x),$$

where g_1, g_2 are arbitrary single-variable function.

Next, suppose a factor $(aD + bD' + c)$ of operator $\mathscr{P}(D, D')$ repeats, say with multiplicity two. In this case, taking

$$(aD + bD' + c)u = v(x, y),$$

we can write

$$(aD + bD' + c)v = (aD + bD' + c)^2 u = 0,$$

which, by the above argument, gives

$$v(x, y) = e^{(-c/a)x} g(bx - ay).$$

Therefore, in this case, we need to solve the linear equation

$$(aD + bD' + c)u = e^{(-c/a)x} g(bx - ay),$$

which has the characteristic system given by

$$\frac{dx}{a} = \frac{dy}{b} = \frac{dz}{e^{(-c/a)x}g(bx - ay) - cz}.$$

Clearly, we can take the first integral curve as $\varphi(bx - ay) = k_1$. For the second integral curve, we need a solution of the differential equation given by

$$\frac{dz}{dx} + \frac{c}{a}z = \frac{e^{(-c/a)x}g(bx - ay)}{a}.$$

For, applying method of *integrating factor*, we obtain

$$e^{(c/a)x}z = \int \left[\frac{g(bx - ay)}{a}\right]dx + k_2 = h(bx - ay) + k_2.$$

Hence, the general solution is given by

$$u(x, y) = e^{(-c/a)x}\left[g(bx - ay) + h(bx - ay)\right], \qquad (6.3.10)$$

where g, h are arbitrary single-variable functions. Of course, we need to repeat the same procedure whenever some factor of $\mathscr{P}(D, D')$ of the form $(aD + bD' + c)$ has multiplicity more than two.

Example 6.18 For the third order homogeneous linear differential equation

$$(D^3 - 4D^2D' + 4DD'^2)u = 0, \quad \text{for } z = u(x, y),$$

we have

$$\mathscr{P}(D, D') = D(D + 2D')^2.$$

Therefore, the general solution is given by

$$u(x, y) = g_1(y) + g_2(y + 2x) + g_3(y + 2x),$$

where g_is are arbitrary single-variable functions.

Example 6.19 For the third order homogeneous linear differential equation

$$(D^3 + 3D^2D' - 18DD' - 6D^2 + 12D'^2 + 9D - 9D')z = 0,$$

the operator $\mathscr{P}(D, D')$ factorises as $(D - D')(D + 2D' - 3)^2$. Therefore, the general solution is given by

$$u(x, y) = g(x + y) + e^{3x}\left[h_1(2x - y) + h_2(2x - y)\right],$$

where g, h_1, h_2 are arbitrary single-variable functions.

When a factor of operator $\mathscr{P}(D, D')$ is not of the form $(aD + bD' + c)$, we take a trial solution of the form

$$u(x, y) = A\,e^{\alpha x + \beta y}, \quad \text{with } A, \alpha, \beta \in \mathbb{R},$$

so that it follows from the direct substitution that $\mathscr{P}(\alpha, \beta) = 0$, and if $\beta_i = f(\alpha_i)$ for some function, then the general solution of the homogeneous equation $\mathscr{P}(D, D')u = 0$ is given as a linear combination of functions of the form

$$A_i \, e^{\alpha_i x + f(\alpha_i) y}, \quad \text{for } i = 1, 2, \ldots, n.$$

Example 6.20 For the differential equation $(2D^2 - D')u = 0$, we have $F(a, b) = 2a^2 - b = 0$ so that $b = 2a^2$ implies that the general solution of the differential equation is given by

$$u(x, y) = \alpha \, e^{a(x + 2ay)}.$$

Similarly, for the differential equation $(D^3 - 3DD' + D' + 2)z = 0$, we have $F(a, b) = a^3 - 3ab + b + 2 = 0$. Therefore, for $b = (a^3 + 2)/3a - 1$, the general solution of the differential equation in this case is given by

$$u(x, y) = \alpha \, e^{ax + by}, \quad \text{where } b = \frac{a^3 + 2}{3a - 1}.$$

Example 6.21 For the homogeneous differential equation

$$(D^3 + D^2 D' + 3DD' + 3D^2 - D'^2 + 2D') z = 0,$$

we have $\mathscr{P}(D, D') = (D + D')(D^2 + 3D - D' + 2)$. Thus, for $(D^2 + 3D - D' + 2)z = u$, we have $(D + D')u = 0$ so that $u(x, y) = \phi(x - y)$. Hence, we obtain the following second order nonhomogeneous linear differential equation

$$(D^2 + 3D - D' + 2)u = \phi(x - y),$$

where ϕ is a smooth function. Solving this differential equation, it follows that the general solution is given by

$$u(x, y) = Ae^{ax + (a^2 + 3a + 2)y} + \frac{1}{7} \int \int \phi(v) dv dv, \quad \text{with } v = x - y.$$

Next, let us discuss how to find a particular integral u_p of a nonhomogeneous differential equation given by

$$\mathscr{P}(D, D')u = g(x, y), \quad \text{for } z = u(x, y).$$

In general, it is difficult to find u_p by using the formula (6.3.5). For example, in the simplest case, when the operator $\mathscr{P}(D, D')$ is of the form

$$\mathscr{P}(D, D') = (D - a_1 D') \cdots (D - a_n D'),$$

we can use simple formula given by

$$\frac{1}{(D - a_k D')} g(x, y) = \int g(x, c - a_k x) \, dx, \quad \text{for } c \in \mathbb{R}. \tag{6.3.11}$$

Therefore, particular integral u_p in this case is given by

$$u_p(x, y) = \frac{1}{\mathscr{P}(D, D')} g(x, y)$$

$$= \frac{1}{(D - a_1 D')(D - a_2 D') \cdots (D - a_n D')} g(x, y)$$

$$= \left(\cdots \left(\cdots \int h \left(\int g(x, c_n - a_n x) \, dx, c_{n-1} - a_{n-1} x \right) dx \cdots \right) \cdots \right)$$

Since integral computation in above expression gets messier by every passing stage, it is impractical to use the above scheme. Fortunately, it is convenient to apply some known formulas directly when the function $g(x, y)$ is a linear combination of functions of some specific types. In this latter situation, the underlying procedure is referred to as the *short method* of finding a particular integral.

To start with, we consider first the case when $g(x, y)$ is of the form

$$g(x, y) = \phi(v), \quad \text{where } v = ax + by \text{ and } a, b \in \mathbb{C}.$$

Then, the Lagrange's formula as given below applies:

$$\frac{1}{\alpha D^2 + \beta DD' + \gamma D'^2} \phi(ax + by) = \frac{1}{\alpha a^2 + \beta ab + \gamma b^2} \iint \phi(v) \, dv dv,$$

provided $\alpha a^2 + \beta ab + \gamma b^2 \neq 0$. The next example illustrates the procedure.

Example 6.22 For the differential equation

$$(2D^2 - 5DD' + 2D'^2)u = 3x + 2y, \quad \text{for } z = u(x, y),$$

the operator $\mathscr{P}(D, D')$ factorises as $(2D - D')(D - 2D')$. Therefore, in this case, the general solution u_h is given by

$$u_h(x, y) = \phi_1(x + 2y) + \phi_2(2x + y).$$

Further, the particular integral u_p is given by

$$u_p(x, y) = \frac{1}{2D^2 - 5DD' + 2D'^2}(3x + 2y)$$

$$= \frac{1}{2 \cdot 3^2 - 5(3 \cdot 2) + 2 \cdot 2^2} \iint v \, dv dv, \quad \text{where } v = 3x + 2y,$$

$$= -\frac{1}{8} \int v^2 \, dv$$

$$= -\frac{1}{24}v^3 = -\frac{1}{24}(3x + 2y)^3.$$

Hence, the complete solution is obtained as

$$u(x, y) = \phi_1(y + 3x) + \phi_2(y - 2x) - (1/36)(x + y)^3,$$

where ϕ_1, ϕ_2 are arbitrary single-variable functions.

The above procedure applies, in particular, when function $g(x, y)$ is one of the following types:

$$e^{ax+by}, \quad \sin(ax + by) \quad \text{or} \quad \cos(ax + by), \quad \text{etc..}$$

We usually apply the following *formulas* directly:

$$\frac{1}{\mathscr{P}(D, D')}e^{ax+by} = \frac{e^{ax+by}}{\mathscr{P}(a, b)}, \quad \text{provided } F(a, b) \neq 0; \tag{6.3.12}$$

$$\frac{1}{\mathscr{P}(D, D')}\sin(ax + by) = \frac{1}{\mathscr{P}(D^2, DD', D'^2)}\sin(ax + by)$$

$$= \frac{1}{\mathscr{P}(-a^2, -ab, -b^2)}\sin(ax + by); \tag{6.3.13}$$

$$\frac{1}{\mathscr{P}(D, D')}\cos(ax + by) = \frac{1}{\mathscr{P}(D^2, DD', D'^2)}\cos(ax + by)$$

$$= \frac{1}{\mathscr{P}(-a^2, -ab, -b^2)}\cos(ax + by). \tag{6.3.14}$$

provided $\mathscr{P}(-a^2, -ab, -b^2) \neq 0$ for Eqs. (6.3.13) and (6.3.14).

Example 6.23 For the nonhomogeneous differential equation given by

$$(D^3 - 2D^2D' - DD'^2 + 2D'^3)u = e^{2x-3y}, \quad \text{for } z = u(x, y),$$

we have $\mathscr{P}(D, D') = (D + D')(D - D')(D - 2D')$. Therefore, the general solution of the above equation is given by

$$u_h(x, y) = \phi_1(x - y) + \phi_2(x + y) + \phi_3(2x + y).$$

Also, in this case, the particular integral u_p is given by

$$u_p(x, y) = \frac{1}{D^3 - 2D^2D' - DD'^2 + 2D'^3} e^{2x-3y}$$

$$= \frac{e^{2x-3y}}{2^3 - 2\,2^2(-3) - 2(-3)^2 + 2(-3)^3}$$

$$= -\frac{1}{40} e^{2x-3y}.$$

Hence, the complete solution of the given differential equation is obtained as

$$u(x, y) = \phi_1(x - y) + \phi_2(x + y) + \phi_3(2x + y) - \frac{1}{40} e^{2x-3y},$$

where ϕ_1, ϕ_2, ϕ_3 are arbitrary single-variable functions.

Furthermore, when function $g(x, y)$ is one of the forms as given by

$$x^m y^n \quad \text{or} \quad e^{ax+by} V(x, y),$$

a particular integral is found by using directly the formulas as given below:

$$\frac{1}{\mathscr{P}(D, D')} x^m y^n = \frac{1}{D^s(1 - F(D, D'))} x^m y^n,$$

$$= \frac{1}{D^s} [x^m y^n + F(D, D')x^m y^n + (F(D, D'))^2 x^m y^n + \dots]; \qquad (6.3.15)$$

$$\frac{1}{\mathscr{P}(D, D')} e^{ax+by} V(x, y) = \frac{e^{ax+by}}{\mathscr{P}(D + a, D' + b)} V(x, y); \qquad (6.3.16)$$

Example 6.24   For the nonhomogeneous differential equation given by

$$(D^3 - 8D'^3)u = x^3 y^3, \quad \text{for } z = u(x, y),$$

the operator $\mathscr{P}(D, D')$ has the factorisation given by $(D - 2D')(D^2 + 2DD' + 4D'^2)$. Therefore, the general solution u_h is given by

$$u_h(x, y) = \phi_1(2x + y) + \phi_2(\alpha x - y) + \phi_3(\bar{\alpha} x - y),$$

where $\alpha = -1 + i\sqrt{3}$. Also, a particular integral u_p is given by

$$u_p(x, y) = \frac{1}{D^3 - 8D'^3} x^2 y^3 = \frac{1}{D^3}\left[1 - 8\frac{D'^3}{D^3}\right]^{-1} x^2 y^3$$

$$= \frac{1}{D^3}\left[1 + 8\frac{D'^3}{D^3} + \cdots\right] x^2 y^3$$

$$= \frac{1}{D^3}[x^2 y^3 + 24\frac{D'^2}{D^3} x^2 y^2] = \frac{1}{D^3}[x^2 y^3 + 48\frac{1}{D^3} x^2]$$

$$= \frac{1}{D^3}[x^2 y^3 + \frac{4}{5} x^5]$$

$$= \frac{1}{60} x^5 y^3 + \frac{2}{375} x^7.$$

Hence, the complete solution is obtained as

$$z = \phi_1(2x + y) + \phi_2(\alpha x - y) + \phi_3(\bar{\alpha} x - y) + \frac{1}{60} x^5 y^3 + \frac{2}{375} x^7,$$

where ϕ_1, ϕ_2, ϕ_3 are arbitrary single-variable functions.

Example 6.25 For the nonhomogeneous differential equation

$$\left(D^3 - 2D^2 D' - DD'^2 + 2D'^3\right) u = e^{x-y} \cos(2x + y), \quad \text{for } z = u(x, y),$$

$\mathscr{P}(D, D') = (D + D')(D - D')(D - 2D')$. So, the general solution u_h is given by

$$u_h(x, y) = \phi_1(x - y) + \phi_2(x + y) + \phi_3(2x + y).$$

Also, a particular integral u_p is given by

$$u_p(x, y) = \frac{1}{D^3 - 2D^2 D' - DD'^2 + 2D'^3} e^{x-y} \cos(2x + y)$$

$$= \frac{e^{x-y}}{D^3 - 2D^2 D' - DD'^2 + 2D'^3 + 5D^2 - 2DD' - 7D'^2 + 3D - 6D'} \cos(2x + y)$$

$$= -\frac{e^{x-y} \sin(2x + y)}{9}.$$

Hence, the complete solution of the given equation is obtained as

$$u(x, y) = \phi_1(x - y) + \phi_2(x + y) + \phi_3(2x + y) - \frac{e^{x-y} \sin(2x + y)}{9},$$

where ϕ_1, ϕ_2, ϕ_3 are arbitrary single-variable functions.

For the cases when a formula of *short method* fails, we apply the modified formulas as given below:

$$\frac{1}{\mathscr{P}(D, D')} e^{ax+by} = \frac{x(e^{ax+by})}{\partial \phi(D, D')/\partial D} \quad \text{or} \quad \frac{y(e^{ax+by})}{\partial \mathscr{P}(D, D')/\partial D'}, \tag{6.3.17}$$

if $\mathscr{P}(a, b) = 0$, where a choice for a particular operator (D or D') is made considering the power of D or D' in the given the expression $\mathscr{P}(D, D')$. Similarly,

$$\frac{1}{\mathscr{P}(D, D')} \sin(ax + by) = \frac{x \sin(ax + by)}{\partial \mathscr{P}(D, D')/\partial D} \quad \text{or} \quad \frac{y \sin(ax + by)}{\partial \mathscr{P}(D, D')/\partial D'}, \quad (6.3.18)$$

if $\mathscr{P}(-a^2, -ab, -b^2) = 0$.

Example 6.26 For the nonhomogeneous differential equation given by

$$\left(2D^3 - 2DD'^2 - D^2D' + D'^3\right) u = e^{x+2y}, \quad \text{for } z = u(x, y),$$

we have $\mathscr{P}(D, D') = (D + D')(D - D')(2D - D')$. Therefore, the general solution u_h is given by

$$u_h(x, y) = \phi_1(x - y) + \phi_2(x + y) + \phi_3(x + 2y).$$

Also, a particular integral u_p is given by

$$\begin{aligned}
u_p(x, y) &= \frac{1}{2D^3 - 2DD'^2 - D^2D' + D'^3} e^{x+2y} \\
&= \frac{1}{2} \frac{1}{3D^2 - D'^2 - DD'} xe^{x+2y} \\
&= -\frac{1}{6} xe^{x+2y}.
\end{aligned}$$

Hence, the complete solution of the given equation is obtained as

$$u(x, y) = \phi_1(x - y) + \phi_2(x + y) + \phi_3(x + 2y) - \frac{1}{6} xe^{x+2y},$$

where ϕ_1, ϕ_2, ϕ_3 are arbitrary single-variable functions.

Example 6.27 For the nonhomogeneous linear differential equation

$$(D^2 - 4D'^2 - 3D + 6D') u = \cos(2x + y), \quad \text{for } z = u(x, y),$$

the differential operator $F(D, D')$ factorises as $(D - 2D')(D + 2D' - 3)$. Therefore, the general solution is given by

$$u_h = \phi_1(2x + y) + e^{3x} \phi_2(2x - y).$$

Also, a particular integral is given by

$$u_p = \frac{1}{D^2 - 4D'^2 - 3D + 6D'} \cos(2x + y)$$

$$= \frac{1}{-4 + 4 - 3D + 6D'} \cos(2x + y)$$

$$= -\frac{1}{3} \frac{D + 2D'}{D^2 - 4D'^2} \cos(2x + y) = -\frac{1}{6} \frac{D + 2D'}{D} x \cos(2x + y)$$

$$= -\frac{1}{6} [D + 2D'][(x/2) \sin(2x + y) + (1/4) \cos(2x + y)]$$

$$= -\frac{1}{6} [2x \cos(2x + y) - (1/2) \sin(2x + y)].$$

Hence, the complete solution of the given equation is obtained as

$$u(x, y) = \phi_1(2x + y) + e^{3x}\phi_2(2x - y) - \frac{1}{6} [2x \cos(2x + y) - (1/2) \sin(2x + y)],$$

where ϕ_1, ϕ_2, ϕ_3 are arbitrary single-variable functions.

Definition 6.3 A nonhomogeneous linear differential equation

$$\mathscr{P}(D, D') u = g(x, y), \quad \text{for } z = u(x, y), \tag{6.3.19}$$

is called a *Cauchy-Euler equation* if each associated homogeneous operator $\mathscr{P}_n(D, D')$ of order n is of the form

$$a_0 x^n \frac{\partial^n}{\partial x^n} + a_1 x^{n-1} y \frac{\partial^n}{\partial x^{n-1} \partial y} + \cdots + a_{n-1} x y^{n-1} \frac{\partial^{n-1}}{\partial x \partial y^{n-1}} + a_n y^n \frac{\partial^n}{\partial y^n},$$

where each $a_k \in \mathbb{R}$.

As in case of a *Cauchy–Euler equation* in one variable, the first step is to transform the equation to a differential equations with constant coefficients. In the present situation, we can use the transformations $(x, y) \mapsto (\xi, \eta)$ given by

$$x = e^\xi \iff \xi(x) = \log |x| \quad \text{and} \quad y = e^\eta \iff \eta(y) = \log |y|.$$

Differentiating the relation given by

$$z = u(x, y) = v(e^\xi, e^\eta),$$

it follows that we have

$$\frac{\partial u}{\partial x} = \frac{1}{x} \frac{\partial v}{\partial \xi} \quad \text{and} \quad \frac{\partial u}{\partial y} = \frac{1}{y} \frac{\partial v}{\partial \eta},$$

which on second differentiation gives

$$x^2 \frac{\partial^2 u}{\partial x^2} = \frac{\partial v}{\partial \xi}\left(\frac{\partial v}{\partial \xi} - 1\right);$$

$$xy \frac{\partial^2 u}{\partial x \partial y} = \frac{\partial v}{\partial \xi}\frac{\partial v}{\partial \eta};$$

$$y^2 \frac{\partial^2 u}{\partial y^2} = \frac{\partial v}{\partial \eta}\left(\frac{\partial v}{\partial \eta} - 1\right); \quad \text{and so on.}$$

Therefore, each operator $\mathscr{P}_n(D, D')$ transforms to

$$a_0 \frac{\partial^n}{\partial \xi^n} + \alpha_1 \frac{\partial^n}{\partial \xi^{n-1} \partial \eta} + \cdots + \alpha_n \frac{\partial^n}{\partial \eta^n},$$

which is a differential equation with constant coefficients. We may write the transformed differential equation in terms of operators given by

$$E \equiv \frac{\partial}{\partial \xi}, \quad E' \equiv \frac{\partial}{\partial \eta}, \quad EE' \equiv \frac{\partial^2}{\partial \xi \partial \eta}, \quad \text{etc..}$$

The next example illustrates the above procedure.

Example 6.28 Consider the nonhomogeneous linear differential equation

$$(x^2 D^2 + 2xy^2 DD' + y^2 D'^2)u = xy \sin\left(\log|x/y|\right), \quad \text{for } z = u(x, y).$$

Taking $x = e^\xi$ and $y = e^\eta$, above equation transforms to the equation given by

$$(E^2 - E + 2EE' + E'^2 - E')v = e^{\xi+\eta}\sin(\xi - \eta), \quad \text{for } z = v(\xi, \eta)$$

which we can also write as

$$(E + E')(E + E' - 1)v = e^{\xi+\eta}\sin(\xi - \eta).$$

Therefore, the general solution $v_h(\xi, \eta)$ of the above equation is given by

$$v_h(\xi, \eta) = \phi_1(\xi - \eta) + e^\xi \phi_2(\xi - \eta),$$

and a particular integral $v_p(\xi, \eta)$ is given by

$$v_p(\xi, \eta) = \frac{1}{E^2 + 2EE' + E'^2 - E - E'} e^{\xi+\eta} \sin(\xi - \eta)$$

$$= e^{u+v} \frac{1}{E^2 + 2EE' + E'^2 + 3E + 3E' + 2} \sin(\xi - \eta)$$

$$= e^{\xi+\eta} \frac{1}{3E + 3E' + 2} \sin(\xi - \eta)$$

$$= e^{\xi+\eta} \frac{3E + 3E' + 2}{9E^2 + 9E'^2 + 6EE' - 4} \sin(\xi - \eta)$$

$$= -e^{\xi+\eta} \frac{3E + 3E' + 2}{16} \sin(\xi - \eta)$$

$$= -e^{\xi+\eta} \frac{3\cos(\xi - \eta) - 3\cos(\xi - \eta) + 2\sin(\xi - \eta)}{16}.$$

Hence, the complete solution of the given equation is obtained as

$$u(x, y) = \psi_1(x/y) + x\psi_2(x/y) - \frac{xy\left[3\cos(x/y) - 3\cos(x/y) + 2\sin(x/y)\right]}{16},$$

where ψ_1, ψ_2 are arbitrary single-variable functions.

6.4 Lagrange–Charpit Method

We discuss here Charpit's method of finding the complete integral of a nonlinear first order differential equation in two variables. Let $\Omega \subseteq \mathbb{R}^2$ be an open set, and consider a nonlinear first order differential equation for a function $u \in C^1(\Omega)$ given by

$$f(x, y; u; p, q) = 0, \quad \text{with } p = u_x \text{ and } q = u_y. \tag{6.4.1}$$

The basic idea of *Lagrange-Charpit method* is to find the complete integral of Eq. (6.4.1), by using a 1-parameter family of first order differential equations of the form

$$g(x, y; u; p, q; a) = 0, \quad \text{where } a \in \mathbb{R}. \tag{6.4.2}$$

Finally, we recover a complete integral of (6.4.1) by solving the above two first order differential equations for unknowns p and q. Suppose we can write

$$p = \phi(x, y, u, a) \quad \text{and} \quad q = \psi(x, y, u, a), \tag{6.4.3}$$

for some functions $\phi, \psi \in C^1(\Omega)$, such that the *differential* given by

$$du = \phi(x, y, u, a)\,dx + \psi(x, y, u, a)\,dy \tag{6.4.4}$$

is an *integrable form*. Then, it is possible to find C^1-functions $\lambda(x, y, z)$ and $\varphi = \varphi(x, y, z)$ such that

$$\lambda[\phi\, dx + \psi\, dy - du] = d\varphi.$$

It follows easily that the above conditions hold if and only if the *Jacobi bracket* $[f, g]$ given by

$$[f, g] = \frac{\partial(f, g)}{\partial(x, p)} + \frac{\partial(f, g)}{\partial(y, q)} + p\, \frac{\partial(f, g)}{\partial(u, p)} + q\, \frac{\partial(f, g)}{\partial(u, q)} \tag{6.4.5}$$

is identically zero (Exercise 6.4). Therefore, if we could somehow find a family of 1-parameter differential equations (6.4.2), then an implicit equation of the form

$$F(x, y;\ u;\ a, b) = 0, \quad \text{for } b \in \mathbb{R}, \tag{6.4.6}$$

as obtained from Eq. (6.4.4), provides a complete integral of Eq. (6.4.1).

Definition 6.4 Two first order differential equations of the form (6.4.1) are said to be **compatible** if they share a common solution over a domain $\Omega \subseteq \mathbb{R}^2$.

Since a solution (6.4.6) of the *Pfaffian* is a common solution of these two equations. So, equivalently, the two equations are *compatible* if they share a one-parameter family of common solutions. Notice that, to obtain the relations (6.4.3) for the derivatives p and q from the two differential equations (6.4.1) and (6.4.2), the Jacobian

$$J[f, g; p, q] = \frac{\partial(f, g)}{\partial(p, q)} = \begin{vmatrix} f_p & f_q \\ g_p & g_q \end{vmatrix} \neq 0, \quad \text{over } \Omega. \tag{6.4.7}$$

So, it is a necessary condition for the compatibility of the two equations. Thus, the two differential equations for functions f and g satisfying the *Jacobian condition* are compatible if and only if the condition (6.4.5) holds. Finally, assuming the two conditions, the final solution of Eq. (6.4.1) is then obtained from (6.4.4). The next example illustrates the point.

Example 6.29 Consider the following system of two differential equations:

$$x p - y q = 0, \quad x u p + y u q = 2xy$$

First, we show that these two equations are *compatible*, by verifying the Jacobian condition (6.4.7). We write

$$f(x, y, u, p, q) = x p - y q$$

and

$$g(x, y, u, p, q) = x u p + y u q - 2xy$$

so that we have

$$f_x = p, \quad f_y = -q, \quad f_p = x, \quad f_q = -y, \quad f_u = 0; \quad \text{and,}$$

$$g_x = up - 2y, \quad g_y = uq - 2x, \quad g_p = xu, \quad g_q = yu, \quad g_u = xp + yq.$$

Then, the Jacobian $J(f, g; p, q) = 2xyu \neq 0$ because $x \neq 0$, $y \neq 0$, and $u \neq 0$. Next, since

$$\frac{\partial(f, g)}{\partial(x, p)} = \begin{vmatrix} f_x & f_p \\ g_x & g_p \end{vmatrix} = \begin{vmatrix} p & x \\ up - 2y & xu \end{vmatrix} = 2xy;$$

$$\frac{\partial(f, g)}{\partial(y, q)} = \begin{vmatrix} f_y & f_q \\ g_y & g_q \end{vmatrix} = \begin{vmatrix} -q & -y \\ uq - 2x & yu \end{vmatrix} = -2xy;$$

$$\frac{\partial(f, g)}{\partial(u, p)} = \begin{vmatrix} f_z & f_p \\ g_z & g_p \end{vmatrix} = \begin{vmatrix} 0 & x \\ xp + yq & xu \end{vmatrix} = -x^2 p - xyq;$$

$$\frac{\partial(f, g)}{\partial(u, q)} = \begin{vmatrix} f_z & f_q \\ g_z & g_q \end{vmatrix} = \begin{vmatrix} 0 & -y \\ xp + yq & uy \end{vmatrix} = xyp + y^2 q,$$

it follows that $[f, g] = 0$. Hence, the given equations are compatible. Next, to determine p and q from the given two equations, observe that

$$xup + yuq = 2xy \implies xp + yq = \frac{2xy}{u}$$

so that, together with the equation $xp - yq = 0$, we get

$$2xp = \frac{2xy}{u} \implies p = \frac{y}{u} = \phi(x, y, u); \quad \text{and,}$$

$$2yq = \frac{2xy}{u} \implies q = \frac{x}{u} = \psi(x, y, u).$$

Substituting these expressions for p and q in Eq. (6.4.4), we obtain

$$du = \frac{y}{u} dx + \frac{x}{u} dy \implies u\, du = y\, dx + x\, dy \implies u^2 = xy + c,$$

where c is a constant, which is a complete solution.

Notice that for the compatibility of Eqs. (6.4.1) and (6.4.2), it is not necessary that every solution of Eq. (6.4.1) must be a solution of Eq. (6.4.2) and vice versa. For example, as one can easily verify that the two conditions (6.4.7) and (6.4.5) hold for the following differential equations:

$$f(x, y, u, p, q) = xp - yq - x = 0$$

and

$$g(x, y, u, p, q) = x^2 p + q - xu = 0$$

So, they are compatible, with $u(x, y) = x + c(1 + xy)$ being a common solution, where c is a constant. Notice that, however, $u(x, y) = x(y + 1)$ is a solution of the first differential equation, but not of the second.

Suppose two differential equations for functions f, $g \in C^1(\Omega)$ are *compatible*. We may expand right side of Eq. (6.4.5) to yield the following linear differential equation in g:

$$f_p \frac{\partial g}{\partial x} + f_q \frac{\partial g}{\partial y} + (p f_p + q f_q) \frac{\partial g}{\partial u} - (f_x + p f_u) \frac{\partial g}{\partial p} - (f_y + q f_u) \frac{\partial g}{\partial q} = 0. \quad (6.4.8)$$

We solve Eq. (6.4.8) by finding the integrals from the following *characteristic system*:

$$\frac{dx}{f_p} = \frac{dy}{f_q} = \frac{du}{p f_p + q f_q} = \frac{dp}{-(f_x + p f_u)} = \frac{dq}{-(f_y + q f_u)}. \quad (6.4.9)$$

These are known as the *Charpit equations* for the differential equation (6.4.1). Hence, given the integral $g(x, y, z, p, q, a)$, the problem reduces to solving for p and q, and finally integrating Eq. (6.4.4). Notice that we do not need all Charpit's equations (6.4.9) to find the integrals, but p or q must occur in the solution obtained from Eq. (6.4.9). The next example illustrates the point.

Example 6.30 For the nonlinear differential equation

$$p^2 x + q^2 y = z, \quad (6.4.10)$$

we have $f(x, y, z, p, q) = p^2 x + q^2 - z$ so that

$$f_p = 2px, \quad f_q = 2qy, \quad f_x = p^2,$$
$$f_y = q^2, \quad f_z = -1$$
$$\Rightarrow p f_p + q f_q = 2(p^2 x + q^2 y),$$
$$-(f_x + p f_z) = -p(p - 1), \quad \text{and}$$
$$-(f_y + q f_z) = -q(q - 1)$$

Thus, Charpit's equations for the given differential equation are as follows:

$$\frac{dx}{2px} = \frac{dy}{2qy} = \frac{dz}{2(p^2 x + q^2 y)} = \frac{dp}{-p(p - 1)} = \frac{dq}{-q(q - 1)},$$

Then,

$$\frac{p^2 dx + 2pxdp}{2p^3 x - 2p^3 x + 2p^2 x} = \frac{q^2 dy + 2qydq}{2q^3 y - 2q^3 y + 2q^2 y}$$

$$\Rightarrow \frac{p^2 dx + 2pxdp}{2p^2 x} = \frac{q^2 dy + 2qydq}{2q^2 y}$$

so that integration gives

$$\log(p^2 x) = \log(q^2 y) + \log(a) \ \Rightarrow \ p^2 x = aq^2 y, \qquad (6.4.11)$$

where a is a constant. Next, solving Eqs. (6.4.10) and (6.4.11) together, we get

$$aq^2 y + q^2 y = z \ \Rightarrow \ q = \left[\frac{z}{(1+a)y} \right]^{1/2},$$

and so

$$p^2 = \frac{az}{x(1+a)} \ \Rightarrow \ p = \left[\frac{az}{x(1+a)} \right]^{1/2}.$$

Finally, solving Eq. (6.4.4) for these values of p and q, we have

$$dz = \left[\frac{az}{x(1+a)} \right]^{1/2} dx + \left[\frac{z}{(1+a)y} \right]^{1/2} dy$$

$$\Rightarrow \left[\frac{(1+a)}{z} \right]^{1/2} dz = \left[\frac{a}{x} \right]^{1/2} dx + \left[\frac{1}{y} \right]^{1/2} dy.$$

Hence, $[(1+a)z]^{1/2} = (ax)^{1/2} + (y)^{1/2} + b$ is the complete solution of the differential equation.

As in case of linear differential equations, in some special cases, the details get simpler.

1. *CASE-I* : For a first order nonlinear differential equation is of the form

$$f(p, q) = 0, \qquad (6.4.12)$$

Charpit's equations are given by

$$\frac{dx}{f_p} = \frac{dy}{f_q} = \frac{du}{f_p x + f_q y)} = \frac{dp}{0} = \frac{dq}{0},$$

where the last two ratios are equivalent to $dp/dt = 0$ and $dq/dt = 0$. Using the first, it follows that $p = a$ is a solution, where a is a constant. Substituting this back into Eq. (6.4.12), we get $q = Q(a)$. Then, integrating the expression $du = a\,dx + Q(a)\,dy$, we obtain $u(x, y) = ax + Q(a)y + b$ as the general solution. Notice that one could also use $dq/dt = 0$ to work with the condition $q = constant$. For some problems, the latter approach leads to simpler computation. For example, suppose $f(p, q) = pq$. Take $dp/dt = 0$ so that $p = a \Rightarrow Q(a) = 1/a$. Hence, $du = a\,dx + Q(a)\,dy \Rightarrow u(x, y) = ax + (y/a) + b$ is the complete solution, where a and b are constants.

Example 6.31 Consider the *eikonal equation* (eikonal in Greek means *the image*) given by

$$u_x^2 + u_y^2 = 1, \quad \text{for some } z = u(x, y) \in C^1(\mathbb{R}^2).$$

We show that the solution of eikonal equation is given by a function $\varphi \in C^1(\mathbb{R}^2)$ such that $\varphi(x, y)$ at each point (x, y) outside a convex closed curve in $\mathbb{R}^2$ equals the distance of the point to the curve.

1. *CASE-II* : Suppose a first order nonlinear differential equation does not involving the independent variables so that it is of the form

$$f(u, p, q) = 0. \tag{6.4.13}$$

Then, Charpit's equations for such a differential equation are given by

$$\frac{dx}{f_p} = \frac{dy}{f_q} = \frac{du}{f_p x + f_q y} = \frac{dp}{-pf_u} = \frac{dq}{-qf_u},$$

so that the last two ratios imply that $p = aq$, where a is a constant. Using $p = aq$ in Eq. (6.4.13), and solving for q, we get $q = Q(u, a) \Leftrightarrow p = aQ(u, a)$. Finally, integrating $du = aQ(u, a)dx + Q(u, a)dy$, we obtain

$$\int \frac{1}{Q(u, a)}\,du = ax + y + b,$$

as the complete solution, where a and b are arbitrary constants.

Example 6.32 Consider the nonlinear differential equation $p^2 u^2 + q^2 = 1$. Substitute $p = aq$, then

$$q^2(a^2 u^2 + 1) = 1 \quad \Rightarrow \quad q = (a^2 u^2 + 1)^{-1/2}.$$

It then follows that

$$p^2 = \frac{1 - q^2}{u^2} = \left(1 - \frac{1}{a^2 u^2 + 1}\right)\frac{1}{u^2} = \frac{a^2}{a^2 u^2 + 1}$$

$$\Rightarrow \quad p = a(a^2 u^2 + 1)^{-1/2}.$$

Hence, integration of $du = a(a^2u^2 + 1)^{-1/2}dx + (a^2u^2 + 1)^{-1/2}dy$ implies that

$$(a^2u^2 + 1)^{1/2}du = adx + dy$$

$$\Rightarrow \ z = \frac{1}{2a}\left[au(a^2u^2 + 1)^{1/2} - \log(au+)a^2u^2 + 1)^{1/2})\right]$$

$$= ax + y + b,$$

is the complete solution, where a and b are constants.

1. *CASE-III* : Suppose a first order nonlinear differential equation is of *separable* form:

$$f(x, p) = g(y, q). \tag{6.4.14}$$

Then, *Charpit's equations* are given by

$$\frac{dx}{f_p} = \frac{dy}{-g_q} = \frac{du}{pf_p - qg_q} = \frac{dp}{-f_x} = \frac{dq}{g_y},$$

so that by using first and second last ratios, we get

$$\frac{dx}{f_p} = \frac{dp}{-f_x} \quad \Rightarrow \quad \frac{dp}{dx} + \frac{f_x}{f_p} = 0 \quad \Rightarrow \quad f_p dp + f_x dx = 0,$$

which gives $f(x, p) = a$, where a is some constant. Similarly, we get $g(y, q) = a$. Finally, we solve equations $f(x, p) = a$, $g(y, q) = a$ for p and q and obtain the complete solution, using Eq. (6.4.4).

Example 6.33 Consider the equation $p^2y(1 + x^2) = qx^2$, which we may write as

$$\frac{p^2(1 + x^2)}{x^2} = \frac{q}{y}.$$

Then, for an arbitrary constant a, we have

$$\frac{p^2(1 + x^2)}{x^2} = a^2 \quad \Rightarrow \quad p = \frac{ax}{\sqrt{1 + x^2}}.$$

Similarly, one gets $q/y = a^2 \Rightarrow q = a^2y$. Substituting these expressions of p and q in (6.4.4), we have

$$du = \frac{ax}{\sqrt{1 + x^2}}dx + a^2y\,dy$$

$$\Rightarrow \ z = a\sqrt{1 + x^2} + \frac{a^2y^2}{2} + b,$$

is the complete solution, where a and b are constants.

1. *CASE-IV* : Suppose the given first order nonlinear differential equation is of *Clairaut's form*:
$$u = px + qy + f(p, q). \tag{6.4.15}$$

Charpit's equations for this form are given by

$$\frac{dx}{x + f_p} = \frac{dy}{y + f_q} = \frac{dz}{xp + yq + pf_p + qf_q} = \frac{dp}{0} = \frac{dq}{0},$$

so that $p = a$ and $q = b$, using the last two ratios, where a and b are constants. It thus follows from Eq. (6.4.15) that $u(x, y) = ax + by + f(a, b)$ is the complete solution.

Example 6.34 Consider the equation $(p + q)(u - xp - yq) = 1$, which we may write as

$$u = xp + yq + \frac{1}{p + q}.$$

Since this later differential equation is in *Clairaut's form*, with $f(p, q) = 1/(p + q)$, it follows that

$$u(x, y) = ax + by + \frac{1}{a + b},$$

the complete solution of the given equation, where a and b are constants.

Exercises 6

6.1 Solve the initial value problem given by

$$y^2 u_{xx} - x^2 u_{tt} = 0;$$
$$u(x, 0) = e^x, \quad u_t(x, 0) = x.$$

6.2 Show that the differential equation $3u_{xx} + 7u_{xy} + 2u_{yy} = 0$ is hyperbolic for all x, y, and hence solve it by the *method of characteristics*.

6.3 Solve the initial value problem: $u_{tt} = e^t$, with $u(x, 0) = e^x$ and $u_t(x, 0) = x^3$.

6.4 Solve the Cauchy problem given by

$$xp + yq = z + 1, \quad \text{with } u(x, y) = x^2 \text{ along } y = x^2.$$

6.5 Let $\Omega \subseteq \mathbb{R}^3$ be a domain, $\varphi \in C^1(\Omega)$ be a first integral of a vector field f, and C be an integral curve of f given by $\gamma(t) = (x(t), y(t), z(t))$, for t in some subinterval $I \subseteq \mathbb{R}$. Show that C lies on some level surface of φ.

6.6 Let $\Omega \subseteq \mathbb{R}^3$ be a domain and C be an integral curve of a vector field $f = (a, b, c)$ defined over Ω, which is given parametrically by a regular curve

$\gamma(t) = \big(x(t), y(t), z(t)\big)$, for t in some subinterval $I \subseteq \mathbb{R}$, with $\gamma \in C^1(I)$. Show that there exists a function $\mu \in C^1(I)$ such that $\mu(t) \neq 0$ over I, and

$$f\big(x(t), y(t), z(t)\big) = \mu(t)\,\gamma'(t).$$

Let $t = t(s)$ be a solution of the differential equation $t'(s) = \mu(t)$ given over a subinterval $J \subseteq \mathbb{R}$ and take

$$\bar{x}(s) = x\big(t(s)\big), \quad \bar{y}(s) = y\big(t(s)\big), \quad \bar{z}(s) = z\big(t(s)\big).$$

Show that, with respect to parametrisation $\bar{\gamma} : J \to \Omega$ given by above relations, the curve $C_{\bar{\gamma}}$ is a characteristic of the system

$$\frac{dx}{ds} = a(x, y, z), \quad \frac{dy}{ds} = b(x, y, z), \quad \frac{dz}{ds} = c(x, y, z).$$

6.7 Use d'Alembert's formula to solve the wave equation

$$u_{tt} = c^2 u_{xx}, \quad -\infty < x < \infty, \quad t > 0, \quad \lim_{x \to \infty} u(x, t),$$

for the initial conditions given by $u(x, 0) = e^{-x^2}$ and $u_t(x, 0) = \sin x$.

6.8 Use Lagrange's method to find the general solution of the semilinear equation given by

$$x^2 p + y^2 q = z(x + y), \quad \text{for a } C^1\text{-function } z = u(x, y).$$

6.9 Use Lagrange's method to find the general solution of the quasilinear equation given by

$$x(y - z)p + y(z - x)q = z(x - y), \quad \text{for a } C^1\text{-function } z = u(x, y).$$

6.10 Use Lagrange's method to find the general solution of the quasilinear equation given by

$$\left(\frac{y^2 + z^2}{y}\right)p + xz\,q = xy, \quad \text{for a } C^1\text{-function } z = u(x, y).$$

6.11 Use Lagrange's method to find the general solution of the quasilinear equation given by

$$(y + z)p + y\,q = x - y, \quad \text{for a } C^1\text{-function } z = u(x, y).$$

6.12 Find the general solution of the differential equation $(D^2 - DD' - 6D'^2)z = x + y$.

6.13 Find the general solution of differential equation

$$(D^3 - 2D^2D' - DD'^2 2 + 2D'^3)z = \cos(3x + y).$$

6.14 Solve the differential equation $(D^3 - 4D^2D' + 4DD'^2)u = \cosh(y - 2x)$.

6.15 Solve the differential equation $(D^3 - 7DD'^2 + 6D'^3)u = e^{x+y} + \sin(x + y)$.

6.16 Solve the differential equation $rq - pt + 3pq = e^x + xy$, where notations have usual meaning.

6.17 Solve the differential equation $r + s + q - z = e^{-x}$, where notations have usual meaning.

6.18 Solve the differential equation $r - q = 2y + \cos x$, where notations have usual meaning.

6.19 Solve the equation $(D^3 - 7DD'^2 - 6D'^3)u = 0$.

6.20 Solve the Euler equation $x^2r - 2xy\,s - 3y^2t + xp - 3y\,q = x^2y\cos(2\log x)$, where notations have usual meaning.

6.21 Solve the Euler equation $x^2r + 2xy\,s + y^2t = x^3y^2$, where notations have usual meaning.

6.22 Solve the differential equation $x(y - u)p + y(x + u)q = (x + y)u$, given that $u = x^2 + 1$ on $y = x$.

6.23 Verify Eq. (6.3.12), using Lagrange's method.

6.24 Find the general solution of the differential equation $y^2p - xy\,q = x(u - 2y)$.

6.25 Show that conditions (6.4.7) and (6.4.5) hold for the differential equations

$$p^2 + q^2 = 1 \quad \text{and} \quad (p^2 + q^2)x = pu.$$

Find a solution of the first equation that not a solution of the second.

6.26 Solve the Cauchy problem given by

$$xp + yq = u + 1, \quad \text{with } u(x, y) = x^2 \text{ along } y = x^2.$$

6.27 Discuss the solution obtained in Theorem 6.1 for $x > 0$, taking the initial condition as $u(0, t) = b(t)$, for $t > 0$.

6.28 Show that the solution of the Cauchy problem given by

$$y_{tt} = c^2y_{xx} + f(x, t), \quad \text{for } x \in \mathbb{R} \text{ and } t > 0;$$
$$y(x, 0) = f(x) \quad \text{and} \quad y_t(x, 0) = g(x), \quad \text{for all } x \in \mathbb{R}, \tag{6.4.16}$$

is the sum of *Dirichlet formula* and the solution as obtained in Theorem 6.2.

6.29 Solve the initial value problem given by

$$y^2u_{xx} - x^2u_{tt} = 0;$$
$$u(x, 0) = e^x, \quad u_t(x, 0) = x.$$

6.30 Show that two first order differential equations given by

$$F(x, y; u; p, q) = 0 \quad \text{and} \quad G(x, y; u; p, q) = 0$$

are compatible if and only if

$$\frac{\partial(F, G)}{\partial(x, p)} + \frac{\partial(F, G)}{\partial(y, q)} + p\,\frac{\partial(F, G)}{\partial(u, p)} + q\,\frac{\partial(F, G)}{\partial(u, q)} \equiv 0$$

or the above equation follows from the implicit relations $F = 0$ and $G = 0$.

6.31 Solve the initial value problem: $u_{tt} = e^t$, with $u(x, 0) = e^x$ and $u_t(x, 0) = x^3$.

Chapter 7
Method of Characteristics

> *I regard as quite useless the reading of large treatise of pure analysis: too large a number of methods pass at once before the eyes. It is in the work of applications that one must study them; one judges their ability there and one apprises the manner of making use of them.*

> *Joseph-Louis Lagrange (1736–1813)*

Recall that the general first order partial differential equation in n-variable is given by

$$F\left(x, u(x), \nabla u\right) = 0, \quad \text{for } x \in \Omega \subset \mathbb{R}^n,$$

where F is a sufficiently smooth function of its arguments. As said earlier in Chap. 3, the theory of ordinary differential equation is the main tool to deal with the related problems. A Cauchy problem for the above type of equation in two variables (x, t) is about finding a surface $S \subset \mathbb{R}^3$ such that S contains a prescribed curve defined by the assigned initial condition. The *method of characteristics* as discussed in this chapter provides an effective geometric procedure to construct such a surface. It produces a unique solution, provided no *singularity* is developed in finite time t. The solution of a Cauchy problem beyond a singular point continues as a *shock wave* that may not even be continuous.

The involved working procedure uses the *local solution* of the associated first order system of initial value problems at each point of the *initial curve*. The same idea is then extended to deal with Cauchy problems for first order differential equations in n variables ([1–4]). The method also applies to Cauchy problems for hyperbolic type second order linear equations (Example 6.7).

© The Author(s), under exclusive license to Springer Nature Singapore Pte Ltd. 2022   305
A. K. Razdan and V. Ravichandran, *Fundamentals of Partial Differential Equations*,
https://doi.org/10.1007/978-981-16-9865-1_7

Gaspard Monge (1746–1818) introduced the concept of characteristics in 1770 while finding the general solution of fully nonlinear first order partial differential equation. However, the related work was published only in 1795. Meanwhile, Lagrange developed similar idea in two influential papers that were published in 1772 and 1779. *William Hamilton* (1805–1865) developed the idea further while working on the nonlinear *eikonal equation* of geometric optics. The term *characteristics* was coined by Hamilton.

7.1   Linear and Semilinear Equations

Let $\Omega \subseteq \mathbb{R}^2$ be an open set. A general first order differential equation for an unknown function $u = u(x, y) \in C^1(\Omega)$ is an implicit relation given by

$$F(x, y; u; p, q) = 0, \quad \text{with } p = u_x \text{ and } q = u_y, \tag{7.1.1}$$

where F is a sufficiently smooth function of variables x, y, u, p, q such that

$$F_p^2 + F_q^2 \not\equiv 0, \quad \text{over } \Omega. \tag{7.1.2}$$

The simplest type of equation (7.1.1) correspond to the case when F is linear with respect to variables u, p, q, and all the coefficients are some functions in $C^1(\Omega)$. Therefore, a linear equation in two independent variables x and y is given by

$$a(x, y)\, p + b(x, y)\, q + c_1(x, y)\, u = f(x, y), \quad \text{for } (x, y) \in \Omega. \tag{7.1.3}$$

In practical terms, the function $f(x, y)$ represents a *source* (or a *sink*). Further, equation (7.1.1) is said to be *semilinear* if F is linear with respect to derivatives p, q, and the coefficients appearing in the principle part are some functions in $C^1(\Omega)$. In this chapter, we may write a first order semilinear equation for a function $u \in C^1(\Omega)$ as

$$a(x, y)p + b(x, y)q = c(x, y, z), \quad \text{for } z = u(x, y). \tag{7.1.4}$$

which reduces to a linear equation when c in the above equation is given by

$$c(x, y, z) = -c_1(x, y)\, u + f(x, y), \quad \text{for }, (x, y) \in \Omega. \tag{7.1.5}$$

Also, equation (7.1.1) is said to be *quasilinear* if F is linear with respect to derivatives p, q, and the coefficients appearing in the principle part are functions of variables x, y and u. Therefore, a *quasilinear equation* in two independent variables x and y is given by

$$a(x, y, z)\, u_x + b(x, y, z)\, u_y = c(x, y, z), \quad \text{for } z = u(x, y). \tag{7.1.6}$$

The above type of differential equation is also known as a *Lagrange equation* because Lagrange was the first to study such types of differential equations. Finally, as said earlier, (7.1.1) is a *fully nonlinear* equation if it is a nonquasilinear equation.

Now, suppose a C^1-function u defined on an open set $D \subseteq \Omega$ is a solution of the equation (7.1.1). For $D_1 = D \times u(D)$, the graph surface Γ_u given by

$$\Gamma_u = \{(x, y, u(x, y)) \in D_1 \mid \text{for } (x, y) \in D\}$$

is called the *integral surface* associated with the solution $u = u(x, y)$. Further, let

$$C_\gamma = \{(\gamma_1, \gamma_2, \gamma_3) : s \in I \subseteq \mathbb{R}\},$$

be a curve on the surface Γ_u such that

$$(\gamma_1')^2 + (\gamma_2')^2 \neq 0, \quad \text{over } I,$$

Recall that a Cauchy problem is about finding a function $u \in C^1(D)$ such that it satisfies the equation (7.1.1), and

$$u(\gamma_1(s), \gamma_2(s)) = \gamma_3(s), \quad \text{for all } s \in I.$$

The next example illustrates the main idea.

Example 7.1 Suppose Γ_u is the integral surface associated with a solution $u = u(x, y) \in C^1(\Omega)$ of linear differential equation $p + y\,q = 0$. That is, we have

$$\Gamma_u = \{(x, y, z) \in \Omega \times \mathbb{R} : z = u(x, y)\}$$

At each point $(x, y, z) \in \Gamma_u$, the normal $\mathbf{n} = (p, q, -1)$ is perpendicular to the *level curve* $u(x, y) = constant$. We also have that, for the vector field $\mathbf{v} := (a, b, 0) = (1, y, 0)$,

$$\mathbf{n} \cdot \mathbf{v} = p + y\,q = 0, \quad \text{over } \Omega.$$

Therefore, at each point $(x, y, z) \in \Gamma_u$, the tangent to the level curve $u = constant$ is parallel to the vector $\mathbf{v}$. Said differently, the level curves of the surface Γ_u are the *integral curves* of the vector field $\mathbf{v}$. In this case, by solving the *characteristic equation* given by $y'(x) = y$, we obtain a 1-parameter family of *characteristics* given by the exponential function $y(x) = ce^x$. Clearly, by varying parameter $c \in \mathbb{R}$, characteristic curves fill the xy-plane completely without any self-intersections. Also, on each one of these curves, the function $u(x, y)$ is constant, because

$$\frac{d(u(x, ce^x))}{dx} = p + ce^x q = p + y\,q = 0.$$

We thus have

$$u(x, ce^x) = u(0, ce^0) = u(0, c)$$

is independent of the variable x. Taking $c = ye^{-x}$, we obtain $u(x, y) = u(0, ye^{-x})$. Hence, the general solution of the given equation is obtained as

$$u(x, y) = \phi(ye^{-x}), \quad \text{for any } \phi \in C^1(\mathbb{R}).$$

We may consider next a homogeneous linear equation such as

$$a(x, y) p + b(x, y) q = 0, \quad \text{for } (x, y) \in \Omega, \tag{7.1.7}$$

where $a, b \in C^1(\Omega)$ are such that $(a, b) \not\equiv (0, 0)$ over Ω. More generally, suppose $u \in C^1(\Omega)$ is a solution of a semilinear equation given by

$$a(x, y) p + b(x, y) q = c(x, y, z), \quad \text{for } z = u(x, y) \in C^1(\Omega), \tag{7.1.8}$$

where $a, b, c \in C(\Omega)$, with $(a, b) \not\equiv (0, 0)$ over Ω. The normal to the integral surface Γ_u at point (x, y, z) is given by

$$\mathbf{n} = (p, q, -1),$$

which is perpendicular to the level curve $u = constant$. Also, by (7.1.8), the vector field $v = (a, b, c)$ at each point $(x, y, z) \in \Gamma_u$ satisfies the equation

$$\mathbf{n} \cdot v = a\,p + b\,q - c = 0.$$

Therefore, at each point (x, y, z), the tangent to the level curve $u = constant$ is in the direction of the field v. Said differently, level curves of the surface Γ_u are the *integral curves* of the vector field v. Hence, integral surface Γ_u of equation (7.1.8) has the property that, at every point (x, y, z), the *tangential plane* defined by the vector field v is given by

$$p(X - x) + q(Y - y) = Z - u, \quad \text{for } (X, Y, Z) \in \mathbb{R}^3.$$

In this case, we also say that the tuple (x, y, z, p, q) is the *surface element* of the integral surface Γ_u supported at the point (x, y, z). In particular, for all pairs (p, q) satisfying the homogeneous equation (7.1.7), we obtain a bundle of planes at the *support point* (x, y, z). That is, for each point $(x, y) \in \Omega$, there exists a 1-parameter family of planes $\Pi(\lambda) := \Pi(x, y; \lambda)$ specified in terms of the derivatives

$$p(\lambda) = p(x, y; \lambda) \quad \text{and} \quad q(\lambda) = q(x, y; \lambda).$$

As the envelope of planes $\Pi(\lambda)$ defines a 1-parameter family of lines given by

$$a(x, y)p(\lambda) + b(x, y)q(\lambda) = 0, \qquad (7.1.9)$$

it follows that the normal $\mathbf{n}(\lambda)$ to the plane $\Pi(\lambda)$ is perpendicular to vector $\boldsymbol{v} = (a, b)$. Now, with reference to equation (7.1.7), suppose an *initial curve* Γ_0 on the integral surface Γ_u is parametrised as given below:

$$\Gamma_0(r) : \quad x = x(r), \quad y = y(r), \quad z = z(r), \quad \text{for } r_0 \le r \le r_1,$$

which passes through a point $x_0 = \big(x(r_0), y(r_0), z(r_0)\big)$. Let Π_{x_0} be the tangent plane at x_0. We may assume that the line on Π_{x_0} given by

$$L(x_0) : \quad \ell(\alpha) = x_0 + \alpha\, x'(r_0), \quad \alpha \in \mathbb{R},$$

coincides with the envelope of planes $\Pi(\lambda)$ at point x_0. For $\Pi_{x_0} = \Pi(\lambda_0)$, consider planes given by

$$\Pi(\lambda_0) : \quad Z - z_0 = (X - x_0)p(\lambda_0) + (Y - y_0)q(\lambda_0);$$
$$\Pi(\lambda_0 + \epsilon) : \quad Z - z_0 = (X - x_0)p(\lambda_0 + \epsilon) + (Y - y_0)q(\lambda_0 + \epsilon).$$

Clearly, at the line of intersection $\ell(\alpha)$, we have

$$(X - x_0)p(\lambda_0) + (Y - y_0)q(\lambda_0) = (X - x_0)p(\lambda_0 + \epsilon) + (Y - y_0)q(\lambda_0 + \epsilon).$$

It thus follows that

$$x'(r_0)p'(\lambda_0) + y'(r_0)q'(\lambda_0) = 0.$$

On the other hand, equation (7.1.9) implies that we have

$$a\big(x(r_0), y(r_0)\big)p(\lambda) + b\big(x(r_0), y(r_0)\big)q(\lambda) = 0,$$

which gives the equation

$$a\, p'(\lambda_0) + b\, q'(\lambda_0) = 0.$$

Therefore, as the parameter r_0 is arbitrary, we obtain

$$\big(x'(r), y'(r)\big) = \frac{x'(r)}{a(x(r), y(r))}\, \big(a(x(r), y(r)), b(x(r), y(r))\big).$$

The above equation is valid as long as $x'(r) \ne 0$ and $a\big(x(r), y(r)\big) \ne 0$. Let

$$t(r) = \int_{r_0}^{r} \frac{x'(s)}{a(x(s), y(s))}\, ds.$$

so that we have $x'(t) = a(x, y)$ and $y'(t) = b(x, y)$. We may write $\boldsymbol{x}(r(t)) := \boldsymbol{x}(t)$. Since $a, b \in C^1(\Omega)$, by Theorem 3.19, the first order system given by

$$x'(t) = a(x, y), \quad x(0) = x_0;$$
$$y'(t) = b(x, y), \quad y(0) = y_0,$$

has a unique integral curve passing through point (x_0, y_0), where the slope at each point is given by the vector $v = (a, b)$.

In general, for equation (7.1.8), the integral curve of the vector field $v = (a, b, c)$ is parametrised as

$$\Gamma(s) : \quad x = x(s), \quad y = y(s), \quad z = z(s), \quad \text{for } s_1 \leq s \leq s_2,$$

where (x, y, z) is a unique (local) solution of the first order system given by

$$\frac{dx}{ds} = a(x(s), y(s)); \tag{7.1.10a}$$

$$\frac{dy}{ds} = b(x(s), y(s)); \tag{7.1.10b}$$

$$\frac{du}{ds} = c(x(s), y(s), z(s)). \tag{7.1.10c}$$

The above first order system of initial value problems is called the *characteristic system* associated with the Cauchy problem for a semilinear equation (7.1.8). Notice that the characteristics $\Gamma(s)$ obtained from the above system provides a parametrisation of the integral surface Γ_u.

Definition 7.1 A plane curve $C(s) \subset \Omega$ is called a *projected characteristic curve* for a semilinear equation (7.1.8) if it satisfies the first two differential equations of the system (7.1.10).

Notice that, along a projected characteristic curve, we have

$$\frac{du}{dx} = u_x + \frac{dy}{dx} u_y$$

$$= u_x + \frac{b(x, y)}{a(x, y)} u_y$$

$$= \frac{c(x, y, z)}{a(x, y)}.$$

Therefore, variation in values $z = u(x, y)$ along the curve $C(s)$ is given by

$$\frac{du}{ds} = \frac{du}{dx} \frac{dx}{ds} = c(x, y, z). \tag{7.1.11}$$

Hence, along the curve $C(s)$, differential equation (7.1.8) reduces to the form

$$\frac{d}{ds}u\big(x(s),\,y(s)\big) = c\big(x(s),\,y(s),\,z(s)\big).\qquad(7.1.12)$$

Example 7.2 Consider the *transport equation* given by

$$u_t + c\,u_x = 0,\quad \text{for } (x,t) \in \Omega \subset \mathbb{R} \times \mathbb{R}^+,\qquad(7.1.13)$$

where c is constant. Suppose a projected characteristics $C(s)$ is parametrised as

$$x = x(s),\quad t = t(s),\quad \text{for } s \in \mathbb{R}.$$

Then, for $u(s) = u(x,t) = u\big(x(s),t(s)\big)$, we have

$$\frac{du}{ds} = u_x\,x'(s) + u_t\,t'(s)$$

which gives variation in values of u along the curve $C(s)$. Comparing with equation (7.1.13), we have $x'(s) = c$ and $t'(s) = 1$, along the curve $C(s)$, which implies that

$$x(s) = c\,s + x_0 \quad \text{and} \quad t(s) = s + t_0, \quad \text{for some } (x_0, t_0) \in C(s).$$

Clearly, we have $du/ds = 0$. That is, u is constant along the curve $C(s)$. Therefore, Γ_u is an integral surface of equation (7.1.13) if and only if the function u is constant along the curve $C(s)$. Hence, the general solution of equation (7.1.13) is given by $u(x,t) = g(x - ct)$, where g is an arbitrary function. Notice that the curve $C(s)$ is a line in xt-plane parametrised by

$$x(s) = c\,s + x_0, \quad t(s) = s + t_0, \quad \text{for } -\infty < s < \infty,$$

where $(x_0, t_0) = (x(0), t(0)) \in C(s)$. Eliminating the parameter s, we obtain $x - ct = x_0 - ct_0 = c_0$. Therefore, we have

$$u(x,t) = g(x - ct), \quad \text{for an arbitrary } g \in C^1(\mathbb{R}).$$

Clearly, the above solution of equation (7.1.13) is far from being unique.

Example 7.3 With notations as in Example 5.1.12, suppose the speed $v = v(x)$ is uniform. Consider the Cauchy problem given by

$$u_t + v\,u_x + v'\,u = 0, \quad \text{with } u(x,0) = u_0.$$

In this case, the *characteristic system* is given by

$$\frac{dx}{ds} = a(x,t) = v(x) \quad \text{and} \quad \frac{dt}{ds} = b(x,t) = 1,$$

so that the associated initial value problem is given by

$$\frac{\mathrm{d}x}{\mathrm{d}t} = v(x), \quad \text{with } x(0) = u(x, 0) = u_0.$$

Suppose a unique solution is given by $x = x(t)$. A solution $u = u(x, t)$ of the given Cauchy problem, along the characteristic $x = x(t)$, can be written as $u = u(x(t), t)$ so that we have

$$\frac{\mathrm{d}u}{\mathrm{d}t} = u_t \frac{\mathrm{d}t}{\mathrm{d}t} + u_x \frac{\mathrm{d}x(t)}{\mathrm{d}t}.$$

Comparing with the given transport equation expressed as

$$-v'(x(t))u = u_t + v(x(t))u_x,$$

it follows that

$$\frac{\mathrm{d}u}{\mathrm{d}t} = -v'(x(t))u(x(t), t),$$

which has a unique solution given by

$$u(x(t), t) = u(u_0, 0) e^{-\int_0^t v'(x(s))\mathrm{d}s}.$$

For $x \in \mathbb{R}$ and $t > 0$, we need to find the value u_0 such that a characteristics starting at the point $(u_0, 0)$ reaches to the point $(x(t), t)$ in time t. For, by using the fact that $u = u(x, t)$ is constant along each characteristics, it follows that a unique solution is given by $u = u(x(t), t)$, which is a *steady state* solution of the given semilinear equation. The same conclusion holds whenever a Cauchy problem is related to a typical *conservation equation*.

In general, for a Cauchy problem related to equation (7.1.8) with initial condition $u(x, 0) = \phi(x)$, the integral surface Γ_u contains the initial curve

$$\Gamma_\phi = \{(x, 0, \phi(x)) \mid (x, 0) \in \Omega\}.$$

We choose a point $(x_0, 0, \phi(x_0))$ on Γ_ϕ to begin the construction of the integral surface of the Cauchy problem along the curve Γ_ϕ. In the first step, we find a characteristic curve satisfying the system (7.1.10), with initial conditions specified by the point $(x_0, 0, \phi(x_0))$. The integral surface is constructed by repeating the same procedure for every point on Γ_ϕ and, subsequently, taking union of all characteristics. Clearly, the surface contains the initial curve Γ_ϕ. This is how we obtain a unique solution of a Cauchy problem, subject to certain conditions as explained later.

In actual practice, we take a parametrisation of the initial curve Γ_ϕ given by

$$\Gamma_\phi(r) : \quad x = x(r, 0), \quad y = y(r, 0), \quad u = u(r, 0), \quad r_0 \le r \le r_1, \qquad (7.1.14)$$

such that the characteristics are obtained from the first order system given by

$$\frac{dx}{ds}(r, s) = a(x(r, s), y(r, s)), \qquad (7.1.15a)$$

$$\frac{dy}{ds}(r, s) = b(x(r, s), y(r, s)), \qquad (7.1.15b)$$

$$\frac{du}{ds}(r, s) = F(x(r, s), y(r, s)) - c(x(r, s), y(r, s))u(x(r, s), y(r, s)), \qquad (7.1.15c)$$

subject to the initial conditions given by

$$x(r, 0) = x_0, \qquad (7.1.16a)$$
$$y(r, 0) = y_0, \qquad (7.1.16b)$$
$$u(r, 0) = \phi(r). \qquad (7.1.16c)$$

Finally, a local solution near Γ_ϕ to the Cauchy problem given by

$$a(x, y)\, p + b(x, y)\, q + c_1(x, y)u = f(x, y),$$
$$u(x, 0) = \phi(x), \qquad (7.1.17)$$

is found by solving r, s in terms of x, y. Notice that we must have

$$J = \det \begin{pmatrix} ccx_r & y_r \\ x_s & y_s \end{pmatrix} = x_r y_s - y_r x_s \ne 0,$$

to solve (r, s) in terms of (x, y). In particular, at $s = 0$, we obtain the condition

$$x'(r)\, b(x(r), y(r)) - y'(r)\, a(x(r), y(r)) \ne 0. \qquad (7.1.18)$$

Therefore, as in Example 7.3, the characteristics are obtained as the 1-parameter family of curves from the system (7.1.15), where the tangent vector at each point is specified by the vector field

$$\mathbf{V} = (a(x, y), b(x, y), c(x, y)), \quad \text{for } (x, y) \in \Omega.$$

Example 7.4 Consider the Cauchy problem given by

$$xu_x - yu_y = u;$$
$$fu(x, x) = x^2, \quad \text{for } x \in \mathbb{R}. \qquad (7.1.19)$$

We may view points on Γ_ϕ as the origin for the parameter s along the characteristics, namely, $s = 0$ on Γ_ϕ, we may parametrise Γ_ϕ as

$$x(r, 0) = r \quad y(r, 0) = r, \quad \text{and} \quad u(r, 0) = r^2$$

such that the characteristics are given by the solution of the first order system of initial value problem given by

$$\frac{dx}{ds}(r, s) = x, \quad \text{with } x(r, 0) = r;$$

$$\frac{dy}{ds}(r, s) = -y, \quad \text{with } y(r, 0) = r;$$

$$\frac{du}{ds}(r, s) = u, \quad \text{with } u(r, 0) = r^2.$$

Therefore, we obtain the characteristic curves given by

$$x(r, s) = re^s, \quad y(r, s) = re^{-s}, \quad u(r, s) = r^2 e^s,$$

Using the first two equations, we have

$$x/y = e^{2s} \quad \Rightarrow \quad s = \ln\sqrt{x/y}$$

and also

$$xy = r^2 \Rightarrow r = \sqrt{xy} \quad (r \geq 0).$$

Substituting the above expressions for r, s in the third equation, a unique solution of the Cauchy problem is obtained as

$$u(x, y) = xy \exp\left(\ln\sqrt{\frac{x}{y}}\right) = xy\sqrt{\frac{x}{y}} = x\sqrt{xy}, \quad \text{for } x \geq 0.$$

The next example illustrates uniqueness aspect for a solution of a first order nonhomogeneous Cauchy problem.

Example 7.5 Consider the Cauchy problem related to a transport equation with a source term, given by

$$u_t(x, t) + c\, u_x(x, t) = g(x, t), \quad \text{for } t > 0; \tag{7.1.20a}$$

$$u(x, 0) = \phi(x), \quad \text{for } -\infty < x < \infty, \tag{7.1.20b}$$

where functions g, $\phi \in C^1(\mathbb{R} \times \mathbb{R}^+)$. If $u_1(x, t)$ is a solution of the homogeneous part of the problem, then we already know that $u_1(x, t) = \phi(x - ct)$. Next, if $u_2(x, t)$ is the solution of the Cauchy problem for the case $\phi(x) = 0$, with $F(x, t) \neq 0$, then we have

$$\partial_t(u_2(x,t)) + c\,\partial_x(u_2(x,t)) = g(x,t), \quad t > 0;$$
$$u_2(x,0) = 0, \quad -\infty < x < \infty.$$

The above nonhomogeneous equation is equivalent to the differential equation

$$\frac{d}{ds}\big[u_2(x(s),t(s))\big] = g(x(s),t(s)),$$

where $x'(s) = c$ and $t'(s) = 1$, so that

$$x(s) = cs + x_0 \quad \text{and} \quad t(s) = s + t_0, \quad \text{with } x_0 = x(0), \ t_0 = (0).$$

So, for arbitrary values $s_2 > s_1$,

$$u_2(x(s_2),t(s_2)) = u_2(x(s_1),t(s_1)) + \int_{s_1}^{s_2} d/ds\{u_2(x(s),t(s))\}ds$$

$$= u_2(x(s_1),t(s_1)) + \int_{s_1}^{s_2} g(x(s),t(s))ds.$$

Thus, taking $s_2 = 0 > s_1 = -t$, then $x_0 = x(s_2)$ and $t_0 = t(s_2)$. That is,

$$x(s_2) = c \cdot 0 + x \quad \text{and} \quad t(s_2) = 0 + t$$
$$x(s_1) = c(-t) + x \quad \text{and} \quad t(s_1) = -t + t = 0$$

and

$$u_2(x,t) = u_2(x - ct, 0) + \int_{-t}^{0} g(x + cs, t + s)ds$$

$$= 0 + \int_{0}^{t} g(x + c(\theta - t), \theta)d\theta,$$

where change of variable $\theta = t + s$ is used in the integral. Therefore,

$$u(x,t) = u_1(x,t) + u_2(x,t) = f(x - ct) + \int_{0}^{t} g(x - c(t - \theta), \theta)d\theta$$

satisfies the Cauchy problem (7.1.20). Notice that for $g \equiv 0$ and

$$\phi(x) = \begin{cases} 1, & x < 1 \\ 2, & x > 1 \end{cases},$$

it follows that

$$u(x, t) = \begin{cases} 1 \; if \; x < 1 + ct \\ 2 \; if \; x > 1 + ct \end{cases}$$

satisfies the Cauchy problem but since the first derivatives of $u(x, t)$ fail to exist at points on the line $x - ct = 1$, it is not a solution in classical sense.

Definition 7.2 For a Cauchy problem for a first order linear equation given by

$$a(x, y) \, p + b(x, y) \, q + c_1(x, y) \, u = f(x, y);$$
$$z|_\Gamma = \phi, \tag{7.1.21}$$

an initial curve Γ_ϕ on the integral surface Γ_u of the equation is called *non-characteristic* if the projection $(\gamma_1(r), \gamma_2(r))$ of Γ_ϕ onto the xy-plane is nowhere tangent to projected characteristic curves. Or, equivalently, if

$$(a(\gamma_1(r), \gamma_2(r)), b(\gamma_1(r), \gamma_2(r))) \cdot (-\gamma_2'(r), \gamma_1'(r)) \neq 0.$$

Therefore, initial curve Γ_ϕ is said to be *characteristic* if the Jacobian given by

$$J = \det \begin{pmatrix} a & b \\ \gamma_1'(r) & \gamma_2'(r) \end{pmatrix} = a \, \gamma_1'(r) - b \, \gamma_2'(r) = 0.$$

For the Cauchy problem given in Example 7.4, we have

$$J = \det \begin{pmatrix} a & b \\ \gamma_1'(r) & \gamma_2'(r) \end{pmatrix} = \det \begin{pmatrix} x & -y \\ 1 & 1 \end{pmatrix} = x + y.$$

It thus follows that, in this case, the Cauchy problem can have a unique solution only when $x + y \neq 0$. Notice that, for $y = -x$ with $x > 0$, we have $u(x, y) = i \, x^2$. Therefore, no real solution exists along the line $y = -x$, with $x > 0$.

Example 7.6 Consider the Cauchy problem given by

$$x \, u_x + 2u_y = 1,$$
$$u(x, 0) = x^2. \tag{7.1.22}$$

For $\phi(x) = x^2$, we parametrise the initial curve Γ_ϕ as

$$x(0, t) = t, \quad y(0, t) = 0, \quad u(0, t) = t^2,$$

so that we have

$$J = \det \begin{pmatrix} a & b \\ \gamma_1'(r) & \gamma_2'(r) \end{pmatrix} = \det \begin{pmatrix} x & 1 \\ 1 & 0 \end{pmatrix} = -1 \neq 0.$$

The characteristics are obtained from the system of initial value problems given by

$$\frac{dx}{ds}(s, t) = x, \quad \text{with } x(0, t) = t; \tag{7.1.23a}$$

$$\frac{dy}{ds}(s, t) = 2, \quad \text{with } y(0, t) = 0; \tag{7.1.23b}$$

$$\frac{du}{ds}(s, t) = 1, \quad \text{with } u(0, t) = t^2. \tag{7.1.23c}$$

Therefore, we have

$$x(s, t) = te^s \qquad y(s, t) = 2s \qquad \text{and} \qquad u(s, t) = s + t^2.$$

Solving s, t in terms of x, y, we obtain

$$s = \frac{y}{2}, \quad \text{and} \quad t = xe^{-s} = xe^{-y/2}.$$

Hence, a unique solution of the given Cauchy problem is obtained as

$$u(x, y) = \frac{y}{2} + x^2 e^{-y}.$$

Referring back to the general solution of equation (7.1.8), suppose the initial condition $u_0 = (x_0, y_0) = q(s)$ is prescribed along an arbitrary projected characteristic curve $C(s) : x = x_0(s), y = y_0(s)$. When $C(s)$ is *noncharacteristic*, Cauchy problem is *well-posed*, and there is a unique function g satisfying the condition

$$q(s) = e^{\alpha(x_0(s), y_0(s))} \big[\beta(x_0(s), y_0(s)) + g(x_0(s), y_0(s)) \big].$$

On the other hand, when $C(s)$ is characteristic, the relation between q and g becomes

$$q(s) = e^{\alpha(x_0(s), y_0(s))} \big[\beta(x_0(s), y_0(s)) + G \big], \tag{7.1.24}$$

where $G = g(k)$ is a constant. Therefore, in this case, Cauchy problem is *ill-posed*.

Example 7.7 Consider the Cauchy problem given by

$$u_x + u_y = 1 - 2u, \quad \text{with } u(x, x) = \cos x.$$

For $\phi(x) = \cos x$, let the initial curve Γ_ϕ be parametrised as

$$x(0, t) = t, \quad y(0, t) = t, \quad u(0, t) = \cos t.$$

It then follows that

$$J = \det \begin{pmatrix} a & b \\ \gamma_1'(r) & \gamma_2'(r) \end{pmatrix} = \det \begin{pmatrix} 1 & 1 \\ 1 & 1 \end{pmatrix} = 0.$$

Therefore, the initial curve is indeed a characteristic curve. The same can be verified directly. For, in this case, the associated characteristic system

$$\frac{dx}{ds}(s, t) = 1, \quad \text{with } x(0, t) = t;$$

$$\frac{dy}{ds}(s, t) = 1, \quad \text{with } y(0, t) = t;$$

$$\frac{du}{ds}(s, t) = 1 - 2u, \quad \text{with } u(0, t) = \cos t,$$

has the solution given by

$$x(s, t) = s + t, \quad y(s, t) = s + t, \quad \text{and} \quad u(s, t) = \frac{e^{2s}}{2}\left[2 \cos t - e^{-2s} + 1\right].$$

It follows from the above equations that the given Cauchy problem has no solution.

More generally, the functions $\alpha(x, y)$ and $\beta(x, y)$ in equation (7.1.24) are determined by the differential equation, so they places a constraint on the given data function $q(x)$. It $q(s)$ is not of this form for any constant G, then there is no solution taking on these prescribed values on C. On the other hand, if $q(s)$ is of this form for some G, then there are infinitely many such solutions, because we can choose for g any differentiable function so that $g(k) = G$.

   In summary, to solve a Cauchy problem for a first order linear or semilinear equation, we solve the characteristic equations (7.1.10), together with the *compatibility equation* (7.1.11). Notice that, for an equation of the form (7.1.4), it is possible to solve the compatibility equation independent of the characteristic equations. However, the same is not true for a first order quasilinear equation.

7.2   Quasilinear Equations

Consider a first order *quasilinear equation* for a function $u = u(x, y)$ given by (7.1.6). In this case, the vector field of coefficient functions is given by

$$\mathbf{V} = \big(a(x, y, u), b(x, y, u), c(x, y, u)\big),$$

which defines the tangent plane at the point (x, y, u) on an integral surface Γ_u of the differential equation (7.1.6). As before, integral curves of the field $\mathbf{V}$ are given by the system:

$$\frac{dx}{ds} = a(x(s), y(s), u(s)) \tag{7.2.1}$$

$$\frac{dy}{ds} = b(x(s), y(s), u(s)) \tag{7.2.2}$$

$$\frac{du}{ds} = c(x(s), y(s), u(s)), \tag{7.2.3}$$

which define the characteristics of the equation (7.1.6). The projected characteristic curves $C(s)$ for this equation are obtained from the first two of these differential equations, and along these characteristics in xy-plane, we have

$$\frac{du}{dx} = u_x + \frac{dy}{dx} u_y = u_x + \frac{b}{a} u_y = \frac{c(x, y, u)}{a(x, y, u)}.$$

Therefore, the variation in $u = u(x, y)$ along the curves $C(s)$ is given by

$$\frac{du}{ds} = \frac{du/dx}{dx/ds} = c(x, y, u). \tag{7.2.4}$$

That is, along the curves $C(s)$, (7.1.6) transforms to the equation

$$\frac{du}{ds} = c(x(s), y(s), u(s)), \tag{7.2.5}$$

which is known as the *compatibility condition*. Now, for a subinterval $I \subseteq \mathbb{R}$, and $i = 1, 2, 3$, let $\gamma_i : I \to \mathbb{R}$ be a continuously differentiable functions such that

$$\left(\gamma_1'\right)^2 + \left(\gamma_2'\right)^2 \neq 0, \quad \text{over } I,$$

and the curve C_γ given by

$$C_\gamma = \left\{(\gamma_1(s), \gamma_2(s), \gamma_3(s)) : \text{ for } s \in I\right\}$$

be a part of the integral surface Γ_φ associated with a solution $\varphi \in C^1(D)$ of the quasilinear equation (7.1.6), where $D \subseteq \Omega$ is an open set. As before, a typical Cauchy problem is about finding a function $u \in C^1(D)$ such that it satisfies the equation, and $u(\gamma_1(s), \gamma_2(s)) = \gamma_3(s)$, for all $s \in I$.

Notice that, as both the characteristics and compatibility equations involve the dependent variable u, the projected characteristics in this case may *intersect* at some point. Further, the solution remains constant along the characteristics only when $c \equiv 0$. Starting from an initial curve specified by assigned initial condition, an integral surface Γ_u of the field $\mathbf{V} = (a, b, c)$ is constructed by using the parametrised curves

$$\Gamma(s) = (x(s), y(s), u(s)), \quad \text{for } -\infty < s_0 \le s \le s_1 < \infty.$$

Therefore, a general solution $u = u(x, y)$ of the equation (7.1.6) defines a smooth surface that is locus of a point $(x, y, u) \in \mathbb{R}^3$ changing directions subject to that all through the normal $\mathbf{N} = (p, q, -1)$ to the surface remains perpendicular to the field $\mathbf{V}$.

For a Cauchy problems related to a quasilinear equation (7.1.6), we adopt the following explicit approach. That is, to find a solution of the Cauchy problem in parametric form as

$$x = x(s, t), \quad y = y(s, t), \quad u = u(s, t),$$

the first step is to obtain a curve $\Gamma(s, t)$ by solving the characteristic equations (7.2.1) – (7.2.3), with an initial curve given by

$$\Gamma(0, t) : \quad u(0, t) = u_0(t) \tag{7.2.6}$$

along the curve

$$x(0, t) = x_0(t), \quad y(0, t) = y_0(t).$$

So, for every $t = t_0$, we obtain a curve $\Gamma(s, t_0)$ linked to the point $\Gamma(0, t_0)$. Finally, at the second step, we find a solution $u = u(x, y)$ by inverting the relations $x = x(s, t)$ and $y = y(s, t)$ to obtain $s = s(x, y)$ and $t = t(x, y)$.

Example 7.8 Consider the Cauchy problem

$$u\, p + q = 2, \quad \text{with } u(x, y) = x, \quad \text{along the line } y = x, \ x \neq 1.$$

In view of given initial data, the initial curve $\Gamma(0, t)$ is given by

$$x(0, t) = t, \quad y(0, t) = t, \quad u(0, t) = t.$$

So, the characteristic equations are given by

$$\frac{dx}{ds}(s, t) = u(s, t), \quad \text{with } x(0, t) = t;$$

$$\frac{dy}{ds}(s, t) = 1, \quad \text{with } y(0, t) = t; \tag{7.2.7}$$

$$\frac{du}{ds}(s, t) = 2, \quad \text{with } u(0, t) = t.$$

The second and the third equations in (7.2.7) imply

$$y(s, t) = s + t \quad \text{and} \quad u(s, t) = 2s + t.$$

Using the above expression for $u(s, t)$, the first equation in (7.2.7) gives

$$x(s, t) = s^2 + ts + t.$$

Solving s, t in terms of x, y, we have

$$x(s, t) = (y - t)^2 + t(y - t) + t = y^2 + t(1 - y) \quad \Rightarrow \quad t = \frac{x - y^2}{1 - y},$$

so that $s = y - t = (y - x)/(1 - y)$. Hence, the solution is obtained as

$$u(x, y) = \frac{2y - y^2 - x}{1 - y}, \quad y \neq 1.$$

More generally, it is possible to find a local solution in the neighbourhood of the initial curve as long as the coefficients a, b, c are continuously differentiable functions. However, as in the case of semilinear equation, the local feasibility of the afore mentioned second step depends on whether or not the initial data lies on a characteristics of the equation. More precisely, to apply the implicit function theorem, we must have that

$$J[x, y; s, t] \neq 0, \quad \text{along } s = 0.$$

We say the initial curve $\Gamma(0, t)$ is *noncharacteristic* if its projection $C(0, t) = (x(0, t), y(0, t))$ is nowhere tangent to projected characteristic curves $C(s, t) = (x(s, t), y(s, t))$. Equivalently, if

$$\det \begin{pmatrix} a(x_0(t), y_0(t), u_0(t)) & x_0'(t) \\ b(x_0(t), y_0(t), u_0(t)) & y_0'(t) \end{pmatrix} \neq 0. \tag{7.2.8}$$

For example, in case of the above example, we have

$$\det \begin{pmatrix} a(x_0(t), y_0(t), u_0(t)) & x_0'(t) \\ b(x_0(t), y_0(t), u_0(t)) & y_0'(t) \end{pmatrix} = \det \begin{pmatrix} t & 1 \\ 1 & 1 \end{pmatrix} \neq 0, \ \forall t.$$

So, for $y \in [1, \infty)$, the problem has a unique solution as obtained above.

The next example illustrates the fact that a Cauchy problem for a quasilinear equation may not have any solution or it may have infinitely many when the initial data is specified along a characteristic curve.

Example 7.9 Consider the general Cauchy problem for the equation (5.1.33):

$$u_t + \alpha(u) u_x = 0, \quad \text{for } x \in [0, \ell] \text{ and } t \geq 0;$$
$$u(x, 0) = \phi(x).$$

In this case, the characteristic equations are given by

$$\frac{dx}{ds}(s, \tau) = \alpha(u(s, \tau)), \quad \text{with } x(0, \tau) = x_0;$$

$$\frac{dt}{ds}(s, \tau) = 1, \quad \text{with } t(0, \tau) = 0;$$

$$\frac{du}{ds}(s, \tau) = 0, \quad \text{with } u(0, \tau) = \phi(x_0).$$

As the function u is not known, the first two of above differential equations cannot be solved. However, we can write the general solution as

$$u(x, t) = \phi(x_0), \quad \text{with } x(t) = x_0 + \alpha(\phi(x_0)) t.$$

Therefore, we obtain

$$u(x, t) = \phi(x - a(u)t), \quad \text{for } (x, t) \in [0, \ell] \times \mathbb{R}^+.$$

Hence, in this case, characteristic curves corresponding to different u in general may cross each other, leading to a *blow up* in finite time. That is, beyond a certain time, solution does not exists in classical sense. In particular, as $\alpha(u) = u$ for Burgers' equation, the characteristic equations are given by

$$\frac{dx}{ds} = u(s), \quad \frac{dt}{ds} = 1, \quad \frac{du}{ds} = 0,$$

so that by taking $\Gamma_\phi \equiv (r, 0, \phi(r))$ we can write

$$x(r, s) = \phi(r) s + r, \quad t(r, s) = s, \quad u(r, s) = \phi(r),$$

with $u(x, t) = \phi(x - ut)$. The line $x(r, s) = \phi(r)s + r$ is the projected characteristic in the xt-plane, passing through the point $(r, 0)$, along which we have $u(r, s) = \phi(r)$. If the two characteristics

$$x_1(r, s) = \phi(r_1) s + r_1 \quad \text{and} \quad x_2(r, s) = \phi(r_2) s + r_2$$

intersect at a point (x, t), with

$$t = -\frac{r_2 - r_1}{\phi(r_2) - \phi(r_1)},$$

then $u(x, t) = \phi(x - ut)$ implies that

$$u_x = \phi'(r)(1 - u_x t) \quad \Rightarrow \quad u_x = \frac{\phi'(r)}{1 + \phi'(r)t}.$$

It thus follows that, for $\phi'(r) < 0$, we have

$$u_x \to \infty \quad \text{as} \quad t \to -\frac{1}{\phi'(r)}.$$

At the smallest such time t_0, so that for the corresponding $r = r_0$, $\phi'(s)$ is minimum (or $-\phi'(s)$ is maximum), the solution $u = u(x, t)$ experiences a *gradient catastrophe*. Therefore, for Burger equation with $\phi(x) = -x$, the maximum time for the existence of a classical solution is $t = 1$.

Theorem 7.1 *Suppose a, b, c are C^1-functions in their arguments, the initial data $x_0, y_0, z_0 \in C^1([s_1, s_2])$, and the initial curve Γ is noncharacteristic. Then there exists a neighbourhood of the projected curve C such that the Cauchy problem has exactly one solution $z = u(x, y)$.*

Proof Left to the reader as exercise.

Therefore, in general, a Cauchy problem for a first order quasilinear equation in two variables is about finding a unique function $u = u(x, y) \in C^1(\Omega)$ in a neighbourhood of projected curve of the assigned initial curve $\phi = \phi(s) : [s_1, s_2] \to \Omega \times \mathbb{R}$ such that it takes prescribed value $z_0(s)$ on ϕ. That is,

$$u(x_0(s), y_0(s)) = z_0(s), \quad \text{for all } s \in [s_1, s_2].$$

An initial curve specified by the given initial condition looks like a spine about which the whole integral surface is *knitted* as a union of characteristics.
Integrating the equation (5.1.32) over an interval $x_0 \le x \le x_1$, we obtain

$$\frac{d}{dt} \int_{x_0}^{x_1} u(x, t)dx + f(u(x_1, t)) - f(u(x_0, t)) = 0, \qquad (7.2.9)$$

which is a typical 1-dimensional conservation law. Let $u = u(x, t)$ be a solution of this equation such that, for some fixed t, u has a *jump discontinuity* at a point $x = \xi(t)$, with $\xi(t) \in C^1$. Suppose all the three functions u, u_x, u_t are continuous upto ξ. Then, for $x_0 < \xi(t) < x_1$, we have

$$\frac{d}{dt}\left(\int_{x_0}^{\xi} u(x, t)dx + \int_{\xi}^{x_1} u(x, t)dx \right) + f(u(x_1, t)) - f(u(x_0, t))$$

$$= \xi'(t)u^-(\xi(t), t) - \xi'(t)u^+(\xi(t), t) + \int_{x_0}^{\xi} u_t(x, t)dx + \int_{\xi}^{x_1} u_t(x, t)dx$$

$$= -\big[f(u(x_1, t)) - f(u(x_0, t))\big],$$

where u^-, u^+ are, respectively, the left and right limits of u taken on sides of the *shock interval* $[x_0, x_1]$. According to *Rankine–Hugoniot jump condition*, as $x_0 \uparrow \xi(t)$ and $x_1 \downarrow \xi(t)$, we have

$$\xi'(t)\big[u^- - u^+\big] + f(u^+) - f(u^-) = 0 \quad \Leftrightarrow \quad \xi'(t) = \frac{f(u^+) - f(u^-)}{u^+ - u^-}.$$

7.3 Fully Nonlinear Equation

Let $U \subset \mathbb{R}^3$ be an open set, and consider a differentiable implicit function $F : U \times \Omega \to \mathbb{R}$ given by

$$F = F(x, y; \, u; \, p, q),$$

where $p, q : \Omega \to \mathbb{R}$ are arbitrary functions. For a fixed arbitrary point $x_0 = (x_0, y_0, u_0) \in U$, let $V(x_0)$ be the set of vectors $v = (p, q, -1)$ such that the relation $F(x_0; p, q) = 0$ holds, and the associated family of tangent planes $\Pi(x_0)$ at the point x_0 are orthogonal to vectors v. Also, if

$$F_q(x_0; \, p, q) \not\equiv 0 \quad \text{over} \quad U,$$

by implicit function theorem we can write $q = g(p)$, where g is a continuous function defined over an interval contained in Ω. Specialising to our needs, let a function $u = u(x, y) \in C^1(\Omega)$ be a solution of a nonlinear differential equation of the form

$$F(x, y; \, u; \, p, q) = 0, \quad \text{with} \quad p = u_x, \quad q = u_y, \tag{7.3.1}$$

with associated integral surface given by

$$\Gamma_u := \big\{(x, y, u(x, y)) \in U : \text{for } (x, y) \in \Omega\big\}.$$

Then, the tangent plane of Γ_u at the point (x_0, y_0, z_0) (with , $(z_0 = u(x_0, y_0))$) is a member of the family $\Pi(x_0)$, with $v = \mathbf{n} = (p, q, -1)$. Also, all possible one-parameter family of tangent planes to the integral surface Γ_u passing through the point (x_0, y_0, z_0), and orthogonal to the normal vector $\mathbf{n}$, are given by

$$(p, q, -1) \cdot (x - x_0, y - y_0, z - z_0) = 0$$
$$\Leftrightarrow \quad z - z_0 = p(x - x_0) + g(p)(y - y_0), \tag{7.3.2}$$

where p is the parameter. We know that, in general, if $S_a \subset \mathbb{R}^3$ is a one-parameter family of regular surface defined by a smooth function $\varphi = \varphi(x, y; a)$, then the envelope $\mathscr{E} : a = f(x, y)$ of the family of surfaces S_a is precisely the union of one-parameter family of (intersection) curves Γ_a given by the following two surfaces:

$$\varphi(x, y; a) = 0 \quad \text{and} \quad \frac{\partial \varphi}{\partial a} = 0.$$

Therefore, we have $\varphi = \varphi(x, y; f(x, y))$. For a quick visualisation, one may take S_a to be the unit sphere with centre at the point $(a, 0, 0)$, i.e., we are considering the situation when $\varphi(x, y; a) = (x - a)^2 + y^2 + z^2 - 1$. In this case, $\mathscr{E}$ is right circular cylinder of radius 1 around the x-axis. In general, along Γ_a, the tangent planes to the surface $\varphi(x, y; a) = 0$ and the envelope $\mathscr{E}$ coincides, and a is constant. So, we have

$$d\varphi = \varphi_x dx + \varphi_y dy \quad \text{and} \quad 0 = \varphi_{ax} dx + \varphi_{ay} dy.$$

In particular, the envelope $\mathscr{E}$ of the family of planes in (7.3.2) is a cone $K(x_0)$ with vertex at x_0, which is known as the *Monge cone*. Clearly, on this cone $K(x_0)$, we have

$$du = p\, dx + q\, dy; \tag{7.3.3}$$

$$0 = dx + \frac{dq}{dq} dy. \tag{7.3.4}$$

Differentiating (7.3.1) with respect to p to also have

$$F_p + f_q \frac{dq}{dq} = 0.$$

On comparing the above equation with (7.3.4), we obtain

$$\frac{dx}{F_p} = \frac{dy}{dF_q}. \tag{7.3.5}$$

In view of (7.3.3), equations defining the cone $K(x_0)$ can be expressed in parametric form as follow:

$$\frac{dx}{ds} = F_p(x, y; u; p, q) \tag{7.3.6}$$

$$\frac{dy}{ds} = F_q(x, y; u; p, q) \tag{7.3.7}$$

$$\frac{du}{ds} = p \frac{dx}{ds} + q \frac{dy}{ds}$$
$$= p\, F_p(x, y; u; p, q) + q\, F_q(x, y; u; p, q) \tag{7.3.8}$$

The description above suggests that, among a field of cones at points $x_0 \in \mathbb{R}^3$, a function $u = u(x, y) \in C^1(\Omega)$ is a solution of the differential equation (7.3.1) if and only if the surface Γ_u is tangent to the cone $K(x_0, y_0, u(x_0, y_0))$, for every $(x_0, y_0) \in \Omega$. By taking

$$z_0 = u(x_0, y_0), \quad p_0 = u_x(x_0, y_0), \quad \text{and} \quad q_0 = u_y(x_0, y_0),$$

the tangent plane of Γ_u at x_0 is the set of points $x = (x, y, z)$ such that

$$z - z_0 = p_0(x - x_0) + q_0(y - y_0)$$

holds. Therefore, using *characteristic system* (7.3.6) – (7.3.8), it follows that x are points on the line

$$\frac{x - x_0}{F_p} = \frac{y - y_0}{F_q} = \frac{z - z_0}{p_0 F_p + q_0 F_q},$$

where both F_p, F_q are evaluated at $(x_0, y_0, z_0, p_0, q_0)$. Said differently, x are points on a line in the direction $(F_p, F_q, p_0 F_p + q_0 F_q)$, passing through the point x_0. Hence, the field of directions defined at each point of Γ_u are the directions lying along a generator of the Monge cone at that point. The *Monge cone*, in general, is a *ruled surface* with generator a line in some tangent plane defined by (7.3.2). Notice that a plane Π belongs to the family $\Pi(x_0)$ if and only if Π is tangent to the cone $K(x_0)$, along a generator of the cone.

Definition 7.3 A surface S is called an *integral surface* of a first order nonlinear differential equation in two variables if, for each $x_0 \in S$, $\Pi(x_0)$ is tangent to the cone $K(x_0)$. Further, since the unique line of tangency between $\Pi(x_0)$ and $K(x_0)$ determines a direction field on S, the *integral curves* of the field are called *characteristic curves*.

In general, such types of curves depend on our choice for a tangent plane $\Pi(x_0)$, and hence on the point p_0, because $q_0 = g(p_0)$. As seen in previous sections, the cone $K(x_0)$ degenerates to a points, for the semilinear equations, and to lines, for the quasilinear case. So, neither of the two require any specification for p_0.

The *characteristic system* of differential equations (7.3.6) – (7.3.8), that suffices to determine the characteristics for a quasilinear equation, is not solvable in this general case unless equations for $p'(s)$ and $q'(s)$ are specified. For that purpose, differentiate

$$F(x, y; u(x, y); p, q) = 0$$

with respect to x to we obtain

$$F_x + p \, F_u + p_x F_p + q_x F_q = 0.$$

So, using the relation $q_x = u_{xy} = p_y$, and equations (7.3.6)–(7.3.8), it follows that

$$F_x + p \, F_u + p_x x'(s) + p_y y'(s) = 0 \quad \Rightarrow \quad \frac{dp}{ds} = -F_x - p \, F_u \qquad (7.3.9)$$

Similarly, by differentiating the relation $F(x, y; u(x, y); p, q) = 0$ with respect to y, we obtain

$$\frac{dq}{ds} = -F_y - q\, F_u. \tag{7.3.10}$$

The above system of five ODEs in variables x, y, u, p, q is said to define the *characteristic system* of the quasilinear equation (6.2.1). Notice that, since

$$\begin{aligned}
\frac{dF}{ds} &= F_x x'(s) + F_y y'(s) + F_u u'(s) + F_p p'(s) + F_q q'(s) \\
&= F_x F_p + F_y F_q + F_u[pF_p + qF_q] - F_p[F_x + pF_u] - F_q[F_y + qF_q] \\
&= 0,
\end{aligned}$$

the smooth function

$$F(s) = F(x(s), y(s); u(s); p(s), q(s))$$

is a *first integral* of the system of five autonomous ODEs, and so if for some s_0 we have

$$F(x(s_0), y(s_0); u(s_0); p(s_0), q(s_0)) = 0,$$

then $F \equiv 0$. Hence, the characteristic system (7.3.6)–(7.3.8) determine the characteristics

$$\Gamma(s) := (\gamma_1(s), \gamma_2(s), \gamma_3(s)), \quad \text{for} \quad s \in I,$$

called the *base characteristic*, for the integral surface of a quasilinear equation. The two ODEs (7.3.9) and (7.3.10) help define the normal vector $\mathbf{n} = (p, q, -1)$. Notice that, if we identify the point $(x_0, p, q) \in \mathbb{R}^5$ with a plane $\Pi(x_0)$ orthogonal to the vector $v = (p, q, -1)$, then various specifications of $(p(s), q(s))$ gives infinitesimal pieces of the tangent planes along the characteristic curves $\Gamma = \Gamma(s)$ and only support of these *strips* are used to construct the integral surface.

Definition 7.4 *(Strip Condition)* Let $C : I \to \mathbb{R}^5$ be a parametrised curve given by

$$C(s) := (\gamma_1(s), \gamma_2(s), \gamma_3(s), \gamma_4(s), \gamma_5(s)), \quad s \in I,$$

be such that, for each $s \in I$, $C(s)$ belongs to a plane $\Pi(s)$ passing through the point $\Gamma(s) = (\gamma_1(s), \gamma_2(s), \gamma_3(s))$ and orthogonal to the vector

$$v(s) = (\gamma_4(s), \gamma_5(s), -1).$$

The curve C is called a **strip** if the tangent vector $\Gamma'(s)$ of the *base curve* $\Gamma(s)$ always lies in the plane $\Pi(s)$. That is, for every $s \in I$,

$$\gamma_3'(s) = \gamma_4(s)\gamma_1'(s) + \gamma_5(s)\gamma_2'(s). \tag{7.3.11}$$

As before, the method tries to solve a fully nonlinear differential equation of the form (7.3.1) along certain suitably chosen curves by reducing the equation to a system of ODEs that can be solved explicitly, atleast locally. So, the first step is to figure out what type of parametrised curves would serve the purpose. In the case of a fully nonlinear differential equation, the main difference arises due to the fact the nonlinearity of the equation constraints the vectors that are orthogonal to the normal vector $\mathbf{n} = (p, q, 1)$ must belong to Monge cone, at every point (x, y, u). So, to solve such a fully nonlinear differential equation, we need to construct a regular surface that is tangent to the Monge cones, everywhere.

For a quasilinear equation, as illustrated in Example 7.8, situation is facilitated by the fact that the Monge cones are actually planes so that every solution of the characteristic system is automatically a strip, called the *characteristic strips*, and so the corresponding integral surface defined by a solution is the *Monge strip*. Furthermore, subject to an assigned initial conditions imposed on p and q, we obtain an *initial characteristic strip*:

$$\Gamma_0 := \Gamma(s_0) = (\gamma_1(s_0), \gamma_2(s_0), \gamma_3(s_0)), \quad \text{for some } s_0 \in I$$

Therefore, starting with a point on the initial curve Γ_0, an integral surface of a quasilinear equation is the locus of a point $(x, y, u(x, y))$ changing directions smoothly subject to the condition that the normal $\mathbf{n}$ all through its motion remains perpendicular to the Monge strips defined by the vector field $\mathbf{V} = (P, Q, R)$. In the present case, a function $u = u(x, y) \in C^1(\Omega)$ satisfies a nonlinear equation if the integral surface Γ_u is the union of certain base curves of the characteristic strips. Therefore, to find the base curves, we solve the characteristic system (7.3.6)–(7.3.8) and then construct the integral surface Γ_u by computing functions p, q using equations (7.3.9) and (7.3.10).

A Cauchy problem related to a fully nonlinear differential equation, as usual, is about finding a solution $u = u(x, y)$ of equation (7.3.1) such that

$$u(\gamma_1(s), \gamma_2(s)) = \varphi(s),$$

where $\varphi = \varphi(s)$ is prescribed by the given Cauchy data on the projected characteristics parametrised as $\gamma = (\gamma_1, \gamma_2)$ in the xy plane. For the existence of a solution, the first condition is that the initial curve

$$\Gamma_0 : \quad x(s) = \gamma_1(s), \ y(s) = \gamma_2(s), \ u(s) = \varphi(s)$$

must be **noncharacteristic**, i.e., at each point on Γ_0, the Monge cone is not tangent to Γ_0. However, since the function F in (7.3.1) only determines a Monge cone along the curve Γ_0, no additional information is available to know which direction to flow along a characteristic. Said differently, there is system of five ODEs to solve, whereas initial values are known only for the characteristic system (7.3.6)–(7.3.8).

A natural way to resolve the issue is to specify initial conditions for p and q in terms of two functions, say $\phi_1, \phi_2 : I \to \mathbb{R}$, along the characteristic Γ_0. Suppose

$$\sigma(s) = (\gamma_1(s), \gamma_2(s), \varphi(s), \phi_1(s), \phi_2(s)), \quad s \in I,$$

is an arbitrary curve, with base curve $\sigma_b(s) = (\gamma_1(s), \gamma_2(s))$, such that $F(\sigma(s)) = 0$, for all $s \in I$. Notice that there is a unique solution $C(s, t)$ of the system (7.3.6)–(7.3.10) passing through each point of the curve $\sigma(s)$, and it is a characteristic strip because F is the *first integral* of the system. Then, the union of the corresponding base curves $\sigma_b(s, t)$ is a surface Γ_u, containing the base curve $\sigma_b(s)$. The following theorem tells us when Γ_u is an integral surface of the differential equation (7.3.1).

Theorem 7.2 *Let $U \subset \mathbb{R}^5$ be a domain, containing the open set Ω, and $F \in C^2(U)$ be given by*

$$F = F(x, y; u(x, y); p, q), \quad \text{with} \quad F_p^2 + F_q^2 \neq 0.$$

For $I = [a, b]$, let $\gamma = (\gamma_1, \gamma_2) \in C^2(I)$, $\varphi \in C^2(I)$, and $\phi_1, \phi_2 \in C^1(I)$, such that, for all $s \in I$,

$$F(\sigma(s)) = F(\gamma_1(s), \gamma_2(s), \varphi(s), \phi_1(s), \phi_2(s)) = 0; \tag{7.3.12}$$

and, the following strip condition holds for all $s \in I$:

$$\frac{d\varphi}{ds} = \phi_1(s)\gamma_1'(s) + \phi_2(s)\gamma_2'(s) \tag{7.3.13}$$

Further, suppose $\sigma_b'(s) = (\gamma_1'(s), \gamma_2'(s))$ and the projection of the characteristic direction $(F_p, F_q, p_0 F_p + q_0 F_q)$ onto xy-plane given by $(F_p(\sigma(s)), F_q(\sigma(s)))$ are linearly independent. That is, we have

$$\gamma_1'(s)F_q(\sigma(s)) \neq \gamma_2'(s)F_p(\sigma(s)), \quad \forall \; s \in I. \tag{7.3.14}$$

Then, there exists a unique solution of equation (7.3.1) satisfying

$$u(\sigma_b(s)) = \varphi(s), \quad p(\sigma_b(s)) = \phi_1(s), \quad \text{and} \quad q(\sigma_b(s)) = \phi_2(s),$$

for all $s \in I$.

Proof Left to the reader as an exercise. For example, see [2], Theorem 2, p.107.

Therefore, as each tangent plane to an integral surface Γ_u must be tangent to the Monge cone, the functions ϕ_1, ϕ_2 must satisfy (7.3.12). And, for these planes to fit together nicely along the initial characteristic Γ_0, the strip condition (7.3.13) must hold. A Cauchy problem has a unique solution if so does the equation (7.3.12), and it has no solution if the equation has no (real) solution. For $(x_0, y_0, u_0) = (\gamma_1(0), \gamma_2(0), \varphi(0))$, if (p_0, q_0) is a solution of the following system of nonlinear equations:

$$F(x_0, y_0, u_0, p_0, q_0) \equiv 0;$$
$$\varphi'(0) = p_0 \gamma_1'(0) + q_0 \gamma_2'(0),$$

obtained by the implicit function theorem, then the condition

$$\det \begin{pmatrix} \gamma_1'(0) & F_p(x_0, y_0, u_0, p_0, q_0) \\ \gamma_2'(0) & F_q(x_0, y_0, u_0, p_0, q_0) \end{pmatrix} \neq 0$$

gives the existence of a solution (ϕ_1, ϕ_2) of the following system of nonlinear equations:

$$F(\gamma_1(s), \gamma_2(s), \varphi(s), \phi_1(s), \phi_2(s)) \equiv 0;$$
$$\varphi'(s) = \phi_1(s)\gamma_1'(s) + \phi_2(s)\gamma_2'(s),$$

in a neighbourhood of the point (x_0, y_0). So, at the first step, we solve these two equations for $\phi_1(s)$ and $\phi_2(s)$. Then, we solve characteristic system of five equations, with parameter t, using the initial conditions

$$x(s, 0) = \gamma_1(s), \quad y(s, 0) = \gamma_2(s), \quad u(s, 0) = \varphi(s),$$
$$p(s, 0) = \phi_1(s), \quad q(s, 0) = \phi_2(s).$$

Finally, we solve $x = X(s, t)$ and $y = Y(s, t)$ for s, t. The following examples illustrate the procedure to construct an integral surface for the aforementioned Cauchy problem.

Example 7.10 Consider the Cauchy problem:

$$p^2 - 3q^2 = u, \quad \text{with } u(x, y) = x^2 \text{ along } y = 0.$$

Here, $F(x, y; u; p, q) = p^2 - 3q^2 - u$, and so the characteristic system is given by

$$x'(t) = 2p, \quad y'(t) = -6q, \quad u'(t) = 2p^2 - 6q^2 = 2u; \qquad (7.3.15)$$
$$p'(t) = p, \quad q'(t) = q. \qquad (7.3.16)$$

In view of given Cauchy data, we can parametrise the initial curve as

$$\Gamma_0 : \quad \gamma_1(s) = s, \quad \gamma_2(s) = 0, \quad \varphi(s) = s^2$$

Now, to complete the initial strip, we solve the system

$$\phi_1^2 - 3\phi_2^2 = s^2; \quad \phi_1 = 2s.$$

There are following two possibilities:

$$\phi_1(s) = 2s, \quad \phi_2(s) = \pm s.$$

First, consider the case when $\phi_2(s) = s$. Then, using (7.3.16), we obtain

$$P(s, t) = 2se^t \quad \text{and} \quad Q(s, t) = se^t.$$

So, it follows from (7.3.15) that

$$X(s, t) = 4s(e^t - 1) + s;$$
$$Y(s, t) = -6s(e^t - 1);$$
$$U(s, t) = s^2 e^{2t}.$$

We can solve the first two equations for s and t and then use the third to obtain

$$u(x, y) = \left(x + \frac{2}{y}\right)^2.$$

Similarly, the case when $\phi_2(s) = -s$ yields

$$u(x, y) = \left(x - \frac{2}{y}\right)^2.$$

Example 7.11 Consider the Cauchy problem

$$pq = u, \quad \text{with} \quad u(x, y) = y^2 \text{ along } x = 0.$$

Here, $F \equiv pq - u$, and so the characteristic system is given by

$$x'(t) = q, \quad y'(t) = q, \quad u'(t) = pq + pq = 2u; \qquad (7.3.17)$$
$$p'(t) = -p, \quad q'(t) = -q. \qquad (7.3.18)$$

In view of given Cauchy data, we may use $\Gamma_0(s) = (0, s, s^2)$. Now, to complete the initial strip, we solve the system

$$\phi_1\phi_2 = s^2; \quad \phi_2 = 2s,$$

which has the solution $\phi_1(s) = s/2$ and $\phi_2(s) = 2s$. Then, using (7.3.18), we obtain

$$P(s, t) = \frac{s}{2}e^{-t} \quad \text{and} \quad Q(s, t) = 2se^{-t}.$$

So, it follows from (7.3.17) that

$$X(s, t) = -2se^{-t} + 2s;$$
$$Y(s, t) = -2se^{-t} + 3s;$$
$$U(s, t) = s^2 e^{2t}.$$

We can solve the first two equations for s and t and then use the third to obtain

$$u(x, y) = \frac{4}{25} \frac{(x + y)^4}{(2y - 3x)^2}.$$

Exercises 7

7.1 Given that the differential equation $p - q = x^2 + y^2$ has the function

$$u(x, y) = a(x + y) + \frac{1}{3}(x^3 - y^3) + b$$

as a complete solution, show that the equation has $F\left(x + y, u - \frac{1}{3}(x^3 - y^3)\right) = 0$ as a general solution.

7.2 Show that $z = axy + b$ is a complete solution of $px = qy$, and hence find the general solution.

7.3 Show that $z = a(x^2 + y^2) + b$ is a complete solution of the differential equation $py - qx = 0$, and hence find the general solution.

7.4 Solve the equation (i) $p^2 + q^2 = x + y$, (ii) $pq = xy$, (iii) $\tan x \ p + \tan y \ q = \tan z$ (v) $9(p^2 z + q^2) = 4$.

7.5 Find a complete solution of (i) $p^2 + q^2 = 10pq$, (ii) $p + q = pq$ (iii) $z^2(p^2 + q^2 + 1) = 1$.

7.6 Find the general solution of the differential equation $u(x, y) = px + qy + \sqrt{1 + p^2 + q^2}$.

7.7 Solve the following initial-boundary value problem:

$$u_t + cu_x = 0, \quad t > 0, \ 0 < x < \ell, \ c > 0;$$
$$u(x, 0) = g(x), \quad 0 < x < R, \quad u(0, t) = f(t), \quad t > 0.$$

7.8 Solve the semilinear equation $x^2 p + y q + xyu = 1$, using method of characteristics.

7.9 Solve the semilinear equation $x \ p - y q + y^2 u = y^2$, using method of characteristics.

7.10 Solve the following transport equation in the presence of a *distributed source* of the contaminant with density per unit time $f(x, t)$ at the time t and the position x, where both f, f_x are continuous:

$$u_t + cu_x = f(x, t), \quad t > 0, \ x \in \mathbb{R};$$
$$u(x, 0) = g(x), \quad g \in C^1(\mathbb{R}).$$

What if, in addition to the transport process, there is also decay proportional to the present density $i.\,e.\ f(x, t) = -u(x, t)$?

7.11 Find the solution of the differential equation $x\, p - y\, q = u$ subject to $u(x, y) = x^2$ along the line $C : y = x$, with $1 \le y \le 2$.

7.12 Find the solution of the differential equation $x\, p + y\, q = 2xy$ subject to $u(x, y) = 2$ along the parabola $C : y = x^2$.

7.13 Suppose u_1 and u_2 are any two solutions of a quasilinear equation of the form (7.1.6) such that the surfaces $u - u_1 = 0$ and $u - u_2 = 0$ intersect along a curve Γ in xyu-space. Show that Γ must be a characteristic curve.

7.14 Supply a proof of Theorem 7.1.

7.15 (A Riemann Problem) Let the positive scalars a, b, c represent the velocities concerning a problem of kinetics in chemistry such that $u = u(t, x)$ is the concentration of the chemical substance at time $t \ge 0$ and at a *height* x (of a tube). For $x \ge 0, \ y \ge 0$, consider the Cauchy problem

$$u_t + u_x = (ae^{-bt} + c)(1 - u),$$

with the initial data given by

$$u(t, 0) = 0, \quad \text{for } t > 0, \quad \text{and} \quad u(0, x) = u_0(x), \quad \text{for } x > 0.$$

Solve the problem separately in domains Ω_1 and Ω_2 in the first quadrant of $\mathbb{R}^2$ partitioned by he line $t = x$.

7.16 Solve the initial value problem $u\, u_x + u_y = 2$, with initial curve given by

$$x_0(s) = s, \quad y_0(s) = 1, \quad z_0(s) = 1 + s, \quad \text{for } 0 < s < 1.$$

7.17 Solve the initial value problem $u\, u_x^2 + u_y^2 = 1 + x$, with initial curve given by

$$x_0(s) = 0, \quad y_0(s) = s, \quad z_0(s) = 1, \quad p_0(s) = 1, \quad q_0(s) = 0, \quad \text{for } s \in \mathbb{R}.$$

7.18 Solve the initial value problem $(x - y)\, u_x + 2y\, u_y = 3x$, with projected curve given by

$$x_0(s) = s, \quad y_0(s) = 1, \quad \text{for } s \in \mathbb{R}.$$

7.19 Solve the initial value problem $u\,u_x^2 + u_y^2 = 1$, with initial curve given by

$$x_0(\theta) = a\cos\theta, \quad y_0(\theta) = a\sin\theta, \quad z_0(\theta) = 1, \quad p_0(\theta) = \cos\theta, \quad q_0(s) = \sin\theta,$$
for $\theta \in [0, 2\pi]$.

7.20 Supply a proof of Theorem 7.2.

References

1. Arnold, V. I., *Lecturers on Partial Differential Equations*, Springer, 2004.
2. Evans, L. C., *Partial Differential Equations*, AMS GTS, Vol.19, 2010.
3. John, F., *Partial Differential Equations*, Springer, 1982.
4. Prasad, P. and Ravindran, R., *Partial Differential Equations*, New Age International (P) Ltd., New Delhi, 1985.

Chapter 8
Separation of Variables

*Profound study of nature is the most fertile source of
mathematical discoveries.*

Joseph Fourier (1768–1830)

The basic idea of *separation of variables* is somewhat similar to the method of
reduction of a homogeneous ordinary differential equation to variables separable
form, which was introduced by *Gottfried Leibniz* (1646–1716) in 1691. An modified version of the same was applied by *Jean d' Alembert* (1717–1783) in 1773 to
formulate the *series solution* of the 1-dimensional wave equation. However, it was
French mathematician and engineer *Joseph Fourier* (1768–1830) who developed the
method more systematically while providing a series solution for some important
initial-boundary value problems related to heat conduction in solids. Fourier's *conjecture* about possibility of an *arbitrary* function to have Fourier series representation
emerged from the need to ensure the existence of a unique solution of any such a
problem, for various types of initial conditions.

The central idea is based on the assumption that the underlying equation has
a solution that can be written as a product of univariate functions of the involved
variables. The method reduces the original problem to solving some familiar types
of boundary value problems for linear ordinary differential equations. Finally, a
series solution of the original IBVP is obtained by using Fourier series techniques.
In general, the method separation of variables depends on the *type* of differential
equation, *nature* of the assigned boundary conditions and also on the coordinates
system chosen to work with. Therefore, the method fails to provide a solution of
for certain important types of nonhomogeneous BVPs. Nonetheless, the method
provides series solution of certain fundamental homogeneous IBVPs.

In this chapter, we apply separation of variables to solve some important initial-
boundary value problems concerning physical phenomena such as vibrations, heat
conduction, fluid flow, acoustic, and electrodynamics. As seen earlier, in most

situations, underlying *differential equation model* is a second order linear differential equations in two independent variables defined over a bounded domain or an unbounded domain, together with some limit condition on the possible solution. The reader may refer texts [1–6], for further details.

8.1 Vibrating String Controversy

For brevity, we may repeat some details related to wave equation as derived earlier in Chap. 2. Suppose a perfectly elastic thin string tied tautly along the x-axis is fixed at points $x = 0$ and $x = \ell$. Let a C^2-function $y(x, t)$ represent the *transverse displacement* of the string at a point $x \in [0, \ell]$ and at time $t > 0$, when distorted from the equilibrium position. It is known that the function $y = y(x, t)$ satisfies the wave equation given by

$$y_{tt}(x, t) = y_{xx}(x, t), \quad x \in [0, \ell] \text{ and } t \geq 0. \tag{8.1.1}$$

Suppose the *initial deflection* is given by a C^2-function $d(x) = y(x, 0)$. Therefore, it is assumed that the function d cannot have a *kink* such as in the case of plucked string. Further, as the string has the ends fixed at points $x = 0$ and $x = \ell$, the function $y = y(x, t)$ satisfies the boundary conditions given by

$$y(0, t) = y(\ell, t) = 0, \quad \text{for all } t \geq 0. \tag{8.1.2}$$

As in the original paper of d'Alembert, it may be assumed that the *initial velocity* is zero, i.e., $v(x) = y_t(x, 0) = 0$. We know that a general solution of Eq. (8.1.1) is given by

$$y(x, t) = F(t + x) + G(t - x), \tag{8.1.3}$$

where F and G are arbitrary univariate C^2-functions given by Eq. (6.1.22). The boundary condition $y(0, t) = 0$ implies that $G \equiv -F$ so that (8.1.3) can be written as $y(x, t) = F(x + t) + F(x - t)$, provided F is an *odd function*. Also, it follows from the boundary condition $y(\ell, t) = 0$ that F is a *periodic function*[1] of period 2ℓ. Subsequently, by using the initial condition $y(x, 0) = d(x)$, d'Alembert was able to conclude that the solution $y(x, t)$ of the above initial-boundary value problem is given by

$$y(x, t) = \frac{1}{2}\big[d(x + t) + d(x - t)\big], \quad \text{for } x \in [0, \ell] \text{ and } t > 0.$$

---

[1] Recall that a function $f : \mathbb{R} \to \mathbb{R}$ is said to be **periodic** if, for some $p > 0$, we have $f(t + p) = f(t)$, for all $t \in \mathbb{R}$. The smallest $p > 0$ satisfying above condition is called the **period** of the function f. For example, sine and cosine are periodic of *period* 2π.

It thus makes sense to write the solution of Eq. (8.1.1) as

$$y(x,t) = \frac{1}{2}[\hat{d}(x+t) + \hat{d}(x-t)],$$

where $\hat{d}$ is the odd periodic extension of function $d : [0, \ell] \to \mathbb{R}$, with period 2ℓ. We will give further details in Sect. 8.2.

Euler entered the debate by publishing two papers in 1748 and 1749 wherein he used argument based on the idea of *infinitesimal displacements* to analyse problems related to vibrating string of uniform density ρ. Taking τ as the fixed tension at each point on the string, he proposed general wave equation given by

$$y_{tt}(x,t) = c^2 \, y_{xx}(x,t), \quad \text{where } c = \sqrt{\frac{\tau}{\rho}}. \tag{8.1.4}$$

The physical constant c represents the *wave speed*. Based on d'Alembert's picture of string vibrations, he found the following solution

$$y(x,t) = d(x+ct) + d(x-ct).$$

However, rejecting d'Alembert's smoothness condition on the deflection function, he took d to be any 2ℓ-periodic smooth function defined over the half-interval $(0, \ell)$, subject to that it is odd around both the points $x = 0$ and $x = \ell$. He proposed that the function d can be specified considering the initial conditions suggested by the equation as given below (Theorem 6.1):

$$y(x,t) = \frac{1}{2}\left[d(x+ct) + d(x-ct) + \frac{1}{c} \int_{x-ct}^{x+ct} v(u)du\right].$$

Further, he proclaimed that $d(x)$ and $v(x)$ can be any curves *drawn by hand* in $[0, \ell]$, extended to $\mathbb{R}$ as odd periodic extensions. However, d'Alembert in his publication of 1761 took strong exception to Euler's assertion about "*any curves drawn by hand*" quoting the reason that the derivative y_{xx} would be undefined in the neighbourhood of a point where d has a *kink*. It was also pointed out that such types of physical arguments are flawed due to fact that the role of *infinitesimals* is indispensable in deriving the governing equation. Euler attempted to defend his assertions in two separate papers published in 1762 and 1765, arguing mainly that small displacements ensures negligible error.

In two delayed papers[2] of 1741, both published in 1751, Daniel Bernoulli found the *closed form expression* for the transversal motions of a vibrating elastic strings,

---

[2] In "*De Vibrationibus et Sono Laminarum Elasticarum*", he discussed the case of a horizontal rod of finite length, with one end fixed; and, in "*De Sonis Multifariis quos Laminae Elasticae Diversimode edunt disquisitiones Mechanico-Geometricae Experimentis Acusticis Illustratae et Confirmatae*", he used a relation between curvature and moment to study wave equation for the case when ends are free.

using infinite series, and also exponential and trigonometric functions. Bernoulli argued that the wave equation has a series solution of the form

$$\sum_{n=1}^{\infty} a_n \sin(\omega_n x) \cos(\omega_n ct), \quad \text{with } \omega_n = \frac{n\pi}{\ell}, \tag{8.1.5}$$

which is true if $d(x)$ is a sum of sine functions. As we know now, it takes arguments of the *distribution theory* to make precise the statement such as the general solution $y = y(x, t)$ is a superposition of an infinite number of harmonics such as $y_n = \sin(\omega_n x) \cos(\omega_n t)$ that represents a *standing wave* of the string.

In 1753, Daniel Bernoulli published a long note confronting both d'Alembert and Euler for obscuring the subject using unverified sophisticated mathematics. He was mainly concerned about how the two have ignored his deeper insight into the vibrating string problems. Referring back to the *principle of superposition*, he remarked that every type of complex acoustic vibration with a well-defined frequency can be produced by superposition of any number of harmonics, as defined in terms of *Taylor vibrations* $y_n = \sin(\omega_n x) \cos(\omega_n t)$. More specifically, the argument was based on following two assertions:

1. Since in an infinite series of the form

$$d(x) = \sum_{n=1}^{\infty} a_n \sin(\omega_n x), \tag{8.1.6}$$

there is a sufficiently large number of constants a_n so that the series (8.1.5) can adapt to any curve, d'Alembert's and Euler's supposedly new solutions are absolutely nothing but infinite sum of Taylor vibrations;
2. The aggregation of these partial modes in a single formula is incompatible with the physical character of the decomposition.

Bernoulli had no idea how to compute the coefficients a_n, so could not possibly have a proof of (1). As we will see shortly, it was Fourier's destiny to do so. In his reply, Euler acknowledged Bernoulli's contribution as the best *physical model*, but rejected the assertion that the superposition of harmonics always preserve the structure of the partial modes. Taking $b_n = \alpha^n$ ($\alpha \in \mathbb{R}$) in (8.1.6), he illustrated the fact that if the number of terms becomes infinite then it is not possible to have a solution of the form (8.1.5), and so concluded that *Bernoulli solutions* make a subclass of his solutions. In continuation to his earlier work, Euler also analysed the (non-homogeneous) equation of the *heavy vibrating string*

$$y_{tt}(x, t) - c\, y_{xx}(x, t) = g(x, t),$$

and obtained a solution of the form

$$y(x, t) = \varphi(x + ct) + \psi(x - ct) - \frac{x(x - \ell)g}{2c^2}.$$

Euler also attempted the problem of vibrations of rectangular and circular membranes, but the complete theory (for arbitrary domains) was developed much later by Denis Poisson in 1829.

Lagrange entered the debate when he published his work on sound propagation in 1759. Avoiding the wave equation-based analysis, he viewed string to be made up of equally spaced n point masses connected by a light chord. Such consideration led him to construct the solution of the problem by using *limit of partial sums*. He solved the system of equations given by

$$\frac{d^2 y_k}{dt^2} = c^2 (y_{k-1} - 2y_k + y_{k+1}), \quad \text{for } k = 1, \ldots, n,$$

allowing $n \to \infty$. With $\omega_n = n\pi/\ell$, he found

$$y(x, t) = \frac{2}{\ell} \int_0^\ell \left(\sum_{n=1}^\infty \sin(\omega_n x) \sin(\omega_n t) \cos(\omega_n ct) \right) d(x) dx$$

$$+ \frac{2}{\pi c} \int_0^\ell \left(\sum_{n=1}^\infty \sin(\omega_n x) \sin(\omega_n t) \cos(\omega_n ct) \right) v(x) dx$$

The above approach led Lagrange to obtain, inadvertently, the formula for the (Fourier) coefficients given by

$$b_n = \frac{2}{\ell} \int_0^\ell d(t) \sin(\omega_n t) dt, \quad (n \geq 1).$$

Notice that the last equation can be written as

$$d(x) = y(x, 0) = \sum_{n=1}^\infty b_n \sin(\omega_n x),$$

by using the assumption that $v(x) = 0$ at $t = 0$.

8.2 Fourier Series

As shown in Example 4.4, if $u = u(x, t) \in C^2([0, \ell] \times \mathbb{R}^+)$ is the temperature in a perfectly insulated thin bar of length ℓ at point $x \in [0, \ell]$ and at time $t \geq 0$, then the function u satisfies the (parabolic) *heat equation*

$$u_t(x, t) = a^2 u_{xx}(x, t). \qquad (8.2.1)$$

The physical constant a representing thermal conductivity can be *normalised* to assume that $a = 1$. We may write

$$u_t(x, t) = L[u], \quad \text{with } L \equiv \frac{\partial^2}{\partial x^2}. \qquad (8.2.2)$$

To use the argument leading to Theorem 3.20, suppose

$$u(x, t) = e^{\lambda t} w(x), \quad \text{for } x \in [0, \ell] \text{ and } t > 0. \qquad (8.2.3)$$

Then, in this case, Eq. (8.2.2) implies that

$$w''(x) = \lambda w(x), \quad \text{for } x \in [0, \ell].$$

Therefore, w is an eigenfunction of the operator L corresponding to the eigenvalue λ. We may consider the three possibilities given by

$$\lambda = -p^2 < 0 \quad \Rightarrow \quad w(x) = \cos px \text{ or } w(x) = \sin px;$$
$$\lambda = 0 \quad \Rightarrow \quad w(x) = 1 \text{ or } w(x) = x;$$
$$\lambda = p^2 > 0 \quad \Rightarrow \quad w(x) = e^{-px} \text{ or } w(x) = e^{px}.$$

Notice that the above procedure is a special case of the method of *separation of variables*. In 1807, Joseph Fourier used the same technique, together with principle of superposition, to conclude that initial-boundary value problem related to Eq. (8.2.1) with auxiliary conditions given by

$$u(0, t) = 0 = u(\ell, t), \quad \text{for all } t > 0; \qquad (8.2.4)$$
$$u(x, 0) = f(x), \quad \text{for all } 0 < x < \ell, \qquad (8.2.5)$$

has a series solution[3] of the form

$$u(x, t) = \sum_{n=1}^{\infty} b_n e^{\omega_n^2 t} \sin(\omega_n x), \quad \text{where } \omega_n = \frac{n\pi}{\ell}. \qquad (8.2.6)$$

Applying the *initial condition* (8.2.5), Eq. (8.2.6) gives

$$f(x) = u(x, 0) = \sum_{n=1}^{\infty} b_n \sin(\omega_n x). \qquad (8.2.7)$$

---

[3] In the original text, Fourier took $\ell = 1$.

Therefore, by using the first of the following *orthogonality property* for sine and cosine functions:

$$\int_0^\ell \sin(\omega_n x) \sin(\omega_m x) = \begin{cases} 0, & \text{if } n \neq m \\ \ell/2, & \text{if } n = m = 1, 2, \ldots \end{cases} ;$$

$$\int_0^\ell \cos(\omega_n x) \cos(\omega_m x) = \begin{cases} 0, & \text{if } n \neq m \\ \ell/2, & \text{if } n = m = 1, 2, \ldots \end{cases} ; \qquad (8.2.8)$$

$$\int_0^\ell \sin(\omega_n x) \cos(\omega_m x) = 0, \quad \text{for all } m, n.$$

it follows that the *Fourier coefficient*

$$b_n = \frac{2}{\ell} \int_0^\ell f(y) \sin(\omega_n x) \mathrm{d}y. \qquad (8.2.9)$$

To make the above equation valid for an arbitrary function f, Fourier wrongly assumed that each (periodic) function can be represented by a *trigonometric series* of the form

$$\frac{a_0}{2} + \sum_{n=1}^{\infty} h_n(x), \qquad (8.2.10)$$

where the function h_n (known as the nth harmonics) is given by

$$h_n(x) = a_n \cos(\omega_n x) + b_n \sin(\omega_n x), \quad \text{for } n \geq 1.$$

Notice that, using the Euler identity

$$e^{i \omega_n x} = \cos(\omega_n x) + i \sin(\omega_n x),$$

we can also write

$$h_n(x) = c_{-n} e^{-i\omega_n x} + c_n e^{i\omega_n x},$$

where $c_n = (a_n - ib_n)/2$ and $c_{-n} = (a_n + ib_n)/2$, with $b_0 = 0$. Thus, the series (8.2.10) can also be expressed as the following *exponential series*:

$$\sum_{n=0}^{\infty} h_n(x) = \sum_{n=-\infty}^{\infty} c_n e^{i\omega_n x}. \qquad (8.2.11)$$

The function $h_n(t)$ satisfies the *periodicity condition*

$$h_n(x + 2\ell) = h_n(x), \quad \text{for all } x \in \mathbb{R}.$$

We come across numerous practical situations such as this when it is important to know: When does a 2ℓ-periodic function $f(x)$ can be represented by a series of the form (8.2.10)?. The main problem boils down to finding conditions so that the series (8.2.10) *converges* in a nice way to ensure that the *sum function* is termwise differentiable and integrable. Even when the series converges in some sense, its *sum* may not define the function $f(x)$.

Example 8.1 Consider the case of heat flow in an insulated circular ring, taking $-\pi < x \leq \pi$ to represent the angular coordinates, so that we have to solve a heat equation based BVP with *periodic boundary conditions* of the form (5.4.13). We thus apply separation of variables to solve the IBVP:

$$u_t = u_{xx}, \quad \text{with } -\pi < x \leq \pi; \tag{8.2.12a}$$

$$u(-\pi, t) = u(\pi, t) \quad \text{and} \quad u_x(-\pi, t) = u_x(\pi, t), \text{ for } t > 0; \tag{8.2.12b}$$

$$u(x, 0) = f(x), \quad \text{for } -\pi < x \leq \pi. \tag{8.2.12c}$$

Here, taking $u(x, t) = e^{\lambda t} w(x)$, we need to solve the BVP

$$w''(x) = \lambda w(x), \quad \text{with } -\pi < x \leq \pi; \tag{8.2.13a}$$

$$w(-\pi) = w(\pi) \quad \text{and} \quad w'(-\pi) = w'(\pi). \tag{8.2.13b}$$

As shown earlier in Chap. 2, the nontrivial solution is obtained for $\lambda = -p^2 < 0$, with associated eigenfunctions given by

$$w_p(x) = a_n \cos px + b_n \sin px, \quad \text{for } p = 1, 2, 3, \ldots .$$

Therefore, for each $n \in \mathbb{Z}^+$, we have that

$$u_n(x, t) = e^{-n^2 t} \cos nx \quad \text{and} \quad \widehat{u}_n(x, t) = e^{n^2 t} \sin nx,$$

are the eigenfunctions of the given initial-boundary value problem. Hence, applying *superposition principle*, the general solution is given by

$$u(x, t) = \frac{a_0}{2} + \sum_{n=1}^{\infty} e^{\omega_n^2 t} \left[a_n \cos(\omega_n x) + b_n \sin(\omega_n x) \right], \tag{8.2.14}$$

where the term $a_0/2$ is added for the eigenfunction $u_0 \equiv 1$, and the coefficients a_n, b_n are determined from the initial condition, and orthogonality.

The series solutions such as (8.2.6) and (8.2.14) were part of Fourier's fundamental work "*Theory of propagation of heat in solid bodies*" that he presented to *Académie des Sciences.* Both Laplace and Lagrange were members of the committee that Académie constituted to review Fourier's work. For reasons as discussed below, the paper was not accepted for publication. In 1811, an improvised manuscript was submitted by Fourier to the Académie for a prize essay. Though the paper was awarded this time, but the publication was denied, once again, quoting the reasons as under:

The manner in which the Author arrives at his equations is not exempt from difficulties, and that his analysis, to integrate them, still leaves something to be desired in the realms of both generality and even rigour.

These comments, mainly due to Lagrange, could be justified at least on the ground that Fourier's work on heat propagation was convincingly so remarkable and counterintuitive: (*a*) In a time when there was no clarity about the concept of continuity, he asserted that an *arbitrary* function (possibly discontinuous) can be expressed as a trigonometric series; (*b*) in sharp contrast to common usage of integrals as *antiderivatives*, he viewed integral as the area under a curve; and, among all reservations, (*c*) the issues related to convergence of proposed series solution were the most controversial because the concept of *convergence* was not fully understood. Also, because he was well aware of the relevance of his ideas to problems related to other differential equations such as the wave equations.

Howsoever controversial these *assertions* of Fourier remained during that period, the success of his ideas in diverse applications ultimately led the best mathematicians of nineteenth century such as Cauchy, Dirichlet, Riemann, Cantor, Weierstrass, Lebesgue, and Hilbert to give precise formulation of fundamental concepts such as sets, functions, continuity, uniform convergence, integration, and many others of advanced mathematics. In fact, Riemann developed his theory of integration specifically to deal with issues related to convergence of Fourier series of a function.[4] The gist of the story is that Fourier's claimed *sinusoid decomposition* shaped the modern analysis, and some other parts of Mathematics, the way we understand today.

The collected work of Fourier appeared finally in 1822 as the celebrated book "*Théorie analytíque de la chaleur*" [7]. Some of the issues concerning convergence of series solution were resolved much later by Dirichlet's (Theorem 8.2). In 1904, a new procedure of summing the Fourier series devised by L. Fejér made it possible to talk about convergence of Fourier series of periodic *integrable* functions. Recall that a function $f : [a, b] \to \mathbb{R}$ is *absolutely integrable* if

$$\|f\|_1 := \int_a^b |f(t)|dt < \infty, \quad \text{with } -\infty \le a < b \le \infty.$$

We write $L^1[a, b]$ for the space of real-valued absolutely integrable functions defined over the interval $[a, b]$. As shown later, Fourier was *almost correct* in asserting that every 2ℓ-periodic function $f \in L^1[-\ell, \ell)$ can be represented by a trigonometric series of the form

---

[4] H. Weber, *The Collected Works of Riemann*, Dover, New York, 1953.

$$S(f)(t) := \frac{a_0}{2} + \sum_{n=1}^{\infty} [a_n \cos(\omega_n t) + b_n \sin(\omega_n t)], \tag{8.2.15}$$

where the constants $a_n = a_n(f)$, $b_n = b_n(f)$ are respectively given by formulas

$$a_n(f) = \frac{1}{\ell} \int_{-\ell}^{\ell} f(x) \cos(\omega_n x) dx; \tag{8.2.16}$$

$$b_n(f) = \frac{1}{\ell} \int_{-\ell}^{\ell} f(x) \sin(\omega_n x) dx, \tag{8.2.17}$$

which are now known as the *Fourier coefficients* of the function $f(t)$. Fourier was the first to use the *orthogonality property* (8.2.8) to derive these *integral formulas* for the coefficients $a_n(f)$, $b_n(f)$. The next theorem proves uniqueness of these coefficients, provided the series (8.2.15) converges uniformly.

Theorem 8.1 (Euler, 1777) *If the series in Eq. (8.2.15) converges uniformly to a function $f(t)$ over the interval $[-\ell, \ell)$ then the coefficients a_n, b_n are given by Eqs. (8.2.16) and (8.2.17).*

Proof Left as exercise for the reader.   □

In what follows, we are dealing with functions $f \in L^1([-\ell, \ell))$, with $f(-\ell) = f(\ell)$, extended as 2ℓ-periodic function to the line $\mathbb{R}$. In this case, since a_n, b_n exist always, we can write the trigonometric series $S(f)$ (8.2.15). However, we will use the notation

$$f(t) \sim \frac{a_0}{2} + \sum_{n=1}^{\infty} [a_n \cos(\omega_n t) + b_n \sin(\omega_n t)].$$

as a reminder of the fact that we have yet to decide in what sense the series on the right represents the function $f(t)$. The issues related to the convergence of the series are discussed in the next part of the section. By Theorem 8.1, we know that a 2ℓ-periodic function $f \in L^1([-\ell, \ell))$ admits a *Fourier series representation* if it is the *uniform limit* of the series (8.2.15).

Definition 8.1 Let $f \in L^1([-\ell, \ell])$ be a 2ℓ-periodic function. The trigonometric series (8.2.15) is called a *Fourier series* of the function f if the coefficients a_n, b_n are, respectively, given by the integrals (8.2.16) and (8.2.17). In this case, we call a_n and b_n the *Fourier coefficients* of f. Further, an exponential series of the form

$$\sum_{-\infty}^{\infty} c_n(f) e^{i\omega_n t}, \tag{8.2.18}$$

is a *complex Fourier series* of the function f when

$$c_n(f) := \frac{1}{2\ell} \int_{-\ell}^{\ell} e^{-i\omega_n x} f(x)\, dx, \qquad (8.2.19)$$

where $c_n(f) = \widehat{f}(n)$ are known as the (complex) Fourier coefficients of f.

Notice that $c_0(f) = \widehat{f}(0)$ is same as the *average* of f over the interval $[-\ell, \ell]$. The next example illustrates how to obtain Fourier series of a periodic functions $f \in L^1([-\ell, \ell])$ of *period* 2ℓ.

Example 8.2 Consider the *square wave* given by the function

$$f(x) = \begin{cases} 0, & \text{for } -\ell < x < -1 \\ 1, & \text{for } -1 < x < 1 \\ 0, & \text{for } 1 < x < \ell \end{cases} , \quad \text{with } \ell > 1.$$

By Eq. (8.2.16), we have

$$a_0 = \frac{1}{\ell} \int_{-\ell}^{\ell} f(x)dx = \frac{1}{\ell} \int_{-1}^{1} dx = \frac{2}{\ell};$$

and, for $n \geq 1$,

$$a_n = \frac{1}{\ell} \int_{-\ell}^{\ell} f(x) \cos(\omega_n x)dx = \frac{1}{\ell} \int_{-1}^{1} \cos(\omega_n x)dx$$

$$= \frac{2}{\ell} \int_{0}^{1} \cos(\omega_n x)dx = \frac{2}{\ell} \Big(\frac{\ell}{2n\pi} \sin(\omega_n x \Big|_0^1$$

$$= \frac{1}{n\pi} \sin(\omega_n).$$

Also, by Eq. (8.2.17), we have

$$b_n = \frac{1}{\ell} \int_{-\ell}^{\ell} f(x) \sin(\omega_n x)dx = \frac{1}{\ell} \int_{-1}^{1} \sin(\omega_n x)dx = 0.$$

Hence, we obtain

$$f(x) \sim \frac{1}{\ell} + \frac{1}{\pi} \sum_{n=1}^{\infty} \frac{1}{n} \sin(\omega_n) \cos(\omega_n x).$$

A Fourier series of a 2ℓ-periodic function $f \in L^1([0, 2\ell])$ is obtained by the same procedure, using the following integral formulas to compute the Fourier coefficients a_n, b_n of the function $f(t)$:

$$a_n = \frac{1}{\ell} \int_0^{2\ell} f(x) \cos(\omega_n x) dx, \tag{8.2.20}$$

and

$$b_n = \frac{1}{\ell} \int_0^{2\ell} f(x) \sin(\omega_n x) dx. \tag{8.2.21}$$

Remark 8.1 Notice that, for any function $f \in L^1([-\ell, \ell])$, with $f(-\ell) = f(\ell)$, the *transformed function* given by

$$g(x) := f\left(\frac{\ell x}{\pi}\right), \quad \text{for } x \in [-\pi, \pi],$$

defines a function $g \in L^1([-\pi, \pi])$, with $f(-\pi) = f(\pi)$. So, if the Fourier series (8.2.15) represents $f(t)$, then the function $g(x)$ is represented by the Fourier series of the form

$$g(x) = \frac{a_0'}{2} + \sum_{n=1}^{\infty} \left[a_n' \cos(nx) + b_n' \sin(nx) \right], \tag{8.2.22}$$

where the Fourier coefficients of $g(x)$ are given by the relations

$$a_n' = \frac{1}{\pi} \int_{-\pi}^{\pi} g(u) \cos(nu) \, du \quad \text{and} \quad b_n' = \frac{1}{\pi} \int_{-\pi}^{\pi} g(u) \sin(nu) du.$$

Clearly, we can reverse the process by using the transformation $x = \pi t / \ell$.

For a 2π-periodic function $f \in L^1([-\pi, \pi])$, we have $\omega_n = n$, and so the series (8.2.15) takes the form

$$\frac{a_0}{2} + \sum_{n=1}^{\infty} \left[a_n \cos(nt) + b_n \sin(nt) \right], \tag{8.2.23}$$

and the coefficients $a_n, b_n \in \mathbb{R}$ are given by the relations

$$a_n(f) = \frac{1}{\pi} \int_{-\pi}^{\pi} f(x) \cos nx \, dx, \qquad (8.2.24)$$

and

$$b_n(f) = \frac{1}{\pi} \int_{-\pi}^{\pi} f(x) \sin nx \, dx. \qquad (8.2.25)$$

Example 8.3 For $t \in \mathbb{R}$, consider the *unit step function* given by

$$h(t) = \begin{cases} 0 & \text{for} \quad -\pi < t < 0 \\ 1 & \text{for} \quad 0 < t < \pi \end{cases}, \quad \text{with} \quad h(t + 2\pi) = h(t).$$

By (8.2.24), we have

$$a_0 = \frac{1}{\pi} \int_0^{\pi} dx = 1,$$

and, for $n \geq 1$,

$$a_n = \frac{1}{\pi} \int_0^{\pi} \cos nx \, dx = \frac{1}{n\pi} \left[\sin nx \right]_0^{\pi} = 0.$$

Also, by (8.2.25), we have

$$\begin{aligned} b_n &= \frac{1}{\pi} \int_0^{\pi} \sin nx \, dx \\ &= -\frac{1}{n\pi} \left[\cos nx \right]_0^{\pi} \\ &= \frac{1 - (-1)^n}{n\pi}, \quad \text{for } n \geq 1, \end{aligned}$$

so that $b_n = 0$, for n even; and, $b_n = 2/n\pi$, for n odd. Therefore, a Fourier series of $h(t)$ is obtained as follows:

$$h(t) \sim \frac{1}{2} + \frac{2}{\pi} \sum_{n=1}^{\infty} \frac{\sin(2n-1)t}{(2n-1)} = \frac{1}{2} + \frac{2}{\pi} \left[\sin t + \frac{\sin 3t}{3} + \cdots \right].$$

Example 8.4 Consider the function $f : [-\pi, \pi] \to \mathbb{R}$ defined by $f(t) = t + t^2$, with $f(t + 2\pi) = f(t)$ for all $t \in \mathbb{R}$. In this case, we have

$$a_0 = \frac{1}{\pi} \int_{-\pi}^{\pi} (x + x^2)dx = \frac{2}{\pi} \int_{0}^{\pi} x^2 dx = \frac{2}{\pi} \left(\frac{x^3}{3} \right) \Big|_0^\pi = \frac{2\pi^2}{3},$$

and, for $n \geq 1$, we have

$$a_n = \frac{1}{\pi} \int_{-\pi}^{\pi} (x + x^2) \cos nx dx = \frac{2}{\pi} \int_{0}^{\pi} x^2 \cos nx dx$$

$$= \frac{2}{\pi} \left(x^2 \frac{\sin nx}{n} + 2x \frac{\cos nx}{n^2} - 2 \frac{\sin nx}{n^3} \right) \Big|_0^\pi$$

$$= \frac{2}{\pi n^2} \left(2\pi(-1)^n \right) = \frac{4(-1)^n}{n^2}.$$

Also, for $n \geq 1$, we have

$$b_n = \frac{1}{\pi} \int_{-\pi}^{\pi} (x + x^2) \sin nx dx = \frac{2}{\pi} \int_{0}^{\pi} x \sin nx dx$$

$$= \frac{2}{\pi} \left(-x \frac{\cos nx}{n} + \frac{\sin nx}{n^2} \right) \Big|_0^\pi$$

$$= \frac{2}{\pi n} \left(\pi(-1)^n \right) = \frac{2(-1)^{n+1}}{n}.$$

Therefore, a Fourier series of $f(t)$ is given by

$$f(t) \sim \frac{\pi^2}{6} + \sum_{1}^{\infty} (-1)^n \left[\frac{4}{n^2} \cos nt - \frac{2}{n} \sin nt \right].$$

Notice that, in all the above worked examples, we have

$$a_n \to 0 \quad \text{and} \quad b_n \to 0, \quad \text{as } n \to \infty.$$

The next lemma says, for any 2π-periodic function $f \in L^1[-\pi, \pi]$, the above assertion holds in general.

Lemma 8.1 (Riemann-Lebesgue Lemma) *Let $f \in L^1[-\pi, \pi]$ be a 2π-periodic function. Then, for any $\lambda \in \mathbb{R}$,*

$$\lim_{\lambda \to \infty} \int_{-\pi}^{\pi} f(t) \sin(\lambda t) dt = 0 = \lim_{\lambda \to \infty} \int_{-\pi}^{\pi} f(t) \cos(\lambda t) dt. \qquad (8.2.26)$$

Proof The assertion is trivial when $f = \chi_{(a,b)}$, with $(a, b) \subset [-\pi, \pi]$. From this follows a proof for the case when f is a *simple function*. That is, if

$$f = \sum_{k=1}^{n} \alpha_i \chi_{(a_k, b_k)}, \quad \text{with } \alpha_i \in \mathbb{R}.$$

And, also for the case when $f = \sum_{k=1}^{n} \chi_{M_k}$, where each set $M_k \subset (-\pi, \pi)$ is measurable. Notice that the only nontrivial part is to prove the assertion for the function $f = \chi_M$, where $M \subset (-\pi, \pi)$ is a measurable set. To complete the proof, let $\varepsilon > 0$ be arbitrary. We find a *simple function* $s(t)$ such that $\int_{-\pi}^{\pi} |f(t) - s(t)| < \varepsilon/2$ so that

$$\left| \int_{-\pi}^{\pi} f(t) \sin(\lambda t) dt \right|$$

$$= \left| \int_{-\pi}^{\pi} [f(t) - s(t) + s(t)] \sin(\lambda t) dt \right|$$

$$\leq \left| \int_{-\pi}^{\pi} [f(t) - s(t)] \sin(\lambda t) dt \right| + \left| \int_{-\pi}^{\pi} s(t) \sin(\lambda t) dt \right|.$$

As the second term on the right is bounded (Exercise 8), the left side integral in Eq. (8.2.26) holds. By taking $\beta = \pi/2$ in Eq. (8.2.27), proof follows for the other integral.

More generally, we have

$$\lim_{|\alpha| \to \infty} \int_{-\pi}^{\pi} f(t) \sin(\alpha t + \beta) dt = 0, \quad \alpha, \beta \in \mathbb{R}. \qquad (8.2.27)$$

Fourier Sine and Cosine Series

The case discussed here is based on the fact that computation of Fourier coefficients gets simpler for functions having some nice geometry over the specified period.

Definition 8.2 Let $\ell > 0$. We say a function $f : [-\ell, \ell] \to \mathbb{R}$ is an **even function** if $f(-t) = f(t)$, for $t \in [-\ell, \ell]$. Also, f is an **odd function** if $f(-t) = -f(t)$, for $t \in [-\ell, \ell]$.

Example 8.5 The functions x^{2n} and $\cos mx$ are even for any $n, m \in \mathbb{N}$; and, the functions x^{2n+1} and $\sin mx$ are odd for any $n, m \in \mathbb{N}$. Some other examples can be constructed by using the fact that the product of any two even or any two odd functions is an even function. We also have that the product of an odd and an even function is always an odd function. In general, for any function $f : [-\ell, \ell] \to \mathbb{R}$, the functions g and h given by

$$g(t) := \frac{f(t) + f(-t)}{2} \quad \text{and} \quad h(t) := \frac{f(t) - f(-t)}{2},$$

for $t \in [-\ell, \ell]$, are, respectively, an even and an odd functions. Notice that we have $f = h + g$, with respect to the *pointwise addition*.

The Fourier series (8.2.15) of a periodic function $f \in L^1([-\ell, \ell])$ of *period 2ℓ* takes a simpler form when f is an *Even function* or an *odd function*. For, if f is an even function, then $f(t) \sin(\omega_n x)$ is an odd function, and $f(t) \cos(\omega_n x)$ is an even function. It thus follows that $b_n = 0$, for $n \geq 1$, and

$$a_n = \frac{2}{\ell} \int_0^{\ell} f(t) \cos(\omega_n t) dt, \quad \text{for } n \geq 0. \tag{8.2.28}$$

So, we obtain a *Fourier cosine series* of the form

$$f(t) \sim \frac{a_0}{2} + \sum_{n=1}^{\infty} a_n \cos(\omega_n t), \quad \text{for } t \in (-\ell, \ell). \tag{8.2.29}$$

Similarly, if f is an odd function, then $f(t) \cos(\omega_n t)$ is an odd function, and $f(t) \sin(\omega_n t)$ is an even function. In this case, $a_n = 0$, for all $n \geq 0$; and, for $n \geq 1$, we have

$$b_n = \frac{2}{\ell} \int_0^{\ell} f(t) \sin(\omega_n t) dt. \tag{8.2.30}$$

Therefore, in this case, we obtain a *Fourier sine series* of the form

$$f(t) \sim \sum_{n=1}^{\infty} b_n \sin(\omega_n t), \quad \text{for } t \in (-\ell, \ell). \tag{8.2.31}$$

Example 8.6 Consider the *sawtooth function* $f : \mathbb{R} \to \mathbb{R}$ given by

$$f(t) = t, \quad \text{for } -\pi < t < \pi,$$

with $f(t + 2k\pi) = f(t)$, for all $k \in \mathbb{Z}$. Clearly, f is a periodic odd function of period 2π. Then the Fourier sine series of f is given by

$$b_1 \sin t + b_2 \sin 2t + b_3 \sin 3t + \cdots,$$

where the Fourier coefficient b_n is given by

$$b_n = \frac{2}{\pi} \int_0^\pi x \sin nx \, dx$$

$$= \frac{2}{\pi} \left(-x \frac{\cos nx}{n} + \frac{\sin nx}{n^2} \right) \Big|_0^\pi = \frac{2}{n}(-1)^{n+1}.$$

Therefore, in this case, we obtain (Fig. 8.1)

$$t \sim 2 \sum_{n=1}^{\infty} \frac{(-1)^{n+1}}{n} \sin nt, \quad \text{for } -\pi < t < \pi.$$

Example 8.7 Consider the *rectangle wave* function given by

$$f(t) = \begin{cases} -a, & \text{for } -\pi < t < 0 \\ a, & \text{for } 0 < t < \pi \end{cases}, \quad \text{with } f(t + 2\pi) = f(t), \ \forall \, t \in \mathbb{R}.$$

As it is an odd periodic function of period 2π, we have $a_n = 0$, for all $n \geq 0$, and the coefficient b_n is obtained as follows:

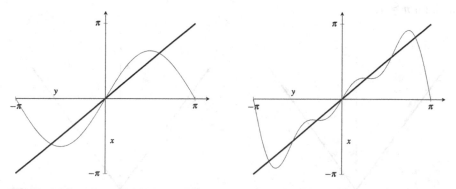

Fig. 8.1 Function $f(t) = t$ approximated by $s_2(f)$ and $s_3(f)$

$$b_n = \frac{2}{\pi} \int_0^{\pi} f(x) \sin(n\,x)\mathrm{d}x = \frac{2a}{\pi} \int_0^{\pi} \sin(n\,x)\mathrm{d}x$$

$$= -\frac{2a}{n\,\pi}[(-1)^n - 1] = \begin{cases} \frac{4a}{n\,\pi}, & \text{when } n \text{ is odd} \\ 0, & \text{when } n \text{ is even} \end{cases}.$$

Therefore, for $0 \neq t \in (-\pi, \pi)$, we have

$$f(t) \sim \sum_{n=1}^{\infty} \frac{4a}{\pi(2n - 1)} \sin((2n - 1)t).$$

in particular, for $a = 1$, we obtain (Fig. 8.2)

$$\frac{\pi}{4} \sim \sin t + \frac{1}{3} \sin 3t + \frac{1}{5} \sin 5t + \cdots.$$

Example 8.8 The 2π-periodic function $f : [-\pi, \pi] \to \mathbb{R}$ given by $f(t) = |t|$ is even, and so its Fourier cosine series is given by

$$\frac{a_0}{2} + \sum_{n=1}^{\infty} a_n \cos nt,$$

where

$$a_0 = \frac{2}{\pi} \int_0^{\pi} |x|\mathrm{d}x = \frac{2}{\pi} \int_0^{\pi} x\mathrm{d}x = \frac{2}{\pi}\left[\frac{x^2}{2}\right]_0^{\pi} = \pi$$

and, for $n \geq 1$,

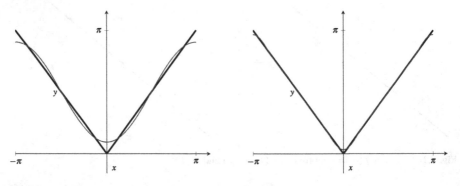

Fig. 8.2 Function $f(t) = |t|$ approximated by $s_1(f)$ and $s_3(f)$

$$a_n = \frac{2}{\pi} \int_0^\pi |x| \cos nx \, dx = \frac{2}{\pi} \int_0^\pi x \cos nx \, dx$$

$$= \frac{2}{\pi} \left(x \frac{\sin nx}{n} + \frac{\cos nx}{n^2} \right) \Big|_0^\pi$$

$$= \frac{2}{\pi n^2} \left((-1)^n - 1 \right) = \begin{cases} -\frac{4}{\pi(2k-1)^2} & (n = 2k - 1) \\ 0 & (n = 2k). \end{cases}$$

Therefore, we can write

$$|t| \sim \frac{\pi}{2} - \frac{4}{\pi} \sum_{k=1}^\infty \frac{\cos(2k - 1)t}{(2k - 1)^2}.$$

Fourier Series over Half-Range

The procedure described in the previous part of the section applies to any integrable function f defined over a half-interval of the form $[0, \ell]$. We consider first the case of computing the Fourier series of the even extension of the function f. Recall that an *even extension* f_e of a function f is obtained by using *reflection of f in y-axis* (see Fig. 8.3). That is, the function $f_e : [-\ell, \ell] \to \mathbb{R}$ is given by

$$f_e(t) = \begin{cases} -f(t), & \text{for } (-)0 \leq t \leq \ell \\ f(-t), & \text{for } -\ell \leq t \leq 0 \end{cases}.$$

Subsequently, we extend f_e to the line $\mathbb{R}$ by taking

$$f_e(t + 2\ell) = f_e(t), \quad \text{for all } t \in \mathbb{R}.$$

The periodic function $f_e \in L^1(\mathbb{R})$ of period 2ℓ as obtained above is called *even periodic extension* of the function f. Notice that, in this case, the Fourier series of the extension f_e contains only cosine terms. For, the Fourier cosine series of the even periodic extension $f_e(t)$ is given by

$$f_e(x) \sim \frac{a_0}{2} + \sum_{n=1}^\infty a_n \cos \omega_n t, \quad \text{where } \omega_n = \frac{n\pi}{\ell}$$

and the coefficient a_n is given by

$$a_n = \frac{2}{\ell} \int_0^\ell f_e(x) \cos \omega_n x \, dx.$$

Fig. 8.3 Even and odd extensions of $f(x) = x^2$, $x \in [0, 1]$

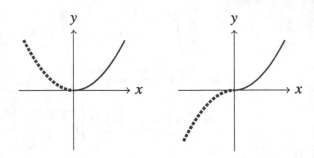

However, as $f_e(t) = f(t)$ for $t \in [0, \ell]$, we obtain

$$f(t) \sim \frac{a_0}{2} + \sum_{n=1}^{\infty} a_n \cos \omega_n t, \quad \text{where} \quad a_n = \frac{2}{\ell} \int_0^{\ell} f(x) \cos \omega_n x \, dx.$$

Therefore, we apply the above technique whenever it is required to have a *Fourier cosine series* of a function f defined over a *half-range*.

On the other hand, to obtain a *Fourier sine series* for an integrable function f defined over an interval $[0, \ell]$, we use odd periodic extension of the function f. Recall that an *odd extension* $f_o : [-\ell, \ell] \to \mathbb{R}$ of a function f is obtained by using the *reflection of f about the origin* (see Fig. 8.3). That is, f_o is given by

$$f_o(t) = \begin{cases} f(t), & \text{for } 0 \le t \le \ell \\ -f(-t), & \text{for } -\ell \le t \le 0 \end{cases}.$$

As in the above case, we extend f_o to the real line $\mathbb{R}$ by taking

$$f_o(t + 2\ell) = f_o(t), \quad \text{for all } t \in \mathbb{R}.$$

The periodic function $f_o \in L^1(\mathbb{R})$ of period 2ℓ as obtained above is called *odd periodic extension* of the function f. The Fourier series of the extension f_o in this case contains only sine terms. For, we know that the Fourier sine series of the odd periodic extension $f_o(t)$ is given by

$$f_o(t) \sim \sum_{n=1}^{\infty} b_n \sin \omega_n t, \quad \text{with} \quad b_n = \frac{2}{\ell} \int_0^{\ell} g_o(t) \sin \omega_n x \, dx.$$

Fig. 8.4 Function with discontinuous periodic extension

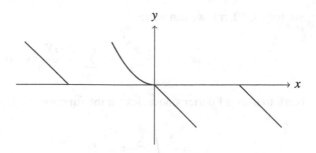

However, as $f_e(t) = f(t)$ for $t \in [0, \ell]$, it follows that

$$f(t) \sim \sum_{n=1}^{\infty} b_n \sin \omega_n t, \quad \text{where } b_n = \frac{2}{\ell} \int_0^{\ell} f(x) \sin \omega_n x \, dx.$$

Therefore, we apply the above technique whenever it is required to have a *Fourier sine series* of a function f defined over a half-range.

Notice that, however, it is most unlikely that the even or odd periodic extension so obtained is continuous on the line $\mathbb{R}$, even when $f \in L^1([0, \ell]) \cap C([0, \ell])$. For example, the *periodic extension* to $\mathbb{R}$ of the function

$$f(t) = \begin{cases} t^2, & \text{for } -1 \le t \le 0 \\ -t, & \text{for } 0 < t \le 1 \end{cases} \tag{8.2.32}$$

defined by $f(t + 2) = f(t)$, $t \in \mathbb{R}$, is *discontinuous* at every integer point (see Fig. 8.4).

Remark 8.2 Subject to some minor *adjustments*, when the periodic even extension f_e or odd extension f_o of a half-range function f satisfies the Dirichlet's conditions of Theorem 8.2, it is possible to write for the function f as a cosines series or a sine series. In any practical situation, the nature of the problem in hand guide us to choose which one of the above two types of Fourier series has to be used.

Example 8.9 Consider the function $f(t) = t^2$, $t \in (0, \pi)$. To have a Fourier cosine series of $f(t)$, we have

$$a_0 = \frac{2}{\pi} \int_0^{\pi} x^2 \, dx = \frac{2\pi^2}{3},$$

and, for $n \ge 1$,

$$a_n = \frac{2}{\pi} \int_0^{\pi} x^2 \cos nx \, dx = \frac{4(-1)^n}{n^2}.$$

So, for $t \in (0, \pi)$, we can write

$$t^2 \sim \frac{\pi^2}{3} + 4 \sum_{n=1}^{\infty} \frac{(-1)^n}{n^2} \cos nt.$$

Next, to have a Fourier sine series for the function $f(t)$, we have

$$b_n = \frac{2}{\pi} \int_0^{\pi} x^2 \sin nx \, dx$$

$$= \frac{2}{\pi} \left[x^2 \frac{\cos nx}{-n} + \frac{2x \sin nx}{n^2} + \frac{2 \cos nx}{n^3} \right]_0^{\pi}$$

$$= \frac{2}{\pi} \left[\frac{\pi^2 (-1)^{n+1}}{n} + \frac{2}{n^3}((-1)^n - 1) \right]$$

$$= \begin{cases} -\frac{2\pi}{n}, & \text{for } n \text{ even} \\ \frac{2\pi}{n} - \frac{8}{n^3 \pi}, & \text{for } n \text{ odd} \end{cases}.$$

So, for $t \in (0, \pi)$, we also have

$$t^2 \sim \sum_{n=1}^{\infty} \left(\frac{2\pi}{(2n-1)} - \frac{8}{(2n-1)^3 \pi} \right) \sin(2n-1)t - \sum_{n=1}^{\infty} \frac{\pi}{n} \sin 2nt.$$

Example 8.10 With $\omega_n = n\pi/\ell$, to write a Fourier cosine series for the function $f(t) = t$, $t \in (0, 2)$, we have

$$a_0 = \int_0^2 x \, dx = 2,$$

and, for $n \geq 1$,

$$a_n = \int_0^2 x \cos \left(\frac{n\pi x}{2} \right) dx = \frac{-4}{n^2 \pi^2}(\cos n\pi - 1) = \frac{4}{n^2 \pi^2}((-1)^{n+1} + 1).$$

So, for $t \in (0, 2)$, we can write

$$t \sim 1 + \frac{8}{\pi^2} \sum_{n=1}^{\infty} \frac{1}{(2n-1)^2} \cos \left(\frac{(2n-1)\pi t}{2} \right).$$

Convergence

Suppose a series $S(f)$ as in (8.2.15) is obtained for a 2ℓ-periodic function $f \in L^1([-\ell, \ell])$, where the coefficients a_n and b_n are computed using Eqs. (8.2.16) and (8.2.17). The next big problem that we need to consider here asks: *Which functions $f(t)$ are nice enough so that $S(f) \to f$ over the interval $[-\ell, \ell]$*? That is, we have to discuss the following two questions :

(a) For what class of functions, does the series $S(f)$ converges in some sense ?
(b) For a function $f(t)$ that satisfies (a), is $S(f) \to f$?

In historical terms, *Gustav Dirichlet*'s (1805–1859) was the first to study rigorously issues related to convergence of Fourier series of a periodic function, with focus mainly on the set of discontinuities[5] of the function. His fundamental paper of 1829 is considered to be first formal publication of mathematical analysis, who proved most periodic functions of practical importance can be expressed as a series of the form (8.2.10). More precisely, for the sequence of partial sum *Fourier series representation*

$$S_k(f)(t) := \frac{a_0}{2} + \sum_{n=1}^{k} [a_n \cos(\omega_n t) + b_n \sin(\omega_n t)] = \sum_{n=-k}^{k} c_n e^{i\omega_n t}, \qquad (8.2.33)$$

of the series (8.2.15), we ask when does the following conditions hold:

1. $S_k(f)(t) \to S(f)(t)$, as $k \to \infty$, $t \in [-\ell, \ell]$;

2. $\lim_{k \to \infty} \sup_{t \in [-\ell, \ell]} |S(f)(t) - S_k(f)(t)| = 0$;

3. $\lim_{k \to \infty} \int_{-\ell}^{\ell} |S(f)(t) - S_k(f)(t)|^2 dt = 0$;

4. $\lim_{k \to \infty} \int_{-\ell}^{\ell} |S(f)(t) - S_k(f)(t)| dt = 0$.

Notice that (1) implies $S(f) \to f$ *pointwise*. The first major result of the section due to Dirichlet is Theorem 8.2, which provides a sufficient condition for (1) to hold for a reasonable class of functions. However, we remark that it is in general true that pointwise convergence is too weak to serve all applications. For example, even it does not preserve *continuity*, i.e. pointwise limit function of a series of continuous function may not be continuous.

The condition (2) implies that the function f is the *uniform limit* of Fourier series $S(f)$. The second major result of the section is Theorem 8.5, which provides a sufficient condition for (2) to hold for a reasonable class of functions. The uniform convergence of the sequence $\langle S_k(f) \rangle$ is particularly crucial when $S_k(f)$ are used to solve a homogeneous IBVP using *sparation of variables*. For example, the solution (8.2.6) of the heat equation makes sense only when the series converge uniformly

---

[5] This approach ultimately led Cantor to *formulate* the definition of an **infinite set**, and to him the definition of a **function**, the way we understand these two notions today.

on the interval $[0, \ell]$ so that termwise first and second derivatives of the series converge uniformly, respectively, to the partial derivative $u_x(x, t)$ and $u_{xx}(x, t)$, and the formula for coefficients b_n is valid.

Definition 8.3 A 2ℓ-periodic function $f : [-\ell, \ell] \to \mathbb{R}$ is said to have a Fourier series representation if (2) holds for the function $f(t)$.

The condition (3) implies $S(f) \to f$ with respect to L^2-*norm*. Recall that a Lebesgue measurable functions $f : \mathbb{R} \to \mathbb{R}$ is said to be *square integrable* if

$$\| f \|_2^2 = \int_{-\infty}^{\infty} |f(x)|^2 dx < \infty, \tag{8.2.34}$$

We write $L^2(\mathbb{R})$ for the space of real or complex-valued square integrable functions defined on $\mathbb{R}$. For $f, g \in L^2(\mathbb{R})$, $\| f - g \|_2$ is called the *mean square distance* between the functions f and g. So, (3) is also known as *mean square convergence*. The third major result of the section is Theorem 8.6, which proves that the condition (3) holds if and only if f is a square integrable function.

The condition (4) implies $S(f) \to f$ with respect to L^1-*norm*. It is now known that, for some 2π-periodic functions in $L^1[-\pi, \pi]$, the series $S(f)$ diverges at every point $t \in [-\pi, \pi]$. Also, for some 2π-periodic continuous functions in $L^1[-\pi, \pi]$, even pointwise convergence fails. The reader may refer [8], for further details.

It is straightforward to verify the following implications:

$$(2) \Rightarrow (1), \quad (2) \Rightarrow (3), \quad (2) \Rightarrow (4), \quad (4) \not\Rightarrow (2).$$

For example, if a sequence of functions $\langle f_n \rangle \in L^2[-\ell, \ell]$ converges uniformly to a function f, then

$$\| f_n - f \|^2 = \int_{-\ell}^{\ell} |f_n(t) - f(t)|^2 dt$$

$$\leq \sup |f_n(t) - f(t)|^2 \int_{-\ell}^{\ell} dt$$

$$= 2\ell \Big(\sup |f_n(t) - f(t)| \Big)^2,$$

implies that $\langle f_n \rangle$ mean square converges to f. The sequence of functions $f_n(t) = \sqrt{n}\, t^n$, $0 \leq t < 1$, converges pointwise to 0, but not in L^2-norm, i.e. (1) $\not\Rightarrow$ (3). Also, using the sequence of functions

$$f_n(t) = \begin{cases} n^{1/3}, & \text{for } 0 \leq t \leq \frac{1}{n} \\ 0, & \text{otherwise} \end{cases},$$

it follows easily that (3) $\not\Rightarrow$ (1). Also, by Cauchy–Schwarz inequality, we have

$$\int_a^b |f(t) - f_n(t)| dt \leq |a - b|^{1/2} \left[\int_a^b (f(t) - f_n(t))^2 dt \right]^{1/2},$$

provided functions f, f_n are nice enough to make integrals exist, so (3) $\Rightarrow$ (4). On the other hand, if functions f_n are uniformly bounded by a constant $M > 0$ and $f_n \to f$ with respect to L^1-norm, then

$$\int_a^b (f(t) - f_n(t))^2 dt \leq 2M \int_a^b |f(t) - f_n(t)| dt,$$

shows that (4) $\Rightarrow$ (3). In fact, it can be shown that under the above assumption we have (1) $\Rightarrow$ (4), and so also (1) $\Rightarrow$ (3).

As it turns out the whole issue related to convergence of a Fourier series $S(f)$ boils down to knowing how much the function $f(t)$ could be recovered from the information about its Fourier coefficients. Also, there arises some other type of convergence issues when a Fourier series is used to approximate physical phenomena.

In technical terms, a nice approximation amounts to ensuring that a sufficiently smooth periodic signal can be analysed in terms of *harmonically related* sinusoids given by the sequence $\langle S_k(f) \rangle$. For example, Fig. 8.5 depicts second and third approximation of the signal $f(t) = t + t^2$ using the Fourier series as obtained in Example 8.4.

Example 8.11 The Fourier series of the *triangular wave* function

$$f(t) = \begin{cases} t, & 0 \leq t \leq 1/2 \\ 1 - t, & 1/2 \leq t \leq 1 \end{cases}, \quad \text{with } f(t + 1) = f(t), \ \forall \, t \in \mathbb{R},$$

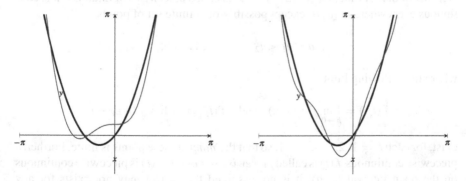

Fig. 8.5 Function $f(t) = t + t^2$ approximated by $s_2(f)$ and $s_3(f)$

is given by

$$\frac{1}{4} - 2 \sum_{n=0}^{\infty} \frac{\cos(2(2n+1)\pi t)}{\pi^2 (2n+1)^2}.$$

In this case, the sequence $\langle S_k(f) \rangle$ of *partial sums* given by

$$S_k(f)(t) = \frac{1}{4} - 2 \sum_{n=0}^{k-1} \frac{\cos(2(2n+1)\pi t)}{\pi^2 (2n+1)^2}$$

converges *very fast* due to the fact that convergence is uniform. Notice that the function f is continuous, but its derivative f' is not, though it is piecewise smooth (Definition 8.4). On the other hand, the *square wave*

$$g(t) = \begin{cases} 1, & 0 \le t \le 1 \\ 0, & 1 < t \le 2 \end{cases}, \quad \text{with } f(t+2) = f(t), \ \forall t \in \mathbb{R},$$

has the Fourier series

$$\frac{2}{\pi} \sum_{n=0}^{\infty} \frac{\sin((2n+1)\pi t)}{(2n+1)},$$

which converges at each point of continuity of the function g. The convergence *at discontinuities* is not that good due to a general failure, known as the *Gibbs phenomenon* (Example 8.13). Notice that, in this case, both g and g' are not continuous, but g is piecewise smooth.

Considering the convergence of Fourier series for functions obtained in this example, and also the construction of Fourier series for *half-range* functions as explained previously, we find it quite natural to base our study working around with functions in the space of *piecewise continuous functions* defined over a bounded interval of the form $[0, \ell]$.

Definition 8.4 A function $f : [a, b] \to \mathbb{R}$ is called *piecewise continuous* if it is continuous everywhere in $[a, b]$ except possibly on a finite set of points

$$a \le t_1 < t_2 < \ldots < t_n \le b,$$

where the following limits

$$f(t_k^+) = \lim_{h \to 0} f(t_k + h) \quad \text{and} \quad f(t_{k+1}^-) = \lim_{h \to 0} f(t_{k+1} - h)$$

exist, for each $k = 1, 2, \ldots, n - 1$, so that the *jump* at these points is finite. Further, a piecewise continuous $f(t)$ is called *piecewise smooth* if $f'(t)$ is piecewise continuous on the open interval (a, b); it is not assumed that $f'(t_k)$ may not exists for any $1 \le k \le n$.

The space of *piecewise continuous* functions $f : [a, b] \to \mathbb{R}$ is denoted by $PC[a, b]$, and $PS[a, b] \subset PC[a, b]$ denotes the space of *piecewise smooth* functions. Clearly, every $f \in PC[a, b]$ is *bounded*. The function $f(t) = \sin(t^{-1})$ is not piecewise continuous, and $f(t) = \sqrt{|t|} \in C^0[-1, 1]$ is not piecewise smooth. However, the function defined in Eq. (8.2.32) is piecewise smooth. Notice that, for $f \in PS[a, b]$, the derivative $f'(t)$ may not exists at the points of *jump discontinuities*. Moreover, since

$$\int\limits_{-\infty}^{\infty} |f(t)| dt = \int\limits_{-\infty}^{\infty} f^+(t) dt + \int\limits_{-\infty}^{\infty} f^-(t) dt,$$

where $f^+(t) = \max(f(t), 0)$ and $f^-(t) = \max(-f(t), 0)$, it follows that

$$\int\limits_{-\infty}^{\infty} f^+(t) dt < \infty \quad \text{and} \quad \int\limits_{-\infty}^{\infty} f^-(t) dt < \infty.$$

This is equivalent, in Lebesgue sense, to the fact that $f(t)$ is integrable. So, every piecewise continuous function defined on a bounded interval is bounded, square integrable, and absolutely integrable. Notice that $C^1[a, b] \subset PS[a, b]$.

Dirichlet Theorem

We prove here Dirichlet's theorem about pointwise convergence of Fourier series for functions satisfying the following condition.

Definition 8.5 *(Dirichlet Conditions)* A function $f : [a, b] \to \mathbb{R}$ is said to satisfy the *Dirichlet conditions* if

1. The function $f(t)$ has finitely many maxima and/or minima in the interval $[a, b]$.
2. The function $f(t)$ has finitely many discontinuities in the interval $[a, b]$.
3. The *jump* at each point of discontinuity is finite.

The condition (1) implies that there are only finitely many points $t_i \in (a, b)$ such that the limits $f'(t_i^+)$ and $f'(t_i^-)$ exist. Also, the conditions (2) and (3) imply that there are only finitely many points $t_i \in [a, b]$ such that the limits $f(t_i^+)$ and $f(t_i^-)$ exist. For example, every continuous function of *bounded variations* satisfies the Dirichlet conditions, and so do every function in the space $PS[a, b]$.

Theorem 8.2 *(Dirichlet Theorem) Let a 2ℓ-periodic function $f \in L^1[-\ell, \ell]$, that satisfies the Dirichlet conditions over the interval $[-\ell, \ell]$, has Fourier coefficients a_n and b_n given by (8.2.16) and (8.2.17). Then the series (8.2.15) converges pointwise to the* mean jump

$$\frac{f(t^+) + f(t^-)}{2}, \quad \text{for all } t \in (-\ell, \ell).$$

In particular, $S(f)(t) = f(t)$ whenever $f(t)$ is continuous at $t \in (-\ell, \ell)$.

An outline proof of the above theorem is as follows: Putting in Eq. (8.2.33) the expressions for the coefficients a_n, b_n, we obtain

$$S_k(f)(t) = \frac{1}{2\ell} \int_{-\ell}^{\ell} f(x)dx + \frac{1}{\ell} \int_{-\ell}^{\ell} f(x) \left[\sum_{n=1}^{k} \cos(\omega_n(t-x)) \right] dx$$

$$= \frac{1}{\ell} \int_{-\ell}^{\ell} f(x) \left[\frac{1}{2} + \sum_{n=1}^{k} \cos(\omega_n(t-x)) \right] dx$$

$$= \frac{1}{\ell} \int_{-\ell}^{\ell} f(x) D_k(t-x) dx, \tag{8.2.35}$$

where the expression

$$D_k(t-x) = \frac{1}{2} + \sum_{n=1}^{k} \cos(\omega_n(t-x)), \quad \text{for } k \geq 1, \tag{8.2.36}$$

is known as the *Dirichlet's kernel*. For any fixed $k \geq 1$, we may write

$$D_k(y) = \frac{1}{2} + \sum_{n=1}^{k} \cos(\omega_n y), \quad \text{for } k \geq 1. \tag{8.2.37}$$

The main argument used in the formal proof given below is based on the fact that though the function $D_k(y)$ has a huge spike when y is near zero, i.e. when x is close to t, but the area under the spike for $D_k(y)$ is approximately 1. So, the integral in (8.2.35) approaches $f(t)$, as $k \to \infty$. For values of x away from t, the function $D_k(x-t)$ is oscillatory, where as $k \to \infty$ these oscillations tend to cancel out in the integral on the right of Eq. (8.2.35) (see Fig. 8.6).

Proof (Proof of Theorem 8.2) In view Remark 8.1, it suffices to prove the theorem for a piecewise smooth function $f : [-\pi, \pi] \to \mathbb{R}$, with $f(t+2\pi) = f(t)$ for all $t \in \mathbb{R}$. In this case, the *Dirichlet's kernel* is given by

$$D_k(y) = \frac{1}{2} + \sum_{n=1}^{k} \cos(n y), \quad \text{for } k \geq 1, \tag{8.2.38}$$

so that the kth partial sums $S_k(f)$ of the Fourier series

Fig. 8.6 Graphs of
Dirichlet's kernel D_4 and D_8

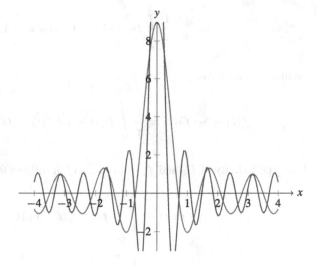

$$\frac{a_0}{2} + \sum_{n=1}^{\infty} [a_n \cos(n\,t) + b_n \sin(n\,t)]$$

can be written as

$$S_k(f)(t) = \frac{1}{\pi} \int_{-\pi}^{\pi} f(x)\, D_k(t - x)\, dx. \qquad (8.2.39)$$

Multiplying Eq. (8.2.38) by $\sin(y/2)$, $y \neq 0$, and using the identity

$$2 \sin \alpha \cos \beta = \sin (\alpha + \beta) - \sin (\alpha - \beta),$$

we obtain the relation

$$\sin \left(\frac{y}{2}\right) D_k(y) = \frac{1}{2} \left\{ \sin \left(\frac{y}{2}\right) + \sum_{n=1}^{k} \sin \left(\left(n + \frac{1}{2}\right)y\right) - \sin \left(\left(n - \frac{1}{2}\right)y\right) \right\}.$$

As the sum on the right is *telescopic*, we obtain closed-form expression of *Dirichlet's kernel* as given by

$$D_k(y) = \frac{\sin \left(\left(k + \frac{1}{2}\right)y\right)}{2 \sin \left(\frac{y}{2}\right)}, \quad \text{for } y \neq 0. \qquad (8.2.40)$$

Clearly, we have $D_k(0) = (2k + 1)/2$. Notice that the points $y = 2m\pi$ are only *removable discontinuities* of the function on the right side of Eq. (8.2.40), so it holds for all $y \in \mathbb{R}$ subject to that limits are taken at points wherever it is not well-defined. Notice that D_k is a 2π-periodic, even function. Also, by definition, we have

$$\frac{1}{\pi} \int_{-\pi}^{\pi} D_k(y)dy = 1, \quad \text{for } k \geq 1, \tag{8.2.41}$$

so that we may write

$$f(t) - S_k(f)(t) = \frac{1}{\pi} \int_{-\pi}^{\pi} D_k(t - x)[f(t) - f(x)]\,dx. \tag{8.2.42}$$

Using periodicity of D_k and f, from (8.2.41), it follows that

$$S_k(f)(t) = \frac{1}{\pi} \int_{-\pi}^{\pi} f(x)D_k(t - x)dx$$

$$= \frac{1}{\pi} \int_{-\pi+b}^{\pi+b} f(x)D_k(t - x)dx$$

$$= \frac{1}{\pi} \int_{-\pi}^{\pi} f(u + b)D_k(t - u - b)\,du$$

$$= \frac{1}{\pi} \int_{-\pi}^{\pi} f(t - x)D_k(x)dx. \tag{8.2.43}$$

A similar manipulation gives

$$S_k(f)(t) = \frac{1}{\pi} \int_{0}^{\pi} \big[f(t - x) + f(t + x)\big]D_k(x)dx. \tag{8.2.44}$$

Therefore, taking

$$h(x) = \frac{f(t) - f(x)}{2\sin((t - x)/2)},$$

we obtain from relations (8.2.40) and (8.2.42)

$$f(t) - S_k(f)(t) = \frac{1}{\pi} \int_{-\pi}^{\pi} \sin\left(\left(k + \frac{1}{2}\right)(t - x)\right) h(x)dx.$$

Now, for any fixed $t \in [-\pi, \pi]$, if f is differentiable at t then

$$\frac{f(t) - f(x)}{t - x} \in PC[-\pi, \pi] \;\Rightarrow\; h(x) \in PC[-\pi, \pi].$$

As said before, as $k \to \infty$, oscillations of the sin function become large, and the difference between its contributions to the integral from the intervals where *it is positive* and where it is *negative* is almost zero. A rigorous proof of this last assertion follows from Lemma 8.1. Therefore,

$$|f(t) - S_k(f)(t)| \;\to\; 0 \quad \text{as } k \to \infty, \; , \forall t \in [-\pi, \pi].$$

Likewise, at a point of *jump discontinuity* $t = t_0$, we may use (8.2.44) to conclude that the assertion of the theorem is equivalent to the statement

$$\lim_{k \to \infty} \frac{1}{\pi} \int_0^\pi \left[f(t_0 - x) + f(t_0 + x) \right] D_k(x) dx = \frac{f(t_0^+) + f(t_0^-)}{2}.$$

However, using the relations

$$\frac{1}{\pi} \int_0^\pi D_k(y) = \frac{1}{\pi} \int_{-\pi}^0 D_k(y) = \frac{1}{2}, \quad k \geq 1, \tag{8.2.45}$$

it is enough to prove the following equalities:

$$\lim_{k \to \infty} \int_0^\pi \left[f(t_0 \pm x) - f(t_0^\pm) \right] D_k(x) dx = 0,$$

which in integral form are given by

$$\lim_{k \to \infty} \frac{1}{\pi} \int_0^\pi \frac{f(t_0 \pm x) - f(t_0^\pm)}{2 \sin(x/2)} \sin\left(k + \frac{1}{2} \right) dx = 0.$$

Therefore, using fact (Exercise 8) that the function

$$g_\pm(x) := \frac{f(t_0 \pm x) - f(t_0^\pm)}{2 \sin(x/2)} \in L^1(0, \pi], \tag{8.2.46}$$

we may apply the generalised form of Riemann–Lebesgue lemma (as in (8.2.27)), and argument as above, to complete the proof of the Dirichlet's theorem. □

In particular, if both f and f' are continuous, it follows that the function $f(t)$ at every point $t \in (-\pi, \pi)$ is represented by the Fourier series

$$\frac{a_0}{2} + \sum_{n=1}^{\infty} \left[a_n \cos(nt) + b_n \sin(nt) \right]. \qquad (8.2.47)$$

The argument used above, together with Lemma 8.1, shows that at a given point $t \in (-\pi, \pi)$ the value of the Fourier series

$$S(f)(t) := \lim_{k \to \infty} \left[\frac{a_0}{2} + \sum_{n=1}^{k} \left[a_n \cos(n\,t) + b_n \sin(n\,t) \right] \right],$$

for any function $f \in L^1[-\pi, \pi]$ satisfying Dirichlet conditions depends only on the *local values* of the function $f(t)$. This is known as the *localisation principle* (Exercise 8).

Remark 8.3   There are other forms of convergences that hold importance in some situations. For example, Féjer proved *Cesáro summability* of Fourier series of an absolutely integrable 2π-periodic function $f : [-\pi, \pi] \to \mathbb{R}$ in terms of the sequence of functions

$$f_n(t) := \frac{1}{n+1} \left(\frac{\sin[(n+1)t/2]}{\sin(t/2)} \right),$$

known as the *Féjer kernels*. It is very instructive to study a connection between this and Dirichlet's approach discussed above. Poisson used Abel's series to develop his version of convergence of complex form of Fourier series. For further details, the reader may refer [6].

Gibbs Phenomenon

The behaviour of a Fourier series at a point of *jump discontinuity* as illustrated in next two worked examples is known as *Gibbs phenomenon*, which states that the value of the *truncated series* at the point *overshoots* by about 9% of the *size of the jump*. This observations was published in 1899 as letters to *Nature* magazine by American scientist J. W. Gibbs (1839 - 1903).

Example 8.12   Consider the *step function*

$$h(t) = \begin{cases} 0 & \text{for } -\pi < t < 0 \\ 1 & \text{for } 0 < t < \pi \end{cases}, \quad \text{with } h(t + 2\pi) = h(t), \ \forall \, t \in \mathbb{R}.$$

Let t be a very small positive number. Now,

$$S_k(h)(t) = \frac{1}{\pi} \int_0^\pi D_k(x - t)\, \mathrm{d}x = \frac{1}{\pi} \int_{t-\pi}^t D_k(y)\, \mathrm{d}y. \qquad (8.2.48)$$

We may use (8.2.45) to write

$$S_k(h)(t) = \frac{1}{2} + \frac{1}{\pi} \int_0^t D_k(y)\,dy - \frac{1}{\pi} \int_{-\pi}^{-\pi+t} D_k(y)\,dy. \tag{8.2.49}$$

For a fixed $a > 0$, take $t_n = a/[k + (1/2)]$. Since $D_k(-\pi) = (-1)^k/2$, for $-\pi \leq y \leq -\pi + t$, we have

$$|D_k(y)| < \frac{1}{2}, \quad \text{for small values of } t.$$

Therefore,

$$\left| \int_{-\pi}^{-\pi+t} D_k(y)\,dy \right| \leq \frac{t_n}{2} < \frac{a}{2n}. \tag{8.2.50}$$

Also, since the function $(1/\sin(y/2)) - (2/y)$ is differentiable, and it vanishes at $y = 0$, for small values of t and

$$\widehat{D}_k(y) = \frac{\sin\left(\left(k + \frac{1}{2}\right)y\right)}{y},$$

we have

$$\left| \int_0^{t_n} D_k(y)\,dy - \int_0^{t_n} \widehat{D}_k(y)\,dy \right| < \frac{t_n}{2} < \frac{a}{2n}. \tag{8.2.51}$$

Notice that, using the substitution $x = (k + 1/2)y$,

$$\int_0^{t_n} \widehat{D}_k(y)\,dy = \int_0^a \frac{\sin x}{x} = \mathrm{Si}(a), \tag{8.2.52}$$

where $\mathrm{Si}(x) = \sin x/x$, which attains its *mximum* at $a = \pi$ and $\mathrm{Si}(\pi) \approx 1.85194.$, known as the *Gibbs constant* It thus follows from the discussion above that

$$\lim_{k\to\infty} S_k(h)(t_n) = \frac{1}{2} + \frac{\mathrm{Si}(\pi)}{\pi} \approx 1.08949,$$

that is, in limiting case, the partial sequence of the Fourier series of the function $h(t)$ overshoots its maximal value by 9%.

Fig. 8.7 Gibbs phenomenon for signal $f(x) = 1$, for $1 < x < \pi$, and 0, otherwise

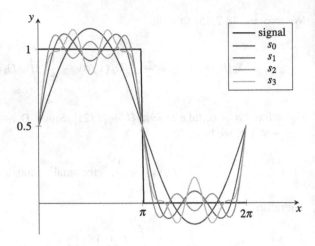

Example 8.13 Consider the *step function* given by

$$f(x) = \begin{cases} -1, & \text{for } -\pi < x < 0 \\ -1, & \text{for } 0 < x < \pi \end{cases}, \quad \text{with } f(x + 2\pi) = f(x), \quad \forall\, x \in \mathbb{R}.$$

As obtained in Example 8.7 (with $a = 1$), the sequence $\langle S_k(f) \rangle$ of *partial sums* of the Fourier series of $f(x)$ is given by

$$S_k(f)(x) = \frac{4}{\pi} \sum_{n=0}^{k-1} \frac{\sin(2n+1)x}{2n+1}$$

$$= \frac{4}{\pi} \operatorname{Im} \left[\sum_{n=1}^{k-1} \frac{e^{i(2n+1)x}}{2n+1} \right], \quad \text{for } x \in (0, \pi).$$

Assuming that the termwise differentiation is possible, we have

$$S_k'(f)(x) = \frac{4}{\pi} \operatorname{Im} \left[\sum_{n=1}^{k-1} i\, e^{i(2n+1)x} \right] = \frac{4}{\pi} \operatorname{Im} \left[i\, e^x \sum_{n=1}^{k-1} e^{i2nx} \right]$$

$$= \frac{4}{\pi} \operatorname{Im} \left[i\, e^x \frac{1 - e^{i2kx}}{1 - e^{i2x}} \right] = \frac{4}{\pi} \operatorname{Im} \left[i\, \frac{1 - e^{i2kx}}{e^{-x} - e^{ix}} \right]$$

$$= \frac{4}{\pi} \operatorname{Im} \left[i\, \frac{e^{i2kx} - 1}{e^{ix} - e^{-ix}} \right]$$

$$= \frac{2}{\pi} \operatorname{Im} \left[\frac{e^{i2kx} - 1}{\sin x} \right] = \frac{2}{\pi} \frac{\sin 2kx}{\sin x}.$$

Therefore, $S_k(f)(x)$ attains its maximum value at $x = \pi/2k$. Also, since we can write

$$S_k(f)(x) = \frac{2}{\pi} \int_0^x \frac{\sin 2ku}{\sin u} du = \frac{2}{\pi} \int_0^{2kx} \frac{\sin t}{2k \sin(t/2k)} dt,$$

it follows that

$$S_k(f)(x) \approx \frac{2}{\pi} \int_0^{2kx} \mathrm{Si}(t) dt, \quad \text{for } k \text{ sufficiently large.}$$

where $\mathrm{Si}(t) = \dfrac{\sin t}{t}$ is the *integral sine function*, with $\mathrm{Si}(\pi) = 1.85194$. Thus, the maximum value of $S_k(f)(x)$ at $x = \pi/2k$ is approximately 1.17898. Thus, as $k \to \infty$, the *overshoots* is 0.18, which is 9% of the jump at $x = 0$ (see Fig. 8.7).

In the original article, Gibbs used the Fourier series of the sawtooth function as obtained in Example 8.6. Significance of this function is due to the fact that we can write any periodic function $f \in PS[-\pi, \pi)$, with $f(-\pi) = f(\pi)$, as the sum of a continuous piecewise smooth function and a finite linear combinations of sawtooth-like functions. So, by Theorem 8.5, Gibbs' phenomenon occurs only due to the second part of the sum. The related details are left as exercise for the reader (Exercise 8).

Uniform Convergence

For a 2π-periodic function $f \in L^1([-\pi, \pi))$, consider the sequence of *exponential polynomials*

$$p_n[f](t) = \sum_{k=-n}^{n} c_k e^{ikt}, \quad \text{for } n \geq 0, \tag{8.2.53}$$

where the coefficients c_k are given by the formula

$$c_k = c_k(f) := \frac{1}{2\pi} \int_{-\pi}^{\pi} e^{-ikx} f(x) dx, \quad \text{for } k \in \mathbb{Z}.$$

A simple adaptation of Theorem 8.1 shows that if the series

$$\lim_{n \to \infty} p_n[f](t) = \sum_{k=-\infty}^{\infty} c_k(f) e^{ikt}, \tag{8.2.54}$$

converges uniformly to the function f then the coefficients c_k can be recovered from this series by using the following orthogonality relation

$$\frac{1}{2\pi} \int\limits_{-\pi}^{\pi} e^{i(k-n)x} dx = \begin{cases} 0, & \text{for } k \neq n \\ 1, & \text{for } k = n \end{cases}. \tag{8.2.55}$$

Recall that the set of $L^2([-\pi, \pi])$ of square integrable real valued function defined on $[-\pi, \pi)$ is a Banach space with respect to L^2-norm given by

$$\|f\|_2^2 := \int\limits_{-\pi}^{\pi} |f(t)|^2 dt < \infty, \quad \text{for } f \in L^2([-\pi, \pi]).$$

Theorem 8.3 *Let* $f \in L^2([-\pi, \pi))$ *be a* 2π*-periodic function with a Fourier series given by*

$$f(t) = \frac{a_0}{2} + \sum_{m=1}^{k} [a_m \cos(m\,t) + b_m \sin(m\,t)], \tag{8.2.56}$$

where $a_n, b_n \in \mathbb{R}$ *are the Fourier coefficients of* f. *Then, for any trigonometric polynomial*

$$\sigma(t) := \frac{c_0}{2} + \sum_{m=1}^{k} [c_m \cos(mt) + d_m \sin(mt)], \quad c_m, d_m \in \mathbb{R}, \tag{8.2.57}$$

the integral

$$I(c_0, \ldots, c_k; d_1, \ldots, d_k) := \int\limits_{-\pi}^{\pi} [f(x) - \sigma(x)]^2 dx \tag{8.2.58}$$

has the least value only if $c_0 = a_0$, $c_m = a_m$ *and* $d_m = b_m$, $1 \leq m \leq k$.

Proof Notice that, by simple calculus argument, the integral I has a *least value* provided for all m we have

$$\int\limits_{-\pi}^{\pi} [f(x) - \sigma(x)]\frac{\partial \sigma(x)}{\partial c_m} dx = \int\limits_{-\pi}^{\pi} [f(x) - \sigma(x)] \cos mx \, dx = 0;$$

$$\int\limits_{-\pi}^{\pi} [f(x) - \sigma(x)]\frac{\partial \sigma(x)}{\partial d_m} dx = \int\limits_{-\pi}^{\pi} [f(x) - \sigma(x)] \sin mx \, dx = 0.$$

It thus follows by using *orthogonality relations* (8.2.8) that

$$\int_{-\pi}^{\pi} f(x)\cos mx\,dx = \frac{c_0}{2}\int_{-\pi}^{\pi}\cos mx\,dx + \pi\,c_m;$$

$$\int_{-\pi}^{\pi} f(x)\sin mx\,dx = \frac{c_0}{2}\int_{-\pi}^{\pi}\sin mx\,dx + \pi\,d_m.$$

These equation proves the assertion. Therefore, the polynomial $\langle S_k(f)\rangle$ gives the *best approximation* of the 2π-periodic function $f(t)$ over the interval $[-\pi, \pi]$. Further,

$$\int_{-\pi}^{\pi} (f-\sigma)^2 dt = \int_{-\pi}^{\pi} (f^2 - 2f\sigma + \sigma^2)dt. \tag{8.2.59}$$

Now, since we have

$$\int_{-\pi}^{\pi} \sigma(t)^2 dt \tag{8.2.60}$$

$$= \int_{-\pi}^{\pi} \left[\frac{c_0}{2} + \sum_{n=1}^{k}[c_n\cos(nt) + d_n\sin(nt)]\right]^2 dt$$

$$= \int_{-\pi}^{\pi} \left[\frac{c_0^2}{4} + \sum_{n=1}^{k}[c_n^2\cos^2(nt) + d_n^2\sin^2(nt)]\right] dt$$

$$= \frac{c_0}{4}\int_{-\pi}^{\pi} dt + \sum_{n=1}^{k}\left(c_n\int_{-\pi}^{\pi} f(t)\cos(nt)dt + d_n\int_{-\pi}^{\pi} f(t)\sin(nt)dt\right)$$

$$= \frac{c_0^2}{4}\cdot 2\pi + \pi\sum_{n=1}^{k}[c_n^2 + d_n^2]$$

$$= \pi\left[\frac{c_0^2}{2} + \sum_{n=1}^{k}[c_n^2 + d_n^2]\right], \tag{8.2.61}$$

This is known as the *Plancherel-Parseval's Identity* for trigonometric polynomial $\sigma(t)$. Similarly, using Eqs. (8.2.16) and (8.2.17), it follows that

$$\int_{-\pi}^{\pi} f(t)\sigma(t)\mathrm{d}t \tag{8.2.62}$$

$$= \int_{-\pi}^{\pi} f(t)\left[\frac{c_0}{2} + \sum_{n=1}^{k}[c_n\cos(nt) + d_n\sin(nt)]\right]\mathrm{d}t$$

$$= \frac{c_0}{2}\int_{-\pi}^{\pi} f(t)\mathrm{d}t + \sum_{n=1}^{k}\left(c_k\int_{-\pi}^{\pi} f(t)\cos(nt)\mathrm{d}t + d_k\int_{-\pi}^{\pi} f(t)\sin(nt)\mathrm{d}t\right)$$

$$= \frac{c_0}{2}\cdot\pi a_0 + \sum_{n=1}^{k}\left[c_k\cdot\pi a_k + d_k\cdot\pi b_k\right]$$

$$= \pi\left[\frac{1}{2}a_0 c_0 + \sum_{n=1}^{k}[a_n c_n + b_n d_n]\right]. \tag{8.2.63}$$

Putting the last two expressions back into Eq. (8.2.59), a simple algebraic manipulation of the resulting expression gives

$$\int_{-\pi}^{\pi}(f - \sigma)^2\mathrm{d}t$$

$$= \int_{-\pi}^{\pi} f^2\mathrm{d}t - 2\int_{-\pi}^{\pi} f\sigma\,\mathrm{d}t + \int_{-\pi}^{\pi}\sigma^2\mathrm{d}t$$

$$= \int_{-\pi}^{\pi} f^2\mathrm{d}t - \pi A + \pi B, \tag{8.2.64}$$

where

$$A = \left[\frac{a_0^2}{2} + \sum_{k=1}^{n}(a_k^2 + b_k^2)\right]; \tag{8.2.65}$$

$$B = \left[\frac{(a_0 - c_0)^2}{2} + \sum_{k=1}^{n}[(a_k - c_k)^2 + (b_k - d_k)^2]\right]. \tag{8.2.66}$$

Notice that only B depends upon σ, and the left side integral in (8.2.64) is *minimum* only when $c_k = a_k$ and $d_k = b_k$, for $0 \leq k \leq n$.   □

Further, since $f(t)$ is given by the Fourier series (8.2.56), $B = 0$, and so the positivity of L^2-norm proves that the following *Bessel's Inequality* holds:

$$\frac{|a_0^2|}{2} + \sum_{n=1}^{\infty} (|a_n|^2 + |b_n|^2) \le \frac{1}{\pi} \int_{-\pi}^{\pi} [f(t)]^2 dt. \tag{8.2.67}$$

The assertion holds also for the complex Fourier series (8.2.18), so

$$\sum_{n=-\infty}^{\infty} |c_n|^2 \le \frac{1}{2\pi} \int_{-\pi}^{\pi} [f(t)]^2 dt. \tag{8.2.68}$$

To prove this, one has to use complex form of the definition of L^2-norm. In particular, by *monotone sequence property* of $\mathbb{R}$, we conclude the convergence of the following series:

$$\sum_{n=1}^{\infty} |a_n|^2, \quad \sum_{n=1}^{\infty} |b_n|^2, \quad \text{and} \quad \sum_{n=-\infty}^{\infty} |c_n|^2.$$

Therefore, $a_n, b_n, c_n \to 0$, as $n \to \infty$. That is, for a 2π-periodic function $f \in L^2[-\pi, \pi]$, the Fourier coefficients in absolute values are always arbitrarily small.

In what follows, it is convenient to view a 2π-periodic function $f : (-\pi, \pi] \to \mathbb{R}$, with $f(-\pi) = f(\pi)$, as a function $f : \mathbb{T} \to \mathbb{R}$, where

$$\mathbb{T} := \frac{\mathbb{R}}{2\pi \mathbb{Z}} = \{(x, y) \in \mathbb{R}^2 \mid x^2 + y^2 = 1\}$$

is the *unit circle*. Conversely, each function $f : \mathbb{T} \to \mathbb{R}$ defines a 2π-periodic function $f : (-\pi, \pi] \to \mathbb{R}$. It thus makes sense to talk about *complex Fourier series* of a function $f \in PC(\mathbb{T})$, having the polynomial $\langle p_n(t) \rangle$ as its nth *partial sum*, with $c_k = c_k(f)$ as given by Eq. (8.2.19). It then follows that the *Dirichlet kernel* can be written as

$$D_k(y) = \sum_{n=-k}^{k} e^{iny} = \frac{e^{i(k+1)y} - e^{-iky}}{e^{iy} - 1}, \quad \text{for } k \ge 0,$$

and the function $g_{\pm}(x) \in PS(\mathbb{T})$ is given by

$$g(x) = \begin{cases} \dfrac{f(t - x) - f(t^+)}{e^{ix} - 1}, & \text{if } -\pi < x < 0 \\[2mm] \dfrac{f(t - x) - f(t^-)}{e^{ix} - 1}, & \text{if } 0 < x \le \pi \end{cases}.$$

Therefore, a proof of the following *complex version* of Dirichlet's theorem can be obtained by imitating the proof of Theorem 8.2.

Theorem 8.4 *Let $f \in PS(\mathbb{T})$, and suppose $c_n = c_n(f)$ are given by (8.2.19), with $\omega_n = n$. Then*

$$\lim_{n \to \infty} c_n = \lim_{n \to \infty} c_{-n} = 0,$$

and we have

$$\lim_{n \to \infty} p_n[f](t) = \frac{f(t^+) + f(t^-)}{2}, \quad \text{for all } t \in \mathbb{T}.$$

Notice that the first assertion in above theorem follows from Lemma 8.1.

By Theorem 8.2, 2π-periodic functions in $PS(\mathbb{T})$ are good for some applications. However, these are not good enough to meet all practical needs such as termwise differentiation and integration of the Fourier series. For example, differentiating the Fourier series of *sawtooth function* as in Example 8.6, we get

$$1 = 2 \sum_{n=1}^{\infty} (-1)^{n+1} \cos nt, \quad \text{for } -\pi < t < \pi,$$

which is certainly *false*. Notice that the series on the right is not even convergent. In fact, by Riemann–Lebesgue lemma, it cannot even be a Fourier series of any function in $PC(\mathbb{T})$. As said before, such type of problem do not arise when the convergence is *uniform*. That is, if *sup metric* is used on the space of 2π-periodic *continuous* functions in $PS(\mathbb{T})$.

Theorem 8.5 *Every continuous, piecewise smooth function in $PS(\mathbb{T})$ can be represented by its Fourier series.*

Proof Let $f(t) \in PS(\mathbb{T})$ be continuous. Then, $f' \in PC(\mathbb{T})$ implies that

$$c_n(f') = \frac{1}{2\pi} \int_{-\pi}^{\pi} f'(x)e^{-inx}dx < \infty. \quad \text{for all } n \in \mathbb{Z}. \tag{8.2.69}$$

Integrating by parts, we obtain

$$c_n(f') = in\, c_n(f), \quad \text{for all } n \in \mathbb{Z}. \tag{8.2.70}$$

Similarly, we can obtain

$$a_n(f') = n\, a_n(f) \quad \text{and} \quad b_n(f') = -n\, b_n(f),$$

for all applicable n. Now, we have

$$\sum_{n=-\infty}^{\infty} |c_n(f)| = |c_0(f)| + \sum_{0 \neq n \in \mathbb{Z}} |c_n(f)|$$

$$= |c_0(f)| + \sum_{0 \neq n \in \mathbb{Z}} \left| \frac{c_n(f')}{in} \right|, \tag{8.2.71}$$

$$\leq |c_0(f)| + \frac{1}{2} \sum_{0 \neq n \in \mathbb{Z}} |c_n(f')|^2 + \frac{1}{2} \sum_{0 \neq n \in \mathbb{Z}} \frac{1}{n^2}, \tag{8.2.72}$$

where the second equality uses (8.2.70), and the last inequality follows by applying the following standard inequality to $a = c_n(f')$ and $b = (in)^{-1}$:

$$|ab| \leq \frac{|a|^2 + |b|^2}{2}, \quad \text{for } a, b \in \mathbb{C}.$$

The first sum in (8.2.72) converges by Bessel's inequality (8.2.68), and the second by usual *p-test*. We thus conclude that

$$\sum_{n=-\infty}^{\infty} |c_n(f)| < \infty. \tag{8.2.73}$$

Finally, to conclude that proof, we prove that the sequence $\langle S_k(f) \rangle$ converges uniformly to f on $\mathbb{T}$. For, let $t \in \mathbb{T}$. Then

$$|S_k(f)(t) - f(t)| = \left| \sum_{n=-k}^{k} c_n(f) e^{int} - \sum_{n=-\infty}^{\infty} c_n(f) e^{int} \right|$$

$$= \left| \sum_{|n|>k} c_n(f) e^{int} \right|$$

$$\leq \sum_{|n|>k} \left| c_n(f) e^{int} \right| = \sum_{|n|>k} |c_n(f)|.$$

Therefore, by completeness property of $\mathbb{R}$, we obtain

$$0 \leq \sup_{t \in \mathbb{T}} |S_k(f)(t) - f(t)| \leq \sum_{|n|>k} |c_n(f)|,$$

which proves the assertion, because the right side series is convergent by (8.2.73). □

Notice that, by (8.2.73), we also have

$$\sum_{n=-\infty}^{\infty} |c_n(f) e^{int}| < \infty.$$

So, for a continuous, piecewise smooth function in $PS(\mathbb{T})$, the Fourier series converges absolutely, as well. Further, for a continuous function $f \in PS(0, \ell)$, Theorem 8.5 applies only to the *even periodic extension* of the function $f(t)$. The problem with the *odd periodic extension* is that it is continuous on $\mathbb{R}$ if and only if $f(0^+) = f(\ell^-)$.

A similar remark applies to ℓ-*periodic extension* of a continuous, piecewise smooth function in $PS(\mathbb{T})$. However, in this case, the problems arise due to points of discontinuities introduced, and also slow rate of decay of the Fourier coefficient

$b_n(f)$. The reader is encouraged to use worked examples given earlier to reflect on these comments.

Mean Square Convergence

Recall that $L^2(\mathbb{T})$ denotes the space of *square integrable* real or complex-valued function defined on $\mathbb{T} = [-\pi, \pi)$, with L^2-*norm* given by (8.2.34):

$$\|f\|_2^2 := \int_{-\pi}^{\pi} |f(t)|^2 dt < \infty, \quad \text{for } f \in L^2(\mathbb{T}).$$

Here, we discuss L^2- *convergence* of the Fourier series $S(f)$ for $f \in L^2(\mathbb{T})$. Notice that, since

$$e^{ikx} \in C(\mathbb{T}) \quad \Rightarrow \quad e^{ikx} \in L^2(\mathbb{T}),$$

so the product $f(t)e^{ikx} \in L^1(\mathbb{T})$. Thus, all the coefficients c_k in (8.2.53) are elements of the space $L^1(\mathbb{T})$. Hence, the polynomial $p_n(t) \in L^2(\mathbb{T})$, for all $n \in \mathbb{Z}$.

By Stone–Weierstrass theorem, the space of polynomials $p(t)$ defined over the interval $[-\pi, \pi]$, satisfying $p(-\pi) = p(\pi)$, is *dense* in $C[-\pi, \pi]$. As mentioned in the concluding part of the previous section, the 2π - *periodic extension* of any such polynomial lies in the class of continuous, piecewise smooth functions defined over the set $\mathbb{T}$, which we denote by $PSC(\mathbb{T})$. Certainly then, the space $PSC(\mathbb{T})$ is dense in $C(\mathbb{T})$. Also, we know that every $f \in L^2(\mathbb{T})$ can be approximated by a sequence of functions in the space $C(\mathbb{T})$. Notice that the function $f(t) = -\ln|2\sin(t/2)| \in L^2(\mathbb{T})$ is 2π - periodic, but $f \notin PC(\mathbb{T})$. It is instructive to obtain the Fourier series of this function. Does $p_n[f](t) \to f(t)$ for all t ?

Theorem 8.6 *For every $f \in L^2(\mathbb{T})$,*

$$\lim_{n \to \infty} \|p_n[f](t) - f(t)\|_2 = 0, \quad \text{for all } t \in \mathbb{T}. \tag{8.2.74}$$

The converse holds too.

Proof Let $f_n \in PSC(\mathbb{T})$ such that

$$\|f_n - f\|_2 \to 0, \quad \text{as } n \to \infty. \tag{8.2.75}$$

Also, by triangle inequality, we have

$$\|p_n[f] - f\|_2 \leq \|p_n[f] - p_n[f_n]\|_2 + \|p_n[f_n] - f_n\|_2 + \|f_n - f\|_2.$$

However, since

$$\| p_n[f] - p_n[f_n] \|_2 = \| p_n[f - f_n] \|_2 \le \| f - f_n \|_2,$$

we obtain

$$\| p_n[f] - f \|_2 \le \| p_n[f_n] - f_n \|_2 + 2 \| f_n - f \|_2.$$

Now, we know that $p_n[f_n]$ converges uniformly to f_n for each $n \in \mathbb{N}$, by Theorem 8.5, and so with respect to L^2-norm as well. Hence, (8.2.75) proves (8.2.75).

Conversely, since we can write

$$f = p_n[f] + (f - p_n[f]),$$

so if $p_n[f] \to f$ with respect to L^2-norm, then from the implication

$$p_n[f] \in L^2(\mathbb{T}) \quad \Rightarrow \quad f - p_n[f] \in L^2(\mathbb{T}), \quad \text{for large enough } n,$$

it follows that $f \in L^2(\mathbb{T})$. □

In particular, we obtain L^2-convergence of the sequence $\langle p_n[f] \rangle$ for a 2π-periodic function $f(t)$ with $f, f' \in PC(\mathbb{T})$.

In 1907, *Riesz-Fischer* proved the following:

If $\{\varphi_n\}$ is an orthonormal system in $L^2(\mathbb{T})$ and $\{c_n\}$ is a sequence in $\ell^2(\mathbb{C})$, then the convergence of the series $\sum_{-\infty}^{\infty} c_n^2$ is a necessary and sufficient condition for the existence of a function f such that

$$\int_{-\pi}^{\pi} f(x)\varphi_n(t) \, dt = c_n, \quad \text{for every } n. \tag{8.2.76}$$

In particular, for $\varphi_n = (1/\pi)e^{-2nt}$, it follows that the series (8.2.19) converges uniformly to $f(t)$, if $f \in L^2(\mathbb{T})$, and the coefficients c_n are given by Eq. (8.2.76).

Application to Infinite Series

Fourier series provides a useful tool to find the sum of important infinite series. The basic idea is to *identify* an appropriate continuously differentiable 2ℓ-periodic function $f(t)$ defined over the interval $[-\ell, \ell]$ so that Theorem 8.2 applies. Then, for $t \in (-\ell, \ell)$, we can write

$$\frac{f(t^+) + f(t^-)}{2} = \frac{a_0}{2} + \sum_{n=1}^{\infty} [a_n \cos(\omega_n t) + b_n \sin(\omega_n t)], \tag{8.2.77}$$

where the Fourier coefficients a_n, b_n are given by Eqs. (8.2.16) and (8.2.17). The sum of given infinite series is subsequently obtained by evaluating the Fourier series of the function $f(t)$ at a suitable point of continuity $t = t_0 \in (-\ell, \ell)$.

Example 8.14 By Theorem 8.2, the Fourier series of 2π-periodic function $f(t)$ obtained in Example 8.4 gives at $t = \pi$

$$\frac{f(\pi^+) + f(\pi^-)}{2} = \frac{(-\pi + \pi^2) + (\pi + \pi^2)}{2} = \pi^2.$$

In particular, it follows that

$$\sum_1^\infty \frac{1}{n^2} = \frac{\pi^2}{6},$$

which is known as the *Basel theorem*. Further, taking $t = 0$, we obtain

$$\sum_{n=1}^\infty \frac{(-1)^{n-1}}{n^2} = \frac{\pi^2}{12}.$$

Example 8.15 By Example 8.6, we know that the Fourier sine series of 2π-periodic function $f(t) = t$, $-\pi < t \leq \pi$, is given by

$$t = 2 \sum_{n=1}^\infty \frac{(-1)^{n+1}}{n} \sin nt, \quad \text{for } -\pi < t < \pi.$$

In particular, taking $t = \pi/2$, we obtain the *Gregory series*

$$\frac{\pi}{4} = 1 - \frac{1}{3} + \frac{1}{5} - \frac{1}{7} + \cdots .$$

On the other hand, the Fourier cosine series of 2π-periodic function $f(t) = |t|$, $t \in [-\pi, \pi]$, obtained in Example 8.8 is given by

$$\frac{\pi}{2} - \frac{4}{\pi} \sum_{k=1}^\infty \frac{\cos(2k - 1)t}{(2k - 1)^2}.$$

In particular, taking $t = 0$, we obtain

$$\sum_{k=1}^\infty \frac{1}{(2k - 1)^2} = \frac{\pi^2}{8}.$$

Example 8.16 For any $f \in L^2([-\pi, \pi])$, notice that equality holds in (8.2.67):

$$\frac{1}{2\ell} \int_{-\pi}^\pi f(t)^2 dt = \frac{a_0^2}{4} + \frac{1}{2} \sum_{n=1}^\infty [a_n^2 + b_n^2], \tag{8.2.78}$$

where a_n and b_n are the Fourier coefficients of the function $f(t)$. Applying this to the function $f(t) = t$, $t \in (0, 2)$, and using Example 8.10, we have

$$\frac{1}{4} \int_0^2 x^2 dx = 1 + \frac{32}{\pi^4} \sum_{n=1}^{\infty} \frac{1}{(2n-1)^4},$$

so that

$$\sum_{n=1}^{\infty} \frac{1}{(2n-1)^4} = \frac{\pi^4}{96}. \tag{8.2.79}$$

Notice that

$$S = \sum_{n=1}^{\infty} \frac{1}{n^4} \quad \Rightarrow \quad \sum_{n=1}^{\infty} \frac{1}{(2n)^4} = \frac{1}{2^4} \sum_{n=1}^{\infty} \frac{1}{n^4} = \frac{S}{16}.$$

Thus, using (8.2.79), we obtain

$$S = \sum_{n=1}^{\infty} \frac{1}{n^4} = \sum_{n=1}^{\infty} \frac{1}{(2n-1)^4} + \sum_{n=1}^{\infty} \frac{1}{(2n)^4}$$

$$= \frac{\pi^4}{96} + \frac{S}{16}$$

$$\Rightarrow \quad S = \sum_{n=1}^{\infty} \frac{1}{n^4} = \frac{\pi^4}{90}.$$

Notice that the infinite series in the above example is a special case of the *Riemann zeta function* $\zeta(s)$ defined as an absolutely convergent series given by

$$\zeta(s) := \sum_{n=1}^{\infty} \frac{1}{n^s}, \quad s \in \mathbb{C}, \text{ with } \mathrm{Re}(s) > 1,$$

which can extended to $\mathbb{C} \setminus \{s = 1\}$ by meromorphic continuation. Most notable conjecture about $\zeta(s)$ is *Riemann Hypothesis*: *Other than the even negative integers, all solutions of the equation* $\zeta(s) = 0$ *have their real part on the line* $x = 1/2$.

8.3 Separation of Variables

Earlier in Chap. 4, we illustrated how basic idea works for a first order differential equation. While dealing with a second order differential equation, say for a function $u = u(x, y) \in C^2$, we need to do a little more. As before, by using a solution of the form $u(x, y) = X(x)Y(y)$ followed by substitution of derivatives back into the equation, we obtain

$$F(X, X', X''; x) = G(Y, Y', Y''; y).$$

Therefore, the method of separation of variables decomposes a differential equation into ordinary differential equations, each involving a separation constant, say k. Subsequently, considering assigned boundary conditions, the suitability of k is determined by solving the ordinary differential equations of the form

$$F(X, X', X''; x) = k \quad \text{and} \quad G(Y, Y', Y''; y) = k.$$

For brevity, let us consider a second order differential equation for a C^2-function $u = u(x, y)$ given by

$$a\, u_{xx} + b\, u_{xy} + c\, u_{yy} = 0, \quad \text{with } a \neq 0 \text{ and } b \neq 4c. \tag{8.3.1}$$

It then follows by taking $u(x, y) = X(x)Y(y)$ that

$$X''Y + \alpha X'Y' + \beta XY'' = 0, \quad \text{where } \alpha = \frac{b}{a}, \text{ and } \beta = \frac{c}{a}.$$

Dividing the above equation by XY, we obtain

$$\frac{X''}{X} + \alpha \frac{X'}{X}\frac{Y'}{Y} + \beta \frac{Y''}{Y} = 0.$$

Choose a constant $k \neq 0$ such that $Y'/Y = -k$. Then, we can write $Y''/Y = k^2$. Therefore, we have to solve the two ordinary differential equations given by

$$X'' - (k\alpha)X' + k^2\beta X = 0 \quad \text{and} \quad Y'' - k^2 Y = 0. \tag{8.3.2}$$

Recall that the general solution of the second equation in (8.3.2) is given by

$$Y(y) = Ce^{kx} + Be^{-kx}, \quad \text{where } C \text{ and } B \text{ are constants.}$$

Further, as the first equation in (8.3.2) has the *discriminant* given by

$$k^2\alpha^2 - 4k^2\beta = k^2(\alpha - 4\beta),$$

it follows that for the two roots given by

$$\lambda_1 = \frac{k}{2}(\alpha + \sqrt{\alpha - 4\beta}) \quad \text{and} \quad \lambda_2 = \frac{k}{2}(\alpha - \sqrt{\alpha - 4\beta}),$$

we can write the solution of the first equation in (8.3.2) as

$$X(x) = Ae^{\lambda_1 x} + Be^{\lambda_2 x},$$

when λ_1, λ_2 are real numbers, or as

$$X(x) = e^{\gamma x}[A\cos(\mu x) + B\sin(\mu x)],$$

when λ_1, λ_2 are complex numbers of the form

$$\lambda_1 = \gamma + i\mu \quad \text{and} \quad \lambda_2 = \gamma - i\mu.$$

Notice that the solutions of the two equations in (8.3.2) when $k = 0$ are given by

$$X(x) = Ax + B \quad \text{and} \quad Y(y) = Cy + D,$$

so that the complete solution of Eq. (8.3.1) can be written as

$$u(x, y) = (Ax + B)(Cy + D), \quad \text{where } A, B, C, \text{ and } D \text{ are parameters.}$$

Otherwise, the complete solution of (8.3.1) can be written as given below:

1. For $k = p^2 > 0$ & $b > 4c$, we have

$$u(x, y) = \left[Ae^{\lambda_1 x} + Be^{\lambda_1 x}\right]\left[Ce^{p^2 y} + De^{-p^2 y}\right].$$

2. For $k = p^2 > 0$ & $b < 4c$, we have

$$u(x, y) = e^{\gamma x}\left[A\cos(\mu x) + B\sin(\mu x)\right]\left[Ce^{p^2 y} + De^{-p^2 y}\right].$$

3. For $k = -p^2$ & $b > 4c$, we have

$$u(x, y) = \left[Ae^{\lambda_1 x} + Be^{\lambda_1 x}\right]\left[C\cos(p^2 y) + D\sin(p^2 y)\right].$$

4. For $k = -p^2 < 0$ & $b < 4c$, we have

$$u(x, y) = e^{\gamma x}\left[A\cos(\mu x) + B\sin(\mu x)\right]\left[C\cos(p^2 y) + D\sin(p^2 y)\right].$$

Hence, the complete solutions of two equations in (8.3.2) are obtained by using the specified boundary conditions. Finally, considering the three cases as given below:

$$k = 0, \quad k = p^2 > 0, \quad \text{and} \quad k = -p^2 < 0,$$

the solution $u = u(x, y)$ of Eq. (8.3.1) is obtained by solving the two resulting boundary value problems for ordinary differential equations. The above procedure usually leads to a sequence $\langle u_n \rangle$ of product solutions given by $u_n(x, y) = X_n(x)Y_n(y)$ that are *superposed by linearity* to write

$$u(x, y) = \sum_{n=0}^{\infty} \alpha_n u_n(x, y) = \sum_{n=0}^{\infty} \alpha_n X_n(x) Y_n(y).$$

It is assumed that the analytical conditions for the (uniform) convergence of the series on the right of the above equation are satisfied by the involved functions. The constants α_n are determined by using the assigned initial conditions and the concept of Fourier series.

Remark 8.4 For an equation involving a function $u = u(x, y, z)$, we take a *product solution* of the form $u(x, y, z) = X(x)Y(y)Z(z)$ so that substituting the derivatives back into the equation result into the three ordinary differential equations in unknowns $X = X(x)$, $Y = Y(y)$, and $Z = Z(z)$, each involving a *separation constant*. Subsequently, assuming that $u \not\equiv 0, \pm\infty$, we apply the assigned boundary conditions to find appropriate values of the *three* separation constants. If the solutions of the resulting boundary value problems for the three ordinary differential equations are, respectively, given by $\langle X_n \rangle$, $\langle Y_n \rangle$, and $\langle Z_n \rangle$, then each function $u_n(x, y, z) = X_n(x)Y_n(y)Z_n(z)$ is a solution of the original boundary value problem. Next, by applying the *principle of superposition*, we write the general solution as an infinite series given by

$$u(x, y, z) = c_1 u_1 + c_2 u_2 + c_3 u_3 + \cdots,$$

Finally, the coefficients c_i are computed by using the Fourier series of the functions as given in the assigned initial conditions. Clearly, the details get messy as we consider initial-boundary value problems for equations in n number of variables, with $n \geq 3$, and also if the order of the equation is more than two.

We begin by applying the above *three-step procedure* to solve the initial-boundary value problem for the 1 -dimensional *homogeneous* wave equation.

Fourier Solution of Wave Equations

Let the *transverse displacement* of a vibrating string of length ℓ at a point $0 \leq x \leq \ell$ and at time $t \geq 0$ be given by a function $y = y(x, t) \in C^2([0, \ell] \times \mathbb{R}^+)$. By Example 4.1, we know that y satisfies the wave equation given by

$$y_{tt}(x, t) = c^2 y_{xx}(x, t), \quad \text{for } 0 \leq x \leq \ell, \text{ and } t \geq 0.$$

The ends of the string are taken to be fixed at points $x = 0$ and $x = \ell$ along x-axis. Suppose the equilibrium state of the string is distorted at time $t = 0$ by a force whose magnitude is represented a function $f(x)$, and the initial velocity be given by a function $g(x)$. Therefore, the displacement function y satisfies the boundary and initial conditions as given below:

$$y(0, t) = 0, \quad \text{and} \quad y(\ell, t) = 0, \quad \text{for } t \geq 0;$$
$$y(x, 0) = f(x), \quad \text{and} \quad y_t(x, 0) = g(x), \quad \text{for } 0 < x < \ell.$$

To solve the above IBVP, we follow the stepwise procedure as given below:

1. We separate variables by taking $y(x, t) == F(x)G(t)$ so that by differentiating $y(x, t)$ twice with respect to variables x and t, and substituting the resulting expressions back into the above wave equation, we obtain

$$\frac{d^2 F}{dx^2} - k\, F = 0 \quad \text{and} \quad \frac{d^2 G}{dt^2} - c^2 k\, G = 0, \tag{8.3.3}$$

where k is the *separation constant*. Next, we determine admissible value of k. For, consider the three possibilities as given below:

a. For $k = 0$, the solutions of the two equations in (8.3.3) are, respectively, given by

$$F(x) = Ax + B \quad \text{and} \quad G(t) = Ct + D.$$

That is, in this case, the function y given by

$$y(x, t) = (Ax + B)(Ct + D) \tag{8.3.4}$$

is the complete integral of the equation $y_{tt} - c^2 y_{xx} = 0$.

b. For $k = \mu^2$ $(\mu > 0)$, the two solutions of equations in (8.3.3) are, respectively, given by

$$F(x) = Ae^{\mu x} + Be^{-\mu x} \quad \text{and} \quad G(t) = Ce^{c\mu t} + De^{-c\mu t},$$

which shows, in this case, the function y given by

$$y(x, t) = (Ae^{\mu x} + Be^{-\mu x})(Ce^{c\mu t} + De^{-c\mu t}) \tag{8.3.5}$$

is the complete integral of the equation $y_{tt} - c^2 y_{xx} = 0$.

c. For $k = -\mu^2$ $(\mu > 0)$, the solutions of two equations in (8.3.3) are, respectively, given by

$$F(x) = A\cos \mu x + B\sin \mu x \quad \text{and} \quad G(t) = C\cos(c\mu t) + D\sin(c\mu t),$$

which shows, in this case, the function y given by

$$y(x, t) = [A\cos \mu x + B\sin \mu x][C\cos(c\mu t) + D\sin(c\mu t)] \tag{8.3.6}$$

is the complete integral of the equation $y_{tt} - c^2 y_{xx} = 0$.

2. We next find precisely which one of the above three solutions are *compatible* with the assigned boundary conditions. Notice that we have

$$y(0, t) = F(0)G(t) = 0 \quad \text{and} \quad y(\ell, t) = F(\ell)G(t) = 0, \quad \forall t.$$

It thus follows that

$$G \equiv 0 \quad \Rightarrow \quad y = FG = 0 \quad \Rightarrow \quad G \not\equiv 0.$$

Then, by equation (8.3.4), we have

$$F(0) = F(\ell) = 0 \quad \Rightarrow \quad A = B = 0$$
$$\Rightarrow \quad F = 0$$
$$\Rightarrow \quad y = FG = 0.$$

Therefore, we conclude that $k \neq 0$. Similarly, the value $k = \mu^2 \, (\mu > 0)$ is not admissible because (8.3.5), together with $F(0) = F(\ell) = 0$, gives

$$F \equiv 0 \quad \Rightarrow \quad y = FG = 0.$$

Consequently, Eq. (8.3.6) together with $F(0) = F(\ell) = 0$ implies $A = 0$, and hence $B \sin(\mu\ell) = 0$. Finally, as we must have $B \not\equiv$ to obtain a nontrivial solution, it follows that

$$\mu_n = \frac{n\pi}{\ell}, \quad \text{for } n = 1, 2, 3, \ldots .$$

Therefore, by setting $B = 1$, we obtain *infinitely many* solutions given by

$$F_n(x) = \sin\left(\frac{n\pi x}{\ell}\right), \quad \text{for } n = 1, 2, 3, \ldots .$$

Hence, for $k = -\mu_n^2 = -(n\pi)^2/\ell^2$, the complete solution of the second equation in (8.3.3) is the function given by

$$G_n(t) = C_n \cos(\lambda_n t) + D_n \sin(\lambda_n t), \quad \text{where } \lambda_n = \frac{n c \pi}{\ell}.$$

We conclude infinitely many solutions of the IBVP are obtained as

$$y_n(x, t) = \left[C_n \cos(\lambda_n t) + D_n \sin(\lambda_n t) \right] \sin\left(\frac{n\pi x}{\ell}\right). \tag{8.3.7}$$

3. The concluding step is about applying the Fourier series to find the complete solution of the given IBVP. Using the *principle of superposition*, we may write

$$y(x, t) = \sum_{n=1}^{\infty} y_n(x, t) = \sum_{n=1}^{\infty} [C_n \cos(\lambda_n t) + D_n \sin(\lambda_n t)] \sin\left(\frac{n\pi x}{\ell}\right),$$
$$\tag{8.3.8}$$

which in view of given initial conditions yields

$$f(x) = y(x, 0) = \sum_{n=1}^{\infty} C_n \sin\left(\frac{n\pi x}{\ell}\right); \quad \text{and,} \qquad (8.3.9)$$

$$g(x) = y_t(x, 0) = \sum_{n=1}^{\infty} \lambda_n D_n \sin\left(\frac{n\pi x}{\ell}\right) \qquad (8.3.10)$$

Now, to obtain the coefficients C_n and $C_n^* = \lambda_n D_n$, we apply *Fourier series*. Notice that Eq. (8.3.9) suggests that we must choose coefficients C_n so that $y(x, 0)$ becomes the *Fourier sine series* of the given function $f(x)$. That is, we must have that

$$C_n = \frac{2}{\ell} \int_0^{\ell} f(x) \sin\left(\frac{n\pi x}{\ell}\right) dx, \quad n = 1, 2, \ldots . \qquad (8.3.11)$$

Similarly, Eq. (8.3.10) suggests that we must choose C_n^* so that $y_t(x, 0)$ becomes the *Fourier sine series* of the given function $g(x)$. That is, we must have that

$$C_n^* = \frac{2}{\ell} \int_0^{\ell} g(x) \sin\left(\frac{n\pi x}{\ell}\right) dx, \quad n = 1, 2, \ldots .$$

That is, using the relation $\lambda_n = (nc\pi)/\ell$,

$$D_n = \frac{2}{nc\pi} \int_0^{\ell} g(x) \sin\left(\frac{n\pi x}{\ell}\right) dx, \quad n = 1, 2, \ldots . \qquad (8.3.12)$$

Hence, the verticalhistorically important particular deflection $y = y(x, t)$ of the vibrating string at a position $0 < x < \ell$ and at a time $t > 0$ is given by the series (8.3.8), where the coefficients are given by integrals (8.3.11) and (8.3.12).

The next two examples illustrate the procedure.

Example 8.17 A historically imporant particular case corresponds to the situation when initially the string is *plucked* at a point *i. e.* the initial defection $y(x, 0)$ is given by a *nondifferentiable triangular curve* of the form

$$f(x) = \begin{cases} \dfrac{2k}{\ell} x & \text{, if } 0 < x < \frac{\ell}{2} \\ \dfrac{2k}{\ell}(\ell - x), & \text{if } \frac{\ell}{2} < x < \ell \end{cases},$$

so that the *initial velocity* $y_t(x, 0)$ is zero. That is, we may have $g(x) = 0$. Clearly then, in this case, $D_n = 0$, and C_n may be computed using the integral (8.3.11). It then follows that Eq. (8.3.8) takes the form

$$y(x, t) = \frac{8k}{\pi^2} \left[\frac{1}{1^2} \sin\left(\frac{\pi x}{c\ell}\right) \cos\left(\frac{\pi t}{\ell}\right) - \frac{1}{3^2} \sin\left(\frac{3\pi x}{c\ell}\right) \cos\left(\frac{3\pi t}{\ell}\right) + \cdots \right],$$

which is the complete solution.

Fourier Solution of Heat Equation

Suppose the *temperature distribution* of a properly insulated bar of length ℓ is given by function $u = u(x, t)$. By Example 4.4, we know that u satisfies the equation

$$u_t(x, t) = a^2 u_{xx}(x, t), \quad \text{for } 0 \le x \le \ell \text{ and } t \ge 0. \tag{8.3.13}$$

As said before, we need to solve this equation subject to that $u = u(x, t)$ satisfies certain specific type of boundary and initial conditions. Let us first consider the case when $u = u(x, t)$ satisfies (8.3.13), Dirichlet boundary conditions of the form

$$u(0, t) = \alpha, \quad \text{and} \quad u(\ell, t) = \beta, \quad \text{for } t > 0, \tag{8.3.14}$$

where α, β are constants, and an initial condition given by

$$u(x, 0) = g(x), \quad \text{for } 0 < x < \ell. \tag{8.3.15}$$

As *nonhomogeneous* boundary conditions are involved, we cannot apply separation of variables directly. However, in this case, the IBVP can be reduced to a homogeneous problem by considering the function

$$v(x) = \frac{\beta - \alpha}{\ell} x + \alpha, \quad \text{for } 0 \le x \le \ell \text{ and } t \ge 0. \tag{8.3.16}$$

Clearly, $v = v(x)$ satisfies the heat equation and also the boundary conditions in (8.3.14). Notice that the function $v = v(x)$ provides the *equilibrium solution* of the given BVP. So, if $w = w(x, t)$ is a solution of the following homogeneous BVP

$$\begin{aligned} w_t(x, t) &= a^2 w_{xx}(x, t); \\ w(0, t) &= 0, \quad \text{and} \quad w(\ell, t) = 0, \quad \text{for } t > 0, \end{aligned} \tag{8.3.17}$$

then the function

$$u(x, t) = v(x) + w(x, t), \quad 0 \le x \le \ell, \ t \ge 0, \tag{8.3.18}$$

provides a complete solution of the given *nonhomogeneous* BVP. The function $w = w(x, t)$ is called the *transient solution* because $w(x, t) \to 0$ as $t \to \infty$. So, the complete solution $u = u(x, t)$ of a given nonhomogeneous IBVP approaches to the *equilibrium temperature* $v = v(x)$, as $t \to \infty$.

It is thus possible to solve a nonhomogeneous IBVP for the heat equation (8.3.13) with Dirichlet boundary conditions of the form (8.3.14) by using a solution of the homogeneous BVP of the type (8.3.17) that also satisfies the *modified initial condition* of the form

$$w(x, 0) = g(x) - v(x) = f(x), \quad \text{for } 0 < x < \ell. \tag{8.3.19}$$

A similar procedure applies if the ends of the bar are insulated so that the boundary conditions are of the Neumann type:

$$u(0, t) = u(\ell, t) = 0, \quad \text{for all } t > 0, \tag{8.3.20}$$

In each case, we can apply separation of variables and Fourier series to solve the IBVP. A stepwise procedure is as described below.

1. First, we apply separation of variables by taking $u(x, t) = F(x)G(t)$. Then, Eq. (8.3.13) yields the following two differential equations:

$$\frac{d^2 F}{dx^2} - k F = 0 \quad \text{and} \quad \frac{dG}{dt} - a^2 k G = 0, \tag{8.3.21}$$

where k is a constant of separation. Consider the following three different possibilities for k.

a. If $k = 0$, the complete solutions of the two differential equations in equation (8.3.21) are $F(x) = Ax + B$ and $G(t) = t + C$. This shows that

$$u(x, t) = (Ax + B)(t + C), \tag{8.3.22}$$

is a complete integral of the given heat equation.

b. If $k = \mu^2 > 0$ $(\mu > 0)$, the complete solutions of the two differential equations in Eq. (8.3.21) are $F(x) = A_1 e^{\mu x} + B_1 e^{-\mu x}$ and $G(t) = C e^{a^2 \mu^2 t}$. This shows that

$$u(x, t) = (A e^{\mu x} + B e^{-\mu x}) e^{a^2 \mu^2 t}, \quad \text{with } A = A_1 C \,\&\, B = B_1 C, \tag{8.3.23}$$

is a complete integral of the given heat equation.

c. If $k = -\mu^2 < 0$ $(\mu > 0)$, the complete solutions of the two differential equations in Eq. (8.3.21) are $F(x) = A_1 \cos \mu x + B_1 \sin \mu x$ and $G(t) = C e^{-a^2 \mu^2 t}$. This shows that

$$u(x, t) = (A \cos \mu x + B \sin \mu x) e^{-a^2 \mu^2 t} \tag{8.3.24}$$

is a complete integral of the give heat equation.

2. Next, as before, we wish to find which one of these three complete integrals of the heat equation are *compatible* with the given homogeneous boundary conditions. Note that

$$u(0, t) = F(0)G(t) = 0 \quad \text{and} \quad u(\ell, t) = F(\ell)G(t) = 0, \quad \forall \, t.$$

Thus, $G \equiv 0 \Rightarrow u = FG = 0 \Rightarrow G \not\equiv 0$. But then, by Eq. (8.3.22),

$$F(0) = F(\ell) = 0 \quad \Rightarrow \quad A = B = 0 \quad \Rightarrow \quad F = 0 \quad \Rightarrow \quad u = FG = 0.$$

Thus, the case $k = 0$ is not admissible. Similarly, the value $k = \mu^2 \, (\mu > 0)$ is not admissible because Eq. (8.3.23), together with $F(0) = F(\ell) = 0$, implies that $F \equiv 0 \Rightarrow u = FG = 0$. Thus, Eq. (8.3.24), together with $F(0) = F(\ell) = 0$, implies that $A = 0$ and $B \sin(\mu \ell) = 0$. Finally, since we must have $B \neq 0$ for a *nontrivial* solution, it follows that $\mu = (n\pi)/\ell$. Setting $B = 1$, we thus obtain the *infinitely many* complete solutions

$$F_n(x) = \sin\left(\frac{n\pi \, x}{\ell}\right), \quad \text{for} \quad n = 1, 2, \dots .$$

For $k = -\mu_n^2 = -(n\pi)^2/\ell^2$, the complete solution of the second equation in (8.3.21) is given by

$$G_n(t) = e^{-\lambda_n^2 t}, \quad \text{where} \quad \lambda_n = \frac{na\pi}{\ell}.$$

Whence, the complete integrals of the heat equation are given by

$$u_n(x, t) = \sin\left(\frac{n\pi \, x}{\ell}\right) e^{-\lambda_n^2 t}, \quad n = 1, 2, \dots . \tag{8.3.25}$$

3. The concluding step is about applying the Fourier series method to find the complete integral of the heat equation (8.3.13). Firstly, using the *principle of superposition*, we write

$$u(x, t) = \sum_{n=1}^{\infty} u_n(x, t) = \sum_{n=1}^{\infty} C_n \sin\left(\frac{n\pi \, x}{\ell}\right) e^{-\lambda_n^2 t}, \tag{8.3.26}$$

so that by the given initial condition

$$f(x) = u(x, 0) = \sum_{n=1}^{\infty} C_n \sin\left(\frac{n\pi \, x}{\ell}\right).$$

In view of Fourier theorem, Eq. (8.3.21) suggests that the coefficients C_n are chosen so that $u(x, 0)$ becomes the *Fourier sine series* of the given function $f(x)$. That is,

$$C_n = \frac{2}{\ell} \int_0^\ell f(x) \sin\left(\frac{n\pi x}{\ell}\right) dx, \quad n = 1, 2, \ldots . \tag{8.3.27}$$

Hence, the distribution of temperature $u = u(x, t)$ in the bar at the position $x > 0$ and time $t > 0$ is given by Eq. (8.3.26), where the coefficients C_n are given by Eq. (8.3.27). Notice that, for Neumann boundary conditions of the type (8.3.20), the formula for coefficients C_n involves cosine function of $n\pi x/\ell$.

The following worked example illustrates the procedure.

Example 8.18 Let $a = \sqrt{3}$, $\ell = 2$, $\alpha = 2$, $\beta = -3$, and $u(x, 0) = x + \sin(\pi x) - (1/2)$, for $0 < x < 2$. Note that, for these values, $\lambda_n = n\pi$ Then, we have to solve the heat equation $u_t - 3u_{xx} = 0$ subject to *modified* initial condition $u(x, 0) = x + \sin(\pi x)$, $0 < x < \ell$, and homogeneous boundary conditions. By Eq. (8.3.27),

$$C_n = \int_0^2 [x + \sin(\pi x)] \sin(n\pi x/2) \, dx$$

$$= \int_0^2 x \sin(n\pi x/2) \, dx + \int_0^2 \sin(\pi x) \sin(n\pi x/2) \, dx$$

$$= \left[1 - (-1)^{n+1}\right] \frac{4}{n\pi} + \left[\frac{\sin((n-2)\pi)}{(n-2)\pi} - \frac{\sin((n+2)\pi)}{(n+2)\pi}\right],$$

$$= \left[1 - (-1)^{n+1}\right] \frac{4}{n\pi}.$$

Thus, $C_n = 0$, for odd n, and $C_n = 4/(n\pi)$, for even n. Hence, the distribution of temperature $u = u(x, t)$ in the bar at the position $x > 0$ and time $t > 0$ is given by

$$u(x, t) = \sin\left(\frac{\pi x}{2}\right) e^{-\pi^2 t}.$$

The same procedure can be adopted while dealing with IBVPs related to heat equations with a *study-state* source term, say $h(x)$. That is, the method applies to solve a nonhomogeneous heat equation of the form

$$u_t(x, t) = a^2 u_{xx}(x, t) + h(x), \quad 0 < x < \ell, \ t > 0, \tag{8.3.28}$$

satisfying the following boundary and initial conditions:

$$u(0, t) = \alpha, \quad \text{and} \quad u(\ell, t) = \beta, \quad \text{for } t > 0; \tag{8.3.29}$$

$$u(x, 0) = f(x), \quad \text{for } 0 < x < \ell. \tag{8.3.30}$$

The method also applies when the boundary condition $u(0, t) = \alpha$ is replaced by the *gradient condition* $u_x(0, t) = \alpha$, subject to that the *equilibrium solution* $v = v(x)$ in this case is of the form

$$v(x) = \alpha(x - \ell) + \beta.$$

The following worked example illustrates the procedure.

Example 8.19 Let the data of the problem be given as follow:

$$a = 2, \quad \ell = 2, \quad u_x(0, t) = \alpha = 3, \quad u(\ell, t) = \beta = -1,$$

and suppose the initial condition is given by

$$u(x, 0) = 3x + \sin^2(\pi x/2) + 7, \quad \text{for } 0 < x < 2.$$

Note that, for the above given values, $\lambda_n = n\pi$. We need to solve the heat equation $u_t - 4u_{xx} = 0$ subject to the *modified initial condition*

$$u(x, 0) = \sin^2(\pi x/2), \quad 0 < x < 2,$$

and satisfying the homogeneous boundary conditions. By Eq. (8.3.27),

$$C_n = \int_0^2 \sin^2(\pi x/2) \sin(n\pi\, x/2)\, dx$$

$$= -\frac{1}{n\pi}[(-1)^n - 1] - \frac{1}{4}\int_0^2 \cos \pi x \sin(n\pi\, x/2)\, dx$$

$$= -\frac{1}{n\pi}[(-1)^n - 1] - \frac{n[(-1)^n - 1]}{\pi(n^2 - 4)}.$$

Thus, $C_n = -8/(n\pi(n^2 - 4))$, for odd n, and $C_n = 0$, for even n. Hence, the distribution of temperature $u = u(x, t)$ in the bar at the position $x > 0$ and time $t > 0$ is given by the series

$$\frac{8}{\pi}\sum_{k=1}^{\infty} \frac{\sin((2k + 1)\pi x/2)}{8k^3 + 12k^2 - 2}\, e^{-(2k+1)^2\pi^2 t} + 3x - 7.$$

The procedure fails for nonhomogeneous Neumann boundary conditions of the form

$$u_x(0, t) = \alpha, \quad \text{and } u_x(\ell, t) = \beta, \quad \text{for } t > 0, \quad \text{with } \alpha \neq \beta.$$

In this case, when the temperature changes at a constant rate, we may consider a function of the form

$$v(x, t) = \gamma t + h(x)$$

for the *equilibrium solution* so that $\gamma = v_t = a^2 v_{xx} = a^2 h''(x)$ implies

$$h(x) = \frac{\gamma}{a^2} x^2 + bx + c,$$

for some constants b, c. Then, by given boundary conditions, we obtain

$$v_x(0, t) = \alpha \quad \text{and} \quad v_x(\ell, t) = \beta, \quad \text{for } t > 0,$$

so that $b = \alpha$ and

$$\beta = v_x(\ell, t) = h'(\ell) = \frac{\gamma \ell}{a^2} + b \implies \gamma = \frac{(\beta - \alpha)a^2}{\ell}.$$

Therefore, with $c = 0$, an *equilibrium solution* is given by

$$v(x, t) = \frac{(\beta - \alpha)}{\ell} a^2 t + \frac{(\beta - \alpha)}{2\ell} x^2 + \alpha x. \tag{8.3.31}$$

Hence, as before, we can apply the method and Fourier series to solve the following homogeneous IBVP:

$$w_t(x, t) = a^2 w_{xx}(x, t), \quad 0 < x < \ell, \quad t > 0;$$
$$w_x(0, t) = \alpha, \quad \text{and} \quad w_x(\ell, t) = \beta, \quad \text{for } t > 0;$$
$$w(x, 0) = f(x) - v(x, 0) = f(x) - \frac{(\beta - \alpha)}{2\ell} x^2 - \alpha x.$$

Finally, a complete solution of the BVP is written as $u(x, t) = v(x, t) + w(x, t)$.

We consider next the case when a solution of heat equation (8.3.13) needs to satisfy the nonhomogeneous boundary conditions of the form

$$u(0, t) = T_0(t), \quad \text{and} \quad u(\ell, t) = T_\ell(t), \quad \text{for } t > 0, \tag{8.3.32}$$

and also an initial condition given by

$$u(x, 0) = f(x), \quad \text{for } 0 < x < \ell. \tag{8.3.33}$$

In this case, the *test function*[6]

$$v(x, t) = \frac{T_\ell(t) - T_0(t)}{\ell} x + T_0(t)$$

---

[6] The function $v = v(x, t)$ in this case may not represent an equilibrium solution.

satisfies the given nonhomogeneous boundary conditions, but it is not a solution of Eq. (8.3.13). In fact, we have

$$v_t - a^2 v_{xx} = 0 \implies v_t(x, t) = \frac{T'_\ell(t) - T'_0(t)}{\ell} x + T'_0(t) = 0.$$

If a function $w = w(x, t)$ gives *transient solution* of the heat equation so that the function

$$u(x, t) = v(x, t) + w(x, t)$$

provides the complete solution of Eq. (8.3.13), and satisfy the boundary conditions (8.3.32), then in view of the following

$$u_t - a^2 u_{xx} = v_t - a^2 v_{xx} + w_t - a^2 w_{xx}$$
$$= \frac{T'_\ell(t) - T'_0(t)}{\ell} x + T'_0(t) + w_t - a^2 w_{xx},$$

the function $w = w(x, t)$ must satisfy the *nonhomogeneous* differential equation

$$w_t(x, t) = a^2 w_{xx}(x, t) - \frac{T'_\ell(t) - T'_0(t)}{\ell} x, \tag{8.3.34}$$

and the following boundary and initial conditions:

$$w(0, t) = 0, \quad \text{and} \quad w(\ell, t) = 0, \quad \text{for } t > 0; \tag{8.3.35}$$
$$w(x, 0) = f(x) - v(x, 0), \quad \text{for } 0 < x < \ell. \tag{8.3.36}$$

Thus, in this case, our search for a complete solution leads to solving a homogeneous IBVP for a nonhomogeneous heat equation of the form

$$w_t - a^2 w_{xx} = f(x, t), \quad 0 < x < \ell, \ t > 0. \tag{8.3.37}$$

Recall that the function $f(x, t)$ here is interpreted as the source of thermal energy. The method of *eigenfunctions expansion* discussed in the next the chapter applies to find a solution of these type of IBVPs.

Laplace Equations
In next few worked examples, method of separation of variables is used to solve some important cases of the Laplace equation.

Example 8.20 We first consider the problem of determination of equilibrium temperature in an infinite horizontal plate, with insulated faces. That is, we solve the Laplace equation

$$\nabla^2 u = u_{xx} + u_{yy} = 0,$$

subject to boundary conditions $u(x, y) \to 0$ as $x \to \infty$, and $u(0, y) = \alpha \cos my$. For, let $u(x, y) = X(x)Y(y)$ so that $u_{xx} = X''Y$ and $u_{yy} = XY''$. With this substitution, the Laplace equation becomes

$$X''Y + XY'' = 0 \quad \text{or} \quad \frac{X''}{X} = -\frac{Y''}{Y}.$$

Writing $X''/X = k$ and $-Y''/Y = k$, we obtain two differential equations given by

$$X'' - kX = 0 \quad \text{and} \quad Y'' + kY = 0. \tag{8.3.38}$$

We consider the following three different possibilities for k.

Case-(i): If $k = 0$, the solutions of (8.3.38) are $X = Ax + B$ and $Y = Cy + D$. This shows that $u(x, y) = (Ax + B)(Cy + D)$ is a complete integral of the Laplace equation.

Case-(ii): If $k = p^2 > 0$ $(p > 0)$, the solutions of (8.3.38) are $X = Ae^{px} + Be^{-px}$ and $Y = C \cos py + D \sin py$. This shows that $u(x, y) = (Ae^{px} + Be^{-px})$ $(C \cos py + D \sin py)$ is a complete integral of the Laplace equation.

Case-(iii): If $k = -p^2 < 0$ $(p > 0)$, the solutions of (8.3.38) are $X = A \cos px + B \sin px$ and $Y = Ce^{py} + De^{-py}$. This shows that $u(x, y) = (A \cos px + B \sin px)$ $(Ce^{py} + De^{-py})$ is a complete integral of the Laplace equation.

So, in this case, a general solution of the Laplace equation is one of the following forms:

$$u(x, y) = (Ax + B)(Cy + D), \tag{8.3.39}$$
$$u(x, y) = (Ae^{px} + Be^{-px})(C \cos py + D \sin py), \tag{8.3.40}$$
$$u(x, y) = (A \cos px + B \sin px)(Ce^{py} + De^{-py}). \tag{8.3.41}$$

Since $u(x, y) \to 0$ as $x \to \infty$, if u is given by (8.3.39) then we must have $A = 0$. So, either $B = 0$ or $Cy + D = 0$, which gives $u(x, y) = 0$. But then, it does not satisfy the other condition. Similarly, if u is given by (8.3.41), then $u(x, y) = 0$. Thus, we are left with the possibility that $u = u(x, y)$ is as in (8.3.40). In this case, the condition $u(x, y) \to 0$ as $x \to \infty$ implies that $A = 0$, and so we have

$$u(x, y) = Be^{-px}(C \cos py + D \sin py).$$

Then, by condition $u(0, y) = \alpha \cos my$, we obtain

$$\alpha \cos my = B(C \cos py + D \sin py),$$

which gives $D = 0$, $BC = \alpha$, and $p = m$. Therefore, the general solution of the BVP is given by the functin

$$u(x, y) = \alpha e^{-mx} \cos m y.$$

Example 8.21 Suppose the boundary conditions are given by

$$u(0, y) = u(\ell, y) = 0, \quad u(x, 0) = 0, \quad u(x, b) = \sin(n\pi x/a).$$

In general, u may have one of the forms given in equations (8.3.39)-(8.3.41). By an argument as used earlier, it follows that u must be of the form (8.3.41). That is,

$$u(x, y) = (A \cos px + B \sin px)(Ce^{py} + De^{-py}).$$

Then, by using boundary condition $u(0, y) = 0$, we conclude $A = 0$. Therefore,

$$u(x, y) = \sin px(Ce^{py} + De^{-py}),$$

where B is taken to be part of the coefficients C and D. Next,

$$u(\ell, y) = 0 \quad \Rightarrow \quad \sin p\ell(Ce^{py} + De^{-py}) = 0,$$

so that we obtain $p\ell = n\pi$, for some integer n. Hence, we have

$$u(x, y) = \sin(n\pi x/\ell)(Ce^{n\pi y/\ell} + De^{-n\pi y/\ell}).$$

Further, by using the condition $u(x, 0) = 0$, it follows that $C + D = 0$. That is,

$$u(x, y) = C \sin(n\pi x/\ell)(e^{n\pi y/\ell} - e^{-n\pi y/\ell})$$
$$= 2C \sin(n\pi x/\ell) \cosh(n\pi y/\ell).$$

Finally, by using the condition $u(x, b) = \sin(n\pi x/\ell)$, we obtain

$$\sin(n\pi x/\ell) = 2C \sin(n\pi x/\ell) \cosh(n\pi b/\ell),$$

which gives $2C = 1/\cosh(n\pi b/\ell)$. The complete solution of the boundary value problem is obtained as

$$u(x, y) = \frac{\sin(n\pi x/\ell) \cosh(n\pi y/\ell)}{\cosh(n\pi b/\ell)}.$$

Example 8.22 Suppose a rectangular plate has vertices at the points $(0, 0)$, $(\ell, 0)$, $(0, h)$, and (ℓ, h). We assume that the equilibrium temperature $u = u(x, y)$ satisfy the boundary conditions

$$u(x, 0) = u(x, h) = 0, \quad u(0, y) = 0, \quad \text{and} \quad u(\ell, y) = f(y),$$

where f is a continuous function satisfying Dirichlet's conditions. As before, Eqs. (8.3.39)–(8.3.41)) provide the complete integral of the Laplace equation. First, consider $u(x, y) = (Ax + B)(Cy + D)$. The condition $u(x, 0) = 0$ gives $Ax + B = 0$ or $D = 0$. In the first case, we get a trivial solution $u(x, y) = 0$. If $D = 0$, then u is given by $u(x, y) = (Ax + B)Cy$. The condition $u(x, h) = 0$ gives $Ax + B = 0$ or $C = 0$. Both conditions lead to trivial solutions. The trivial solution does not satisfy the boundary condition $u(\ell, y) = f(y)$.

If $u(x, y) = (A \cos px + B \sin px)(Ce^{py} + De^{-py})$, the condition $u(x, 0) = 0$ gives $C + D = 0$ and the condition $u(x, h) = 0$ gives $Ce^{ph} + De^{-ph} = 0$. This shows that $C = D = 0$ and again u is a trivial solution. Next, if

$$u(x, y) = (Ae^{px} + Be^{-px})(C \cos py + D \sin py),$$

then the boundary condition $u(x, 0) = 0$ shows that either u is a trivial solution or $C = 0$. In the latter case, u takes the form

$$u(x, y) = (Ae^{px} + Be^{-px}) \sin py.$$

Using the condition $u(x, h) = 0$, we see that either u must be trivial or $\sin ph = 0$. In the latter case, we have $p = n\pi/h$ where n is an integer. The condition $u(0, y) = 0$ gives $A + B = 0$ and so u must be of the form

$$u(x, y) = (Ae^{n\pi x/h} - Be^{-n\pi x/h}) \sin(n\pi y/h) = 2A \sinh(n\pi x/h) \sin(n\pi y/h)$$

or, for each integer n,

$$u_n(x, y) = a_n \sinh(n\pi x/h) \sin(n\pi y/h)$$

is the solution of the Laplace equation satisfying the boundary conditions $u(x, 0) = u(x, b) = 0$, $u(0, y) = 0$. The numbers $p = n\pi/h$ are called the *eigenvalues* of the problem and the functions $u_n(x, y)$ are the corresponding *eigenfunctions*.

By extending the *principle of superposition* for infinite sum, we see that $u(x, y)$ given by the following *eigenfunction expansion*

$$u(x, y) = \sum_{n=1}^{\infty} u_n(x, y) = \sum_{n=1}^{\infty} a_n \sinh(n\pi x/h) \sin(n\pi y/h)$$

satisfy the Laplace equation and the boundary conditions $u(x, 0) = u(x, h) = 0$, $u(0, y) = 0$. Using $u(\ell, y) = f(y)$, we get

$$f(y) = \sum_{n=1}^{\infty} a_n \sinh(n\pi \ell/h) \sin(n\pi y/h).$$

The Fourier coefficients $a_n \sinh(n\pi \ell/h)$ are given by

$$a_n \sinh(n\pi \ell/h) = \frac{2}{h} \int_0^h f(y) \sin(n\pi y/h) dy.$$

Therefore, the function u given by

$$u(x, y) = \sum_{n=1}^{\infty} a_n \sinh(n\pi x/h) \sin(n\pi y/h),$$

with a_n given by

$$a_n = \frac{2}{h \sinh(n\pi \ell/h)} \int_0^h f(y) \sin(n\pi y/h) dy$$

is the solution to our problem.

Example 8.23 Here we consider steady-state case of a 2 - dimensional heat equation for a plate bounded by the lines $x = 0$, $x = \pi$ and $y = 0$, $y = \pi$. Let the edge $y = 0$ be maintained at a temperature $u(x, 0) = f(x)$, and all other edges be maintained at zero temperature. So, the problem is to solve the Laplace equation subject to the conditions

$$u(0, y) = u(\pi, y) = u(x, \pi) = 0, \quad \text{and} \quad u(x, 0) = f(x).$$

As before, the complete integral of the Laplace equation by separation of variables is given by

$$u_n(x, y) = \frac{\sinh n(y - \pi)}{\cosh(n\pi)} \sin(nx), \quad n \in \mathbb{N}.$$

Therefore, by principle of superposition, the function

$$u(x, y) = \sum_{n=1}^{\infty} u_n(x, y) = \sum_{n=1}^{\infty} a_n \frac{\sinh n(y - \pi)}{\sinh(n\pi)} \sin(nx).$$

is also a solution of the Laplace equation satisfying the boundary conditions $u(0, y) = u(\pi, y) = u(x, \pi) = 0$.

We now determine coefficients a_n so that the function $u(x, y)$ satisfies the *Dirichlet condition* $u(x, 0) = f(x)$. For, using Fourier theorem, the constants a_n are given by

$$a_n \left(\frac{-\sinh(n\pi)}{\cosh(n\pi)} \right) = \frac{2}{\pi} \int_0^\pi f(x) \sin nx \, dx$$

$$\implies a_n = -\frac{2\cosh(n\pi)}{\pi \sinh(n\pi)} \int_0^\pi f(x) \sin nx \, dx$$

It thus follows that the complete solution of the boundary value problem is given by

$$u(x, y) = \frac{2}{\pi} \sum_{n=1}^\infty \int_0^\pi f(\xi) \sin(n\xi) \frac{\sinh n(y - \pi)}{\sinh(n\pi)} \sin(nx) \, d\xi.$$

Example 8.24 We solve here BVP formulated for the Laplace equation in Example 5.24. For simplicity, suppose the edge $y = 0$ is maintained at a constant temperature $u = u_0 \neq 0$. The infinite plate is bounded by the lines $x = 0$, $x = \pi$ and $y = 0$, where the two vertical edges are maintained at zero temperature. Assuming that $u(x, y) \to 0$ as $y \to \infty$, we find the equilibrium temperature of the plate at point (x, y). Thus, we solve the Laplace equation, together with the boundary conditions given by

$$u(0, y) = u(\pi, y) = 0, \quad \text{and} \quad u(x, 0) = u_0.$$

As before, the general solutions of the Laplace equation are given by (8.3.39)-(8.3.41). If

$$u(x, y) = (Ax + B)(Cy + D),$$

then by condition $u(0, y) = 0$, we have $B = 0$ or $Cy + D = 0$. The second case leads to the trivial solution $u(x, y) = 0$, which does not satisfy the boundary condition $u(x, 0) = u_0$. . And, if $B = 0$ then $u(x, y) = Ax(Cy + D)$, so the boundary condition $u(\pi, y) = 0$ gives $A = 0$ or $Cy + D = 0$. Both of these lead to the trivial solutions. Next, if

$$u(x, y) = (Ae^{px} + Be^{-px})(C \cos py + D \sin py),$$

then the boundary condition $u(0, y) = 0$ implies $A + B = 0$ or $C \cos py + D \sin py = 0$. The latter condition leads to the trivial solution. Hence, we must have

$$u(x, y) = A(e^{px} - e^{-px})(C \cos py + D \sin py).$$

In this case, the boundary condition $u(\pi, y) = 0$ shows $A = 0$ or $C \cos py + D \sin py = 0$. Both lead to the trivial solution. Finally, if

$$u(x, y) = (A \cos px + B \sin px)(Ce^{py} + De^{-py}),$$

then the condition $u(x, y) \to 0$ as $y \to \infty$ shows that $C = 0$. Hence,

$$u(x, y) = (A \cos px + B \sin px)e^{-py}.$$

The condition $u(0, y) = 0$ gives $A = 0$. Thus, $u(x, y) = Be^{-py} \sin px$. The boundary condition $u(\pi, y) = 0$ shows that $\sin p\pi = 0$ or p is an integer n. Thus, for each integer n, the function

$$u_n(x, y) = a_n e^{-ny} \sin nx$$

is a solution of the Laplace equation satisfying $u(0, y) = u(\pi, y) = 0$, and $u(x, y) \to 0$ as $y \to \infty$. Therefore,

$$u(x, y) = \sum_{n=1}^{\infty} u_n(x, y) = \sum_{n=1}^{\infty} a_n e^{-ny} \sin nx$$

is also a solution of the Laplace equation satisfying $u(0, y) = u(\pi, y) = 0$, and $u(x, y) \to 0$ as $y \to \infty$. We now determine a_n so that the condition $u(x, 0) = u_0$ is satisfied. This condition gives $u_0 = \sum_{n=1}^{\infty} a_n \sin nx$. Therefore, the constants a_n are given by

$$a_n = \frac{2}{\pi} \int_0^{\pi} u_0 \sin nx dx = \frac{2u_0}{n\pi}(1 - \cos n\pi) = \begin{cases} \frac{4u_0}{n\pi} & (n \text{ is odd}) \\ 0 & (n \text{ is even.}) \end{cases}$$

Using this, we see that

$$u(x, y) = \sum_{n=1}^{\infty} u_n(x, y) = \frac{4u_0}{\pi} \sum_{n=1}^{\infty} \frac{\sin(2n-1)x}{2n-1} e^{-(2n-1)y}$$

is the complete solution of the given BVP.

Helmholtz Wave Equation

The *stationary wave* equation

$$\nabla^2 u + k^2 u = 0, \ \ k \geq 0, \tag{8.3.42}$$

is known as the *Helmholtz wave equation*. See Appendix A.2 for details. Since $k = 0$ gives the Laplace equation, we may assume that $k > 0$.

To find the general solution of Helmholtz wave equation by method of separation of variables, let $u(x, y) = X(x)Y(y)$ so that $u_{xx} = X''Y$ and $u_{yy} = XY''$. With this substitution, Eq. (8.3.42) becomes

$$X''Y + XY'' + k^2 XY = 0 \ \ \text{or} \ \ \frac{X''}{X} = -\frac{Y''}{Y} - k^2.$$

Writing $X''/X = \ell$ and $Y''/Y + k^2 = -\ell$, where ℓ is the separation constant, we get

$$X'' - \ell X = 0 \quad \text{and} \quad Y'' + (k^2 + \ell)Y = 0. \tag{8.3.43}$$

As before, we consider the following three different possibilities for ℓ.

Case-(i): If $\ell = 0$, the solutions of (8.3.43) are $X = Ax + B$ and $Y = C \cos ky + D \sin ky$. This shows that

$$u(x, y) = (Ax + B)(C \cos ky + D \sin ky)$$

is a complete integral of the Helmholtz equation.

Case-(ii): If $\ell = p^2 > 0 \, (p > 0)$, the solutions of (8.3.43) are $X = Ae^{px} + Be^{-px}$ and $Y = C \cos(\sqrt{k^2 + p^2}y) + D \sin(\sqrt{k^2 + p^2}y)$. This shows that

$$u(x, y) = (Ae^{px} + Be^{-px})(C \cos(\sqrt{k^2 + p^2}y) + D \sin(\sqrt{k^2 + p^2}y))$$

is a complete integral of the Helmholtz equation.

Case-(iii): If $\ell = -p^2 < 0 \, (p > 0)$, we need to consider three subcases: $k = -\ell^2$, $k < -\ell^2$ and $k > -\ell^2$. In each of these three cases, the solution, respectively, is given by

$$u(x, y) = (A \cos px + B \sin px)(Cy + D),$$

$$u(x, y) = (A \cos px + B \sin px) \left(Ce^{\sqrt{k^2 + p^2}y} + De^{-\sqrt{k^2 + p^2}y} \right)$$

$$u(x, y) = (A \cos px + B \sin px)(C \cos(\sqrt{k^2 + p^2}y) + D \sin(\sqrt{k^2 + p^2}y)).$$

Notice that Eq. (8.3.42) written in the form $\nabla^2 u = -k^2 u$ implies that a solution $u = u(x, y)$ of the Helmholtz wave equation is an *eigenfunction* of the Laplace operator ∇^2, with *eigenvalue* $-k^2$. We will pursue this approach in the last section of the chapter to discuss an alternative method of finding a solution $u = u(x, y)$ of the Helmholtz wave equation.

Exercises 8

8.1 Suppose a (stationary) sine wave associated with the displacement $y = y(x, t)$ of a vibrating string at time $t > 0$ has amplitude α, frequency $\nu = 2\pi\omega$, and the wavelength $\lambda = 2\pi/k$, so that we can write the function $y(x, t)$ as

$$y(x, t) = \alpha \sin (kx - \omega t). \tag{8.3.44}$$

Give a mathematical explanation to prove that a high-pitch sound is produced when a low-density string is stretched tightly, and a low-pitch sound is produced when the density of string material is high.

8.2 Supply a proof of Theorem 8.1.

8.3 Compute the Fourier series of the *half-sine wave* function given by

$$f(t) = \begin{cases} 0, & -\pi < t < 0 \\ \sin t, & 0 < t < \pi \end{cases}$$

8.4 Obtain the Fourier series expansion of $f(x) = \cos ax$ in the interval $(-\pi, \pi)$.

8.5 Compute the Fourier series of the function $f : [-\pi, \pi] \to \mathbb{R}$ given by $f(x) = x \cos x$, with $f(x + 2\pi) = f(x)$ for all $x \in \mathbb{R}$.

8.6 Expand $f(x)$ in a Fourier series where

$$f(x) = \begin{cases} -\pi & -\pi < x < 0 \\ +x & 0 < x < \pi. \end{cases}$$

Also find the sum of the Fourier series at the point of discontinuity.

8.7 Compute the Fourier series of the function $f : [-2, 2] \to \mathbb{R}$ given by $f(x) = e^{-x}$, viewed in natural manner as a 4-periodic function.

8.8 Compute Fourier series of the function $f(x) = x \sin x$, $x \in [0, 2\pi]$, with $f(x + 2\ell) = f(x)$.

8.9 Compute the complex Fourier series of the *square wave* given by

$$f_\ell(t) = \begin{cases} 0, & \text{for } -\ell < t < -1 \\ 1, & \text{for } -1 < t < 1 \\ 0, & \text{for } 1 < t < \ell \end{cases}, \quad \text{with } f_\ell(x + 2\ell) = f_\ell(x), \ \forall \, x \in \mathbb{R}.$$

8.10 Compute the complex Fourier series of the function $f(x) = e^{-|x|}$, $x \in [-\ell, \ell]$, with $f(x + 2\ell) = f(x)$ for $x \in \mathbb{R}$.

8.11 For $a \in \mathbb{R}$, compute the complex Fourier series of the 2π-periodic function $f(x) = e^{-iax}$, $x \in (-\pi, \pi]$. What if $a \in \mathbb{Z}$?

8.12 Compute the Fourier series of the function $f : [0, 2\pi] \to \mathbb{R}$ given by $f(x) = e^{-x}$, with $f(x + 2\pi) = f(x)$ for all $x \in \mathbb{R}$.

8.13 Express the function $f(x) = \dfrac{(\pi - x)^2}{4}$ as a Fourier series over the interval $(0, 2\pi)$. Hence, find the sum of the infinite series $\sum_0^\infty \frac{1}{n^2}$.

8.14 Express the function $f(x) = 2x - x^2$, of period 3 over the interval $(0, 3)$, as a Fourier series. What if the above function is of period 2 over the interval $(0, 2)$?

8.15 Discuss the pointwise, uniform, and also L^2-convergence of the Fourier sine series for each functions as given below: (a) $f(x) = \pi x - x^2$; (b) $g(x) = x^2 + 1$; (c) $h(x) = x^3$, for $x \in [0, \pi]$.

8.16 Discuss Gibbs' phenomenon using Fourier series of *sawtooth function* as obtained earlier in Example 8.6.

8.17 Verify Lemma 8.1 for (i) $f = \chi_{(a,b)}$, with $(a, b) \subset [-\pi, \pi]$; (ii) $f = \sum_{k=1}^{n} \chi_{(a_k, b_k)}$; and, (iii) $f = \sum_{k=1}^{n} \chi_{M_k}$, where each $M_k \subset (-\pi, \pi)$ is a measurable set.

8.18 Let $f : [-\pi, \pi] \to \mathbb{R}$, with $f(t + 2\pi) = f(t)$ for all $t \in \mathbb{R}$, be an absolutely integrable function such that for some $\varepsilon > 0$, $f(t) = 0$, for all $t \in (t_0 - \varepsilon, t_0 + \varepsilon)$. Prove that $S(f)(t_0) = 0$. Hence deduce that if another such function $g : [-\pi, \pi] \to \mathbb{R}$ such that $f(t) = g(t)$, for all $t \in (t_0 - \varepsilon, t_0 + \varepsilon)$, then either both limits do not exist or $S(f)(t_0) = S(g)(t_0)$, if any one of these exists.

8.19 Prove the assertion (8.2.46).

8.20 Prove the assertion (8.2.68) for a function $f \in PC(\mathbb{T})$.

8.21 Using Fourier series of a suitable function, evaluate the infinite sum as given below:

$$\sum_{n=1}^{\infty} \frac{1}{1 + n^2} = \frac{1}{2} + \frac{1}{5} + \frac{1}{10} + \cdots .$$

8.22 Solve the homogeneous initial-boundary value problem:

$$y_{tt} - 4 y_{xx} = 0, \quad \text{for } 0 < x < 2 \text{ and } t \geq 0;$$
$$y(0, t) = 0 = y(2, t), \quad \text{for } t > 0;$$
$$y(x, 0) = \sin \pi x, \quad y_t(x, 0) = \cos \pi x$$

8.23 Solve the homogeneous initial-boundary value problem:

$$u_t - k u_{xx} = 0, \quad \text{for } 0 < x < \ell \text{ and } t > 0;$$
$$u(0, t) = 0 = u(\ell, t), \quad \text{for } t > 0;$$
$$u(x, 0) = 3 \sin(\pi x/\ell) - \sin(3\pi x/\ell)$$

8.24 Solve the nonhomogeneous initial-boundary value problem:

$$u_t - k u_{xx} = \alpha u(x, t), \quad \text{for } 0 < x < \ell \text{ and } t > 0;$$
$$u(0, t) = 0 = u(\ell, t), \quad \text{for } t > 0;$$
$$u(x, 0) = f(x), \quad \text{with } \alpha > 0.$$

8.25 Solve the homogeneous initial-boundary value problem:

$$u_{tt} - c^2 u_{xx} = 0, \quad \text{for } 0 < x < \ell \text{ and } t > 0;$$
$$u(0, t) = 0 = u(\ell, t), \quad \text{for } t > 0;$$
$$u(x, 0) = f(x), \quad \text{and } u_t(x, 0) = h(x), \quad \text{for } 0 < x < \ell.$$

8.26 Solve the homogeneous initial-boundary value problem:

$$u_{tt} - u_{xx} = 0, \quad \text{for } 0 < x < \pi \text{ and } t > 0;$$
$$u(0, t) = 0 = u(\ell, t), \quad \text{for } t > 0;$$
$$u(x, 0) = 1, \quad \text{and } u_t(x, 0) = 0, \quad \text{for } 0 < x < \pi.$$

8.27 Solve the homogeneous initial-boundary value problem:

$$u_{tt} - c^2 u_{xx} = 0, \quad \text{for } 0 < x < \ell \text{ and } t > 0;$$
$$u_x(0, t) = 0 = u_x(\ell, t), \quad \text{for } t > 0;$$
$$u(x, 0) = f(x), \quad \text{and} \quad u_t(x, 0) = h(x), \quad \text{for } 0 < x < \ell.$$

References

1. Cain G, Meyer GH (2006) Separation of variables for partial differential equations—an Eigen function approach. Chapman & Hall/CRC
2. Strauss, W. A. (2007). *Partial differential equations. An introduction*. London: Wiley.
3. Haberman, R. (1987). *Elementary applied partial differential equations* (Vol. 2). Engelwood Cliffs: Prentice Hall.
4. Pivato M (2010) Linear partial differential equations and Fourier theory. Cambridge University Press, Cambridge
5. Dym, H., & McKean, H. P. (1972). *Fourier series and integrals*. New York: Academic.
6. Körner TW (1988) Fourier analysis. Cambridge University Press, Cambridge
7. Fourier J (1955) The analytical theory of heat (trans. by A. Freeman). Dover, New York
8. Champeney, D. C. (1987). *A handbook of Fourier theorems*. New York: Cambridge University Press.

Chapter 9
Method of Eigenfunctions Expansion

He who seeks for methods without having a definite problem in mind seeks in the most part in vain.

David Hilbert (1862–1943)

As discussed in the previous chapter, the method of *separation of variables* has been found quite effective in solving some homogeneous partial differential equations with homogeneous boundary conditions. Recall that the method provides a *series solution* of the given problem in terms of eigenfunctions of some boundary value problem for ordinary differential equation, and the involved coefficients are determined by using Fourier series of the function given by the assigned initial condition. Further, for some nonhomogeneous problems, it is possible to use the same procedure by transforming the original problem to a feasible boundary value problem. However, the method fails for many nonhomogeneous boundary value problems related to some important linear partial differential equations.

In most such situations, we can use eigenvalues and eigenfunctions of the differential operator $\mathscr{D} = D^2 + \lambda$ to solve the associated boundary value problem. Therefore, we are led to use Sturm–Liouville theory to obtain an approximate solution of the problem expressed as *eigenfunctions expansion*. More precisely, Sturm–Liouville theory provides requisite foundation to use the fact that a *generalised Fourier series* of the form

$$\alpha_1 \phi_1(x) + \alpha_2 \phi_2(x) + \alpha_3 \phi_3(x) + \cdots$$

exists for a functions f satisfying certain special type of boundary value problem for a second order homogeneous ordinary differential equation, with mixed type of homogeneous boundary conditions, known as the *Sturm-Liouville equations* (Theorem 3.17). The functions $\phi_n(x)$ are orthogonal in the sense of Definition 3.22, and the coefficients α_n characterise the function f.

© The Author(s), under exclusive license to Springer Nature Singapore Pte Ltd. 2022   403
A. K. Razdan and V. Ravichandran, *Fundamentals of Partial Differential Equations*,
https://doi.org/10.1007/978-981-16-9865-1_9

The central theme of the present chapter is to apply *generalised Fourier series*, as offered by Sturm–Liouville theory, to solve some important types of boundary value problems. The underlying procedure is known as the *method of eigenfunctions expansion*. The main source for the content presented here are the standard texts [1–3].

9.1   Generalised Fourier Series

As before, we consider linear partial differential equations in two independent variables, where both in certain situations may be the space variables such as in potential problems. Most problems considered in this chapter have the same basic equation:

$$\mathscr{L}\big[u(x,t)\big] = f(x,t), \quad \text{for } x \in (0,\ell] \text{ and } 0 < t \le T, \tag{9.1.1}$$

where $\mathscr{L}$ a linear partial differential operator, possibly with variable coefficients. We shall first discuss the general approach and, subsequently, examine how the same applies to solve boundary value problems related to

$$[\textbf{Heat Equation}] \quad \mathscr{L}\big[u(x,t)\big] = u_{xx} - \frac{1}{k}u_t = f(x,t);$$

$$[\textbf{Wave Equation}] \quad \mathscr{L}\big[u(x,t)\big] = u_{xx} - \frac{1}{c^2}u_{tt} = f(x,t);$$

$$[\textbf{Poisson's Equation}] \quad \mathscr{L}\big[u(x,t)\big] = u_{xx} + u_{yy} = f(x,y).$$

The last equation is also known as the *potential equation* that is characteristic feature of eigenfunctions expansion generalisation of the separation of variables as needed to accommodate the *source terms*. Recall that, as found in the previous chapter, the *series solution* of homogeneous wave equation

$$y_{xx} - \frac{1}{c^2}y_{tt} = 0, \quad \text{for } 0 < x < \ell \text{ and } t > 0,$$

subject to the *homogeneous* boundary conditions

$$y(0,t) = y(\ell,t) = 0, \quad \text{for } t \ge 0,$$

can be expressed as

$$y(x,t) = \sum_{n=1}^{\infty} \alpha_n(t)\phi_n(x), \tag{9.1.2}$$

where the eigenfunctions

$$\phi_n(x) = \sin(\lambda_n x), \quad \text{for } x \in [0, \ell],$$

are solutions of the second order homogeneous BVP given by

$$X'' + \lambda^2 X = 0;$$
$$X(0) = X(\ell) = 0,$$

and the coefficients $\alpha_n(t)$ are given by

$$\alpha_n(t) = C_n \cos(\lambda_n t) + D_n \sin(\lambda_n t), \quad \text{for } t \geq 0.$$

The above coefficients are obtained by using the assigned initial conditions and Fourier series techniques. We also know that the set of functions

$$\left\{ \phi_n \in C^\infty[-\ell, \ell] \mid n \geq 1 \right\}$$

forms an *orthogonal system*.

Definition 9.1 A solution of the form (9.1.2) is called an *eigenfunctions expansion*, where $\phi_n(x)$ are *eigenfunctions* associated with the *eigenvalues*

$$\lambda_n = \omega_n c, \quad \text{with } \omega_n = \frac{n\pi}{\ell}, \quad n \geq 1,$$

that define the *spectrum* of the above initial-boundary value problem.

Remark 9.1 Recall that we used the above approach in previous chapter to solve many different types of homogeneous initial-boundary value problems. In practical terms, the eigenfunctions $\phi_n(x)$ represent the nth *normal mode* having the frequency $\lambda_n/(2\pi) = (nc)/2\ell$ cycles per unit time. The mode for $n = 1$ is called the *fundamental mode*, and the rest are known as the *overtones*. For $c = 1$, using relations $\sin(\omega_n x) = 0$ at $x = \ell, 2\ell, \ldots$, it follows from Eq. (9.1.2) that the nth mode has $(n - 1)$ points (*nodes*) on the string remain fixed all through the motion.

As said earlier, separation of variables fails when the usual way of transformation yields a *nonhomogeneous* initial-boundary value problem, with a *source term*. More generally, for $0 < x < \ell$ and $t > 0$, we may consider nonhomogeneous initial-boundary value problem for heat equations as given below:

$$u_{xx} - \frac{1}{c^2} u_t = f(x, t); \tag{9.1.3}$$

$$a_1 u(0, t) + b_1 u_x(0, t) = 0; \tag{9.1.4}$$

$$a_2 u(\ell, t) + b_2 u_x(\ell, t) = 0; \tag{9.1.5}$$

$$u(x, 0) = g(x), \tag{9.1.6}$$

and suppose a solution $u(x, t)$ is given by a *generalised Fourier series* of the form

$$u(x, t) = \sum_{n=1}^{\infty} \alpha_n(t)\, \phi_n(x), \tag{9.1.7}$$

where $\phi_n(x)$ are the orthogonal eigenfunctions obtained by solving the associated homogeneous initial-boundary value problem

$$u_{xx} - \frac{1}{c^2}\, u_t = 0, \quad 0 < x < \ell, \quad t > 0; \tag{9.1.8}$$

$$a_1 u(0, t) + b_1 u_x(0, t) = 0; \tag{9.1.9}$$

$$a_2 u(\ell, t) + b_2 u_x(\ell, t) = 0; \tag{9.1.10}$$

$$u(x, 0) = g(x). \tag{9.1.11}$$

Furthermore, suppose the *generalised* Fourier coefficients $\alpha_n(t)$ are found by solving certain homogeneous ordinary differential equations, using the given initial condition $u(x, 0) = g(x)$. Now, we know that the solution of the differential equation (9.1.8), satisfying the boundary conditions (9.1.9) and (9.1.10), is given by

$$u(x, t) = \sum_{n=1}^{\infty} A_n e^{-\lambda_n^2 a^2 t} \phi_n(x),$$

where λ_n and $\phi_n(x)$ are eigenvalues and eigenfunctions of the SL-BVP:

$$X'' + \lambda^2 X = 0, \quad 0 < x < \ell, \quad t > 0;$$
$$a_1 X(0) + b_1 X'(0) = 0; \quad a_2 X(\ell) + b_2 X'(\ell) = 0.$$

Notice that, in the absence of a source term, it is expected that the temperature $u(x, t)$ decreases over a period of time, and so appearance of the damping term $e^{-\lambda_n^2 a^2 t}$ in last equation makes sense. However, when there is a source of heat given by continuous *time-dependent* function $f(x, t)$, the temperature $u(x, t)$ would not decrease with time, and so our solution must contain a term different from the function $e^{-\lambda_n^2 a^2 t}$.

9.2   Nonhomogeneous Boundary Value Problems

Recall that, while attempting to solving a boundary value problem for the homogeneous heat equation (9.1.3) with nonhomogeneous boundary conditions of the form

$$u(0, t) = T_0(t) \quad \text{and} \quad u(\ell, t) = T_\ell(t), \quad \text{for } t > 0, \tag{9.2.1}$$

we are led to consider an initial-boundary value problem of the form

$$w_t - a^2 w_{xx} = f(x, t), \quad \text{for } 0 < x < \ell, \quad t > 0;$$
$$w(0, t) = 0 \quad \text{and} \quad w(\ell, t) = 0, \quad \text{for } t > 0; \tag{9.2.2}$$
$$w(x, 0) = g(x), \quad \text{for } 0 < x < \ell$$

The time-dependent function $f(x, t)$ in above problem is interpreted as the *source term*. We discuss here application of the method of *eigenfunctions expansion* in solving such types of initial-boundary value problems.

We first recall that the solutions of the homogeneous differential equation associated with Eq. (9.2.2) are the *orthogonal eigenfunctions* given by

$$w_n(x, t) = \sin(\lambda_n x) e^{-\lambda_n^2 a^2 t},$$

for the eigenvalues $\lambda_n = n\pi/\ell$, $n = 1, 2, \ldots$. In general, to solve the nonhomogeneous equation of the form (9.2.2) by the method of eigenfunction expansion, we use Theorem 3.17 to write the unknown solution $w(x, t)$ as the series

$$w(x, t) = \sum_{n=1}^{\infty} \alpha_n(t) \phi_n(x) \tag{9.2.3}$$

in terms of the orthogonal eigenfunctions obtained as solutions of the associated homogeneous BVP, where the functions $\alpha_n(t)$ are the generalised Fourier coefficients. Note that the function $w(x, t)$ (for a fixed t) is a piecewise smooth function, and so it admits an eigenfunction expansion. For example, the eigenfunction

$$\phi_n(x) = \sin\left(\frac{n\pi x}{\ell}\right)$$

defines the generalised Fourier series expansion of the function $w_n(x, t)$, with the generalised Fourier coefficients given by

$$\alpha_n(t) = e^{-\lambda_n^2 t}, \quad \text{with } \lambda_n = n\pi/\ell, \quad n = 1, 2, \ldots.$$

In view of our wave equation-related discussion above, it follows that the Fourier coefficients $\alpha_n(t)$ of a solution $u(x, t)$ in general may not be of the exponential form.

Indeed, construction of the series (9.2.3) implies that the function $w(x, t)$ satisfies the homogeneous boundary conditions. And, it would satisfy the initial condition if

$$g(x) = w(x, 0) = \sum_{n=1}^{\infty} \alpha_n(0) \phi_n(x).$$

But then, by the second assertion of Theorem 3.17, the generalised Fourier coefficients in the last equation must be given by the relation

$$\alpha_n(0) = \frac{\int_0^\ell g(x)\phi_n(x)\mathrm{d}x}{\int_0^\ell [\phi_n(x)]^2\mathrm{d}x}, \tag{9.2.4}$$

because the weight function $r(x) = 1$ in this case. It thus remains to find suitable coefficients $\alpha_n(t)$ so that the function $w(x, t)$ in Eq. (9.2.3) solves the nonhomogeneous equation (9.2.2). To do so, we shall use here the method of *direct substitution*. For, using Eq. (9.2.3), we have

$$w_t(x, t) = \sum_{n=1}^\infty \alpha_n'(t)\phi_n(x); \quad \text{and,}$$

$$w_{xx}(x, t) = \sum_{n=1}^\infty \alpha_n(t)\phi_n''(x) = -\sum_{n=1}^\infty \lambda_n\alpha_n(t)\phi_n(x).$$

Substituting these in Eq. (9.2.2), we obtain

$$\sum_{n=1}^\infty [\alpha_n'(t) + a^2\lambda_n\alpha_n(t)]\phi_n(x) = f(x, t). \tag{9.2.5}$$

Since the series on the left is a generalised Fourier series expansion of the function $f(x, t)$, and eigenfunctions $\phi_n(x)$ are orthogonal, the second assertion of Theorem 3.17 provides the following first order ODE for the determination of the generalised Fourier coefficients $\alpha_n(t)$:

$$\alpha_n'(t) + a^2\lambda_n\alpha_n(t) = \frac{\int_0^\ell f(x, t)\phi_n(x)\mathrm{d}x}{\int_0^\ell [\phi_n(x)]^2\mathrm{d}x} = f_n(t), \tag{9.2.6}$$

where the functions $f_n(t)$, as before, are the Fourier coefficients of the function $f(x, t)$ i.e.

$$f(x, t) = \sum_{n=1}^\infty f_n(t)\phi_n(x). \tag{9.2.7}$$

Observe that the initial condition needed for the complete solution of ODE (9.2.6) is provided by the generalised Fourier coefficient as in (9.2.4). Finally, using the integrating factor method, the complete solution of ODE (9.2.6) is given by

$$\alpha_n(t) = \mathrm{e}^{-\lambda_n a^2 t} \int_0^t f_n(t)\mathrm{e}^{\lambda_n a^2 t}\mathrm{d}t + \alpha_n(0)\mathrm{e}^{-\lambda_n a^2 t}. \tag{9.2.8}$$

Note that, when $f(x, t) = 0$, this procedure is exactly the same as the method of separation of variables.

The next two examples illustrate the procedure for the method of direct substitution. We may start with a nonhomogeneous differential equation, with homogeneous boundary conditions.

Example 9.1 We solve the nonhomogeneous equation

$$u_t - 4u_{xx} = t \sin x, \quad 0 < x < 2, \quad t > 0,$$

satisfying the homogeneous boundary conditions, with $u(x, 0) = g(x)$. As explained above, the general idea is to write the source function $f(x, t) = t \sin x$ as

$$t \sin x = \sum_{n=1}^{\infty} f_n(t)\phi_n(x),$$

and find the nth response function $u_n(x, t) = T_n(t)\phi_n(x)$, for each one of the component function $f_n(t)\phi_n(x)$, $n \geq 1$. First, we obtain the eigenfunctions $\phi_n(x)$ as solution of the associated homogeneous BVP by separation of variables. We already know that $\phi_n(x) = \sin(n\pi x/2)$ for the eigenvalue $\lambda_n = n\pi$, $n = 1, 2, \ldots$. It then follows that

$$t \sin x = \sum_{n=1}^{\infty} f_n(t) \sin(n\pi x/2),$$

Next, we use orthogonality of the functions $\phi_n(x)$, it follows that

$$\int_0^2 t \sin x \sin(m\pi x/2) dx = \sum_{n=1}^{\infty} f_n(t) \int_0^2 \sin(m\pi x/2) \sin(n\pi x/2) dx$$
$$= \frac{1}{2} f_m(t),$$

which implies that

$$f_n(t) = 2 \int_0^2 t \sin x \sin(n\pi x/2) dx = \begin{cases} , & \text{for } n = 2/\pi \\ , & \text{for } n \neq 2/\pi \end{cases}.$$

Next, in order to find the solution

$$u(x, t) = \sum_{n=1}^{\infty} u_n(x, t) = \sum_{n=1}^{\infty} T_n(t) \sin(n\pi x/2),$$

we replace the source function $f(x, t) = t \sin x$ in the given equation by the series expansion obtained above. It then follows from the discussion above that the functions $T_n(t)$ must satisfy the nonhomogeneous initial value problem

$$T_n'(t) + (n\pi)^2 T_n(t) = f_n(t);$$

$$T_n(0) = 2 \int_0^2 g(x) \sin(n\pi x/2) = \tau_n \quad \text{(say)},$$

which can be solved by method of integrating factor. Thus, we have

$$T_n(t) = e^{-n^2\pi^2 t} \int_0^2 e^{n^2\pi^2 u} f_n(u) du + \tau_n e^{-n^2\pi^2 t}.$$

Hence,

$$u(x,t) = \sum_{n=1}^{\infty} T_n(t) \sin\left(\frac{n\pi x}{2}\right)$$

$$= \sum_{n=1}^{\infty} \tau_n e^{-(n\pi)^2 t} \sin\left(\frac{n\pi x}{2}\right) + \sum_{n=1}^{\infty} \sin\left(\frac{n\pi x}{2}\right) \int_0^2 e^{-(n\pi)^2(t-u)} f_n(u) du,$$

where the second term represents the contribution due to heat source.

Example 9.2 We solve the *nonhomogeneous* equation

$$u_t - u_{xx} = e^{-2t} \sin 2x, \quad 0 < x < \pi, \quad t > 0,$$

satisfying the *nonhomogeneous* boundary conditions $u(0, t) = 1$, $u(\pi, t) = 0$, with $u(x, 0) = f(x)$. Taking $v(x, t) = 1 - (x/\pi)$, and $w(x, t) = u(x, t) - v(x, t)$, it follows that the function $w(x, t)$ satisfies the nonhomogeneous equation

$$w_t - w_{xx} = e^{-2t} \sin 2x, \tag{9.2.9}$$

has the homogeneous boundary conditions, and $w(x, 0) = f(x) + 1 - (x/\pi)$. Clearly, the eigenfunctions of the associated homogeneous BVP are given by $\phi_n(x) = \sin nx$, and so we may write

$$w(x, t) = \sum_{n=1}^{\infty} \alpha_n(t) \sin nx$$

Substituting in Eq. (9.2.9), we obtain

$$\sum_{n=1}^{\infty} [\alpha_n'(t) + n\alpha_n(t)] \sin nx = e^{-2t} \sin 2x,$$

so the generalised Fourier coefficients $\alpha_n(t)$ are given by the ODE

$$\alpha'_n(t) + n\alpha_n(t) = \begin{cases} 0, & \text{for } n \neq 2 \\ e^{-2t}, & \text{for } n = 2 \end{cases}.$$

This can be solved directly to give

$$\alpha_n(t) = \alpha_2(0)e^{-2t}, \quad \text{where } \alpha_2(0) = \frac{1}{\pi} + \frac{2}{\pi} \int_0^\pi f(x) \sin 2x \, dx.$$

9.3 Poisson Equations

Consider the *Poisson equation*

$$\nabla^2 u = u_{xx} + u_{yy} = g(x, y). \tag{9.3.1}$$

As said before, this equation also arise in a study of problems related to equilibrium temperature distribution $u = u(x, y)$ for a system having a time-independent source $g(x, y)$ of the thermal energy. In general, Poisson equation-related initial-boundary value problems are usually nonhomogeneous with nonhomogeneous boundary conditions. As a preparation to apply the method of eigenfunction expansion, we decompose the solution $u = u(x, y)$ of the main problem to write

$$u(x, y) = v(x, y) + w(x, y),$$

where $v = v(x, y)$ represents the equilibrium solution due to the forcing function $g(x, y)$, and the function $w = w(x, y)$ represents part of the solution due to nonhomogeneous boundary conditions. The initial-boundary value problems for the function $w(x, y)$ is a Laplace equation with nonhomogeneous boundary conditions, which we solved earlier for bounded domain cases by the method of separation of variables, and for unbounded domain cases by applying Fourier transforms.

We would now explain how method of eigenfunction expansion can be used to obtain the part $v = v(x, y)$ of the solution of the Poisson equation (9.3.1). So, we consider first the homogeneous Poisson equation for the function $v = v(x, y)$ such that $v = 0$ on the boundary. To begin with, let us take the case of the Poisson equation in v defined over a rectangle with vertices at $(0, 0)$, $(\ell, 0)$, $(0, b)$, and (ℓ, b). As before, using the homogeneous boundary conditions at $x = 0$ and $x = \ell$, the solution of homogeneous Laplace equation $\nabla^2 v = 0$ as the generalised Fourier series is given by the eigenfunction expansion

$$v(x, y) = \sum_{n=1}^{\infty} \beta_n(y) \sin\left(\frac{n\pi x}{\ell}\right), \tag{9.3.2}$$

so that we can write the Poisson equations for the function v as

$$\sum_{n=1}^{\infty} \frac{d^2 \beta_n}{dy^2} \sin\left(\frac{n\pi x}{\ell}\right) + \frac{\partial^2 v}{\partial x^2} = g. \qquad (9.3.3)$$

Next, differentiating twice equation (9.3.2) term-wise with respect to x, we can write the last equation

$$\sum_{n=1}^{\infty} \left[\frac{d^2 \beta_n}{dy^2} - \left(\frac{n\pi}{\ell}\right)^2 \beta_n \right] \sin\left(\frac{n\pi x}{\ell}\right) = g. \qquad (9.3.4)$$

Then, by Fourier sine integral theorem,

$$\frac{d^2 \beta_n}{dy^2} - \left(\frac{n\pi}{\ell}\right)^2 \beta_n = \frac{2}{\ell} \int_0^\ell g(x, y) \sin\left(\frac{n\pi x}{\ell}\right) dx = g_n(y), \qquad (9.3.5)$$

where the function $g_n(y)$ is related to g by the relation

$$g(x, y) = \sum_{n=1}^{\infty} g_n(y) \sin\left(\frac{n\pi x}{\ell}\right). \qquad (9.3.6)$$

The conditions on the boundaries $y = 0$ and $y = b$ turn ODE (9.3.5) into a nonhomogeneous *boundary value problem*, with boundary conditions

$$\beta_n(0) = 0 \quad \text{and} \quad \beta_n(b) = 0. \qquad (9.3.7)$$

Note that this is in sharp contrast to the time-dependent nonhomogeneous differential equations wherein the coefficients satisfy a 1-dimensional IVP, which are comparatively easier to solve.

There are many different ways to solve the nonhomogeneous *boundary value problem* (9.3.5), with the homogeneous boundary conditions (9.3.7). For example, the method of variation of parameters gives

$$\beta_n(y) = \sinh\frac{n\pi(b-y)}{\ell} \int_0^y g_n(\lambda) \sinh\frac{n\pi\lambda}{\ell} d\lambda$$

$$+ \sinh\frac{n\pi y}{\ell} \int_0^y g_n(\lambda) \sinh\frac{n\pi(b-\lambda)}{\ell} d\lambda \qquad (9.3.8)$$

Exercises 9

9.1 Consider the initial value problem for the heat equation:

$$u_t(x, t) = k\,u_{xx}(x, t), \quad \text{for } 0 < x < 1 \ \text{ and } \ t > 0;$$
$$u(x, 0) = f(x), \quad \text{for } f \in L^2(0, 1);$$

with boundary conditions given by

$$u(0, t) = u(1, t) = 0 \ \text{ or } \ u_x(0, t) = u(1, t) = 0 \ \text{ or } \ u_x(0, t) = u_x(1, t) = 0.$$

Find the eigenvalues and eigenfunctions in each of the above three cases. Further, determine which problem has the solution reaching to the steady-state faster than others. Are the steady-states same in all the three cases?

9.2 Solve the initial-boundary value problem:

$$u_t(x, t) - u_{xx}(x, t) = 1, \quad \text{for } 0 < x < \pi \ \text{ and } \ t > 0;$$
$$u(x, 0) = 0;$$

with boundary condition given by

$$u(0, t) = u(\pi, t) = 0.$$

9.3 Solve the initial-boundary value problem:

$$u_t(x, t) = u_{xx}(x, t), \quad \text{for } 0 < x < 1 \ \text{ and } \ t > 0;$$
$$u(x, 0) = 1;$$

with boundary condition given by

$$u(0, t) = e^{-t} \ \text{ and } \ u_x(1, t) = 0.$$

9.4 Solve the initial-boundary value problem:

$$u_t(x, t) - u_{xx}(x, t) = 2, \quad \text{for } 0 < x < 1 \ \text{ and } \ t > 0;$$
$$u(x, 0) = \sin 2\pi x;$$

with periodic boundary condition given by

$$u(0, t) = u(1, t) \ \text{ and } \ u_x(0, t) = u_x(1, t).$$

9.5 Solve the initial-boundary value problem:

$$u_t(x, t) - u_{xx}(x, t) = xt, \quad \text{for } 0 < x < 1 \quad \text{and} \quad t > 0;$$
$$u(x, 0) = 2x - (x^2)/2;$$

with periodic boundary condition given by

$$u_x(0, t) = 2 \quad \text{and} \quad u_x(1, t) = 1.$$

9.6 Consider the initial value problem for the heat equation with convection:

$$u_t(x, t) = k u_{xx}(x, t) + d u_x(x, t), \quad \text{for } 0 < x < 1 \quad \text{and} \quad t > 0;$$
$$u(x, 0) = f(x), \quad \text{for } f \in L^2(0, 1);$$

with boundary condition given by

$$u(0, t) = u(1, t) = 0.$$

Let $u(x, t) = e^{ax+bt} w(x, t)$, and choose a, b such that $w(x, t)$ solves

$$w_t(x, t) = k w_{xx}(x, t), \quad \text{for } 0 < x < 1 \quad \text{and} \quad t > 0;$$
$$w(x, 0) = e^{-ax} f(x), \quad \text{for } f \in L^2(0, 1);$$

with boundary condition given by

$$w(0, t) = w(1, t) = 0.$$

Find the function $u = u(x, t)$. Compare the solutions for the cases when $d = 0$ and $d \neq 0$.

9.7 Suppose $u = u(x, t)$ solves the initial-boundary value problem

$$u_t(x, t) = u_{xx}(x, t) + d u_x(x, t), \quad \text{for } 0 < x < 1 \quad \text{and} \quad t > 0;$$
$$u(x, 0) = 0, \quad \text{for } 0 < x < 1;$$
$$u(0, t) = T_0 \quad \text{and} \quad u(1, t) = T_1, \quad \text{for } t > 0.$$

Use the relation $v(x, t) = u(x, t) - T_0(1 - x) - T_1 x$ to convert this to a problem in $v(x, t)$ with homogeneous boundary conditions, and hence solve the problem for $v(x, t)$. Use the same to obtain an expression for $u(x, t)$.

9.8 A radioactive substance is placed at one end of a thin tube of length ℓ, and the other end is sealed. Suppose a detector is placed at the sealed end. Assume that initially the tube has no such substance and as it diffuses down the tube, some amount of it escapes through the sides of the tube due to leakage. The situation above can be studied using a model given by

$$u_t(x, t) = k\,u_{xx}(x, t) - d u_x(x, t), \quad \text{for } 0 < x < \ell \ \text{ and } \ t > 0;$$
$$u(x, 0) = 0, \quad \text{for } 0 < x < \ell;$$
$$u(0, t) = a_0 \ \text{ (for } t > 0) \ \text{ and } \ u_x(\ell, t) = 0, \quad \text{for } t > 0.$$

Find $u(\ell, t)$, for $t > 0$.

9.9 Solve the initial-boundary value problem:

$$u_t(x, t) = u_{xx}(x, t), \quad \text{for } 0 < x < 1 \ \text{ and } \ t > 0;$$
$$u(x, 0) = 1;$$

with boundary condition given by

$$u(0, t) = e^{-t} \ \text{ and } \ u_x(1, t) = 0.$$

9.10 Consider the initial value problem for the wave equation with dispersion:

$$u_{tt}(x, t) = c^2\,u_{xx}(x, t) - d^2 u(x, t), \quad \text{for } 0 < x < 1 \ \text{ and } \ t > 0;$$
$$u(x, 0) = f(x), \quad \text{for } f \in L^2(0, 1) \ \text{ and } \ u_t(x, 0) = 0;$$

with boundary condition given by

$$u(0, t) = u(1, t) = 0.$$

Expand the solution of the above problem in series of the form

$$u(x, t) = \sum_{n=1}^{\infty} u_n(t)\varphi_n(x).$$

Find the ordinary differential equation solved by the function u_n, and hence both u_n and u.

9.11 Solve the initial-boundary value problem:

$$u_{tt}(x, t) = u_{xx}(x, t), \quad \text{for } 0 < x < 1 \ \text{ and } \ t > 0;$$
$$u(x, 0) = \sin\left(x\pi/2\right) \ \text{ and } \ u_t(x, 0) = 0;$$

with boundary condition given by

$$u(0, t) = 0 \ \text{ and } \ u_x(1, t) = 0.$$

9.12 Solve the initial-boundary value problem:

$$u_{tt}(x, t) - u_{xx}(x, t) = tx(x - 1), \quad \text{for } 0 < x < 1 \ \text{ and } t > 0;$$
$$u_t(x, 0) = \sin\left(x\pi\right);$$

with boundary condition given by

$$u(0, t) = u(1, t) = 0.$$

9.13 Consider the problem:

$$u_{xx}(x, y) + u_{yy}(x, y) = 0, \quad \text{for } 0 < x, \ y < 1;$$
$$u(0, y) = f(y), \quad \text{for } 0 < y < 1;$$
$$u_x(1, y) = 0, \quad \text{for } 0 < y < 1;$$
$$u(x, 0) = 0, \quad \text{for } 0 < x < 1;$$
$$u_x(x, 1) = g(x), \quad \text{for } 0 < x < 1.$$

Split the above problem into two subproblems, find the eigenfunctions for each problem, and hence solve the both.

9.14 Solve the initial-boundary value problem:

$$u_{xx}(x, y) + u_{yy}(x, y) = 1 - u_y, \quad \text{for } 0 < x, \ y < 1;$$
$$u(0, y) = f(y), \quad \text{for } 0 < y < 1;$$
$$u_x(1, y) = u(1, y) = 0, \quad \text{for } 0 < y < 1;$$
$$u(x, 0) = u(x, 1) = 0, \quad \text{for } 0 < x < 1.$$

References

1. Cain, G., & Meyer, G. H. (2006). *Separation of variables for partial differential equations—An eigenfunction approach*. Chapman & Hall/CRC.
2. Haberman, R. (1987). *Elementary applied partial differential equations* (2nd ed.). Prentice Hall.
3. McOwen, R. (2002). *Partial differential equations—Methods and applications* (2nd ed.). Pearson.

Chapter 10
Fourier Transforms

Profound study of nature is the most fertile source of
mathematical discoveries.

Joseph Fourier (1768–1830)

Let $I \subset \mathbb{R}$ be an interval, and p be a real or complex *parameter*. Suppose $K(p, t)$ is
a nice function (possibly complex-valued) such that the integral given by

$$\mathscr{I}[\phi](p) := \int\limits_I K(p, t)\phi(t)\,dt, \qquad \text{for} \quad \phi \in PC(I),$$

exists in Lebesgue sense. Then the correspondence $\phi \mapsto \mathscr{I}(\phi)$ is a linear map defined
on a class of functions in the space $PC(I)$, and we say $\mathscr{I}$ is the *integral transform* with
the *kernel function* $K(p, t)$. In particular, the Laplace transform $\mathscr{L}[\phi]$ is an integral
transform, with kernel function given by $K(p, t) = e^{-st}$ ($s \in \mathbb{C}$), that is defined on
the space of real-valued functions $f \in PC([0, \infty))$ of *exponential order*.[1] That is,
we have

$$F(s) = \mathscr{L}[f] := \int\limits_0^\infty e^{-st} f(t)\,dt, \qquad \text{for} \quad f \in PC([0, \infty)).$$

The function $F(s)$ is called the *Laplace transform* of the function f. Further, the
Fourier transform defined over $\mathbb{R}$ is an integral transform with kernel function given
by $K(p, t) = e^{-ipt}$ ($p \in \mathbb{R}$). We may write

---

[1] That is, for some positive numbers α, M and T, we have $\left| f(t) \right| \leq Me^{\alpha t}$, for all $t \geq T$.

© The Author(s), under exclusive license to Springer Nature Singapore Pte Ltd. 2022   417
A. K. Razdan and V. Ravichandran, *Fundamentals of Partial Differential Equations*,
https://doi.org/10.1007/978-981-16-9865-1_10

$$\widehat{f}(p) = F[f] := \int\limits_{-\infty}^{\infty} e^{-ipt} f(t)\, dt, \quad \text{for} \quad f \in L^1(\mathbb{R}).$$

It was introduced by Fourier as a tool to solve certain important IBVP related to heat equation [1]. Michel Plancherel coined the term *transformé de Fourier* in 1915. Notice that, for the existence of the Fourier integral, it is natural to take function f satisfy the *Dirichlet's conditions* on every finite subinterval of the line $\mathbb{R}$, and also it must be *absolutely integrable*. In all what follows in this chapter, a function satisfying such two conditions is called an *admissible function*. For example, every piecewise smooth function on $\mathbb{R}$ is admissible. However, the nasty function such as $f(t) = \sin(t^{-1})$, $t \in \mathbb{R}\backslash\{0\}$ is not an admissible function.

As both Fourier and Laplace transforms act nicely on derivatives, and also on the initial-boundary conditions as specified in various situations, the two find application in many different areas of physics and engineering. A convenience to handle the original problem is made possible due to the fact that the *transformed problem* is an algebraic or ordinary differential equation, with specified initial conditions. In both cases, the solution of the original problem is obtained subsequently by applying the associated *inverse transform*.

In the present book, we discuss basic properties and applications of the two transforms in solving initial-boundary value problems defined over a unbounded spatial domains such as the interval $[0, \infty)$ or $\mathbb{R}$. As the Fourier transform is defined over $\mathbb{R}$, it applies to initial value problems for functions that are known to be bounded in the neighbourhoods of $\pm\infty$. In this chapter, we give the solution of some important initial value problems. On the other hand, the Laplace transform is more suitable to deal with initial-boundary value problems over the half-line $x \geq 0$. We give solutions of some such types of problems in the next chapter.

10.1 Introduction

Many different types of integral transforms, for various suitable choices of kernel functions, have been applied for over two centuries to solve some interesting problems in physics and engineering. An early appearance of application of a type of integral transform could be traced back to the work of Euler[2] (1763), which was published in 1769 as part of his book *"Institutiones Calculi Integralis"*. However, it is more reasonable to attribute the origin to the work of *Simon Laplace* (1749–1827) on probability theory, and also to the fundamental work of *Joseph Fourier* (1768–1830) related to heat conduction in solids. Not withheld by the criticism by some mathematicians, applications of the Laplace transform to problems in electrical engineering were popularised by British engineer *Oliver Heaviside* (1850–1925). The reader may refer ([2, 3, 5–7]) for a brief historical introduction of integral transforms, and also as source books for further study.

---

[2] Michael Deakin, Archive for History of Exact Sciences, Vol. 33, No. 4 (1985), 307–319.

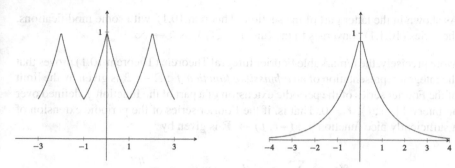

Fig. 10.1 Function $g(t) = e^{-|t|}$ as the limit of the sawtooth function

Fourier transforms have long been proved to be a useful tool in many application areas such as signal and image processing, systems and control, and also ubiquitous differential and difference equations (Appendix A.4). Particularly so when it is just too hard to do computations in the time domain, because the *transformed problem* admits all sort of manageable analytical tools in the frequency domain. Fourier transform is also applicable in many other areas such as quantum mechanics, genetics, analysis of bio-signals and computed tomography. The abstract theory of Fourier transforms is an attractive field of study in harmonic analysis that was developed initially by Norbert Wiener and Raymond Paley in their fundamental paper published in 1933.

Example 10.1 As depicted in Fig. 10.1, the 2ℓ-periodic extension of the *sawtooth function* given by

$$g_\ell(t) = e^{-|t|}, \quad \text{for } -\ell < t < \ell,$$

approaches to the *aperiodic* function $g(t) = e^{-|t|}$ ($t \in \mathbb{R}$) as $\ell \to \infty$. In this case, the exponential Fourier series of the function g_ℓ is given by

$$\frac{e^{-\ell}}{2} \sum_{n=-\infty}^{\infty} \frac{1}{\ell^2 + (2n-1)^2 \pi^2} e^{i(2n-1)\pi x/\ell}. \tag{10.1.1}$$

On the other hand, the Fourier transform of the limit function g is given by

$$\widehat{g}(p) = \int_{-\infty}^{\infty} e^{-ipt} e^{-|t|} dt$$

$$= \int_{-\infty}^{0} e^{(1-ip)t} dt + \int_{0}^{\infty} e^{-(ip+1)t} dt$$

$$= \frac{1}{1-ip} + \frac{1}{1+ip} = -\frac{2}{1+p^2}.$$

As shown in the latter part of the section (Theorem 10.1), with some modifications, the series (10.1.1) converges to the function $\widehat{g}(p)$, as $\ell \to \infty$.

More precisely, the remarkable Fourier Integral Theorem (Theorem 10.1) proves that the integral representation of an *admissible function* $f : \mathbb{R} \to \mathbb{R}$ is given by the limit of the Fourier series of the periodic extension of a part of the function f defined over an interval $[-\ell, \ell)$, $\ell > 0$. That is, if the Fourier series of the periodic extension of a sufficiently nice function $f : (-\ell, \ell) \to \mathbb{R}$ is given by

$$f(t) = \sum_{n=-\infty}^{\infty} c_n e^{i\omega_n x}, \qquad \text{with} \quad \omega_n = \frac{n\pi}{\ell},$$

then the (discrete) exponentials given by

$$e^{i\omega_n t} = \cos \omega_n t + i \sin \omega_n t, \qquad \text{for} \quad n \in \mathbb{Z},$$

approaches to the (continuous) exponentials given by

$$e^{ipt} = \cos pt + i \sin pt, \qquad \text{for} \quad p \in \mathbb{R},$$

as $\ell \to \infty$. The resulting *integral representation* is thus obtained as

$$\widehat{f}(p) = \int_{-\infty}^{\infty} e^{ipt} f(t) \, dt.$$

Therefore, in this case, the *weight factor* $\widehat{f}(p)$ is obtained by integrating over $\mathbb{R}$ the product of the function f and the *conjugate* of the exponential function e^{-ipt}. That is precisely the way (see (8.2.19)) we find the Fourier coefficients c_n by using the relation

$$c_n = \frac{1}{2\pi} \int_{-\infty}^{\infty} e^{-i\omega_n t} f(t) \, dt, \qquad \text{for} \quad n \in \mathbb{Z}.$$

Example 10.2 Consider the case of the *square wave* given by

$$f_\ell(t) = \begin{cases} 0, & \text{for } -\ell < t \le -1 \\ 1, & \text{for } -1 < t < 1 \quad, \quad \ell > 1. \\ 0, & \text{for } 1 \le t < \ell \end{cases}$$

As before, the periodic extension $\tilde{f}_\ell : \mathbb{R} \to \mathbb{R}$ of f_ℓ has the period 2ℓ. Notice that the function $\tilde{f}_\ell$ satisfies the Dirichlet's conditions over every finite interval and $\tilde{f} \in L^1(\mathbb{R})$. As depicted in Fig. 10.2, the function f_ℓ approach to the absolutely integrable function given by

Fig. 10.2 Limit of the unit square function

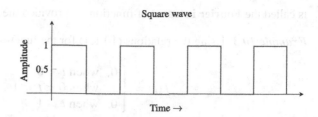

$$f(t) = \begin{cases} 1, & \text{for } -1 < t < 1 \\ 0, & \text{otherwise} \end{cases}, \quad \text{for } t \in \mathbb{R},$$

as $\ell \to \infty$. We shall see, the function $\tilde{f}_\ell$ admits a Fourier series representation in the interval $(-\ell, \ell)$, $\ell > 1$ (Theorem 8.2). That is, the trigonometric polynomial $S_n(t)$ given by

$$S_n(t) = \frac{a_0}{2} + \sum_{k=1}^{n} \left[a_k \cos(\omega_k t) + b_k \sin(\omega_k t) \right], \quad \text{for } n \geq 1,$$

has the function f_ℓ as the *uniform limit* over the interval $(-\ell, \ell)$, where a_k and b_k are the Fourier coefficients of f_ℓ. It follows easily that the Fourier series of f is given by

$$\frac{1}{\pi} \sum_{n=-\infty}^{\infty} e^{i\omega_n x} \frac{\sin \omega_n}{n}, \tag{10.1.2}$$

which converges to the function

$$\widehat{f}(p) = \int_{1}^{1} e^{-ipt} dt = 2 \frac{\sin p}{p}, \quad \text{as } \ell \to \infty.$$

Definition 10.1 For a function $f : \mathbb{R} \to \mathbb{R}$, the integral given by

$$I(t) = \int_{0}^{\infty} \left[A(p) \cos(p\,t) + B(p) \sin(p\,t) \right] dp, \tag{10.1.3}$$

where the functions $A(p)$ and $B(p)$ are respectively given by

$$A(p) = \frac{1}{\pi} \int_{-\infty}^{\infty} f(x) \cos(p\,x) dx \quad \text{and} \quad B(p) = \frac{1}{\pi} \int_{-\infty}^{\infty} f(x) \sin(p\,x) dx \tag{10.1.4}$$

is called the **Fourier integral** of function f, provided the above integrals exist.

Example 10.3 Using the relations (10.1.4) for the function

$$f(t) = \begin{cases} 0, & \text{when } t < 0 \\ t, & \text{when } 0 < t < 1 \ , \\ 0, & \text{when } t > 1 \end{cases}$$

it follows that the integrals $A(p)$ and $B(p)$ are given by

$$A(p) = \frac{1}{\pi} \int_{-\infty}^{\infty} f(t) \cos(pt) \, dt = \frac{1}{\pi} \int_{0}^{1} t \cos(pt) \, dt$$

$$= \frac{1}{\pi} \left[\frac{t \sin(pt)}{p} + \frac{\cos(pt)}{p^2} \right]_{0}^{1}$$

$$= \frac{1}{\pi} \left[\frac{p \sin(p) + \cos(p) - 1}{p^2} \right]; \quad \text{and}$$

$$B(p) = \frac{1}{\pi} \int_{-\infty}^{\infty} f(t) \sin(pt) \, dt = \frac{1}{\pi} \int_{0}^{1} t \sin(pt) \, dt$$

$$= \frac{1}{\pi} \left[-\frac{t \cos(pt)}{p} + \frac{\sin(pt)}{p^2} \right]_{0}^{1}$$

$$= \frac{1}{\pi} \left[\frac{\sin(p) - p \cos(p)}{p^2} \right]$$

Substituting for $A(p)$ and $B(p)$ in Eq. (10.1.3), we obtain

$$I(t) = \frac{1}{\pi} \int_{0}^{\infty} \frac{\cos p(1-t) + p \sin p(1-t) - \cos pt}{p^2} \, dp.$$

Observe that integral fails to converge at $t = 1$. However, we do have

$$\frac{1}{2} \left[I(1^+) + I(1^-) \right] = \frac{0+1}{2} = \frac{1}{2}.$$

In general, the main problem is to know when do we can have $I \rightarrow f$ over $\mathbb{R}$. As proved in the next theorem, a necessary condition for the above to hold is that f is an *admissible function*. That is, as said earlier, $f \in L^1(\mathbb{R})$ and it satisfies the *Dirichlet's conditions* over every *finite* subinterval of $\mathbb{R}$.

Theorem 10.1 (Fourier Integral Theorem) *For every* admissible function $f : \mathbb{R} \rightarrow \mathbb{R}$, *the Fourier integral* (10.1.3) *converges to* $(1/2)[f(t^+) + f(t^-)]$, *for all* $t \in \mathbb{R}$.

Proof Let $\ell > 0$ be arbitrary, and consider the part $f : (-\ell, \ell) \to \mathbb{R}$. By Theorem 8.2, we can write

$$\frac{1}{2}[f(t^-) + f(t^+)] = \frac{a_0}{2} + \sum_{n=1}^{\infty} [a_n \cos(\omega_n t) + b_n \sin(\omega_n t)], \qquad (10.1.5)$$

for all $t \in (-\ell, \ell)$, where $\omega_n = n\pi/\ell$, and a_n, b_n are respectively given by

$$a_n = \frac{1}{\ell} \int_{-\ell}^{\ell} f(t) \cos(\omega_n t) dt;$$

$$b_n = \frac{1}{\ell} \int_{-\ell}^{\ell} f(t) \sin(\omega_n t) dt.$$

Recall that the series (10.1.5) equals the function $f(t)$ at each point of continuity $t \in (-\ell, \ell)$. To obtain an *integral representation* for the function f, it is reasonable to take the limit $\ell \to \infty$. For, by replacing the coefficients a_n and b_n in the series (10.1.5) by above integrals, it follows that the *average* $(1/2)[f(t^+) + f(t^-)]$ equals the series

$$\frac{1}{2\ell} \int_{-\ell}^{\ell} f(t)\, dt + \frac{1}{\ell} \sum_{n=1}^{\infty} \left[\int_{-\ell}^{\ell} f(u) \cos(\omega_n u) du \right] \cos(\omega_n t)$$

$$+ \frac{1}{\ell} \sum_{n=1}^{\infty} \left[\int_{-\ell}^{\ell} f(u) \sin(\omega_n u) du \right] \sin(\omega_n t),$$

for all $t \in (-\ell, \ell)$. Now, since the *adjacent values* of $\omega_n = n\pi/\ell$ are obtained by setting $\Delta n = 1$, i.e., we have

$$\frac{\ell}{\pi} \Delta \omega_n = \frac{\ell}{\pi} [\omega_{n+1} - \omega_n] = 1,$$

by multiplying each term of the series (10.1.5) by $(\ell/\pi)\Delta\omega_n$, we may write the above series as

$$\frac{\Delta \omega_n}{2\pi} \int_{-\ell}^{\ell} f(t)\, dt + \frac{1}{\pi} \sum_{n=1}^{\infty} \left[\int_{-\ell}^{\ell} f(u) \cos(\omega_n u) du \right] \cos(\omega_n t) \Delta \omega_n$$

$$+ \frac{1}{\pi} \sum_{n=1}^{\infty} \left[\int_{-\ell}^{\ell} f(u) \sin(\omega_n u) du \right] \sin(p_n t) \Delta \omega_n.$$

Therefore, as $\ell \to \infty$, $\Delta\omega_n \to d\omega$, and we have

$$f \in L^1(\mathbb{R}) \quad \Rightarrow \quad \frac{\Delta\omega_n}{2\pi} \int_{-\ell}^{\ell} f(t)\,dt \to 0,$$

and the two *Riemann sums* given by

$$\sum_{n=1}^{\infty} \left[\int_{-\ell}^{\ell} f(u) \cos(\omega_n u)\,du \right] \cos(\omega_n t) \Delta\omega_n \quad \text{and}$$

$$\sum_{n=1}^{\infty} \left[\int_{-\ell}^{\ell} f(u) \sin(\omega_n u)\,du \right] \sin(\omega_n t) \Delta\omega_n$$

approach respectively to the *Riemann integrals*

$$\int_{0}^{\infty} \left[\int_{-\infty}^{\infty} f(u) \cos(\omega u)\,du \right] \cos(\omega t)\,d\omega;$$

$$\int_{0}^{\infty} \left[\int_{-\infty}^{\infty} f(u) \sin(\omega u)\,du \right] \sin(\omega t)\,d\omega.$$

Hence, as $\ell \to \infty$, Eq. (10.1.3) holds, with $A(\omega)$ and $B(\omega)$ given by the relations (10.1.4). As before, $\frac{1}{2}[f(t^+) + f(t^-)] = f(t)$ at each point of continuity $t \in \mathbb{R}$. This completes the proof of the theorem. $\qquad\qquad\square$

It is now easy to derive the *exponential form* of Theorem 10.1 by using the relations (10.1.4). For, we may write

$$\frac{1}{2}[f(t^+) + f(t^-)] = \frac{1}{\pi} \int_{0}^{\infty} \int_{-\infty}^{\infty} f(u) \cos p(t-u)\,du\,dp$$

$$= \frac{1}{\pi} \int_{-\infty}^{\infty} f(u) \left[\int_{0}^{\infty} \cos[p(t-u)]\,dp \right] du$$

$$= \frac{1}{2\pi} \int_{-\infty}^{\infty} f(u) \left[\int_{-\infty}^{\infty} \cos(p(t-u))\,dp \right] du;$$

Adding the *zero term*

$$\frac{i}{2\pi} \int\limits_{-\infty}^{\infty} f(u) \left[\int\limits_{-\infty}^{\infty} \sin(p(t-u))\,dp \right] du,$$

to the right side of the last equation, we obtain

$$\frac{1}{2} \left[f(t^+) + f(t^-) \right]$$

$$= \frac{1}{2\pi} \int\limits_{-\infty}^{\infty} \int\limits_{-\infty}^{\infty} e^{ip(t-u)} f(u)\,du\,dp \tag{10.1.6}$$

$$= \frac{1}{2\pi} \int\limits_{-\infty}^{\infty} e^{ipt} \left[\int\limits_{-\infty}^{\infty} e^{-ipu} f(u)\,du \right] dp, \quad \forall\, t \in \mathbb{R}.$$

The relation (10.1.6) is known as the *Fourier inversion formula*. In practical terms, the last equation says the signal f can be resolved continuously into infinite number of harmonic components with varying frequency $p/2\pi$ and amplitude given by the integral

$$\frac{1}{2\pi} \int\limits_{-\infty}^{\infty} e^{-ipu} f(u)\,du. \tag{10.1.7}$$

Remark 10.1 The integral representation of an admissible continuous function in *exponential form* (10.1.6) was known to French mathematician *Augustine Louis Cauchy* (1816). The term *theoremé de Fourier* was coined by French mathematician *Joseph Liouville* in 1831.

The next example illustrates a practical use of Fourier integral (Definition 10.1).

Example 10.4 Consider a Cauchy problem for heat equation as given below:

$$\begin{aligned} u_t &= c^2\, u_{xx}, && \text{for } x \in \mathbb{R} \text{ and } t \geq 0; \\ u(x,0) &= f(x), && \text{for } x \in \mathbb{R}. \end{aligned} \tag{10.1.8}$$

A solution of the problem by separation of variables can be obtained as

$$u_p(x,t) = \left[A(p) \cos px + B(p) \sin px \right] e^{-c^2 p^2 t},$$

where $p > 0$ is not necessarily an integer multiple of a fixed real number. For the general solution of the problem, we may take

$$u(x, t) = \int_0^\infty u_p(x, t)\,\mathrm{d}p$$

$$= \int_0^\infty \left[A(p)\cos px + B(p)\sin px\right]e^{-c^2 p^2 t}\,\mathrm{d}p,$$

provided the integral on the right side exists. Further, as we have

$$f(x) = u(x, 0) = \int_0^\infty \left[A(p)\cos px + B(p)\sin px\right]\mathrm{d}p,$$

it is required to compute the integrals $A(p)$ and $B(p)$ as given by (10.1.4).

10.2   Fourier's Transform Pair

We may start with the next definition.

Definition 10.2 The *Fourier transform* of a function $f \in L^1(\mathbb{R})$, denoted by $\widehat{f}(p)$, is the integral given by

$$\widehat{f}(p) = F[f] = \int_{-\infty}^\infty e^{-ipx} f(x)\,\mathrm{d}x. \qquad (10.2.1)$$

The *inverse Fourier transform* $F^{-1}[g(p)]$ of a function $g(p) \in L^1(\mathbb{R})$ is given by

$$g(t) = F^{-1}[g(p)] = \frac{1}{2\pi} \int_{-\infty}^\infty e^{ipt} g(p)\,\mathrm{d}p. \qquad (10.2.2)$$

We may also write the integral $\widehat{f}(p)$ in Eq. (10.2.1) as

$$\widehat{f}(p) := \lim_{\alpha \to \infty} \int_{-\alpha}^\alpha e^{-ipt} f(t)\,\mathrm{d}t, \quad p \in \mathbb{R}, \qquad (10.2.3)$$

provided the limit exists. Further, Eq. (10.1.6) takes the form

$$\frac{1}{2}\left[f(t^+) + f(t^-)\right] = \frac{1}{2\pi} \int_{-\infty}^\infty e^{ipt} \widehat{f}(p)\,\mathrm{d}p, \quad \forall \, t \in \mathbb{R}. \qquad (10.2.4)$$

The dual transforms pair F and F^{-1} are *mutually invertible* if $f \in C(\mathbb{R})$.

Remark 10.2 The above two transforms can be defined in a variety of ways. For example, the constants 1 and $1/2\pi$ appearing before the integral signs in Eqs. (10.2.1) and (10.2.2), respectively, could be replaced by any two constants a, b such that $ab = 1/(2\pi)$. A simple way to get rid of the *normalising factor* $1/2\pi$ is to lump the constant 2π to the parameter p so that the transform pair can alternatively be defined as

$$\widehat{f}(p) = F[f(t)] = \int_{-\infty}^{\infty} e^{-i\,2\pi\,p\,t} f(t)\,dt;$$

$$f(t) = F^{-1}[F(p)] = \int_{-\infty}^{\infty} e^{i\,2\pi\,p\,t}\,\widehat{f}(p)\,dp$$

Some authors use the above two equations to define the Fourier transform pair. Further, the concept is not limited to functions of time variable alone. Indeed, by replacing t by a vector $\boldsymbol{x} \in \mathbb{R}^n$, we can define a *multivariable version* of Fourier transform using n-fold integral over $\mathbb{R}^n$. However, to follow a unified approach, we may continue refer t as a time variable.

In general, as for the function $\widehat{f} : \mathbb{R} \to \mathbb{C}$, we have

$$\widehat{f}(-p) = [\widehat{f}(p)]^*, \tag{10.2.5}$$

it follows in particular that $\widehat{f}(p)$ is real if $f(t)$ is an *even function*; and, $\widehat{f}(p)$ is purely imaginary if $f(t)$ is an *odd function*. Moreover, as $f \in L^1(\mathbb{R})$, the function $\widehat{f}$ is bounded. Indeed, we have

$$|\widehat{f}(p)| \le \|f\|_1, \quad \text{for all } p, \quad \Rightarrow \quad \|\widehat{f}(p)\|_\infty \le \|f\|_1.$$

The next theorem proves that the correspondence given by

$$f \mapsto \widehat{f} : L^1(\mathbb{R}) \to L^\infty(\mathbb{R})$$

is a linear transformation.

Theorem 10.2 (LINEARITY) *For admissible functions $f(t)$ and $g(t)$, and scalars α, β,*

$$F[\alpha\,f(t) + \beta\,g(t)] = \alpha\,F[f(t)] + \beta\,F[g(t)].$$

Proof Left as an easy exercise for the reader. □

In practical terms, *linearity property* of the Fourier transform holds significance due to fact that operations performed in *time domain* lead to proportional changes in

operations performed in *frequency domain*. Indeed, it is more convenient to deal
with the latter type of data. Further, as we have

$$\left| \widehat{f}(p+h) - \widehat{f}(p) \right| = \left| \int_{-\infty}^{\infty} \left[e^{-i(p+h)t} - e^{-ipt} \right] f(t) \, dt \right|$$

$$\le \int_{-\infty}^{\infty} \left| e^{-iht} - 1 \right| f(t) \, dt$$

$$\to \quad 0, \quad \text{as} \quad h \to 0,$$

it follows by *dominated convergence theorem* that the function $\widehat{f}$ is uniformly continuous. Notice that right side integral in above inequality is dominated by the function
$2|f| \in L^1(\mathbb{R})$. We thus conclude that the correspondence $f \mapsto \widehat{f}$ gives a linear map
$: L^1(\mathbb{R}) \to C(\mathbb{R})$. The extension of *Riemann-Lebesgue lemma* for Fourier transform,
as given by the next theorem, establishes the fact that the space $C(\mathbb{R})$ can be replaced
by the space $C_0(\mathbb{R})$ of continuous function *vanishing at infinity*.

Theorem 10.3 (Riemann–Lebesgue Lemma) *For any* $f \in L^1(\mathbb{R})$,

$$\widehat{f}(p) \to 0 \quad as \quad p \to \pm\infty.$$

Proof The central idea used here is same as given earlier in the case of the Fourier
series of a nice periodic function. For $f = \chi_{[a,b]}$ and $p \ne 0$, we have

$$\int_a^b e^{-ipt} dt = \frac{1}{ip} \left[e^{-iap} - e^{-ibp} \right] \to 0, \quad \text{as} \quad p \to \pm\infty,$$

which proves the assertion if f is a step function, i.e. when f is a linear combination
of simple functions. Next, as such types of functions are *dense* in the space $L^1(\mathbb{R})$,
for any given $f \in L^1(\mathbb{R})$ it is possible to find a sequence of simple functions $\langle s_n \rangle$
such that

$$\left\| s_n - f \right\|_1 \to 0 \quad \text{as} \quad n \to \infty.$$

That is, as $n \to \infty$, we have $\widehat{s_n} \to \widehat{f}$ uniformly over $\mathbb{R}$. Finally, to complete the
proof, we use

$$\left| \widehat{f}(p) \right| \le \left| \widehat{s_n}(p) - \widehat{f}(p) \right| + \left| \widehat{s_n}(p) \right|,$$

which suggests to take firstly n large enough so that the first term on the right side is
small, followed by taking $|p|$ sufficiently large for such an n. $\square$

Hence, it follows that the correspondence

$$f \mapsto \widehat{f} : L^1(\mathbb{R}) \to C_0(\mathbb{R})$$

is a *linear contraction*, where the norm on the space $C_0(\mathbb{R})$ is given by

$$\|f\|_\infty := \sup_{t \in \mathbb{R}} |f(t)|.$$

Next, suppose both f and $tf(t)$ are in $L^1(\mathbb{R})$. We may write

$$\widehat{f}'(p) = \lim_{h \to 0} \frac{1}{h} \int_{-\infty}^{\infty} \left[e^{-i(p+h)t} - e^{-ipt} \right] f(t) \, dt$$

$$= \lim_{h \to 0} \int_{-\infty}^{\infty} \frac{e^{-iht} - 1}{h} e^{-ipt} f(t) \, dt,$$

where the function inside the integral converges pointwise to

$$(-it)e^{-ipt} f(t) \quad \text{as} \quad h \to 0,$$

and for small values of $|h|$, it follows easily that

$$\left| \frac{e^{-iht} - 1}{h} e^{-ipt} f(t) \right| \le |tf(t)|.$$

Therefore, by *dominated convergence theorem*, we obtain

$$\widehat{f}'(p) = \int_{-\infty}^{\infty} (-it\, f(t)) e^{-ipt} dt = -i\,\widehat{(tf(t))}(p). \tag{10.2.6}$$

In actual practices, when a function $f : [0, \infty) \to \mathbb{R}$ is a *continuous time signal* such as voltage difference, current, displacement, radiated electric field, etc., the Fourier transform $\widehat{f} = F[f]$ represents the *frequency spectrum* of the signal f. The next two examples illustrate the fact that there is an inverse relationship between the width of a signal $f(t)$ and the width of its frequency spectrum $\widehat{f}(p)$, concentrated solely at the *zero frequency*.

Example 10.5 Consider the *unit pulse* represented by a *box function* given by

$$f(t) = \begin{cases} 1, & \text{for } |t| < a \\ 0, & \text{for } |t| > a > 0 \end{cases}$$

The Fourier transform $\widehat{f}$ of f is given by

$$\widehat{f}(p) = \int\limits_{-\infty}^{\infty} e^{-ipt} f(t)\, dt$$

$$= \int\limits_{-a}^{a} e^{-ipt}\, dt = \frac{1}{ip}\left(e^{ipa} - e^{-ipa}\right)$$

$$= \frac{2}{p}\sin(ap) = 2a\,\mathrm{Si}(ap), \quad p \neq 0, \tag{10.2.7}$$

where $\mathrm{Si}(x) = \sin x / x$, $x \neq 0$, is the *Dirichlet's sine function*. Notice that the box function $f(t)$ can also be expressed as

$$f(t) := H(t + a) - H(t - a), \quad a > 0,$$

where H is the *Heaviside step function*.

Example 10.6 The Fourier transform of the *damped oscillations* represented by the *sinc function*

$$f(t) = \frac{\sin(at)}{t}, \quad -\infty < t < \infty, \ (a > 0)$$

is given by

$$\widehat{f}(p) = \int\limits_{-\infty}^{\infty} e^{-ipt}\, \frac{\sin(at)}{t}\, dt$$

$$= \int\limits_{-\infty}^{\infty} [\cos(pt) - \sin(pt)]\frac{\sin(at)}{t}\, dt$$

$$= \int\limits_{-\infty}^{\infty} \frac{\cos(pt)\sin(at)}{t}\, dt - \int\limits_{-\infty}^{\infty} \frac{\sin(pt)\sin(at)}{t}\, dt$$

$$= 2\int\limits_{0}^{\infty} \frac{\cos(pt)\sin(at)}{t}\, dt$$

$$= \int\limits_{0}^{\infty} \frac{\sin[(p+a)t] + \sin[(a-p)t]}{t}\, dt$$

Now, since $(p + a)$ and $(a - p)$ are positive for $|p| < a$, we have

$$\widehat{f}(p) = \frac{\pi}{2} + \frac{\pi}{2} = \pi, \quad \text{for } |p| < a.$$

On the other hand, since $(p + a)$ or $(a - p)$ is negative for $|p| > a$, it follows that

$$\widehat{f}(p) = \pm \frac{\pi}{2} \pm \frac{\pi}{2} = 0, \quad \text{for } |p| > a.$$

Therefore, we obtain

$$\widehat{f}(p) = \begin{cases} \pi, & |p| < a \\ 0, & |p| > a > 0 \end{cases}.$$

The behaviour of the *finite wavetrain* represented by the function

$$f(t) = \cos(p_0 t) = \text{Re}(e^{ip_0 t}), \quad \text{for } 0 < t < T,$$

is analogous to situation described in the above example. For, we have

$$\widehat{f}(p) = \text{Re} \int_{-\infty}^{\infty} e^{-i(p-p_0)t} \, dt$$

$$= \text{Re} \int_{0}^{T} e^{-i(p-p_0)t} \, dt$$

$$= \frac{1}{(p - p_0)} \text{Re} \left[ie^{-i(p_0-p)T} - i \right]$$

$$= \frac{\sin[(p_0 - p)T]}{p - p_0}.$$

Therefore, shorter the duration of the wavetrain, more is its spectra contaminated by other frequencies beyond the fundamental one.

Example 10.7 For $a \in \mathbb{R}$, the function $f(t) = e^{-a|t|}$ represents a two-sided *exponential signal* of rate a. By definition, we have

$$\widehat{f}(p) = \int_{-\infty}^{\infty} e^{-ipt} e^{-a|t|} \, dt$$

$$= \int_{-\infty}^{0} e^{-ipt} e^{at} \, dt + \int_{0}^{\infty} e^{-ipt} e^{-at} \, dt$$

$$= \frac{1}{a - ip} + \frac{1}{a + ip} = \frac{2a}{a^2 + p^2}.$$

In particular, when $a > 0$, the transform of *decaying signal* is given by

$$\widehat{f}(p) = \frac{1}{a+ip} = \frac{a-ip}{a^2+p^2}.$$

Therefore, by using the fact $f \in C(\mathbb{R})$, it follows from Eq. (10.2.4) that

$$e^{-a|t|} = \frac{a}{\pi} \int_{-\infty}^{\infty} \frac{e^{ipt}}{a^2+p^2} \, dp$$

$$= \frac{2a}{\pi} \int_{0}^{\infty} \frac{\cos pt}{a^2+p^2} \, dp,$$

because $g(p) = \sin(pt)/(a^2+p^2)$ is an *odd functions*. Hence,

$$e^{-a|t|} = \frac{2a}{\pi} \int_{0}^{\infty} \frac{\cos pt}{a^2+p^2} \, dp,$$

is an integral representation of the two-sided exponential function.

Example 10.8 By *inversion formula*, and Example 10.5, we can write the unit pulse represented by *box function* as

$$f(t) = \frac{1}{2\pi} \int_{-\infty}^{\infty} e^{ipt} \left[\frac{2}{p} \sin(pa) \right] dp$$

$$= \frac{1}{\pi} \left[\int_{-\infty}^{\infty} \frac{\cos(pt)\sin(pa)}{p} \, dp + i \int_{-\infty}^{\infty} \frac{\sin(pt)\sin(pa)}{p} \, dp \right]$$

$$= \frac{2}{\pi} \int_{0}^{\infty} \frac{\cos(pt)\sin(pa)}{p} \, dp, \quad \text{for } t \neq \pm a,$$

because $\frac{\sin(pt)\sin(pa)}{p}$ is an *odd function* with respect to the variable p. Observe that $f(\pm a) = 1/2.$

The next example shows that there are functions such as *Gaussian pulse*

$$\gamma_\varepsilon(t) = \frac{\sqrt{2}}{\sqrt{\pi}} e^{-\frac{t^2}{2\varepsilon^2}}, \quad \varepsilon > 0,$$

that have the *spectra* same as the input function. Said differently, the Gaussian pulse is a *self-dual waveform*.

Example 10.9 We use here a procedure, known as the *completing the square*, to compute the Fourier transform of the Gaussian function given by

$$\gamma(t) = \frac{1}{t_0\sqrt{2\pi}} \, e^{-\frac{t^2}{2t_0^2}}, \quad -\infty < t < \infty.$$

For, by definition, we have

$$\widehat{\gamma}(p) = \frac{1}{t_0\sqrt{2\pi}} \int_{-\infty}^{\infty} e^{-i\,p\,t} \, e^{-t^2/2t_0^2} \, dt$$

$$= \frac{1}{t_0\sqrt{2\pi}} \int_{-\infty}^{\infty} e^{-\left(\frac{t^2}{2t_0^2} + i\,p\,t\right)} \, dt$$

$$= \frac{1}{t_0\sqrt{2\pi}} \int_{-\infty}^{\infty} e^{-\left\{\left(\frac{t}{\sqrt{2}\,t_0} + \frac{i\,p\,t_0}{\sqrt{2}}\right)^2 + \frac{(p\,t_0)^2}{2}\right\}} \, dt$$

$$= \frac{e^{-p^2 t_0^2/2}}{t_0\sqrt{2\pi}} \int_{-\infty}^{\infty} e^{-\left(\frac{t}{\sqrt{2}\,t_0} + \frac{i\,p\,t_0}{\sqrt{2}}\right)^2} \, dt$$

$$= \frac{e^{-p^2 t_0^2/2}}{\sqrt{\pi}} \int_{-\infty}^{\infty} e^{-z^2} \, dz \quad \left(\text{taking } z = \frac{t}{\sqrt{2}\,t_0} + \frac{i\,p\,t_0}{\sqrt{2}}\right)$$

$$= e^{-p^2 t_0^2/2}.$$

The last equality follows by using the fact $\int_{-\infty}^{\infty} e^{-z^2} \, dz = \sqrt{\pi}$. Notice that, for small t_0, $\gamma(t)$ is sharply peaked but $\widehat{\gamma}(p)$ is flattened, and vice versa. The same assertion holds in general. Therefore, the spectra of a Gaussian pulse are again a Gaussian pulse given by

$$\widehat{\gamma}_\varepsilon(p) = 2\varepsilon \, e^{-p^2 \varepsilon^2/2},$$

where the width of transform is reciprocal of the width of the input function (Fig. 10.3).

The next two remarks are important to our subsequent discussion.

1. Like Fourier series, the integral representation of an *admissible* continuous function displays the *Gibb's phenomenon* at points of discontinuities. For example, the integral representation

$$\lim_{\alpha \to \infty} \int_0^\alpha \frac{\cos(p\,t)\sin(p)}{p} \, dp$$

Fig. 10.3 Gaussian pulse is
a self-dual waveform

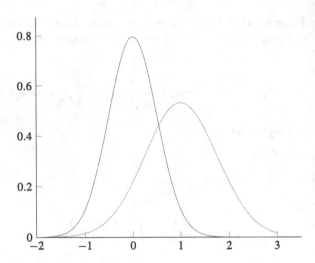

of the function $(\pi/2) f(t)$ oscillates near its discontinuities, where $f(t)$ is as in
Example 10.5 (with $a = 1$). As $\alpha \to \infty$, these oscillations do not disappear, but
they get accumulated around the points ± 1.

2. It therefore makes sense to say that the Fourier transform is to a aperiodic function
   what a Fourier series is to a periodic function. In what follows, we will come
   across many illustrations to gradually fall in with this assertion.

A Simple Application

We have seen earlier how Fourier series of a periodic function can be used to find
the sum of some interesting infinite series. In the same way, it is possible to compute certain important improper integrals by using Fourier transforms of a suitable
(aperiodic) function. The next to examples illustrates the procedure.

Example 10.10 By Example 10.5, we have

$$\int_0^\infty \cos(p\,t) \frac{\sin(a\,p)}{p}\,dp = \begin{cases} \pi/2, & |t| < a \\ 0, & |t| > a \end{cases},$$

so that taking $t = 0$ we obtain the *Dirichlet integral*

$$\int_0^\infty \frac{\sin(a\,x)}{x}\,dx = \frac{\pi}{2}, \quad \text{for all } a > 0.$$

Next, to compute the integral

$$\int_0^\infty \frac{\sin^2(t/2)}{t^2}\,dt,$$

we use the Fourier transform of the *triangular function*

$$f(t) = \begin{cases} a - |t|, & |t| < a \\ 0, & |t| > a > 0 \end{cases}.$$

Example 10.11 Here, with $f(t)$ as above, we have

$$\widehat{f}(p) = \int_{-\infty}^\infty e^{-ipt} f(t)\,dt = \int_{-a}^0 (a+t)e^{-ipt}\,dt + \int_0^a (a-t)e^{-ipt}\,dt$$

$$= \left[\frac{(a+t)e^{-ipt}}{-ip} + \frac{e^{-ipt}}{p^2} \right]_{-a}^0 + \left[\frac{(a-t)e^{-ipt}}{-ip} - \frac{e^{-ipt}}{p^2} \right]_0^a$$

$$= \left[\frac{a}{-ip} + \frac{1}{p^2} - \frac{e^{ipa}}{p^2} \right] + \left[-\frac{e^{-ipa}}{p^2} - \frac{a}{-ip} + \frac{1}{p^2} \right]$$

$$= \frac{2(1 - \cos pa)}{p^2}, \quad \text{for } p \neq 0.$$

It thus follows from (10.2.4) that

$$f(t) = \frac{1}{2\pi} \int_{-\infty}^\infty e^{ipt} \frac{2(1 - \cos pa)}{p^2}\,dp$$

$$= \frac{1}{\pi} \int_{-\infty}^\infty \frac{1 - \cos pa}{p^2} \cos pt\,dp$$

$$= \frac{2}{\pi} \int_{-\infty}^\infty \frac{\sin^2(pa/2)}{p^2} \cos pt\,dp.$$

In particular, for $t = 0$ and $a = 1$, we obtain

$$\int_0^\infty \frac{\sin^2(p/2)}{p^2}\,dp = \frac{\pi}{2}.$$

Fourier Sine and Cosine Transforms

The Fourier integral of an even or odd function takes a simpler form. As in the case of a Fourier series of a periodic function, the symmetry of such types of functions leads directly to conclusion that some involved integral vanishes over a subinterval of $\mathbb{R}$. It is important to note that the Fourier transform of a function don't change its *symmetry type*.

Example 10.12 Consider the function $f : \mathbb{R} \to \mathbb{R}$ given by

$$f(t) = \begin{cases} 0, & -\infty < t < -\pi \\ -1, & -\pi < t < 0 \\ 1, & 0 < t < \pi \\ 0, & \pi < t < \infty \end{cases}.$$

Clearly, we have $f(-t) = -f(t)$, which implies that $f(t)\cos(\omega t)$ is an odd function and $f(t)\sin(\omega t)$ is an even function. It thus follows that

$$A(\omega) = \frac{1}{\pi} \int\limits_{-\infty}^{\infty} f(t)\cos(\omega t)\mathrm{d}t = 0;$$

$$B(\omega) = \frac{1}{\pi} \int\limits_{-\infty}^{\infty} f(t)\sin(\omega t)\mathrm{d}t$$

$$= \frac{2}{\pi} \int\limits_{0}^{\pi} \sin(\omega t)\mathrm{d}t = \frac{2}{\pi\omega}\left(1 - \cos\pi\omega\right)$$

Therefore, we obtain

$$f(t) = \int\limits_{0}^{\infty} \left[A(\omega)\cos(\omega t) + B(\omega)\sin(\omega t)\right]\mathrm{d}\omega$$

$$= \frac{2}{\pi} \int\limits_{0}^{\infty} \frac{1 - \cos\pi\omega}{\omega} \sin(\omega t)\,\mathrm{d}\omega, \quad \text{for } t \neq -\pi.$$

Notice that integral on the right converges to the value $-1/2$ at $t = -\pi$.

More generally, by Eqs. (10.1.3) and (10.1.4), we have that

$$\frac{1}{2}\left[f(t^{+}) + f(t^{-})\right] = \frac{2}{\pi} \int\limits_{0}^{\infty} \cos(pt)\left[\int\limits_{0}^{\infty} f(t)\cos(pt)\mathrm{d}t\right]\mathrm{d}p, \qquad (10.2.8)$$

when f is an *even* admissible function; and, also

$$\frac{1}{2}\left[f(t^+) + f(t^-)\right] = \frac{2}{\pi} \int_0^\infty \sin(p\,t) \left[\int_0^\infty f(t)\sin(p\,t)dt\right] dp, \qquad (10.2.9)$$

when f is an *odd* admissible function. As said earlier, for each one of the above two equations, the integral on the right side equals f at every point of continuity $t \in \mathbb{R}$. In addition, altering the assumptions of Theorem 10.1 suitably, the above two equations prove respectively the *sine integral theorem* and the *cosine integral theorem*. Fourier used cosine integral as the definition of the Fourier transform of a function.

Definition 10.3 In view of (10.2.8) and (10.2.9), the Fourier *cosine transform* $F_c[p]$ and the Fourier *sine transform* $F_s[p]$ of an admissible function $f : [0, \infty)$ are respectively defined by

$$F_c[p] = \int_0^\infty \cos(p\,t)f(t)\,dt \qquad \text{and} \qquad F_s[p] = \int_0^\infty \sin(p\,t)f(t)\,dt. \qquad (10.2.10)$$

It is easy to verify that the following relations hold:

$$F_s[tf(t)] = -\frac{d}{dp}\left(F_c[p]\right) \qquad \text{and} \qquad F_c[tf(t)] = \frac{d}{dp}\left(F_s[p]\right). \qquad (10.2.11)$$

The *derivative property* of Fourier cosine and sine integrals, as stated in the next theorem, is a useful result that would prove to be very helpful in the sequel.

Theorem 10.4 *Suppose* $f,\ f',\ f'' : (0, \infty) \to \mathbb{R}$ *are absolutely integrable continuous function such that both* $f,\ f'$ *approaches zero as* $t \to \infty$. *Then we have*

$$F_c[f''] = -f'(0) - p^2 F_c[f] \qquad (10.2.12)$$
$$F_s[f''] = pf(0) - p^2 F_s[f] \qquad (10.2.13)$$
$$F[f''] = -p^2\,F[f], \qquad (10.2.14)$$

provided in the last equation $f \to 0$ *as* $|t| \to \infty$.

Proof For the first of these equations, we have

$$F_c[f''] = \int_0^\infty \cos pt\, f''(t)\,dt = f'(t)\cos pt\Big|_0^\infty + p \int_0^\infty \sin pt\, f'(t)\,dt,$$

which in view of assumption $t \to \infty \Rightarrow f'(t) \to 0$ implies that

$$F_c[f''] = -f'(0) + p \left\{ [f(t)\sin pt\Big|_0^\infty - p \int_0^\infty \cos pt\, f(t)\,dt \right\}.$$

Hence, using the assumption $t \to \infty \Rightarrow f(t) \to 0$, we obtain Eq. (10.2.12). The proof of the other two equations follow on similar lines.   □

Example 10.13 Let $f(t) = e^{-kt}$, $k > 0$. Then,

$$f'(t) = -ke^{-kt}, \quad f''(t) = k^2 e^{-kt} \quad \Rightarrow \quad f'(0) = -k.$$

Thus, we have that both f, $f' \to 0$ as $t \to \infty$, and so using the relation $F_c[f''(t)] = -f'(0) - p^2 F_c[f(t)]$, it follows that

$$F_c[k^2 e^{-kt}] = k - p^2 F_c[e^{-kt}] \quad \Rightarrow \quad F_c[e^{-kt}] = \frac{k}{k^2 + p^2}.$$

Similarly, it follows that $F_s[e^{-kt}] = \frac{p}{k^2+p^2}$.

Example 10.14 We show that

$$\int_0^\infty \left(\frac{1 - \cos(\pi p)}{p} \right) \sin(p t)\, dp = \begin{cases} \pi/2, & 0 < t < \pi \\ 0, & t > \pi \end{cases},$$

For, since the left side of the equation is the Fourier sine transform $F_s\left[\frac{1-\cos(\pi p)}{p}\right]$, and so Eq. (10.2.9) suggests that we apply Fourier *sine integral theorem* to the function

$$f(t) = \begin{cases} \pi/2, & 0 < t < \pi \\ 0, & t > \pi \end{cases},$$

so that

$$\int_0^\infty f(t) \sin(p t)\, dt = \frac{\pi}{2} \int_0^\pi \sin(p t)\, dt = \frac{\pi}{2} \left(\frac{1 - \cos(\pi p)}{p} \right).$$

Substituting this in Eq. (10.2.9) gives the desired equality.

A similar argument implies that the solution of the *integral equation*

$$\int_0^\infty f(t) \cos(p t)\, dp = \begin{cases} 1, & 0 < t < \pi \\ 0, & t > \pi \end{cases}$$

is the function $f(t) = \frac{2}{\pi} \frac{\sin \pi t}{t}$.

Example 10.15 We use here *Feynman's trick* to compute the Fourier cosine transform of the even function

$$f(t) = \frac{1}{1 + t^2}, \quad t \in \mathbb{R}.$$

For, suppose

$$I(p) = F_c[f(t)] = \int_0^\infty \frac{\cos(pt)}{1+t^2} \, dt \qquad (10.2.15)$$

so that

$$I(0) = \left[\tan^{-1}(t) \right]_0^\infty = \frac{\pi}{2}. \qquad (10.2.16)$$

Differentiating (10.2.15) with respect to p, we have

$$I'(p) = \frac{dI}{dp} = -\int_0^\infty \frac{t \sin(pt)}{1+t^2} \, dt$$

$$= -\int_0^\infty \frac{t^2 \sin(pt)}{t(1+t^2)} \, dt$$

$$= -\int_0^\infty \frac{(t^2+1-1) \sin(pt)}{t(1+t^2)} \, dt$$

$$= -\int_0^\infty \frac{\sin(pt)}{t} \, dt + \int_0^\infty \frac{\sin(pt)}{t(1+t^2)} \, dt$$

$$= -\frac{\pi}{2} + \int_0^\infty \frac{\sin(pt)}{t(1+t^2)} \, dt. \qquad (10.2.17)$$

Again, differentiating (10.2.17) with respect to p, we obtain

$$\frac{d^2 I}{dp^2} = \int_0^\infty \frac{\cos(pt)}{1+t^2} \, dt = I(p)$$

$$\Rightarrow I(p) = c_1 e^p + c_2 e^{-p} \quad \text{so that} \quad I'(p) = c_1 e^p - c_2 e^{-p}$$

So, using (10.2.16) and (10.2.17), it follows that

$$c_1 = 0 \quad \text{and} \quad c_2 = \frac{\pi}{2}.$$

Hence, $I(p) = (\pi/2) e^{-p}$.

Notice that, by (10.2.11), we have

$$-F_s \left[\frac{t}{1+t^2} \right] = \frac{d}{dp} \left(F_c \left[\frac{1}{1+t^2} \right] \right) = \frac{\pi}{2} e^{-p},$$

which implies

$$F_s\left[\frac{t}{1+t^2}\right] = -\frac{\pi}{2}e^{-p}.$$

Definition 10.4 Using Eqs. (10.2.8) and (10.2.9), the Fourier *inverse cosine trans-form* F_c^{-1} of the function $F_c[p]$ and the Fourier *inverse sine transform* F_s^{-1} of the function $F_s[p]$ are defined by

$$f(t) = F_c^{-1}[F_c[p]] = \frac{2}{\pi}\int_0^\infty \cos(p\,t)\,F_c[p]\,dp; \tag{10.2.18}$$

$$f(t) = F_s^{-1}[F_s[p]] = \frac{2}{\pi}\int_0^\infty \sin(p\,t)\,F_s[p]\,dp. \tag{10.2.19}$$

10.3 Transforms of Generalised Functions

As said earlier, most properties about the transform pair are *continuous version* of the corresponding properties of Fourier series. However, theory of Fourier transforms is important by its own right, where a concept from Fourier series is mostly taken as a point of reference. In fact, a lot more can be said about transform pair that have no analogy to Fourier series. For example, unlike the complex Fourier coefficients c_n, we can't recover the coefficients $A(p)$ and $B(p)$ in Eq. (10.1.3) from the function $f(t)$ by using the usual *orthogonality* argument. Here, we have

$$\int_{-\infty}^\infty e^{i(p-p')t}\,dt = \lim_{\alpha\to\infty}\int_{-\alpha}^\alpha e^{i(p-p')t}\,dt$$

$$= 2\lim_{\alpha\to\infty}\frac{\sin[\alpha(p-p')]}{(p-p')},$$

where the function on the right side peaks at $p = p'$, with the height 2α and the width π/α. Since the peak becomes infinitely high and narrow while the integral under the curve remains constant ($= 2\pi$), the function on the right side of the last equation gives the Dirac *delta distribution*, as $\alpha\to\infty$. Therefore, we obtain

$$\int_{-\infty}^\infty e^{i(p-p')t}\,dt = 2\pi\delta(p-p'). \tag{10.3.1}$$

In general, for any $f \in C^0(\mathbb{R})$, we have

$$\int\limits_{-\infty}^{\infty} \int\limits_{-\infty}^{\infty} e^{i(p-p')t} f(p)\, dt\, dp = 2\pi \int\limits_{-\infty}^{\infty} \delta(p-p')\, f(p)\, dp.$$

We can use this relation to establish a connection between the Fourier transform of a *periodic* function to its Fourier series.

Theorem 10.5 *For a periodic signal* $f(t)$ *defined over* $\mathbb{R}$ *and of fundamental frequency* p_0, *satisfying the Dirichlet's conditions, the spectra is given by the series of discrete impulses* $\delta(p - kp_0)$, *with weights* $2\pi c_k$, $k \in \mathbb{Z}$.

Proof Suppose $f(t)$ is a 2ℓ-periodic function defined over $[-\ell, \ell]$ with *fundamental* frequency $p_0 = \pi/\ell$. We can express it as

$$f(t) = \sum_{k=-\infty}^{\infty} c_k\, e^{ikp_0 t}, \quad t \in \mathbb{R}, \tag{10.3.2}$$

where the coefficients c_k are given by

$$c_k = \frac{1}{2\ell} \int\limits_{-\ell}^{\ell} e^{-ikp_0 t} f(t)\, dt.$$

Then, Eq. (10.3.1) implies that

$$\widehat{f}(p) = \sum_{-\infty}^{\infty} c_k \int\limits_{-\infty}^{\infty} e^{i(kp_0 - p)t}\, dt$$

$$= 2\pi \sum_{-\infty}^{\infty} c_k\, \delta(kp_0 - p), \tag{10.3.3}$$

which proves the assertion. Equation (10.3.3) represents a periodic discrete **impulse train** of the function $f(t)$.

Also, the following result has no parallel for Fourier series.

Theorem 10.6 *If $f(t)$ and $g(t)$ are two admissible functions, then*

$$\int_{-\infty}^{\infty} f(p)\widehat{g}(p)dp = \int_{-\infty}^{\infty} \widehat{f}(p)g(p)dp.$$

Proof Left as a simple exercise for the reader.

Though Dirac was the first to use the generalised functions such as $\delta(t - t_0)$ in his study, but a mathematical foundation of such type of functions as *distributions* was provided by the fundamental work of Sobolev and Schwartz [4]. The Heaviside step function

$$H(t) = \begin{cases} 1, & \text{for } t > 0 \\ 0, & \text{for } t < 0 \end{cases}$$

and the Dirac delta *function* $\delta(t) = H'(t)$, also known as the *impulse function*, are simple examples of *generalised functions* in the sense that each is real-valued function having utmost countable number of discontinuities in $\mathbb{R}$. These are also known as the *tempered distributions*. Notice that Fourier transforms of *generalised functions* such as $H(t)$ and $\delta(t)$ do not exists as formal functions. For example, direct evaluation gives

$$\widehat{H}(p) = \int_{0}^{\infty} e^{-ipt}dt = \lim_{\alpha \to \infty} \frac{1 - e^{-ip\alpha}}{ip}$$

$$= \lim_{\alpha \to \infty} 2e^{-i(p\alpha/2)} \frac{\sin(p\alpha/2)}{p},$$

which is not a function in usual sense. However, as we will see shortly, the integral in Eq. (10.2.3) exists as a *distribution*. Here, we may use relation (10.3.1) to find the Fourier transforms of some important *generalised functions*, and also of periodic functions such as sine and cosine.

Taking $g(t) = e^{-ipt}$ in Theorem A.2, we obtain $F[\delta(t)] = 1$, On the other hand, it follows from Eq. (10.3.1) that the Fourier transform of the constant function $\mathbf{1}(t) = 1$ ($\forall t \in \mathbb{R}$) is given by

$$\widehat{\mathbf{1}}(p) = 2\pi\delta(p).$$

It is possible in certain situations to compute the Fourier transform of a function that is not absolutely integrable by applying a simple *repair*.

Example 10.16 Consider the *signum function*

$$s(t) = \begin{cases} -1, & \text{for } t < 0 \\ 0, & \text{for } t = 0 \\ 1, & \text{for } t > 0 \end{cases},$$

which is clearly not absolutely integrable. However, its *repair* $e^{-a|t|}s(t)$, for small $a > 0$, is indeed absolutely integrable. So, we may define

$$
\begin{aligned}
\widehat{s}(p) &= \lim_{a \to 0} \left[-\int_{-\infty}^{0} e^{(a-ip)t}\, dt + \int_{0}^{\infty} e^{-(a+ip)t}\, dt \right] \\
&= \lim_{a \to 0} \left(-\frac{1}{a - ip} + \frac{1}{a + ip} \right) \\
&= \lim_{a \to 0} \frac{-2ip}{a^2 + p^2} \\
&= \frac{2}{ip}, \quad \text{for } p \neq 0; \quad \text{and,} \quad \widehat{s}(p) = 0, \quad \text{otherwise.}
\end{aligned}
$$

Next, we compute the Fourier transform of the Heaviside *step function* $H(t)$. In view of the definition given above, since $H(t - t_0)$ stands for the step function with unit pulse at t_0, so

$$
H(t - t_0) = \begin{cases} 1, & \text{for } t > t_0 \\ 0, & \text{for } t < t_0 \end{cases}.
$$

Thus, the expression $f(t)\{H(t - a) - H(t - b)\}\, (a < b)$ represents the part of the function $f(t)$ for $t \in (a, b)$. This explains why step function $H(t)$ is used to deal with piecewise continuous (*forcing*) functions.

Example 10.17 Viewing the step function $H(t)$ as the limit of the one-sided *exponential decaying functions* f_a $(a > 0)$, where

$$
f_a(t) = \begin{cases} e^{-at}, & \text{for } t \geq 0 \\ 0, & \text{for } t < 0 \end{cases},
$$

it follows that

$$
\widehat{f_a}(\omega) = \frac{1}{a + i\omega} = \frac{a}{a^2 + \omega^2} - i\, \frac{\omega}{a^2 + \omega^2}.
$$

So, Eq. (10.3.1) implies that

$$
\frac{a}{a^2 + \omega^2} \to \pi\, \delta(\omega) \quad \text{and} \quad -i\, \frac{\omega}{a^2 + \omega^2} \to \frac{1}{i\omega},
$$

as $a \to 0$. Thus, we obtain

$$
\widehat{H}(\omega) = \pi\, \delta(\omega) + \frac{1}{i\omega}.
$$

Likewise, it follows that for the *negative* time unit step

$$H^-(t) = \begin{cases} 1, & \text{for } t < 0 \\ 0, & \text{for } t > 0 \end{cases},$$

we may take $\widehat{H}^-(\omega) = \pi\, \delta(\omega) - \dfrac{1}{i\omega}$.

We can also compute the Fourier transform of a periodic signals by allowing impulses in $\widehat{f}(p)$.

Example 10.18 Notice that in view of Eq. (10.3.2) it suffices to consider the sinusoidal components separately. First, let $f(t) = \cos(p_0 t)$ $(t \in \mathbb{R})$, then

$$\widehat{f}(p) = \int_{-\infty}^{\infty} e^{-ipt} \cos(p_0 t)\, dt$$

$$= \frac{1}{2} \int_{-\infty}^{\infty} e^{-ipt} [e^{ip_0 t} + e^{-ip_0 t}]\, dt$$

$$= \frac{1}{2} \int_{-\infty}^{\infty} e^{-i(p-p_0)t}\, dt + \frac{1}{2} \int_{-\infty}^{\infty} e^{-i(p+p_0)t}\, dt$$

$$= \pi \delta(p - p_0) + \pi \delta(p + p_0),$$

using Eq. (10.3.1). Likewise, it follows that for $g(t) = \sin(p_0 t)$ $(t \in \mathbb{R})$, we have

$$\widehat{g}(p) = -i\pi \delta(p - p_0) + i\pi \delta(p + p_0).$$

It is important to observe that the sinusoids $\sin(p_0 t)$, $\cos(p_0 t)$ have the same *amplitude spectra* but different *phase spectra*. In actual application, to do the spectral analysis, these are plotted in two different planes.

10.4   Fundamental Properties

As said before, we need transform pair in Sect. 10.5 to solve some IBVPs defined over an *infinite domain*. To serve that purpose, we discuss here some properties to facilitate the procedure. Transform pair share most properties of usual integral transforms. Most of the properties stated here for transform pair hold good for the Fourier cosine and sine transfers.

Example 10.19 Observe that the *signum function*

$$s(t) = \begin{cases} -1, & \text{for } t < 0 \\ 0, & \text{for } t = 0 \\ 1, & \text{for } t > 0 \end{cases}$$

can be written as

$$s(t) = -1 + 2H(t)\,(t \neq 0),$$

where $H(t)$ is the Heaviside function. So, by linearity and Example 10.17, we obtain

$$\widehat{s}(p) = -2\pi\,\delta(p) + 2\pi\,\delta(p) + \frac{2}{ip} = \frac{2}{ip}, \quad p \neq 0.$$

Theorem 10.7 (Scaling) *For* $f \in L^1(\mathbb{R})$ *and scalar* $\alpha \neq 0$,

$$\widehat{f}(p) = F[f(t)] \quad \Rightarrow \quad F[f(\alpha t)] = 1/\alpha\,\widehat{f}\left(\frac{p}{\alpha}\right).$$

Proof By definition,

$$F[f(\alpha t)] = \int_{-\infty}^{\infty} e^{-ipt} f(\alpha t)\,dt$$

$$= \frac{1}{\alpha} \int_{-\infty}^{\infty} e^{-i(p/\alpha)u} f(u)\,du, \quad \text{taking } u = \alpha t,$$

$$= \frac{1}{\alpha}\,\widehat{f}\left(\frac{p}{\alpha}\right). \qquad \square$$

In practical terms, the property says that compression of the time scale corresponds to expansion of the frequency scale. As the transform expands horizontally the other one not only contracts horizontally but keeps growing vertically so that the area beneath it stays constant. A particular interesting case arises when $f(t)$ is periodic or an impulse.

Example 10.20 We know that $F[e^{-|t|}] = 2/(1 + p^2)$, and so

$$F[e^{-|t/a|}] = \frac{2|a|}{1 + a^2 p^2}, \quad a \neq 0.$$

Notice that the case $\alpha = -1$ of scaling property leads to the *time-reversal property* of the Fourier transforms. That is,

$$f(t) = g(-t) \quad \Rightarrow \quad \widehat{f}(p) = \widehat{g}(-p).$$

Example 10.21 Using Example 10.9, it follows that for $a > 0$,

$$F[e^{-a t^2}] = F[e^{-(\sqrt{a} t)^2}] = \sqrt{\frac{\pi}{a}}\, e^{-p^2/4a}.$$

Theorem 10.8 (Shifting) *For $f \in L^1(\mathbb{R})$,*

$$\widehat{f}(p) = F[f(t)] \quad \Rightarrow \quad F[f(t - \alpha)] = e^{-ip\alpha}\, \widehat{f}(p).$$

Proof By definition,

$$F[f(t - \alpha)] = \int\limits_{-\infty}^{\infty} e^{-ipt} f(t - \alpha)\, dt$$

$$= e^{-ip\alpha} \int\limits_{-\infty}^{\infty} e^{-ip u)} f(u)\, du, \quad \text{taking } u = t - \alpha$$

$$= e^{-ip\alpha}\, \widehat{f}(p). \qquad \square$$

The practical aspect of this property is that if a time-continuous signal $f(t)$ is shifted by an amount α in the positive direction, then there is no changes in *amplitude*, but each spectra component is delayed in phase by an amount proportional to p. So, higher the frequency the greater would be the change in *phase angle*, where the proportionality constant describing the linear change in the phase is $2\pi\alpha$.

Example 10.22 Consider the *sinusoidal signal* $f(t) = \cos(p_0 t + \phi)$. We write $f(t) = \cos(p_0[t + \phi/p_0])$ so that time-shift property and Example 10.18 implies

$$\widehat{f}(p) = \pi e^{ip\phi/p_0} \left[\delta(p - p_0) + \delta(p + p_0)\right].$$

Theorem 10.9 (Symmetry) *For $f \in L^1(\mathbb{R})$,*

$$F[f(t)] = \widehat{f}(p) \quad \Rightarrow \quad F[\widehat{f}(t)] = 2\pi f(-p).$$

Proof By Fourier inversion formula,

$$f(t) = \frac{1}{2\pi} \int\limits_{-\infty}^{\infty} e^{ipt} \widehat{f}(p)\, dp = \frac{1}{2\pi} \int\limits_{-\infty}^{\infty} e^{ixt} \widehat{f}(x)\, dx.$$

It thus follows that

$$2\pi f(-p) = \int\limits_{-\infty}^{\infty} e^{-ixp}\,\widehat{f}(x)\,dx$$

$$= \int\limits_{-\infty}^{\infty} e^{-ipt}\,\widehat{f}(t)\,dt$$

$$= F[\widehat{f}(t)]. \qquad \Box$$

Example 10.23 We show that the Fourier transform of the function $f(t) = (1 + a^2 t^2)^{-1}$ is given by $(\pi/|a|)e^{-|p/a|}$. For, by Example 10.20, we know that

$$F\left[\frac{1}{2|a|}e^{-|t/a|}\right] = \frac{1}{1 + a^2 p^2} = g(p),$$

and so the assertion holds by Theorem 10.9.

Theorem 10.10 (Modulation) *For $f \in L^1(\mathbb{R})$,*

$$F[e^{i\omega t} f(t)] = \widehat{f}(p - \omega) \quad and \quad F[e^{-i\omega t} f(t)] = \widehat{f}(p + \omega).$$

Proof By definition,

$$F[e^{i\omega t} f(t)] = \int\limits_{-\infty}^{\infty} e^{-ipt} e^{i\omega t} f(t)\,dt$$

$$= \int\limits_{-\infty}^{\infty} e^{-i(p-\omega)t} f(t)\,dt,$$

$$= \widehat{f}(p - \omega).$$

Likewise, we have $F[e^{-i\omega t} f(t)] = \widehat{f}(p + \omega)$.

In particular, using the identity $e^{i\omega t} = \cos(\omega t) + i \sin(\omega t)$, we have

$$F[\cos(\omega t) f(t)] = \frac{1}{2} F\left[e^{i\omega t} f(t)\right] + \frac{1}{2} F\left[e^{-i\omega t} f(t)\right]$$

$$= \frac{1}{2}\left[\widehat{f}(p - \omega)) + \widehat{f}(p + \omega)\right]; \quad \text{and,}$$

$$F[\sin(\omega t) f(t)] = \frac{1}{2i} F\left[e^{i\omega t} f(t)\right] - \frac{1}{2i} F\left[e^{-i\omega t} f(t)\right]$$

$$= \frac{1}{2i}\left[\widehat{f}(p - \omega)) - \widehat{f}(p + \omega)\right].$$

In the *amplitude modulation* (*AM*) mode of transmitting, with ω being the *carrier frequency*, an arbitrary signal $f(t)$ is *modulated* as $e^{i\omega t} f(t)$, where the peak amplitude of $f(t)$ is assumed to be far smaller than ω. The modulation property of Fourier transform states that the information about the spectrum of $f(t)$ suffices, with the spectrum of the modulation $e^{i\omega t} f(t)$ is now centred at ω.

Example 10.24 We write the *pulsed cosine*

$$f(t) = \begin{cases} \cos t, & \text{for } |t| \leq a \\ 0, & \text{for } |t| > 0 \end{cases}$$

as the function $f(t) = \cos t\, g(t)$, where $g(t)$ is the rectangular pulse of width $a > 0$. Then, by Example 10.5,

$$\widehat{f}(p) = \frac{1}{2} [\widehat{g}(p-1)) + \widehat{g}(p+1)] = \frac{\sin(a(p-1))}{p-1} + \frac{\sin(a(p+1))}{p+1}.$$

In contrast to amplitude modulation, a *frequency modulation* (FM) such as

$$f(f) = \begin{cases} \alpha_1, & \text{for } |t| < \ell/2 \\ 0, & \text{for } |t| > \ell/2 \end{cases}$$

transmits information by *instantaneous change* in its carrier frequency α_1, which is expressed by

$$g(t) = e^{i[\int_{-\infty}^{t} f(u)du - \alpha_1(\ell/2)] + i\alpha t}$$

$$= \begin{cases} e^{-i\alpha_1(\ell/2)} e^{i\alpha t}, & \text{for } t < -\ell/2 \\ e^{-i\alpha_1 t} e^{i\alpha t}, & \text{for } -\ell/2 < t < \ell/2 \\ e^{i\alpha_1(\ell/2)} e^{i\alpha t}, & \text{for } t > \ell/2 \end{cases} \qquad (10.4.1)$$

Theorem 10.11 (Differentiation) *Suppose $f^{(k)}$, $1 \leq k < n$, are continuous, $f^{(n)}$ is piecewise continuous such that each is an admissible function, and $\lim_{t\to\infty} f^{(k)}(t) = \lim_{t\to-\infty} f^{(k)}(t) = 0$, for all $0 \leq k < n$. Then*

$$\widehat{f}(p) = F[f(t)] \Rightarrow F[f^{(n)}(t)] = (i\,p)^n\, \widehat{f}(p), \quad \text{for all } n \geq 1.$$

Proof For $n = 1$, we have

$$F[f'](p) = \int_{-\infty}^{\infty} e^{-ipt} f'(t)\, dt$$

$$= e^{-ipt} f(t)\Big|_{-\infty}^{\infty} + ip \int_{-\infty}^{\infty} e^{-ipt} f(t)\, dt$$

$$= ip\widehat{f}(p).$$

The proof can be completed by induction.

The operation of *differentiation* in the time domain corresponds to *multiplication* by the frequency, which makes it easier to solve some differential equations in the frequency domain.

Example 10.25 By Example 10.23, we know that

$$F\left[\frac{1}{1+a^2t^2}\right] = \frac{\pi}{|a|}e^{-|p/a|},$$

and so

$$F\left[\frac{t}{(1+a^2t^2)2}\right] = \frac{i\pi p}{2|a|^3}e^{-|p/a|}.$$

Example 10.26 Consider the *generalised function*

$$r(t) = \max(t, 0) = \begin{cases} 0, & \text{for } t < 0 \\ t, & \text{for } t \geq 0 \end{cases},$$

known as the *ramp function*. Clearly, it is differentiable except at $t = 0$. Taking $r'(0) = 1$, it follows that the step function $H(t) = r'(t)$. Thus,

$$\widehat{r}(p) = \frac{1}{ip}\widehat{H}(p) = \frac{1}{ip}[\pi\,\delta(p) + \frac{1}{ip}] = -\frac{i\pi}{p}\,\delta(p) - \frac{1}{p^2}.$$

Example 10.27 Consider the *absolute-value* function

$$\alpha(t) = \begin{cases} -1, & \text{for } t < 0 \\ 1, & \text{for } t \geq 0 \end{cases},$$

which is everywhere differentiable except at $t = 0$. Using the fact that the signum function $s(t)$ of Example 10.19 is the *generalised derivative* of the function $\alpha(t)$, it follows that

$$\widehat{\alpha}(p) = \frac{1}{ip}\widehat{s}(p) = \frac{1}{ip}\frac{2}{ip} = -\frac{2}{p^2}.$$

Notice that, since the absolute-value function $\alpha(t)$ equals the sum $r(t) + r(-t)$, we can also use Example 10.26 to compute the Fourier transform of $\alpha(t)$.

Theorem 10.12 (Integration) *Suppose* $f(t) = \int_{-\infty}^{t} g(u)du$, *then*

$$\widehat{f}(p) = \frac{\widehat{g}(p)}{ip} + \pi\,\widehat{g}(0)\,\delta(p).$$

Proof Left as an exercise for the reader.

Theorem 10.13 (Multiplication by t) *If $t^n f(t) \in L^1(\mathbb{R})$ for $n \geq 1$, then*

$$\widehat{f}(p) = F[f(t)] \quad \Rightarrow \quad F[t^n f(t)] = (i)^n \widehat{f}^{(n)}(p).$$

Proof Differentiating with respect to p the relation

$$\widehat{f}(p) = \int\limits_{-\infty}^{\infty} e^{-ipt} f(t)\,dt,$$

and using the Lagrange rule, we obtain

$$\widehat{f}'(p) = -i \int\limits_{-\infty}^{\infty} e^{-ipt}[tf(t)]\,dt = -i\, F[t\, f(t)].$$

This proves the assertion for $n = 1$. The proof can be completed by induction on n.

Example 10.28 Let us compute $\widehat{f}(p) = F[e^{-t^2}]$, using this property: For, since $(e^{-t^2})' = -2te^{-t^2}$, we have

$$-\frac{ip}{2}\widehat{f}(p) = F[te^{-t^2}] \quad \Rightarrow \quad \widehat{f}(p) = \frac{2i}{p}F[te^{-t^2}].$$

It thus follows that

$$\widehat{f}(p) = -\frac{2}{p}\widehat{f}'(p) \quad \Rightarrow \quad \widehat{f}(p) = Ae^{-\int (p/2)dp} = e^{-p^2/4},$$

which is the same as got in Example 10.9.

The *energy E_s* and the *power P* of a signal f are respectively given by

$$E_s(f) = \int\limits_{-\infty}^{\infty} |f(t)|^2 dt;$$

$$P(f) = \lim_{\ell \to \infty} \frac{1}{\ell} \int\limits_{-\ell}^{\ell} |f(t)|^2 dt.$$

The energy $E_s(f)$ is related to actual energy E by the relation

$$E = \frac{E_s(f)}{Z(f)} = \frac{1}{Z(f)} \int\limits_{-\infty}^{\infty} |f(t)|^2 dt,$$

where $Z(f)$ represents the magnitude of the *load driven by the signal* f. For example, suppose $f(t)$ stands for the potential of an electrical signal (in volts) propagating through a transmission line, then

$$Z = \frac{\text{amplitude of voltage}}{\text{amplitude of current}},$$

is known as the *characteristic* impedance of the transmission line.

Theorem 10.14 (Plancherel theorem) *For a signal* $f \in L^1(\mathbb{R})$,

$$E_s(f) = \frac{1}{2\pi} \int_{-\infty}^{\infty} |\widehat{f}(p)|^2 dp. \tag{10.4.2}$$

Proof Using the fact

$$[\widehat{f}(-p)]^* = F[f^*(t)] = \int_{-\infty}^{\infty} e^{-ipt} f^*(t)\, dt,$$

it follows that

$$E_s(f) = \int_{-\infty}^{\infty} |f(t)|^2 dt = \int_{-\infty}^{\infty} f(t) f^*(t)\, dt$$

$$= \int_{-\infty}^{\infty} f(t)\, F^{-1}[(\widehat{f}(-p))^*]\, dt$$

$$= \int_{-\infty}^{\infty} f(t) \left[\frac{1}{2\pi} \int_{-\infty}^{\infty} e^{ipt} (\widehat{f}(-p))^*\, dp \right] dt$$

$$= \int_{-\infty}^{\infty} f(t) \left[\frac{1}{2\pi} \int_{-\infty}^{\infty} e^{-ipt} (\widehat{f}(p))^*\, dp \right] dt$$

$$= \frac{1}{2\pi} \int_{-\infty}^{\infty} (\widehat{f}(p))^* \left[\frac{1}{2\pi} \int_{-\infty}^{\infty} e^{-ipt} f(t)\, dt \right] dp$$

$$= \frac{1}{2\pi} \int_{-\infty}^{\infty} (\widehat{f}(p))^*\, \widehat{f}(p)\, dp$$

$$= \frac{1}{2\pi} \int_{-\infty}^{\infty} |\widehat{f}(p)|^2 dp. \qquad \square$$

Notice that as $|\widehat{f}(p)|^2$ represents the *spectral energy density* of a time-continuous signal $f(t)$, the Parseval identity (10.4.2) answers the following practical question: *Having the output data as a time-continuous signal $f(t)$, is it possible to find its energy in terms of the spectrum $\widehat{f}(p)$?* The next example illustrates the point.

Example 10.29 The function given by

$$f(t) = \begin{cases} 0, \text{ for } t < 0 \\ e^{-t/\ell} \sin p_0 t, \text{ for } t > 0 \end{cases},$$

represents physical quantities such as *displacement* of a damped harmonic oscillator; the *radiation* of an electric field; the *current* in some communication equipment; etc., with *power* proportional to $|f(t)|^2$. Also, the *total energy* $E_s(f)$ is proportional to $\int_0^\infty |f(t)|^2 dt$, which according to Theorem 10.14 is given by

$$\int_0^\infty |\widehat{f}(p)|^2 dp,$$

where, in this case, we have

$$\widehat{f}(p) = \int_{-\infty}^\infty e^{-ipt} f(t)\, dt = \int_0^\infty e^{-[ip+(1/\ell)]t} \sin p_0 t\, dt$$

$$= \frac{1}{2}\left[\frac{1}{p + p_0 - (i/\ell)} - \frac{1}{p - p_0 - (i/\ell)} \right].$$

In particular, when f represents an electric field radiated wave, the magnitude $|\widehat{f}(p)|^2$ is a measure of the energy radiating per unit frequency interval.

The relation (10.4.2) is a special case of a very powerful theorem of harmonic analysis: *Fourier transform is a unitary operator on the space $L^2(\mathbb{R})$.*

Theorem 10.15 *If f, $g \in L^1(\mathbb{R})$ are complex-valued time-continuous signals, then*

$$\int_{-\infty}^\infty f(t)g^*(t)dt = \int_{-\infty}^\infty \widehat{f}(p)[\widehat{g}(p)]^* dp. \tag{10.4.3}$$

Proof Imitate the proof of the Theorem 10.14.

The relation (10.4.2) also helps to evaluate some intractable integrals.

Example 10.30 It follows from relation (10.2.7) that the Fourier transform of rect-angular function f is the function $\widehat{f}(p) = 2\sin(ap)/p$, so by Eq. (10.4.2)

$$\frac{1}{2\pi} \int_{-\infty}^{\infty} |\widehat{f}(p)|^2 \mathrm{d}p = \int_{a}^{a} (1)^2 dt = 2a.$$

We thus obtain

$$\int_{0}^{\infty} \frac{\sin^2(ap)}{p^2} = \frac{\pi}{2} a,$$

which is same as we got earlier in Example 10.11.

Convolution Theorem

In physical terms, the convolution theorem proves that the *convolution product* of two signals corresponds to the ordinary multiplication of their *spectra*. Therefore, a linear time-invariant system can be expressed in relatively simpler form when subject to convoluted signals (such as a *filter*).

Definition 10.5 The **convolution product** of two real or complex-valued functions f, g, denoted by $f * g$, is given by

$$h(t) = (f * g)(t) = \int_{-\infty}^{\infty} f(x)g(t - x)\,\mathrm{d}x, \quad t \in \mathbb{R},$$

provided the integral exists.

For example, we have

$$f, g \in L^1(\mathbb{R}) \quad \Rightarrow \quad f * g \in L^1(\mathbb{R}).$$

In geometric terms, the integral

$$\int_{-\infty}^{\infty} f(x)g(t - x)\,\mathrm{d}x = \int_{-\infty}^{\infty} f(t - x)g(x)\,\mathrm{d}x$$

is viewed as the limiting value of the *Riemann sum*

$$\sum_{n=-\infty}^{\infty} f(t - x_n)g(x_n)\Delta x_n = \sum_{n=-\infty}^{\infty} f_n(t)g(x_n)\Delta x_n,$$

where $f_n(t) = f(t - x_n)$ is a translation of $f(t)$ by an amount x_n along the x-axis. Thus, the Riemann sum becomes a linear combination of translations of $f_n(t)$, with coefficients $g(x_n)\Delta x_n$. Hence, the convolution product $f * g$ could be interpreted as the continuous *superposition of translates*[3] of the function $f(t)$. Also, for a fixed g, the mapping

$$h = - * g : f \mapsto f * g$$

defines a *functional* on a space containing functions f. So, the product $f * g$ in approximation theory is viewed as the weighted average of the function $f(t)$, with $g(t - x)$ being the *weight function* for any fixed t.

Example 10.31 For $t > 0$, and positive constants $a \neq b$, let $f(t) = e^{-at}$ and $g(t) = e^{-bt}$ be two truncated *decay functions*. Then, since $g(t - u) = e^{-b(t-u)} = 0$ for $t - u \leq 0$ i.e., for $u \geq t$,

$$(f * g)(t) = e^{-at} * e^{-bt}$$

$$= \int_{-\infty}^{\infty} e^{-au} e^{-b(t-u)} \, du$$

$$= e^{-bt} \int_{0}^{t} e^{-(a-b)u} \, du$$

$$= \frac{e^{-bt}[1 - e^{-(a-b)t}]}{a - b}$$

$$. = \frac{e^{-bt} - e^{-at}}{a - b}.$$

Theorem 10.16 (Convolution Theorem) *Let* $f, g \in L^1(\mathbb{R})$. *If* $\widehat{f}(p) = F[f(t)]$ *and* $\widehat{g}(p) = F[g(t)]$, *then*

$$F[(f * g)(t)] = \widehat{f}(p) \cdot \widehat{g}(p).$$

Proof By Fubini's theorem, we have

---

[3] This geometric interpretation of the convolution product explains its German name *Faltung*, meaning *folding*. Further, it is easy to verify that the *convolution product* as a binary operation is *commutative*, *associative*, and *distributes* over the (pointwise) addition of functions.

$$F[(f * g)(t)] = \int\limits_{-\infty}^{\infty} \left[\int\limits_{-\infty}^{\infty} f(u)g(t - u)\, du \right] e^{-ipt}\, dt$$

$$= \int\limits_{-\infty}^{\infty} \int\limits_{-\infty}^{\infty} f(u)g(t - u)e^{-ipt}\, dt\, du$$

$$= \int\limits_{-\infty}^{\infty} f(u)e^{-ipu} \left[\int\limits_{-\infty}^{\infty} g(t - u)e^{-ip(t-u)}\, dt \right] du$$

$$= \int\limits_{-\infty}^{\infty} f(u)e^{-ipu} \left[\int\limits_{-\infty}^{\infty} g(z)e^{-ipz}\, dz \right] du$$

$$= F(p) \int\limits_{-\infty}^{\infty} f(u)e^{-ipu}\, du = \widehat{f}(p) \cdot \widehat{g}(p).$$

The convolution theorem is proved.

Example 10.32 For two known functions $g, h : \mathbb{R} \to \mathbb{R}$, it is possible to solve the following integral equation for the function f:

$$f(t) = h(t) + \int\limits_{-\infty}^{\infty} f(u)g(t - u)\, du = h(t) + (f * g)(t).$$

For, applying the Fourier transform and using Theorem 10.16, we have

$$\widehat{f}(p) = \frac{\widehat{h}(p)}{1 - \widehat{g}(p)},$$

which implies that

$$f(t) = \frac{1}{2\pi} \int\limits_{-\infty}^{\infty} \frac{\widehat{h}(p)}{1 - \widehat{g}(p)} e^{ipt}\, dp,$$

using the Fourier inverse transform (Definition 10.2).

A simple imitation of proof of Theorem 10.16 yields the following result.

Theorem 10.17 If $\widehat{f}(p) = F[f(t)]$ and $\widehat{g}(p) = F[g(t)]$, then

$$F[f(t) \cdot g(t)] = \frac{1}{2\pi} \int\limits_{-\infty}^{\infty} \widehat{f}(\lambda)\widehat{g}(p - \lambda)\, d\lambda = \frac{1}{2\pi}(\widehat{f} * \widehat{g})(\omega).$$

The above theorem says that if $\widehat{f}(p)$ is known, and $h(t) = f(t) \cdot g(t)$ is the same *signal* as $f(t)$ but turned off until $t = 0$, then

$$\widehat{h}(p) = \frac{1}{2\pi} \int\limits_{-\infty}^{\infty} \frac{\widehat{f}(p-\lambda)}{\lambda} \, d\lambda + \frac{\widehat{f}(p)}{2}.$$

There is a multitude of applications wherein *convolution* product of two function is an indispensable tool. For example, if X_1 and X_2 are two random (or even *fuzzy*) variables with probability distributions functions $f(x)$ and $g(x)$ (representing *fuzziness* of a process), then the convolution product $f * g$ is the probability distribution of the random variable $X_1 + X_2$.

Inverse Fourier Transform

The process of *Fourier synthesis* combines the contributions of all different frequencies in the spectrum of a signal to recover the original function of the time. This is achieved by applying the Fourier *inverse transform* as defined in (10.1.6) to the spectra. In view of Fourier inversion formula, computations related to Fourier inverse transforms are straight forward reversal of the procedure adopted to obtain the spectrum of a time-continuous signal. The details about some important *Fourier pairs* are summarised in the following table.

WAVEFORM	$\longleftrightarrow$	SPECTRA
1	$\longleftrightarrow$	$2\pi\delta(p)$
Unit pulse	$\longleftrightarrow$	$\text{sinc}(p)$
$\dfrac{1}{a\sqrt{2\pi}} e^{-(t^2/2a^2)}$	$\longleftrightarrow$	$e^{-(p^2a^2/2)}$
$H(t)$	$\longleftrightarrow$	$\pi\delta(p) + \dfrac{1}{ip}$
signum	$\longleftrightarrow$	$-2ip^{-1}$
$\sin(at)$	$\longleftrightarrow$	$-i\pi\delta(p-a) + i\pi\delta(p+a)$
$\cos(at)$	$\longleftrightarrow$	$\pi\delta(p-a) + \pi\delta(p+a)$
$\text{sinc}(t)$	$\longleftrightarrow$	Unit pulse
$e^{-at}, a > 0$	$\longleftrightarrow$	$(a+ip)^{-1}$
$r(t)$ (ramp function)	$\longleftrightarrow$	$\dfrac{\pi}{ip}\delta(p) - \dfrac{1}{p^2}$

Notice that Eq. (A.3.5) is used to compute the Fourier inverse transform of a spectra $g(p)$ involving Dirac's delta function. The properties of Fourier transforms stated in the previous section could be used to compute the Fourier inverse transforms of some other important spectra. For example, the *symmetry property* of Fourier transforms implies that

$$F[(1 + a^2 t^2)^{-1}] = (\pi/|a|)e^{-|p/a|} \quad \text{(Example 10.23)}$$

The method of *partial fractions* and Theorem 10.16, whenever applicable, are two most frequently used methods to compute the inverse transforms. Finally, when all else fails, one may also use *contour integration* to write

$$f(t) = \frac{1}{2\pi} \oint_C e^{itz} \widehat{f}(z) dz - \frac{1}{2\pi} \int_{C_R} e^{itz} \widehat{f}(z) dz,$$

where contours C and C_R need to be chosen suitably.

10.5 Applications to Partial Differential Equations

As said earlier, the Fourier transform over $\mathbb{R}$ and Fourier sine or cosine transform over $[0, \infty)$ provide powerful tools to solve many different types of linear initial-boundary value problems in physics and engineering. Such transforms convert a problem about linear partial differential equations to a problem about linear ordinary differential equations so that, if $y(t) = u(x, t)$ is the solution of the latter, then its transform $\widehat{y}(p)$ is a solution of the corresponding algebraic equation. We can solve for $\widehat{y}(p)$, and then apply the respective inverse Fourier transform to obtain the solution $y(t)$. The following notations are used while dealing with 1-dimensional cases of linear initial-boundary value problems:

$$\bar{u}(p, t) = \int_{-\infty}^{\infty} e^{-ipx} u(x, t) \, dx;$$

$$\bar{u}_c(p, t) = \int_0^{\infty} \cos(px) u(x, t) \, dx; \quad \text{and}$$

$$\bar{u}_s(p, t) = \int_0^{\infty} \sin(px) u(x, t) \, dx.$$

Further, assuming that both $u(x, t), u_x(x, t) \to 0$ as $x \to \infty$, it follows from Theorem 10.4 that we have

$$F_s[u_{xx}(x, t)] = p\, u(0, t) - p^2 \bar{u}_s(p, t); \tag{10.5.1}$$

$$F_c[u_{xx}(x, t)] = -u_x(0, t) - p^2 \bar{u}_c(p, t). \tag{10.5.2}$$

In addition, when both $u(x, t)$ and $u_x(x, t)$ approach to 0 as $|x| \to \infty$, we also have

$$\overline{u}(p,t) = F\big[u_{xx}(x,t)\big] = -p^2\,\overline{u}(p,t), \qquad (10.5.3)$$

Notice that the relation (10.5.1) implies that Fourier sine transform applies only when $u(0,t)$ is known. Similarly, the relation (10.5.2) implies that Fourier cosine transform applies only when the *gradient condition* $u_x(0,t)$ is known.

To begin with, we first apply Fourier transform to convert a full *time-dependent* wave equation of the form

$$u_{tt} - c^2\,u_{xx} = f(x,t)$$

to the *Helmholtz wave equation* given by

$$\nabla^2 u + k^2 u = -\frac{1}{c^2}\,f(x,t), \qquad \text{for } k \geq 0, \qquad (10.5.4)$$

For, suppose $u = u(x,t)$ and $f = f(x,t)$ are sufficiently smooth functions such that

$$u(x,t) = \frac{1}{2\pi} \int\limits_{-\infty}^{\infty} e^{ipt}\,\overline{u}(p,t)\,\mathrm{d}p;$$

$$f(x,t) = \frac{1}{2\pi} \int\limits_{-\infty}^{\infty} e^{ipt}\,\overline{f}(p,t)\,\mathrm{d}p,$$

where the (temporal) Fourier coefficients $\overline{u}(p,t)$ and $\overline{f}(p,t)$, for each *fixed* p, define functions of x. Notice that the *linearity* of the operators $L \equiv \partial_t^2 - c^2\,\partial_x^2$ allows one to write the wave equation in terms of these (temporal) Fourier coefficients. Now, using the eigenvalue relation for the *Fourier kernel* $\partial_t^2(e^{ipt}) = -p^2 e^{ipt}$, we have

$$0 = -\partial_t^2 u + c^2\,\nabla^2 u + f(x,t)$$

$$= -\partial_t^2 \frac{1}{2\pi} \int\limits_{-\infty}^{\infty} e^{ipt}\overline{u}(p,t)\,\mathrm{d}p + \frac{c^2}{2\pi}\nabla^2 \int\limits_{-\infty}^{\infty} e^{ipt}\overline{u}(p,t)\,\mathrm{d}p + f(x,t)$$

$$= \frac{1}{2\pi} \int\limits_{-\infty}^{\infty} \Big[-\partial_t^2(e^{ipt})\overline{u}(p,t) + c^2\,e^{ipt}\nabla^2\overline{u}(p,t) + e^{ipt}\overline{f}(p,t)\Big]\,\mathrm{d}p$$

$$= \frac{1}{2\pi} \int\limits_{-\infty}^{\infty} \Big[p^2\overline{u}(p,t) + c^2\,\nabla^2\overline{u}(p,t) + \overline{f}(p,t)\Big]e^{ipt}\,\mathrm{d}p, \qquad (10.5.5)$$

which implies that if $f(x,t)$ is known (so that $\overline{f}(p,t)$ is known), then it is possible to find a solution of the original equation by setting $\overline{u}(p,t)$ to be a solution of the (10.5.4), i.e., we obtain

$$\nabla^2 u + k^2 u = -\frac{1}{c^2} f(x, t), \quad \text{with} \quad k = \frac{p}{c}.$$

It is easy to show that the Fourier transform in (10.5.5) gives only a necessary condition, which suffice if only solutions of an initial-boundary value problem is sought.

Heat Equations

We first consider the case when heat conduction is taking place in a semi-infinite rod. For $x \geq 0$ and $t > 0$, let $u(x, t)$ be the temperature in an insulated *semi-infinite* rod. Suppose the initial temperature distribution of the rod is given by the function $f(x)$, $x \in (0, \infty)$, and the end $x = 0$ is maintained at temperature $g(t)$, $t > 0$. Recall that u is the solution of the 1-dimensional initial-boundary value problem

$$
\begin{aligned}
u_t &= c^2 u_{xx}, & x \geq 0 \ \text{ and } \ t \geq 0; \\
u(0, t) &= g(t), & \text{for } \ t > 0; \\
u(x, 0) &= f(x), & \text{for } \ x \in [0, \infty).
\end{aligned}
\tag{10.5.6}
$$

For convenience, we may take $c = 1$. Notice that, by applying sine transform to heat equation (10.5.6), we have

$$
\begin{aligned}
\frac{\mathrm{d}}{\mathrm{d}t} (\bar{u}_s(p, t)) &= \int_0^\infty \sin(px) \frac{\partial(u(x, t))}{\partial t} \, \mathrm{d}x \\
&= F_s[u_t] = F_s[u_{xx}] \\
&= p\, u(0, t) - p^2 \bar{u}_s(p, t) \\
\Rightarrow \quad \frac{\mathrm{d}}{\mathrm{d}t} (\bar{u}_s(p, t)) + p^2 \bar{u}_s(p, t) &= p\, u(0, t).
\end{aligned}
$$

Therefore, by the method of *integrating factor*, we obtain

$$\bar{u}_s(p, t) = e^{-p^2 t} \int e^{p^2 t} p\, u(0, t) \mathrm{d}t + A\, e^{-p^2 t}.$$

Hence, when $u(0, t) = g(t)$ is known, we may use the relation

$$\bar{u}_s(p, t) = p\, e^{-p^2 t} \int e^{p^2 t} g(t) \, \mathrm{d}t + A\, e^{-p^2 t}, \tag{10.5.7}$$

where A is determined using the initial temperature $u(x, 0) = f(x)$. Finally, applying the inverse sine transform, the complete solution is obtained. The next example illustrates the procedure.

Example 10.33 Suppose the end $x = 0$ is maintained at a constant temperature, say u_0, and the initial temperature distribution is given by the function $f(x) = u(x, 0) = e^{-x}$. Then, by Eq. (10.5.7), we obtain

$$\bar{u}_s(p, t) = \frac{u_0}{p} + A\, e^{-p^2 t}, \tag{10.5.8}$$

and so we have

$$A = \bar{u}_s(p, 0) - \frac{u_0}{p}$$

$$= \int_0^\infty \sin(p\, x) u(x, 0)\, dx - \frac{u_0}{p}$$

$$= \int_0^\infty e^{-x} \sin(p\, x)\, dx - \frac{u_0}{p}$$

$$= \frac{p}{1 + p^2} - \frac{u_0}{p} = \frac{(1 - u_0)p^2 - u_0}{p(1 + p^2)}.$$

Substituting this expression for A in Eq. (10.5.8), we obtain

$$\bar{u}_s(p, t) = \frac{u_0}{p} + \frac{p}{1 + p^2}\, e^{-p^2 t} - \frac{u_0}{p}\, e^{-p^2 t}.$$

Applying inverse sine transform, we have

$$u(x, t) = F_s^{-1}[\bar{u}_s(p, t)]$$

$$= u_0 + F_s^{-1}\left[\frac{p e^{-p^2 t}}{1 + p^2}\right] - u_0 F_s^{-1}\left[\frac{e^{-p^2 t}}{p}\right]$$

$$= u_0 + \frac{2}{\pi} \int_0^\infty \frac{p \sin px}{1 + p^2}\, e^{-p^2 t}\, dp - \frac{2u_0}{\pi} \int_0^\infty \frac{\sin px}{p}\, e^{-p^2 t}\, dp$$

which gives the complete solution of 1-dimensional heat equation, in this case. Notice that $u(x, t)$ satisfy the initial condition trivially, and the boundary condition $u(x, 0) = e^{-x}$, by Exercise 10.1.

We next solve an initial-boundary value problem with respect to *Neumann condition*. That is, the assigned boundary condition is given as the gradient $u_x(0, t)$. Then, in this case, we apply sine transform to the heat equation (10.5.6) so that we have

$$\frac{d}{dt}(\bar{u}_c(p, t)) = \int_0^\infty \cos(px)\frac{\partial(u(x, t))}{\partial t}\,dx$$

$$= F_c[u_t] = F_c[u_{xx}] = -u_x(0, t) - p^2\bar{u}_s(p, t)$$

$$\Rightarrow \frac{d}{dt}(\bar{u}_c(p, t)) + p^2\bar{u}_s(p, t) = -u_x(0, t)$$

$$\Rightarrow \bar{u}_c(p, t) = -e^{-p^2t}\int e^{p^2t}u_x(0, t)\,dt + A\,e^{-p^2t}.$$

Therefore, given that $u_x(0, t)$ is known, we will use

$$\bar{u}_c(p, t) = -e^{-p^2t}\int e^{p^2t}u_x(0, t)\,dt + A\,e^{-p^2t}, \tag{10.5.9}$$

where A is determined by using $u(x, 0) = f(x)$ as the specified initial temperature distribution. Finally, the complete solution is obtained by applying the inverse cosine transform. The next example illustrates the procedure.

Example 10.34 Suppose $u_x(0, t) = u_0$ $(t > 0)$, and $u(x, 0) = e^{-x}$ be the initial temperature distribution. Then, by Eq. (10.5.9), we have

$$\bar{u}_c(p, t) = -\frac{u_0}{p^2} + A\,e^{-p^2t}, \tag{10.5.10}$$

and so we obtain

$$A = \bar{u}_c(p, 0) + \frac{u_0}{p^2}$$

$$= \int_0^\infty \cos(px)u(x, 0)\,dx + \frac{u_0}{p^2}$$

$$= \int_0^\infty e^{-x}\cos(px)\,dx$$

$$= -\frac{1}{1 + p^2} + \frac{u_0}{p^2}.$$

Finally, substituting this expression A in Eq. (10.5.10) and applying inverse sine transform, we obtain

$$u(x, t) = F_c^{-1}[\bar{u}_c(p, t)]$$

$$= \frac{2u_0}{\pi}\int_0^\infty \frac{[e^{-p^2t} - 1]\cos px}{p^2}\,dp - \frac{2}{\pi}\int_0^\infty \frac{e^{-p^2t}\cos px}{1 + p^2}\,dp,$$

which is the complete solution of the BVP in this case. Notice that $u(x, t)$ satisfy the initial condition by Exercise 10.1, and also the gradient condition $u(x, 0) = e^{-x}$.

We consider next the case when heat conduction is taking place in an infinite rod. In this case, Eq. (10.5.3) helps to find the temperature $u(x, t)$ at $x \in \mathbb{R}$ and at time $t > 0$, subject to that $u(x, 0) = f(x)$, $x \in \mathbb{R}$, is the initial temperature distribution. Observe that by applying Fourier transform to Eq. (10.5.6), we have

$$-p^2 \overline{u}(p, t) = F[u_{xx}(x, t)] = \frac{1}{c^2} F[u_t(x, t)]$$

$$= \frac{1}{c^2} \int_{-\infty}^{\infty} e^{-ipx} \frac{\partial u(x, t)}{\partial t} \, dx$$

$$= \frac{1}{c^2} \frac{d}{dt} (\overline{u}(p, t)),$$

so that

$$\frac{d\overline{u}(p, t)}{dt} + c^2 p^2 \overline{u}(p, t) = 0 \implies \overline{u}(p, t) = A e^{-c^2 p^2 t}.$$

Also, since

$$A = \overline{u}(p, 0) = \int_{-\infty}^{\infty} e^{-ipx} u(x, 0) \, dx = \int_{-\infty}^{\infty} e^{-ipx} f(x) \, dx,$$

we have

$$\overline{u}(p, t) = \int_{-\infty}^{\infty} e^{-(i\lambda + c^2 pt)p} f(\lambda) \, d\lambda.$$

Therefore, it follows that

$$u(x, t) = F^{-1}[\overline{u}(p, t)]$$

$$= \frac{1}{2\pi} \int_{-\infty}^{\infty} e^{ixp} \left[\int_{-\infty}^{\infty} e^{-(i\lambda + c^2 pt)p} f(\lambda) \, d\lambda \right] dp$$

$$= \frac{1}{2\pi} \int_{-\infty}^{\infty} f(\lambda) \left[\int_{-\infty}^{\infty} e^{-c^2 p^2 t} e^{-i(\lambda - x)p} dp \right] d\lambda \qquad (10.5.11)$$

We may write

$$g(x,t) = \int\limits_{-\infty}^{\infty} e^{-c^2 p^2 t} e^{-ipx} dp = \int\limits_{-\infty}^{\infty} e^{-v^2 t} e^{-ipx} dp,$$

where $v = cp$. So, the integral in Eq. (10.5.11) contains $g(x - \lambda)$. By Example 10.9, we know that

$$\int\limits_{-\infty}^{\infty} e^{-ipx} e^{-v^2 t} dp = \frac{\sqrt{\pi}}{t\sqrt{2}} e^{-p^2/8t^2}$$

The next example illustrate the procedure.

Example 10.35 We find the temperature distribution $u(x,t)$ in an insulated *infinite rod*, where the initial temperature is given by

$$u(x,0) = \begin{cases} u_0, & \text{for } |x| < a \\ 0, & \text{for } |x| > a \end{cases}.$$

For, applying the Fourier transform to Eq. (10.5.6), we have

$$-p^2 \overline{u}(p,t) = F[u_{xx}(x,t)] = F[u_t(x,t)]$$

$$= \int\limits_{-\infty}^{\infty} e^{-ipx} \frac{\partial u(x,t)}{\partial t} dx$$

$$= \frac{d}{dt} (\overline{u}(p,t)),$$

so that

$$\frac{d\overline{u}(p,t)}{dt} + p^2 \overline{u}(p,t) = 0 \implies \overline{u}(p,t) = A e^{-p^2 t}.$$

And, since

$$A = \overline{u}(p,0) = \int\limits_{-\infty}^{\infty} e^{-ipx} u(x,0) dx$$

$$= u_0 \int\limits_{-a}^{a} e^{-ipx} dx$$

$$= -\frac{2 u_0 \sin ap}{p},$$

it follows that

$$u(x, t) = F^{-1}[\bar{u}(p, t)]$$

$$= -\frac{u_0}{\pi} \int\limits_{-\infty}^{\infty} e^{(ix-pt)p} \frac{\sin(ap)}{p} \, dp$$

$$= -\frac{u_0}{\pi} \int\limits_{-\infty}^{\infty} e^{-p^2 t} \left[\frac{\cos(px) \sin(ap)}{p} \right] dp$$

$$= -\frac{2 u_0}{\pi} \int\limits_{0}^{\infty} e^{-p^2 t} \frac{\cos(px) \sin(ap)}{p} \, dp$$

$$= \frac{u_0}{\pi} \int\limits_{0}^{\infty} e^{-p^2 t} \frac{[\sin((a+x)p) + \sin((a-x)p)]}{p} \, dp$$

$$= \frac{u_0}{\pi} \int\limits_{0}^{\infty} e^{-v^2} \frac{\left[\sin\left((a+x) \frac{v}{\sqrt{t}}\right) + \sin\left((a-x) \frac{v}{\sqrt{t}}\right) \right]}{\sqrt{t}} \frac{dv}{v}$$

$$= \frac{u_0}{2} \left\{ erf\left(\frac{a+x}{2\sqrt{t}}\right) + erf\left(\frac{a-x}{2\sqrt{t}}\right) \right\},$$

where erf is the *error function* defined by

$$erf(x) := \frac{2}{\sqrt{\pi}} \int\limits_{0}^{x} e^{-u^2} \, du \tag{10.5.12}$$

Notice that the following relation is used in the third equality:

$$\int\limits_{-\infty}^{\infty} e^{-p^2 t} \left[\frac{\sin(px) \sin(ap)}{p} \right] dp = 0.$$

Finally, we discuss heat flow over a circular membrane. In this case, it is more appropriate to use complex Fourier series. We are dealing with heat equation for the temperature function $u(r, \theta)$ defined over a circular disc given by

$$T = \left\{ (\rho, \theta) \mid 0 \leq \rho \leq r \text{ and } -\pi < \theta \leq \pi \right\}.$$

The discussion is facilitated by the fact that the *exponential map* given by

$$\frac{\mathbb{R}}{2\pi \mathbb{Z}} \ni \theta + 2\pi n \mapsto (\rho, \theta) \in T = \{z \in \mathbb{C} \mid |z| = \rho\},$$

is a *diffeomorphism*, for each $\rho > 0$. Thus, a 2π-periodic function over $\mathbb{R}$ can be viewed as a function $f(\theta) : T \to \mathbb{R}$ with $f(-\pi) = f(\pi)$. Complex Fourier series decomposes the angular dependence of a function $u(r, \theta)$. For example, we may consider the problem related to analysis of vibrating circular membranes.

Example 10.36 Let $0 < r \leq 1$, and suppose $u(r, \theta)$ denote the *temperature* at any point (r, θ) of the unit disc $T = \{z \in \mathbb{C} \mid |z| = 1\}$, where the temperature at the boundary is given by

$$u(1, \theta) = f(\theta), \quad f \text{ being a continuous function.} \qquad (10.5.13)$$

We know that the temperature $u(r, \theta)$ in steady state is given by the Laplace equation

$$\frac{\partial^2 u}{\partial r^2} + \frac{1}{r}\frac{\partial u}{\partial r} + \frac{1}{r^2}\frac{\partial^2 u}{\partial \theta^2} = 0. \qquad (10.5.14)$$

Observe that, for $c_k = \alpha_k r^k$,

$$p_k(\theta) = \sum_{n=-k}^{k} \alpha_k \, [re^{i\theta}]^k$$

is a solution of Eq. (10.5.14), but need not satisfy the boundary condition (10.5.13). Here, applying the *complex version* of Theorem 8.2, it follows that the complex Fourier series

$$\sum_{n=-\infty}^{\infty} \alpha_k \, e^{i\,k\theta}$$

represents the function $f(\theta)$ on the boundary $(r = 1)$, provided $f(\theta)$ is *Lipschutz continuous*. The uniform convergence in this case is taken in the sense of Theorem 8.4. Hence, the series

$$u(r, \theta) = \sum_{n=-\infty}^{\infty} \alpha_k \, [re^{i\theta}]^k$$

is the final solution of the *Dirichelet's problem* over unit disc, where α_k are given by the formula

$$\alpha_k = \frac{1}{2\pi} \int_{-\pi}^{\pi} e^{-i\,k\theta} f(\theta) \mathrm{d}\theta.$$

Wave Equations

Suppose a function $y = y(x, t) \in C^2(\mathbb{R})$ represents the deflection at point x and time $t \geq 0$ of a perfectly elastic infinite string of uniform density. We know that $y = y(x, t)$ satisfies the 1-dimensional *wave equation*

$$y_{tt} = c^2 y_{xx}, \quad -\infty < x < \infty \text{ and } t \geq 0. \tag{10.5.15}$$

Suppose the assigned initial conditions are given by

$$y(x, 0) = f(x) \quad \text{and} \quad y_t(x, 0) = 0. \tag{10.5.16}$$

Applying Fourier transform to Eq. (10.5.15), we have

$$-p^2 \bar{y}(p, t) = F[y_{xx}(x, t)] = \frac{1}{c^2} F[y_{tt}(x, t)]$$

$$= \frac{1}{c^2} \int_{-\infty}^{\infty} e^{-ipx} \frac{\partial^2 y(x, t)}{\partial t^2} \, dx$$

$$= \frac{1}{c^2} \frac{d^2}{dt^2} \bar{y}(p, t),$$

so that

$$\frac{d^2 \bar{y}(p, t)}{dt^2} + c^2 p^2 \bar{y}(p, t) = 0$$

$$\Rightarrow \quad \bar{y}(p, t) = A(p) \cos cpt + B(p) \sin cpt.$$

Now, using the initial conditions (10.5.16), we obtain

$$A(p) = \bar{y}(p, 0)$$

$$= \int_{-\infty}^{\infty} e^{-ipx} y(x, 0) \, dx = \int_{-\infty}^{\infty} e^{-ipx} f(x) \, dx;$$

$$B(p) = 0.$$

It thus follows that

$$\bar{y}(p, t) = \int_{-\infty}^{\infty} e^{-ipx} \cos(cpt) f(x) \, dx$$

$$= \frac{1}{2} \left[\int_{-\infty}^{\infty} e^{-ip(x-ct)} f(x) \, dx + \int_{-\infty}^{\infty} e^{-ip(x+ct)} f(x) \, dx \right]$$

$$= \frac{1}{2} \left[\int_{-\infty}^{\infty} e^{-ip\lambda} f(\lambda + ct) \, d\lambda + \int_{-\infty}^{\infty} e^{-ip\lambda} f(\lambda - ct) \, d\lambda \right]$$

$$= \frac{1}{2} \left(e^{ipt} F[f(\lambda + ct)] + e^{-ipt} F[f(\lambda - ct)] \right),$$

Hence, we have

$$
\begin{aligned}
y(x,t) &= F^{-1}[\overline{y}(p,t)] \\
&= \frac{1}{4\pi}\left(F^{-1}\{e^{ipt}F[f(\lambda+ct)]\} + F^{-1}\{e^{-ipt}F[f(\lambda-ct)]\}\right) \\
&= \frac{1}{4\pi}\left[f(\lambda+c't)+f(\lambda-c't)\right], \quad \text{with } c' = c+1. \quad (10.5.17)
\end{aligned}
$$

This is precisely what we obtained earlier by using the *method of characteristics*. Similarly, we can solve the IVP if the assigned initial conditions are given by

$$
y(x,0) = 0 \quad \text{and} \quad y_t(x,0) = g(x). \quad (10.5.18)
$$

Next, we consider the following initial value problem related to a nonhomogeneous wave equation of a system, without *dampness*:

$$
\begin{aligned}
y_{tt} &= c^2 y_{xx} + f(x,t), \quad -\infty < x < \infty \text{ and } t \ge 0; \\
y(x,0) &= f(x) \quad \text{and} \quad y_t(x,0) = g(x), \quad x \in \mathbb{R}.
\end{aligned} \quad (10.5.19)
$$

We will assign boundary conditions subject to that $y \in L^1(\mathbb{R})$ is piecewise smooth. As before, we first solve the corresponding homogeneous equation. For, if $y(x,t) = X(x)T(t)$ then by separation of variables we have

$$
\frac{d^2 X}{dx} + \lambda X(x) = 0,
$$

so that with $\lambda > 0$ the two independent solutions are given by

$$
X_1(x) = e^{i\sqrt{\lambda}x} \quad \text{and} \quad X_2(x) = e^{-i\sqrt{\lambda}x}.
$$

Thus, with $\lambda = p^2$ $(p \ne 0)$, we need
    We next consider the Cauchy problem for the 1-dimensional wave equation

$$
y_{tt} = c^2 y_{xx}, \quad \text{for } -\infty < x < \infty \text{ and } t \ge 0, \quad (10.5.20)
$$

with the initial conditions given by

$$
y(x,0) = f(x) \quad \text{and} \quad y_t(x,0) = g(x). \quad (10.5.21)
$$

In this case, by applying the Fourier transform to Eq. (10.5.20), we have

$$-p^2 \overline{y}(p, t) = F[y_{xx}(x, t)] = \frac{1}{c^2} F[y_{tt}(x, t)]$$

$$= \frac{1}{c^2} \int_{-\infty}^{\infty} e^{-ipx} \frac{\partial^2 y(x, t)}{\partial t^2} \, dx$$

$$= \frac{1}{c^2} \frac{d^2}{dt^2} \overline{y}(p, t),$$

so that

$$\frac{d^2 \overline{y}(p, t)}{dt^2} + c^2 p^2 \overline{y}(p, t) = 0$$

$$\Rightarrow \quad \overline{y}(p, t) = A(p) e^{icpt} + B(p) e^{-icpt}.$$

Moreover, by applying the Fourier transform to Eq. (10.5.21), we have

$$\overline{y}(p, t) = F(p) \quad \text{and} \quad \overline{y}_t(p, 0) = G(p).$$

Therefore, $A + B = F(p)$ and $A - B = (1/icp)G(p)$. Solving for A and B, we obtain

$$\overline{y}(p, t) = \frac{F(p)}{2} \left[e^{icpt} + e^{-icpt} \right] + \frac{G(p)}{2icp} F(p) \left[e^{icpt} - e^{-icpt} \right]. \qquad (10.5.22)$$

Hence, by applying the inverse Fourier transform, we obtain

$$y(x, t) = \frac{1}{2} \left[\frac{1}{2\pi} \int_{-\infty}^{\infty} \left(e^{ip(x+ct)} + e^{-ip(x-ct)} \right) F(s) \, dp \right]$$

$$\frac{1}{2c} \left[\frac{1}{2\pi} \int_{-\infty}^{\infty} \left(e^{ip(x+ct)} + e^{-ip(x-ct)} \right) \frac{G(s)}{ip} \, dp \right], \qquad (10.5.23)$$

it thus follows that

$$y(x,t) = \frac{1}{2}\left[f(x-ct) + f(x+ct)\right] + \frac{1}{2c}\frac{1}{2\pi}\int_{-\infty}^{\infty} G(p)\,dp \int_{x-ct}^{x+ct} e^{ip\xi}\,d\xi$$

$$= \frac{1}{2}\left[f(x-ct) + f(x+ct)\right] + \frac{1}{2c}\int_{x-ct}^{x+ct} d\xi\left[\frac{1}{2\pi}\int_{-\infty}^{\infty} e^{ip\xi} G(p)\,dp\right]$$

$$= \frac{1}{2}\left[f(x-ct) + f(x+ct)\right] + \frac{1}{2c}\int_{x-ct}^{x+ct} g(\xi)\,d\xi, \qquad (10.5.24)$$

which is the *d'Alembert's formula*. In particular, for $f(x) = e^{-x^2}$ and $g \equiv 0$, the formula gives

$$y(x,t) = \frac{1}{2}\left[e^{-(x-t)^2} + fe^{-(x+t)^2}\right].$$

Laplace Equations

Consider steady-state case of a 2-dimensional heat equation for a semi-infinite plate bounded by the lines $x = 0$, $x = \ell$ and $y = 0$. Let the edge $y = 0$ be maintained at a temperature $u(x,0) = f(x)$, and the temperatures of the two vertical edges $u(0, y)$ and $u(\ell, y)$ be given by $T_1(y)$ and $T_2(y)$, respectively. So, the problem is to solve the Laplace equation

$$u_{xx} + u_{yy} = 0, \quad 0 < x < \ell, \ y > 0; \qquad (10.5.25)$$

subject to the conditions

$$u(0, y) = T_1(y), \quad u(\ell, y) = T_2(y), \quad \text{and} \quad u(x, 0) = f(x). \qquad (10.5.26)$$

Further, we may assume that $u(x, y) \to 0$ as $y \to \infty$, across the range $0 < x < \ell$. As did before, it is convenient to solve Eq. (10.5.25) for the following two types of boundary conditions, with $\lim_{y\to\infty} u_1(x, y) = 0$:

$$v(0, y) = T_1(y), \quad v(\ell, y) = T_2(y), \quad \text{and} \quad v(x, 0) = 0; \qquad (10.5.27)$$

$$w(0, y) = 0, \quad w(\ell, y) = 0, \quad \text{and} \quad w(x, 0) = f(x), \qquad (10.5.28)$$

so that $u(x, t) = v(x, y) + w(x, y)$.

We first use separation of variables to solve the Laplace equation $w_{xx} + w_{yy} = 0$, satisfying the boundary conditions (10.5.28). For, as in Example 8.20, the complete integrals of the Laplace equation for w are given by (8.3.39)–(8.3.41). First, let us consider the complete integral $w(x, y) = (Ax + B)(Cy + D)$. The condition $w(0, y) = 0$ gives $B = 0$ or $Cy + D = 0$. Clearly, in the second case, we get a trivial solution $w(x, y) = 0$. And, if $B = 0$, then w is given by $w(x, y) = Ax(Cy + D)$. Now, the condition $w(\ell, y) = 0$ gives $A = 0$ or $Cy + D = 0$. Both these conditions

lead to trivial solutions $w(x, y) = 0$, which does not satisfy the boundary condition
$w(x, 0) = f(x)$. Next, if

$$w(x, y) = (Ae^{px} + Be^{-px})(C \cos py + D \sin py),$$

then the boundary condition $w(0, y) = 0$ leads to $A + B = 0$ or $C \cos py + D \sin py = 0$. The later condition gives that w must be a trivial solution. So, we write

$$w(x, y) = A(e^{px} - e^{-px})(C \cos py + D \sin py).$$

Now, the boundary condition $w(\ell, y) = 0$ shows $A = 0$ or $C \cos py + D \sin py$;
both lead to trivial solutions. Finally, if the function w is of the form

$$w(x, y) = (A \cos px + B \sin px)(Ce^{py} + De^{-py}),$$

the condition $u(x, y) \to 0$ as $y \to \infty$ implies that $C = 0$ and so

$$w(x, y) = D(A \cos px + B \sin px)e^{-py}.$$

Then, using the condition $w(0, y) = 0$, we obtain $A = 0$, and so $w(x, y) = \alpha \sin px$ e^{-py}, with $\alpha = BD$. The boundary condition $w(\ell, y) = 0$ shows that $\sin p\ell = 0$ *i.e.*, $p = (n\pi/\ell)$. Thus, for each eigenvalue $p = (n\pi/\ell)$, the *eigenfunction*

$$w_n(x, y) = \alpha_n e^{-ny} \sin\left(\frac{n\pi x}{\ell}\right)$$

is a solution of the Laplace equation for w. Therefore, by extended principle of superposition for infinite sums, the following *eigenfunctions expansion*

$$w(x, y) = \sum_{n=1}^{\infty} w_n(x, y) = \sum_{n=1}^{\infty} \alpha_n e^{-ny} \sin\left(\frac{n\pi x}{\ell}\right) \tag{10.5.29}$$

also satisfies the Laplace equation with boundary conditions

$$w(0, y) = w(\ell, y) = 0, \quad \text{and} \quad w(x, y) \to 0 \text{ as } y \to \infty.$$

Using the boundary condition $w(x, 0) = f(x)$, it follows that

$$f(x) = \sum_{n=1}^{\infty} \alpha_n \sin\left(\frac{n\pi x}{\ell}\right),$$

and so the coefficients α_n could be determined by the relation

$$\alpha_n = \frac{2}{\pi} \int_0^\ell f(x) \sin\left(\frac{n\pi x}{\ell}\right) dx, \tag{10.5.30}$$

using Fourier theorem. Hence, the solution of the BVP for w is given by (10.5.29), where the coefficients α_n are given by Eq. (10.5.30).

Next, we use sine transform to solve the Laplace equation $v_{xx} + v_{yy} = 0$ satisfying the boundary conditions (10.5.27). For, as before, applying Fourier sine transform in y, we have

$$\begin{aligned}
p\, v(x, 0) - p^2\, \bar{v}_s(x, p) &= F_s[v_{yy}(x, y)] \\
&= -F_s[v_{xx}(x, y)] \\
&= -\int_{-\infty}^{\infty} \sin py\, \frac{\partial^2 v}{\partial x^2}\, dy \\
&= -\frac{d^2}{dx^2} \bar{v}_s(x, p),
\end{aligned}$$

so that, using the condition $v(x, 0) = 0$, we obtain

$$\frac{d^2 \bar{v}_s(x, p)}{dx^2} - p^2 \bar{v}_s(x, p) = 0,$$

which has the solution given by

$$\bar{v}_s(x, p) = A(p) \sinh px + B(p) \sinh p(\ell - x). \tag{10.5.31}$$

Notice that the first two boundary conditions in Eq. (10.5.27) give

$$\bar{v}_s(0, p) = \int_0^\infty \sin py\, T_1(y) dy$$

and

$$\bar{v}_s(\ell, p) = \int_0^\infty \sin py\, T_2(y) dy.$$

It thus follows that

$$B(p) \sinh p\ell = \bar{v}_s(0, p) = \int_0^\infty \sin py \, T_1(y) dy;$$

$$A(p) \sinh p\ell = \bar{v}_s(\ell, p) = \int_0^\infty \sin py \, T_2(y) dy.$$

Using these equations, the final solution for the Laplace equation in $v = v(x, y)$ can be obtained by applying inverse sine transform to the sum

$$\frac{\sinh px}{\sinh p\ell} \int_0^\infty \sin py \, T_2(y) dy + \frac{\sinh p(\ell - x)}{\sinh p\ell} \int_0^\infty \sin py \, T_1(y) dy. \qquad (10.5.32)$$

We now use Fourier transform to solve the Dirichlet's problem in the *upper-half plane*. That is, we consider the Laplace equation

$$u_{xx} + u_{yy} = 0, \qquad \text{for} \quad -\infty < x < \infty \quad \text{and} \quad y \geq 0, \qquad (10.5.33)$$

with boundary conditions given by

$$\lim_{|x| \to \infty} u(x, y) = \lim_{y \to \infty} u(x, y) = 0, \quad \text{and} \quad u(x, 0) = f(x). \qquad (10.5.34)$$

Applying Fourier transform with respect to variable x, we obtain

$$\frac{d^2 \bar{u}(p, y)}{dy^2} - p^2 \bar{u}(p, y) = 0, \qquad (10.5.35)$$

with boundary conditions given by

$$\bar{u}(p, 0) = F(p) \quad \text{and} \quad \bar{u}(p, y) \, to \, 0 \quad \text{as} \quad y \to \infty. \qquad (10.5.36)$$

The solution of the above *transformed problem* is given by

$$\bar{u}(p, y) = F(p) e^{-|p|y}, \qquad \text{for} \quad y \geq 0, \qquad (10.5.37)$$

Finally, the convolution theorem implies that the solution of the original problem is given by

$$u(x, y) = \int_{-\infty}^\infty f(\xi) g(x - \xi) \, d\xi, \qquad (10.5.38)$$

where the function g is given by

$$g(x) = F^{-1}[, e^{-|p|y}] = \frac{y}{\pi(x^2 + y^2)}. \tag{10.5.39}$$

Therefore, it follows from (10.5.38) that

$$u(x, y) = \frac{y}{\pi} \int_0^\infty \frac{f(\xi)\, d\xi}{(x - \xi)^2 + y^2}, \quad \text{for } y \geq 0, \tag{10.5.40}$$

which is well-known *Poisson integral formula* for the upper-half plane.

Finally, we apply *Fourier cosine transform* to solve the Dirichlet's problem in the quarter-plane. That is, we consider the Laplace equation

$$u_{xx} + u_{yy} = 0, \quad x > 0, \ y > 0, \tag{10.5.41}$$

with the boundary conditions given by

$$u(0, y) = g(y), \ u_y(x, 0) = f(x), \quad \text{and} \quad \lim_{x \to \infty} u(x, y) = \lim_{y \to \infty} u(x, y) = 0. \tag{10.5.42}$$

In this case, it is more convenient to write the solution of Eq. (10.5.41) as $u(x, t) = v(x, y) + w(x, y)$, where the functions $v = v(x, y)$ and $w = w(x, y)$ are harmonic in the quarter domain $x > 0$, $y > 0$ and satisfy respectively the boundary conditions

$$v(0, y) = g(y), v_y(x, 0) = 0, \quad \text{and} \quad \lim_{x \to \infty} v(x, y) = \lim_{y \to \infty} v(x, y) = 0; \tag{10.5.43}$$

$$w(0, y) = 0, w_y(x, 0) = f(x), \quad \text{and} \quad \lim_{x \to \infty} w(x, y) = \lim_{y \to \infty} w(x, y) = 0. \tag{10.5.44}$$

As the solution method for both the problems involving v and w is similar, it suffices to solve the Laplace equation $v_{xx} + v_{yy} = 0$, with the boundary conditions given by (10.5.43). For, by applying Fourier cosine transform in y, we have

$$-v_y(x, 0) - p^2 \bar{v}_c(x, p) = F_c[v_{yy}(x, y)]$$
$$= -F_c[v_{xx}(x, y)]$$
$$= -\int_{-\infty}^\infty \cos py \frac{\partial^2 v}{\partial x^2}\, dy$$
$$= -\frac{d^2}{dx^2} \bar{v}_c(x, p),$$

so that, by using the condition $v_y(x, 0) = 0$, we obtain

$$\frac{d^2 \bar{v}_c(x, p)}{dx^2} - p^2 \bar{v}_c(x, p) = 0,$$

which has the solution given by

$$\bar{v}_c(x, p) = \alpha(p) e^{-px} + \beta(p) e^{px}, \quad x > 0, \quad p > 0. \tag{10.5.45}$$

The boundary condition $\lim_{x \to \infty} v(x, y) = 0$ implies that $\beta(p) = 0$, and so we obtain $\bar{v}_c(x, p) = \alpha(p) e^{-px}$. Notice that the first boundary condition in Eq. (10.5.43) gives

$$\bar{v}_c(0, p) = \int_0^\infty \cos py \, g(y) dy.$$

It thus follows that

$$\alpha(p) = \bar{v}_c(0, p) = \int_0^\infty \cos py \, g(y) dy.$$

Using the above equation, we obtain

$$\bar{v}_c(x, p) = e^{-px} \int_0^\infty \cos py \, g(y) dy.$$

Hence, the final expression for the harmonic function $v = v(x, y)$ is obtained by applying inverse cosine transform to the last equation.

Exercises 10

10.1. For the function $f(t) = e^{-at} \ (t > 0)$, show that

$$e^{-at} = \frac{2}{\pi} \int_0^\infty \frac{a \cos \omega t}{a^2 + \omega^2} d\omega = \frac{2}{\pi} \int_0^\infty \frac{\omega \sin \omega t}{a^2 + \omega^2} d\omega.$$

10.2. Verify Fourier Integral Theorem for the time-continuous signal

$$\sigma_\ell(t) = \begin{cases} 0, & t < 0 \\ \dfrac{t}{\ell}, & t \in [0, \ell] \\ 1, & t \geq \ell \end{cases}.$$

10.3. Verify Fourier Integral Theorem for the *sawtooth function*

$$f_a(t) = \begin{cases} 0, & \text{for } t < 0 \\ \dfrac{a}{b}t, & \text{for } 0 < t < b \\ 0, & \text{for } t > b \end{cases}, \quad a > 0.$$

10.4. Represent the *hat function*

$$\phi(t) = \begin{cases} 1 + \dfrac{t}{a}, & -a \le t \le 0 \\ 1 - \dfrac{t}{a}, & 0 \le t \le a \\ 0, & t \notin [-a, a] \end{cases}$$

as a sum of step functions. Hence, or otherwise, show that its Fourier integral is given by

$$\widehat{\phi}(p) = \frac{2}{\alpha\, p^2}[1 - \cos(\alpha\, p)], \quad p \ne 0, \quad \text{with } \widehat{\phi}(0) = 0.$$

10.5. Compute the integrals

$$\text{(i)} \int_0^\infty \frac{\omega^3 \sin(\omega t)}{\omega^4 + 4}\, dt \quad \text{and} \quad \text{(ii)} \int_0^\infty \frac{(\omega^2 + 2) \cos(\omega t)}{\omega^4 + 4}\, dt.$$

[Hint: Use Fourier inversion formula, with $f(t) = e^{-t} \cos t$.]

10.6. Compute the Fourier integral of the function

$$f(t) = \begin{cases} 1 - t^2, & |t| > 1 \\ 0, & |t| > 1 \end{cases},$$

and hence evaluate the integral

$$\int \left(\frac{t \cos t - \sin t}{t^3} \right) \cos(t/2)\, dt.$$

10.7. Compute the Fourier integral of the time-continuous *sinus function*

$$f(t) = \begin{cases} \sin t, & t \in [0, \pi] \\ 0, & t \notin [0, \pi] \end{cases}.$$

10.8. Compute the Fourier transform of the function

$$f(t) = \begin{cases} e^{3t}, & t < 0 \\ e^{-t}, & t < 0 \end{cases}$$

10.9. Compute $\widehat{\delta}(p)$, using the fact that the *generalised* derivative of the signum function is $2\delta(t)$.

10.10. Compute the Fourier integral $\widehat{f}(p)$ of the function $f(t) = \cos(2t)e^{-t^2}$], $t \in \mathbb{R}$.

10.11. Compute the Fourier integral $\widehat{f}(p)$ of the function $f(t) = te^{-a|t|}$, $t \in \mathbb{R}$.

10.12. Compute the Fourier integral of the function $g(t)$ in Eq. (10.4.1).

10.13. For the *rectangular functions*

$$f(t) = \begin{cases} 1, & |t| < a \\ 0, & |t| > a \end{cases},$$

show that $f * f$ is the *triangular function*. Hence, verify Theorem 10.16 for $f * f$.

10.14. Show that

$$e^{-t} H(t) * e^{-2t} H(t) = [e^{-t} - e^{-2t}]H(t),$$

and hence verify Theorem 10.16 for

$$f(t) = e^{-t} H(t) \quad \text{and} \quad g(t) = e^{-2t} H(t).$$

10.15. Compute $F_s[e^{-|t|}]$, and hence evaluate the integral $\int\limits_{0}^{\infty} \frac{x \sin mt}{1+t^2} \, dt$. [Hint: Use Eq. (10.2.9)]

10.16. Compute $F_c[e^{-t^2/2}]$, and hence $F_s[te^{-t^2/2}]$.

10.17. Compute $F_c[1/\sqrt{t}]$ and $F_s[1/\sqrt{t}]$. [Hint: Consider $F_c[1/\sqrt{t}] + i F_s[1/\sqrt{t}]$ and use *Gamma function* to find $F_c[1^{n-1}]$, and then take $n = 1/2$]

10.18. Solve the integral equation

$$\int\limits_{0}^{\infty} f(t) \cos pt \, dt = \begin{cases} 1 - p, & 0 \leq p \leq 1 \\ 0, & p > 1 \end{cases}.$$

10.19. Show that, for any function $f : \mathbb{R} \to \mathbb{R}$, $(1/2)[F_c(p) - i F_s(p)] = \widehat{f}(p) H(t)$, where H is the Heaviside step function.

10.20. Show that, for any function $f : (0, \infty) \to \mathbb{R}$, $F_c(p) = F[f(t) + f(-t)]$ and $-i F_s(p) = F[f(t) - f(-t)]$.

10.21. Show that, for any $f \in L^1(\mathbb{R})$, the series $\sum_{n=-\infty}^{\infty} f(t + 2n\ell)$ converges absolutely for almost all $t \in (-\ell, \ell)$, and its sum $g \in L^1(-\ell, \ell)$ satisfies the periodicity condition $g(t + 2\ell) = g(t)$, for all $t \in \mathbb{R}$. Further, if c_n is the nth Fourier coefficient of g, then we have

$$c_n = \frac{1}{2\ell} \int\limits_{-\ell}^{\ell} e^{-inx} g(x) \, dx = \frac{1}{2\ell} \int\limits_{-\ell}^{\ell} e^{-inx} f(x) \, dx = \frac{1}{2\ell} F(n).$$

10.22. Find the temperature distribution $u(x, t)$ in an insulated *semi-infinite* rod, with the end at $x = 0$ kept at temperature zero and the initial temperature given by the function

$$f(x) = \begin{cases} 1, & 0 \le x \le 1 \\ 0, & x > 1 \end{cases}.$$

10.23. Find the temperature distribution $u(x, t)$ in an insulated *semi-infinite* rod, where the gradient at the end $x = 0$ is constant, say $u_x(0, t) = -\alpha$, and the initial temperature is given by a function $f(x)$ $(x \ge 0)$. Subsequently, write the complete solution for the following special case:

$$u_x(0, t) = 0; \quad \text{and,} \quad f(x) = \begin{cases} x, & 0 \le x \le 1 \\ 0, & x > 1 \end{cases}.$$

10.24. (*Cauchy Problem for Diffusion Equation*) Find a solution the initial value problem for the 1-dimensional diffusion equation $u_t = k\, u_{xx}$ with the initial condition given by $u(x, 0) = f(x)$, for $x \in \mathbb{R}$ and $t > 0$. The symbol k is as usual the diffusivity constant. What if $f(x) = aH(x)$, for some constant a, or $f(x) = \delta(x)$?

10.25. (*One Dimensional Diffusion Equation on a Half-Line*) Solve the initial-boundary value problem for the 1-dimensional diffusion equation $u_t = k\, u_{xx}$, with the initial condition given by $u(x, 0) = 0$, for $x \in \mathbb{R}$, and the boundary condition $u(0, t) = f(t)$ or $u_x(0, t) = g(t)$, for $t > 0$.

10.26. Solve the nonhomogeneous equation $u_t - u_{xx} = \cos 2t$, $0 < x < 1$, $t > 0$, satisfying the nonhomogeneous boundary conditions $u(0, t) = 0$, $u(2, t) = 1$, with $u(x, 0) = x^2 \cos \pi x - 2x$.

10.27. Solve the nonhomogeneous equation $u_t - 4u_{xx} = t \sin(x/2) - \sin t$, $0 < x < \pi$, $t > 0$, satisfying the nonhomogeneous boundary conditions $u(0, t) = \cos t$, $u(\pi, t) = 0$, with $u(x, 0) = 2$.

10.28. Use the Poisson integral formula (10.5.40) to find the solution of the Laplace equation for a dipole source at $(x, y) = (\xi, 0)$. Further, for $f(x) = \tau_0 H(a - |x|)$, show that the curves in the upper-half plane for which the steady-state temperature is constant are given by the family $x^2 + y^2 - \alpha y = a^2$.

10.29. (*Neumann's Problem in upper-half plane*) Find a solution of the Laplace equation $u_{xx} + u_{yy} = 0$, with boundary conditions given by $u(x, 0) = 0$, $u(0, y) = a$, for $0 < x, y < \infty$, and also $\nabla u \to 0$, as $r = \sqrt{x^2 + y^2} \to \infty$.

10.30. (*Laplace Equation in the Quarter Plane*) Solve the Laplace equation $u_{xx} + u_{yy} = 0$, with boundary condition given by $u_y(x, 0) = f(x)$, for $x \in \mathbb{R}$.

10.31. Use convolution theorem for Fourier transform to show that the solution
of the Laplace equation (10.5.33) in the half-plane, satisfying the boundary
conditions (10.5.34), is given by

$$u(x, y) = \frac{1}{2\pi} \int\limits_{-\infty}^{\infty} \frac{2y}{(x - \lambda)^2 + y^2} f(\lambda) d\lambda.$$

References

1. Fourier, J. (1955). *The analytical theory of heat* (trans.: A. Freeman). Dover.
2. Debnath, L., & Bhatta, D. (2015). *Integral transforms and their applications* (3rd ed.). CRC Press; Taylor & Francis.
3. Cartwright, M. (1990). *Fourier methods for mathematicians, scientists, and engineers*. Ellis Horwood.
4. Strichartz, R. C. (2003). *A guide to distribution theory and Fourier transforms*. World Scientific Publishing Co. Ltd.
5. Hanna, J. R., & Rowland, J. H. (1990). *Fourier series, transforms, and boundary value problems* (2nd ed.). Wiley.
6. Walker, J. S. (1988). *Fourier analysis*. Oxford University Press.
7. Zygmund, A. (2002). *Trigonometric series* (3rd ed.). Cambridge University Press.

Chapter 11
Laplace Transform

*Why should I refuse a good dinner simply because I don't
understand the digestive processes involved.*

Oliver Heaviside (1850–1925)

As said earlier in the previous chapter, the Laplace transform $\mathscr{L}$ introduced by *Simon Laplace* is an integral transform, with kernel function given by $K(p, t) = e^{-st}$ ($s \in \mathbb{C}$), defined on a class of real-valued functions $f \in PC([0, \infty))$ of *exponential order*, say $\alpha > 0$. More precisely, we have

$$\mathscr{L}[f] = F(s) := \int_0^\infty e^{-st} \phi(t) \, dt, \quad \text{for} \quad f \in PC([0, \infty)) \quad \text{and} \quad s > \alpha,$$

where the integral on the right side is taken in the following sense:

$$\int_0^\infty \phi(t) \, dt := \lim_{a \to \infty} \int_0^a \phi(t) \, dt.$$

In this case, we say the function $F(s)$ is the *Laplace transform* of the function f (Theorem 11.1).

The Laplace transform $\mathscr{L}$ is linear operators that acts nicely on derivatives, and so it transforms a linear ordinary differential equation into an algebraic equation, and a linear partial differential equation into a linear ordinary differential equation. In each case, we have a convenient *transformed problem* to deal with. The solution of the original problem is subsequently obtained by applying the inverse Laplace transform. The Laplace transform technique also provides a nice alternative in some situations

A. K. Razdan and V. Ravichandran, *Fundamentals of Partial Differential Equations*,
https://doi.org/10.1007/978-981-16-9865-1_11

wherein the method of separation of variables fails to apply. In this chapter, we discuss properties of the Laplace transform that help solve some important initial-boundary value problems over the half-line $x \geq 0$. The Laplace transform is particularly more suitable to deal with initial value problems.

11.1 Basic Theorems and Examples

Recall that a function $\phi : [a, b] \to \mathbb{R}$ is called *piecewise continuous* if it is continuous at all points of the interval $[a, b]$ except for a finite number of points

$$a = t_0 < t_1 < t_2 < \cdots < t_n = b.$$

That is, ϕ has finite number of discontinuities at points t_i such that the *jump* at each t_i is finite (Definition 8.4). Further, we say a function $\phi : [0, \infty) \to \mathbb{R}$ is piecewise continuous if it is piecewise continuous in $[0, a]$, for all $a > 0$. The space of piecewise continuous defined over the half-line $[0, \infty)$ is written as $PC[0, \infty)$.

Example 11.1 The function given by

$$f(t) = \begin{cases} t^2, & 0 < t \leq 1, \\ 1 + t, & 1 < t < \infty \end{cases}$$

belongs to $PC[0, \infty)$, whereas the function given by

$$f(t) = \begin{cases} (1 - t)^{-1}, & 0 \leq t < 1, \\ 1 + t & 1 \leq t < \infty, \end{cases}$$

·is not a piecewise continuous because $f(t) \to \infty$ as $t \to 1^-$ (see Fig. 11.1).

Definition 11.1 A function $f : [0, \infty) \to \mathbb{R}$ is said to be of *exponential order* α if there exists positive numbers M and T such that

$$|f(t)| \leq Me^{\alpha t}, \quad \text{for } t \geq T.$$

Since for a function $f : [0, \infty) \to \mathbb{R}$ bounded by M, we can write

$$|f(t)| \leq M \leq Me^t, \quad \text{for all } t \geq 0,$$

it follows that every bounded function is of exponential order 1. In particular, a continuous function $f : [0, c] \to \mathbb{R}$ extended periodically to $[0, \infty)$, so that it is bounded over the interval $[0, c]$ and so over $[0, \infty)$, is of exponential order 1. Also, since

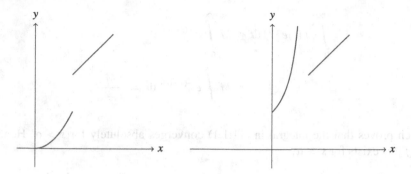

Fig. 11.1 Example and a nonexample of piecewise continuous functions

$$e^t = \sum_{k=0}^{\infty} \frac{t^k}{k!} \geq \frac{t^k}{k!} \implies t^k \leq k!e^t, \quad \text{for all } t \geq 0,$$

so that we can write

$$p(t) = \sum_{k=0}^{n} a_k t^k \leq \left(\sum_{k=0}^{n} k!a_k\right) e^t = Me^t,$$

it follows that every polynomial $p(t) = \sum_{k=0}^{n} a_k t^k$ is of exponential order 1, where $M := \sum_{k=0}^{n} k!a_k$. Notice that, however, the function $f : [0, \infty) \to \mathbb{R}$ defined by $f(t) = e^{t^2}$ is not of exponential order.

Theorem 11.1 (Existence of Laplace transform) *Let $f \in PC[0, \infty)$ be of exponential order α. Then the* Laplace transform *of the function f given by*

$$\mathscr{L}[f](s) := \int_0^{\infty} e^{-st} f(t)\mathrm{d}t \tag{11.1.1}$$

exists for all $s > \alpha$.

Proof As the function $f : [0, \infty) \to \mathbb{R}$ is of exponential order α, for some positive number M, we can write

$$|f(t)| \leq Me^{\alpha t}, \quad \text{for all } t \geq 0.$$

It thus follows that, for $t \geq 0$,

$$|e^{-st} f(t)| \leq Me^{-(s-\alpha)t}.$$

Therefore, for $s > \alpha$, we have

$$\int_0^\infty |f(t)e^{-st}|\,dt \le M \int_0^\infty e^{-(s-\alpha)t}\,dt$$

$$\le M \int_0^\infty e^{-(s-\alpha)t}\,dt = \frac{M}{s-\alpha},$$

which proves that the integral in (11.1.1) converges absolutely for $s > \alpha$. Hence, $\mathscr{L}[f](s)$ exists for $s > \alpha$. □

The condition stated in above theorem is not necessary. For example, the function $f : [0, \infty) \to \mathbb{R}$ given by $f(t) = t^a$, with $-1 < a < 0$, approaches ∞ as $t \to 0^+$. Therefore, f is not piecewise continuous on the interval $[0, \infty)$. However, as shown later in Example 11.3, we have

$$\mathscr{L}[t^a] = \frac{\Gamma(a+1)}{s^{a+1}}, \quad \text{for all } s > 0.$$

Further, in view of Corollary 11.1, functions such as

$$\frac{s}{s+1}, \quad \sin(as), \quad \text{and} \quad \frac{s^2}{1+s^2},$$

cannot be Laplace transform of any function in $PC[0, \infty)$.

Corollary 11.1 *Let $f \in PC[0, \infty)$ be of exponential order α, with $\mathscr{L}[f](s) = F(s)$. Then $F(s) \to 0$ as $s \to \infty$.*

Proof With notations as in Theorem 11.1, we have

$$|F(s)| \le \int_0^\infty e^{-st}|f(t)|\,dt \le \frac{M}{s-\alpha}, \quad \text{for } s > \alpha.$$

Therefore, $|F(s)| \to 0$ as $s \to \infty$. Hence, $F(s) \to 0$ as $s \to \infty$. □

Example 11.2 For any $a \in \mathbb{C}$, the Laplace transform of e^{at} is given by

$$\mathscr{L}[e^{at}] = \int_0^\infty e^{-st} e^{at}\,dt = \int_0^\infty e^{-(s-a)t}\,dt$$

$$= -\frac{e^{-(s-a)t}}{s-a} \bigg|_0^\infty = \frac{1}{s-a}$$

provided $s > a$. In particular, using the relations

$$\cos(at) = \frac{e^{iat} + e^{-iat}}{2} \quad \text{and} \quad \sin(at) = \frac{e^{iat} - e^{-iat}}{2i},$$

and Theorem 11.2, it follows that

$$\mathscr{L}[\cos(at)] = \frac{1}{2}\left(\mathscr{L}[e^{iat}] + \mathscr{L}[e^{-iat}]\right)$$

$$= \frac{1}{2}\left(\frac{1}{s - ia} + \frac{1}{s + ia}\right)$$

$$= \frac{s}{s^2 + a^2};$$

$$\mathscr{L}[\sin(at)] = \frac{1}{2i}\left(\mathscr{L}[e^{iat}] - \mathscr{L}[e^{-iat}]\right)$$

$$= \frac{1}{2i}\left(\frac{1}{s - ia} - \frac{1}{s + ia}\right)$$

$$= \frac{a}{s^2 + a^2}.$$

Notice that, for $s > 0$, we have

$$\left|e^{-st}\cos(at)\right| \le e^{-st} \to 0, \quad \text{as } t \to \infty.$$

Therefore, $e^{-st}\cos(at) \to 0$ as $t \to \infty$. Similarly, $e^{-st}\sin(at) \to 0$ as $t \to \infty$.

One may also, alternatively, use the well-known formulas as given below:

$$\int \cos(at)e^{bt}\,dt = \frac{e^{bt}}{a^2 + b^2}\left(b\cos(at) + a\sin(at)\right);$$

$$\int \sin(at)e^{bt}\,dt = \frac{e^{bt}}{a^2 + b^2}\left(b\sin(at) - a\cos(at)\right).$$

The *gamma function* used in the next example is defined as

$$\Gamma(x) := \int_0^\infty u^{x-1}e^{-u}\,du, \quad \text{for } 0 \le x < \infty.$$

In particular, we have $\Gamma(n + 1) = n!$, for any $n \in \mathbb{N}$.

Example 11.3 Let $a > -1$ be any real number. By definition, the Laplace transform of the *power function* $f(t) = t^a$ $(t \ge 0)$ is given by

$$\mathscr{L}[t^a] = \int_0^\infty e^{-st} t^a \, dt$$

$$= \frac{1}{s^{a+1}} \int_0^\infty e^{-u} u^{(a+1)-1} \, du$$

$$= \frac{\Gamma(a+1)}{s^{a+1}}, \quad \text{for } s > 0.$$

In particular, for any non-negative integer n, we have

$$\mathscr{L}[t^n] = \frac{n!}{s^{n+1}}.$$

Further, using the fact that $\Gamma(1/2) = \sqrt{\pi}$, we also have

$$\mathscr{L}[t^{1/2}] = \frac{1}{2} \left(\frac{\pi}{s^3} \right)^{1/2} \tag{11.1.2}$$

and

$$\mathscr{L}[t^{-1/2}] = \frac{\sqrt{\pi}}{\sqrt{s}}. \tag{11.1.3}$$

In this chapter, we write the *Heaviside step function* as $u(t)$. Recall that u takes value 1 over the interval $[0, \infty)$, and it is 0, otherwise. More generally, the function $u_a :$ $[0, \infty) \to \mathbb{R}$ that takes unit step at $t = a$, with $0 < a < \infty$, is called the (delayed) *unit step function*. That is,

$$u_a(t) = \begin{cases} 0, & \text{for } t < a \\ 1, & \text{for } t \geq a \end{cases}, \quad \text{for } t \in [0, \infty).$$

In practical applications, the function $u_a(t)$ is used to *switch on* an arbitrary function $f(t)$ (such as a time-series) at point $t = a$. That is, we have

$$u_a(t) f(t) = \begin{cases} 0, & \text{for } t < a \\ f(t), & \text{for } t \geq a \end{cases}$$

Example 11.4 The Laplace transform of the function u_a is given by

$$\mathcal{L}[u_a(t)] = \int_0^\infty e^{-st} f(t) \, dt$$

$$= \int_a^\infty e^{-st} \, dt$$

$$= \frac{e^{-as}}{s},$$

for all $s > 0$.

Notice that, for $s = \alpha + i\beta$, we have

$$\left| \mathcal{L}[\phi](s) \right| \leq \int_0^\infty e^{-\alpha t} |\phi(t)| \, dt = \mathcal{L}[|\phi|](\mathrm{Re}(s)),$$

it thus follows that $\beta = \mathrm{Im}(s)$ has no influence on the convergence of the Laplace transform. In fact, if the integral $\mathcal{L}[\phi](s_0)$ is divergent for some $s_0 \in \mathbb{C}$ then the same holds for all complex numbers s such that $\mathrm{Re}(s) < \mathrm{Re}(s_0)$. Hence, the *region of convergence* of the Laplace transform is of the form $\mathrm{Re}(s) \geq \alpha_0$, for some $\alpha_0 \in \mathbb{R}$. If $\phi(t) = 0$ for $t < 0$, and the imaginary axis lies in the *region of convergence* of the Laplace transform $\mathcal{L}[\phi](s)$, then it can be viewed as the Fourier transform only when an exponential weighting has been applied to the function $\phi(t)$; i.e. we have $\mathcal{L}[\phi](s) = \widehat{\phi}(i\,s)$. For example, since

$$g(t) = \begin{cases} e^{-t}, & \text{for } t \geq 0 \\ 0, & \text{for } t < 0 \end{cases} \quad \Rightarrow \quad \mathcal{L}[g](s) = (1+s)^{-1},$$

which has the set $\{s \in \mathbb{C} \mid \mathrm{Re}(s) > -1\}$ as its *region of convergence*, so the Fourier transform of $g(t)$ is given by $\widehat{g}(s) = (1 + i\,s)^{-1}$. On the other hand,

$$f(t) = \begin{cases} e^t, & t \geq 0 \\ 0, & t < 0 \end{cases} \quad \Rightarrow \quad \mathcal{L}[f](s) = (s - 1)^{-1},$$

which has the *region of convergence* $\{s \in \mathbb{C} \mid \mathrm{Re}(s) > 1\}$, and so the Fourier transform $\widehat{f}(s)$ of the function $f(t)$ does not exists.

11.2 Properties of Laplace Transform

Theorem 11.2 (Linearity property) *If the Laplace transforms of the functions f, g : $[0, \infty) \to \mathbb{R}$ exist, then the Laplace transform of $\alpha f + \beta g$ exists for any two real number α, β and*

$$\mathcal{L}[\alpha f(t) + \beta g(t)] = \alpha \mathcal{L}[f(t)] + \beta \mathcal{L}[g(t)].$$

Proof The Laplace transform is linear follows from the linearity of the integral:

$$\mathcal{L}[af(t) + bg(t)] = \int_0^\infty [af(t) + bg(t)]e^{-st}\,dt$$

$$= a\int_0^\infty f(t)e^{-st}\,dt + b\int_0^\infty g(t)e^{-st}\,dt$$

$$= a\mathcal{L}[f(t)] + b\mathcal{L}[g(t)]. \quad \square$$

Example 11.5 Using linearity of Laplace transform, we have

$$\mathcal{L}[(1+t)^2] = \mathcal{L}\left[1 + 2t + t^2\right]$$

$$= \frac{1}{s} + \frac{2}{s^2} + \frac{2}{s^3}$$

$$= \frac{s^2 + 2s + 2}{s^3}.$$

Example 11.6 Using linearity of Laplace transform, we have

$$\mathcal{L}[\cosh(at)] = \mathcal{L}\left[\frac{1}{2}(e^{at} + e^{-at})\right]$$

$$= \frac{1}{2}\left(\mathcal{L}(e^{at}) + \mathcal{L}(e^{-at})\right)$$

$$= \frac{1}{2}\left(\frac{1}{s-a} + \frac{1}{s+a}\right)$$

$$= \frac{s}{s^2 - a^2}.$$

Similarly,

$$\mathcal{L}[\sinh(at)] = \mathcal{L}\left[\frac{1}{2}(e^{at} - e^{-at})\right]$$

$$= \frac{1}{2}\left(\frac{1}{s-a} - \frac{1}{s+a}\right)$$

$$= \frac{a}{s^2 - a^2}.$$

Example 11.7 Since $\sin^2 t = (1 - \cos 2t)/2$, we have

$$\mathcal{L}[\sin^2(at)] = \frac{1}{2} (\mathcal{L}[1 - \cos(2at)])$$

$$= \frac{1}{2} (\mathcal{L}[1] - \mathcal{L}[\sin(2at)])$$

$$= \frac{1}{2} \left(\frac{1}{s} - \frac{s}{s^2 + 4a^2} \right)$$

$$= \frac{2a^2}{s(s^2 + 4a^2)}.$$

Also, using $\cos^2(at) = 1 - \sin^2(at)$,

$$\mathcal{L}[\cos^2(at)] = \mathcal{L}[1 - \sin^2(at)]$$

$$= \frac{1}{s} - \frac{2a^2}{s(s^2 + 4a^2)}$$

$$= \frac{s^2 + 2a^2}{s(s^2 + 4a^2)}.$$

Theorem 11.3 (First Shifting Theorem) *If the Laplace transform* $F(s) := \mathcal{L}[f(t)]$ *of a function* $f : [0, \infty) \to \mathbb{R}$ *exists, for all* $s > \alpha$, *then the Laplace transform of* $e^{at} f(t)$ *exists for* $s > a + \alpha$ *and*

$$\mathcal{L}[e^{at} f(t)] = F(s - a).$$

Proof Since $F(s) = \mathcal{L}[f(t)]$ exist for $s > k$, we have, for $s - a > k$,

$$\mathcal{L}[e^{at} f(t)] = \int_0^\infty e^{-st} e^{at} f(t) \, dt = \int_0^\infty e^{-(s-a)t} f(t) \, dt = F(s - a). \quad \Box$$

Example 11.8 For $b > -1$, we have

$$\mathcal{L}[t^b] = \frac{\Gamma(b + 1)}{s^{b+1}}$$

and therefore

$$\mathcal{L}[e^{at} t^b] = \frac{\Gamma(b + 1)}{(s - a)^{b+1}}.$$

Also,

$$\mathcal{L}[t^b \cosh(at)] = \frac{1}{2}\left(\mathcal{L}[t^b e^{at} + t^b e^{-at}]\right)$$

$$= \frac{\Gamma(b+1)}{2}\left(\frac{1}{(s-a)^{b+1}} + \frac{1}{(s-a)^{b+1}}\right).$$

Example 11.9 Since

$$\mathcal{L}[\sin(bt)] = \frac{b}{s^2 + b^2} \quad \text{and} \quad \mathcal{L}[\cos(bt)] = \frac{s}{s^2 + b^2}$$

we have

$$\mathcal{L}[e^{at}\sin(bt)] = \frac{b}{(s-a)^2 + b^2} \quad \text{and} \quad \mathcal{L}[e^{at}\sin(bt)] = \frac{s-a}{(s-a)^2 + b^2}.$$

The next theorem helps in transforming functions with discontinuities.

Theorem 11.4 (Second Shifting Theorem) *If the Laplace transform of $f : [0, \infty) \to$
$\mathbb{R}$ exists, then the Laplace transform of $f(t-a)u_a(t)$ exists and*

$$\mathcal{L}[f(t-a)u_a(t)] = e^{-as}\mathcal{L}[f(t)],$$

where u_a is the (delayed) unit step function.

Proof From the definition of Laplace transform, we have

$$\mathcal{L}[u_a(t)f(t-a)] = \int_0^{\infty} e^{-st} u_a(t) f(t-a)\, dt$$

$$= \int_a^{\infty} e^{-st} f(t-a)\, dt$$

and, using the substitution $t - a = u$,

$$\mathcal{L}[u_a(t)f(t-a)] = e^{-as}\int_0^{\infty} e^{-su} f(u)\, du$$

$$= e^{-as}\mathcal{L}[f(t)]. \quad \square$$

This property help us to find the LT of functions *switched on* at point $t = a$.

Example 11.10 Suppose the function $g : [0, \infty) \to \mathbb{R}$ is given by

$$g(t) = \begin{cases} 0, & \text{for } 0 \le t \le 2 \\ 3, & \text{for } t > 2 \end{cases}$$

Writing it as $g(t) = 3\,u_2(t)$, it follows from Theorem 11.4 that

$$\mathscr{L}[g(t)] = \mathscr{L}[3u_2(t)] = e^{-2s}\mathscr{L}[3] = 3e^{-2s}/s.$$

Theorem 11.5 *If the Laplace transform of $f : [0, \infty) \to \mathbb{R}$ exists for $s > \alpha$, and $F(s) = \mathscr{L}f(t)$, then the Laplace transform of $f(at)$ exists for all $s > a\alpha$, and we have*

$$\mathscr{L}[f(at)] = \frac{1}{a}F\left(\frac{s}{a}\right).$$

Example 11.11 As shown earlier, we have

$$\mathscr{L}[\sin(t)] = \frac{1}{s^2 + 1}.$$

it thus follows that

$$\mathscr{L}[\sin(at)] = \frac{1}{a}\frac{1}{(s/a)^2 + 1} = \frac{a}{s^2 + a^2}.$$

Theorem 11.6 (Transform of Derivative) *If $f : [0, \infty) \to \mathbb{R}$ is a differentiable function, with the derivative f' piecewise continuous on every finite interval in $[0, \infty)$ and of exponential order, then the Laplace transform of f' exists, and we have*

$$\mathscr{L}[f'(t)] = s\mathscr{L}[f(t)] - f(0).$$

Proof We prove this in the case when f' is continuous in $[0, \infty)$. In this case,

$$\int_0^a e^{-st} f'(t) = e^{-st} f(t)\big|_0^a + s\int_0^a e^{-st} f(t)\,dt$$

$$= e^{-sa} f(a) - f(0) + s\int_0^a e^{-st} f(t)\,dt \qquad (11.2.1)$$

As f' is of exponential order, so is f, and so we can find constants M and $\alpha > 0$ such that

$$|f(t)| \le Me^{\alpha t} \quad (t \ge 0).$$

The above implies that, for $s > \alpha$,

$$|e^{-st} f(t)| \le Me^{-(s-\alpha)t} \to 0 \quad (\text{as } t \to \infty).$$

Using the above, and by letting $a \to \infty$ in (11.2.1), we obtain

$$\mathscr{L}[f'(t)] = s\mathscr{L}[f(t)] - f(0). \quad \Box$$

Example 11.12 When $f(t) = e^{at}$, we have $f'(t) = ae^{at}$ and hence the formula

$$\mathscr{L}[f'(t)] = s\mathscr{L}[f(t)] - f(0)$$

becomes

$$a\mathscr{L}[e^{at}] = s\mathscr{L}[e^{at}] - 1$$

and this gives

$$\mathscr{L}[e^{at}] = \frac{1}{s-a}.$$

Theorem 11.7 *Let $f : [0, \infty) \to \mathbb{R}$ and its derivatives f', f'', ..., $f^{(n-1)}$ be continuous on $[0, \infty)$. If $f^{(n)}$ is piecewise continuous on every finite interval in $[0, \infty)$, and of exponential order $\alpha > 0$, then the Laplace transform of $f^{(n)}$ exists, and we have*

$$\mathscr{L}[f^{(n)}(t)] = s^n \mathscr{L}[f(t)] - s^{n-1} f(0) - s^{n-2} f'(0) - \cdots - f^{(n-1)}(0),$$

for all $s > \alpha$.

Proof For $n = 1$, the result becomes $\mathscr{L}[f'(t)] = s\mathscr{L}[f(t)] - f(0)$. For $n = k$, this result is

$$\mathscr{L}[f^{(k)}(t)] = s^k \mathscr{L}[f(t)] - s^{k-1} f(0) - s^{k-2} f'(0) - \cdots - f^{(k-1)}(0)$$

and using this we get

$$\begin{aligned}
\mathscr{L}[f^{(k+1)}(t)] &= \mathscr{L}[(f^{(k)}(t))'] \\
&= s\mathscr{L}[f^{(k)}(t)] - f^{(k)}(0) \\
&= s\left(s^k \mathscr{L}[f(t)] - s^{k-1} f(0) - \cdots - f^{(k-1)}(0)\right) - f^{(k)}(0) \\
&= s^{k+1} \mathscr{L}[f(t)] - s^k f(0) - s^{k-2} f'(0) - \cdots - f^{(k)}(0).
\end{aligned}$$

By mathematical induction, the result follows for all n. $\Box$

Example 11.13 When $f(t) = \cos(at)$, we have $f'(t) = -a\sin(at)$ and $f''(t) = -a^2 \cos(at)$. In this case, the formula

$$\mathscr{L}[f''(t)] = s^2 \mathscr{L}[f(t)] - sf(0) - f'(0)$$

becomes

$$-a^2 \mathscr{L}[\cos(at)] = s^2 \mathscr{L}[\cos(at)] - s,$$

which in turn gives

$$\mathcal{L}[\cos(at)] = \frac{s}{s^2 + a^2}.$$

Example 11.14 When $f(t) = \cosh(at)$, we have $f'(t) = a\sinh(at)$ and $f''(t) = a^2\cosh(at)$. In this case, we get

$$a^2\mathcal{L}[\cosh(at)] = s^2\mathcal{L}[\cosh(at)] - s$$

and this gives

$$\mathcal{L}[\cosh(at)] = \frac{s}{s^2 - a^2}.$$

Theorem 11.8 *If* $f \in PC([0, \infty))$ *is of exponential order, and* $F(s) = \mathcal{L}f(t)$, *then*

$$\mathcal{L}[t^n f(t)] = (-1)^n F^{(n)}(s), \quad \text{for all } n \geq 1.$$

Proof Differentiating under the integral sign, we have

$$F'(s) = \frac{d}{ds} \int_0^\infty e^{-st} f(t) \, dt$$

$$= \int_0^\infty \frac{\partial}{\partial s} \left(e^{-st} f(t) \right) dt$$

$$= \int_0^\infty e^{-st}(-tf(t)) \, dt$$

$$= \mathcal{L}[-tf(t)].$$

The general formula follows by repeating the same procedure. □

Example 11.15 For $f(t) = 1$, we have $F(s) = 1/s$ and the formula $\mathcal{L}[t^n f(t)] = (-1)^n F^{(n)}(s)$ becomes

$$\mathcal{L}[t^n] = (-1)^n \frac{d^n}{ds^n} \frac{1}{s} = \frac{n!}{s^{n+1}}.$$

Example 11.16 Let $f(t) = \sin(bt)$. Then

$$F(s) = \frac{b}{s^2 + b^2} \quad \text{and so } F'(s) = -\frac{2bs}{(s^2 + b^2)^2}.$$

Hence, using $\mathscr{L}[tf(t)] = -F'(s)$, we get

$$\mathscr{L}[t\sin(bt)] = \frac{2bs}{(s^2+b^2)^2}.$$

It follows from the above relation that

$$\mathscr{L}[te^{at}\sin(bt)] = \frac{2b(s-a)}{((s-a)^2+b^2)^2}.$$

In particular, we have

$$\mathscr{L}[te^{-t}\sin t] = \frac{2(s+1)}{(s^2+2s+2)^2}.$$

Example 11.17 Let $f(t) = \cos(bt)$. Then

$$F(s) = \frac{s}{s^2+b^2} \quad \text{and so} \quad F'(s) = \frac{b^2-s^2}{(s^2+b^2)^2}.$$

Hence, using $\mathscr{L}[tf(t)] = -F'(s)$, we get

$$\mathscr{L}[t\cos(bt)] = \frac{s^2-b^2}{(s^2+b^2)^2}.$$

and so

$$\mathscr{L}[te^{at}\cos(bt)] = \frac{(s-a)^2-b^2}{((s-a)^2+b^2)^2}.$$

In particular,

$$\mathscr{L}[te^{-t}\cos t] = \frac{s(s+2)}{(s^2+2s+2)^2}.$$

Example 11.18 Let $f(t) = \sinh(bt)$. Then

$$F(s) = \frac{b}{s^2-b^2} \quad \text{and so} \quad F'(s) = -\frac{2bs}{(s^2-b^2)^2}.$$

Hence, using $\mathscr{L}[tf(t)] = -F'(s)$, we get

$$\mathscr{L}[t\sinh(bt)] = \frac{2bs}{(s^2-b^2)^2}.$$

From this, it follows that

$$\mathscr{L}[te^{at}\sinh(bt)] = \frac{2b(s-a)}{((s-a)^2-b^2)^2}.$$

Theorem 11.9 *If* $f \in PC([0, \infty))$ *is of exponential order and* $\mathscr{L}[f(t)] = F(s)$, *then*

$$\mathscr{L}\left[\frac{f(t)}{t}\right] = \int_s^\infty F(\tau)d\tau.$$

Proof By definition, $F(s) = \int_0^\infty e^{-st} f(t)dt$ and so

$$\int_s^\infty F(\tau)d\tau = \int_{\tau=s}^\infty \int_{t=0}^\infty e^{-\tau t} f(t)dt d\tau$$

$$= \int_{t=0}^\infty \int_{\tau=s}^\infty e^{-\tau t} f(t)d\tau dt$$

$$= \int_{t=0}^\infty e^{-\tau s} \frac{f(t)}{t} dt$$

$$= \mathscr{L}\left[\frac{f(t)}{t}\right]. \quad \square$$

Example 11.19 As seen earlier, we have

$$\mathscr{L}[e^{at} - e^{bt}] = \frac{1}{s-a} - \frac{1}{s-b}.$$

It thus follows from the above theorem that

$$\mathscr{L}\left[\frac{e^{at} - e^{bt}}{t}\right] = \int_s^\infty \left(\frac{1}{s-a} - \frac{1}{s-b}\right)$$

$$= \log\left(\frac{s-a}{s-b}\right)\Big|_s^\infty$$

$$= \log\left(\frac{s-b}{s-a}\right).$$

Example 11.20 Since

$$\mathscr{L}[\sin(at)] = \frac{a}{s^2 + a^2}$$

we have

$$\mathscr{L}\left[\frac{\sin(at)}{t}\right] = \int_s^\infty \frac{a}{s^2 + a^2}\,ds$$

$$= \tan^{-1}(s/a)\big|_s^\infty$$

$$= \frac{\pi}{2} - \tan^{-1}(s/a)$$

$$= \cot^{-1}(s/a).$$

Example 11.21 Since

$$\mathscr{L}[\cos(at)] = \frac{s}{s^2 + a^2}$$

we have

$$\mathscr{L}\left[\frac{\cos(at) - \cos(bt)}{t}\right] = \int_s^\infty \left(\frac{s}{s^2 + a^2} - \frac{s}{s^2 + b^2}\right)ds$$

$$= \frac{1}{2}\log\left(\frac{s^2 + a^2}{s^2 + b^2}\right)\bigg|_s^\infty$$

$$= \frac{1}{2}\log\left(\frac{s^2 + b^2}{s^2 + a^2}\right).$$

In particular, we have

$$\mathscr{L}\left[\frac{1 - \cos(at)}{t}\right] = \log\frac{\sqrt{s^2 + a^2}}{s}.$$

Since $\sin^2(at) = (1 - \cos(2at))/2$, we have

$$\mathscr{L}[\sin^2(at)] = \frac{1}{2}\log\frac{\sqrt{s^2 + 4a^2}}{s} = \frac{1}{4}\log\frac{s^2 + 4a^2}{s^2}.$$

Theorem 11.10 *If $f \in PC[0, \infty)$ is of exponential order, then*

$$\mathscr{L}\left[\int_0^t f(t)dt\right] = \frac{\mathscr{L}[f(t)]}{s}.$$

Proof Since the function f is piecewise continuous, the function $g : [0, \infty) \to \mathbb{R}$ defined by

$$g(t) = \int_0^t f(\tau)\,d\tau$$

is continuous. Since f is of exponential order, there are constants M, and α satisfying

$$|f(t)| \le Me^{\alpha t} \quad (t \ge 0).$$

Hence,

$$|g(t)| \le M \int_0^t e^{\alpha \tau}\, d\tau = \frac{M}{\alpha}(e^{\alpha t} - 1) \le \frac{M}{\alpha}e^{\alpha t}.$$

Therefore, the function g is continuous and of exponential order. Hence, Laplace transform of g exists. Moreover, we have $g'(t) = f(t)$, with $g(0) = 0$. By Theorem 11.6, we obtain

$$\mathcal{L}[g'(t)] = s\mathcal{L}[g(t)] - g(0).$$

It thus follows that

$$\mathcal{L}[f(t)] = s\mathcal{L}[g(t)] \quad \text{or} \quad \mathcal{L}\left[\int_0^t f(t)dt\right] = (\mathcal{L}[f(t)])/(s). \quad \square$$

Example 11.22 By using the relation $\mathcal{L}[\cos(at)] = s/(s^2 + a^2)$, we have

$$\frac{1}{s^2 + a^2} = \mathcal{L}\left[\int_0^t \cos(at)dt\right]$$

$$= \mathcal{L}\left[\frac{\sin(at)}{a}\right].$$

It thus follows that

$$\mathcal{L}[\sin(at)] = \frac{a}{s^2 + a^2}.$$

Theorem 11.11 (Initial Value Theorem) *If $f : [0, \infty) \to \mathbb{R}$ piecewise continuous and is of exponential order, the Laplace transform of its derivative f' exist, and $F(s) = \mathcal{L}f(t)$, then*

$$\lim_{s \to \infty} sF(s) = \lim_{t \to 0+} f(t)$$

provided the limit $\lim_{t \to 0+} f(t)$ exists.

Proof Since

$$sF(s) - f(0) = \mathcal{L}[f'(t)] = \int_0^\infty e^{-st} f'(t)dt$$

we have

$$\lim_{s \to \infty} s F(s) - f(0) = 0. \quad \square$$

Theorem 11.12 (Final Value Theorem) *If the Laplace transforms of* $f : [0, \infty) \to$ $\mathbb{R}$, *its derivative* f' *exist and* $F(s) = \mathscr{L} f(t)$ *then*

$$\lim_{s \to 0} s F(s) = \lim_{t \to \infty} f(t)$$

provided the two limits exit.

Proof Since

$$s F(s) - f(0) = \mathscr{L}[f'(t)] = \int_0^\infty e^{-st} f'(t) dt$$

we have

$$\lim_{s \to 0} s F(s) - f(0) = \int_0^\infty f'(t) dt = \lim_{t \to \infty} f(t) - f(0)$$

and so $\lim_{s \to 0} s F(s) = \lim_{t \to \infty} f(t)$. $\square$

Example 11.23 Let $f : [0, \infty) \to \mathbb{R}$ be given by $\cos at$. Then its Laplace transform is given by $F(s) = (s)/((s^2 + a^2))$.

$$\lim_{s \to \infty} s F(s) = \lim_{s \to \infty} (s^2)/(s^2 + a^2) = \lim_{s \to \infty} \frac{1}{1 + a^2/s^2} = 1 = \lim_{t \to 0} f(t).$$

Also, we have $\lim_{s \to 0} s F(s) = 0$ but the limit $\lim_{t \to \infty} f(t)$ does not exist.

Periodic functions

Theorem 11.13 *If* $f : [0, \infty) \to \mathbb{R}$ *is piecewise continuous and periodic with period* T, *then*

$$\mathscr{L}[f(t)] = \frac{1}{1 - e^{-sT}} \int_0^T e^{-st} f(t) dt.$$

Proof We have

$$F(s) = \int_0^\infty f(t)e^{-st}\, dt$$

$$= \sum_{n=0}^\infty \int_{nT}^{(n+1)T} e^{-st} f(t)\, dt$$

and using the substitution $u = t - nT$ we have

$$F(s) = \sum_{n=0}^\infty \int_0^T e^{-su-snT} f(u + nT)\, du$$

$$= \sum_{n=0}^\infty e^{-snT} \int_0^T e^{-su} f(u)\, du$$

$$= \left(\sum_{n=0}^\infty (e^{-sT})^n \right) \int_0^T e^{-su} f(u)\, du$$

$$= \frac{1}{1 - e^{-sT}} \int_0^T e^{-st} f(t)\, dt. \quad \square$$

Example 11.24 The function $f : [0, \infty)$ defined by $f(t) = \sin(at)$ is periodic of period $2\pi/a$ as

$$f(t + 2\pi/a) = \sin(a(t + 2\pi/a)) = \sin(at + 2\pi) = \sin(at) = f(t).$$

Therefore, we have

$$\mathscr{L}[\sin(at)] = \frac{1}{1 - e^{-2\pi s/a}} \int_0^{2\pi/a} e^{-st} \sin(at)\, dt$$

$$= \frac{1}{1 - e^{-2\pi s/a}} \frac{e^{-st}}{a^2 + s^2} \Big(-s \sin(at) - a \cos(at) \Big) \Big|_0^{2\pi/a}$$

$$= \frac{a}{s^2 + a^2} \frac{1 - e^{-2\pi s/a}}{1 - e^{-2\pi s/a}}$$

$$= \frac{a}{s^2 + a^2}.$$

Example 11.25 Suppose $f : [0, \infty) \to \mathbb{R}$ be periodic extension of the function given by

$$f(t) = t, \quad \text{for } 0 \le t < 1,$$

with $f(t + 1) = f(t)$, for all $t \ge 0$. The above function is known as a *sawtooth function*. As the period of f is $T = 1$, the Laplace transform of f is given by

$$
\begin{aligned}
F(s) &= \frac{1}{1 - e^{-s}} \int_0^1 t e^{-st}\, dt \\[2mm]
&= -\frac{1}{1 - e^{-s}} \left. \frac{e^{-st}(st + 1)}{s^2} \right|_0^1 \\[2mm]
&= \frac{1 - e^{-s}(1 + s)}{s^2(1 - e^{-s})} \\[2mm]
&= \frac{1}{s^2} - \frac{e^{-s}}{s(1 - e^{-s})}.
\end{aligned}
$$

Recall the definition of *error function* $\mathrm{erf}(x)$ as given in (10.5.12). Clearly, $\mathrm{erf}(x)$ is an odd function, and there is a natural *complementary error function* defined by

$$\mathrm{erfc}(x) := \frac{2}{\sqrt{\pi}} \int_x^{\infty} e^{-u^2}\, du \tag{11.2.2}$$

such that we have the relation

$$\mathrm{erf}(x) + \mathrm{erfc}(x) = 1. \tag{11.2.3}$$

We leave verification of the formulas, as given below, as an exercise for the reader.

$$\mathscr{L}\left[\frac{e^{-a^2/(4t)}}{\sqrt{\pi t}}\right] = \frac{e^{-a\sqrt{s}}}{\sqrt{s}}; \tag{11.2.4a}$$

$$\mathscr{L}\left[\frac{a e^{-a^2/(4t)}}{2\sqrt{\pi t^3}}\right] = e^{-a\sqrt{s}}; \tag{11.2.4b}$$

$$\mathscr{L}[\mathrm{erf}(t)] = \frac{e^{s^2/4}\,\mathrm{erfc}(s/2)}{s}; \tag{11.2.4c}$$

$$\mathscr{L}\left[\mathrm{erfc}\left(\frac{a}{2\sqrt{t}}\right)\right] = \frac{e^{-a\sqrt{s}}}{s}; \tag{11.2.4d}$$

$$\mathscr{L}\left[2\sqrt{t/\pi}\,e^{-a^2/(4t)} - a\left\{\mathrm{erfc}\left(\frac{a}{2\sqrt{t}}\right)\right\}\right] = \frac{e^{-a\sqrt{s}}}{s\sqrt{s}}. \tag{11.2.4e}$$

11.3   Inverse Laplace Transform

For a given complex variable function $F = F(s)$, a function $f : \mathbb{R} \to \mathbb{C}$ is called
the *inverse Laplace transform* of F if $F = \mathscr{L}[f]$. In this case, we also write $f(t) = \mathscr{L}^{-1}[F(s)]$. In general, there is no guarantee that such a function f may exists.
However, it is unique when it does, provided $F \not\equiv 0$. A necessary condition for the
existence of f is that $F(s) \to 0$ as $|s| \to \infty$. By using some basic results from
complex analysis, it is possible to give a *general formula* such as given in the next
theorem.

Theorem 11.14 (Contour Integration Method) *Suppose a function* $f \in PC[0, \infty)$
is of exponential order $\alpha > 0$. *If* $f' \in PC[0, \infty)$ *then* $\mathscr{L}[f(t)] = F$ *exists, and is
analytic on the domain* $\mathrm{Re}(s) > \alpha$. *Conversely, if* $\beta > \alpha$, *then we have*

$$f(t) = \mathscr{L}^{-1}[F(s)] = \frac{1}{2\pi i} \int_{\beta-i\infty}^{\beta+i\infty} e^{st} F(s)\, ds, \quad \text{for } t \in [0, \infty).$$

Proof Left for the reader as an exercise.   □

The next theorem proves that the inverse Laplace transform is a linear operator.

Theorem 11.15 (Linearity property). *If* F *and* G *are Laplace transforms of some
functions, then for any two real number* α, β, *we have*

$$\mathscr{L}^{-1}[\alpha F(s) + \beta G(s)] = \alpha \mathscr{L}^{-1}[F(s)] + \beta \mathscr{L}^{-1}[G(s)].$$

Proof Let F and G are Laplace transforms of the functions f and g. Then, from

$$\mathscr{L}[\alpha f(t) + \beta g(t)] = \alpha \mathscr{L}[f(t)] + \beta \mathscr{L}[g(t)] = \alpha F(s) + \beta G(s),$$

we get, by definition of inverse Laplace transform,

$$\mathscr{L}^{-1}[\alpha F(s) + \beta G(s)] = \alpha f(t) + \beta g(t)$$
$$= \alpha \mathscr{L}^{-1}[F(s)] + \beta \mathscr{L}^{-1}[G(s)]. \quad □$$

Theorem 11.16 *If* F *is the Laplace transform of a function* f, *then*

$$\mathscr{L}^{-1}[F(s + a)] = e^{-at} \mathscr{L}^{-1}[F(s)].$$

Proof By replacing a by $-a$ in first shifting theorem (Theorem 11.3), we get

$$F(s) := \mathscr{L}[f(t)] \Rightarrow \mathscr{L}[e^{-at} f(t)] = F(s + a)$$

and by inverting we have

$$\mathcal{L}^{-1}[F(s+a)] = e^{-at} f(t) = e^{-at} \mathcal{L}^{-1}[F(s)]. \quad \square$$

Example 11.26 We find the inverse Laplace transform of the function $F(s) = 1/s(s+1)^2$. For, we have

$$\mathcal{L}[t] = \frac{1}{s^2} \implies \mathcal{L}[te^{-t}] = \frac{1}{(s+1)^2}$$

Therefore, for $f(t) = te^{-t}$, we have $F(s) = 1/(s+1)^2$. Hence,

$$\frac{1}{s(s+1)^2} = \frac{F(s)}{s} \implies \mathcal{L}^{-1}\left(\frac{1}{s(s+1)^2}\right) = \int_0^t \tau e^{-\tau} d\tau = 1 - (t+1)e^{-t}$$

Theorem 11.17 *If $F(s)$ is the Laplace transform of f, then*

$$\mathcal{L}[-tf(t)] = F'(s), \quad and \quad \mathcal{L}^{-1}[F'(s)] = -tf(t). \tag{11.3.1}$$

Example 11.27 We find the inverse Laplace transform of the function

$$F(s) = \ln\left(\frac{s-a}{s-b}\right)$$

For, let $\mathcal{L}[f(t)] = F(s)$ so that $\mathcal{L}[tf(t)] = -F'(s)$. Therefore,

$$\mathcal{L}[tf(t)] = \frac{1}{s-b} - \frac{1}{s-a} = \mathcal{L}[e^{bt} - e^{at}]$$

$$\implies \quad f(t) = \frac{e^{bt} - e^{at}}{t}.$$

Alternatively, notice that

$$\mathcal{L}[f(t)] = \ln\left(\frac{s-a}{s-b}\right)$$

$$= \int_s^\infty \frac{1}{s-b} dp - \int_s^\infty \frac{1}{s-a} dp$$

$$= \mathcal{L}\left(\frac{e^{bt}}{t}\right) - \mathcal{L}\left(\frac{e^{at}}{t}\right)$$

Hence, it follows that

$$\mathscr{L}[f(t)] = \mathscr{L}\left(\frac{e^{bt} - e^{at}}{t}\right) \implies f(t) = \frac{e^{bt} - e^{at}}{t}.$$

There do exist simple methods to find the inverse Laplace transform when the function $F(s)$ is the Laplace transform of a known function $f(t)$. In what follows, we discuss some commonly used analytical methods[1] those help find the inverse Laplace transform. We first discuss some simple cases as given below:

$$\mathscr{L}[t^a] = \frac{\Gamma(a+1)}{s^{a+1}}, \quad a > -1,$$

$$\implies \mathscr{L}^{-1}\left[\frac{1}{s^{a+1}}\right] = \frac{t^a}{\Gamma(a+1)}.$$

In particular, it follows that

$$\mathscr{L}^{-1}\left[\frac{1}{s}\right] = 1, \quad \mathscr{L}^{-1}\left[\frac{1}{s^2}\right] = t, \quad \text{and} \quad \mathscr{L}^{-1}\left[\frac{1}{s^n}\right] = \frac{t^{n-1}}{(n-1)!}.$$

We also have

$$\mathscr{L}[e^{at}] = \frac{1}{s-a} \implies \mathscr{L}^{-1}\left[\frac{1}{s-a}\right] = e^{at}.$$

$$\mathscr{L}[\sin(at)] = \frac{a}{s^2 + a^2} \implies \mathscr{L}^{-1}\left[\frac{a}{s^2 + a^2}\right] = \sin(at)$$

$$\mathscr{L}[\cos(at)] = \frac{s}{s^2 + a^2} \implies \mathscr{L}^{-1}\left[\frac{s}{s^2 + a^2}\right] = \cos(at)$$

$$\mathscr{L}[\sinh(at)] = \frac{a}{s^2 - a^2} \implies \mathscr{L}^{-1}\left[\frac{a}{s^2 - a^2}\right] = \sinh(at)$$

$$\mathscr{L}[\cosh(at)] = \frac{s}{s^2 - a^2} \implies \mathscr{L}^{-1}\left[\frac{s}{s^2 - a^2}\right] = \cosh(at)$$

$$\mathscr{L}[e^{at}\sin(bt)] = \frac{b}{(s-a)^2 + b^2} \implies \mathscr{L}^{-1}\left[\frac{b}{(s-a)^2 + b^2}\right] = e^{at}\sin(bt),$$

$$\mathscr{L}[e^{at}\sin(bt)] = \frac{s-a}{(s-a)^2 + b^2} \implies \mathscr{L}^{-1}\left[\frac{s-a}{(s-a)^2 + b^2}\right] = e^{at}\sin(bt).$$

---

[1] In some more complex problems, especially related to control theory, we also need to use numerical techniques.

In general, we apply one of the following four to find the inverse Laplace transform:

1. Method of *partial fraction decomposition*: If $F(s) = p(s)/q(s)$, where p, q are polynomials such that $\deg(p) < \deg(q)$, then we express $F(s)$ as the sum of terms with known inverse Laplace transform.
2. The *Convolution Theorem* (Theorem 11.18).
3. *Contour integration of the Laplace inversion integral*: The main point in the proof of Theorem 11.14 is that construction of the inverse Laplace transform $f(t)$ of a complex variable function $F(s)$ involves integration along the so called *Bromwich Contour*, which contains $\mathrm{Re}(s) = \beta$ and limit varies from $-\infty$ to ∞. The details depends on the type of singularities the function $F(s)$ has, and the *Cauchy Residue Theorem* takes care of the rest. Notice that if f is discontinuous then the values given by the complex integral are *averages* as used in the definition of functions in $PC[0, \infty)$. The reader may refer [1], for further details.
4. *Heaviside's Expansion Theorem*: If $F(s) = p(s)/q(s)$, where p, q are polynomials such that $\deg(q) \geq \deg(p)$, then

$$\mathscr{L}^{-1}\left[\frac{p(s)}{q(s)}\right] = \sum_{k=1}^{n} \frac{p(\alpha_k)}{\alpha_k} e^{t\alpha_k}, \qquad (11.3.2)$$

where α_k are the distinct roots of the equations $q(s) = 0$.

Example 11.28 To find the inverse Laplace transform of $1/s(s+1)$, we have

$$\frac{1}{s(s+1)} = \frac{1}{s} - \frac{1}{s+1}.$$

Using linearity of the inverse transform, we get

$$\mathscr{L}^{-1}\left[\frac{1}{s(s+1)}\right] = \mathscr{L}^{-1}\left[\frac{1}{s}\right] + \mathscr{L}^{-1}\left[\frac{1}{s+1}\right] = 1 - e^{-t}.$$

Example 11.29 We show that

$$\mathscr{L}^{-1}\left[\frac{s}{s^2 - a^2}\right] = \cosh(at).$$

As we can write

$$\frac{s}{s^2 - a^2} = \frac{1}{2}\left(\frac{1}{s-a} + \frac{1}{s+a}\right),$$

by using the linearity, we obtain

$$\mathscr{L}^{-1}\left[\frac{s}{s^2 - a^2}\right] = \frac{1}{2}\left(\mathscr{L}^{-1}\left[\frac{1}{s-a}\right] + \mathscr{L}^{-1}\left[\frac{1}{s+a}\right]\right)$$

$$= \frac{e^{at} + e^{-at}}{2}$$

$$= \cosh(at).$$

An elegant way of finding inverse Laplace transform of a function h is given by the *convolution theorem*, provided it is possible to write the function h as a convolution product $f * g$, for some function f and g, and the inverse Laplace transform $\mathscr{L}^{-1}[f]$ and $\mathscr{L}^{-1}[g]$ are known.

Definition 11.2 The *convolution product* of two functions $f, g : [0, \infty) \to \mathbb{R}$, denoted by $f * g$, is given by

$$(f * g)(t) = \int_0^t f(\tau)g(t - \tau)\,d\tau, \quad \text{for } t \geq 0.$$

As said earlier, the functions f and g are "*folded together*" to obtain $h = f * g$. It is easy to verify that if functions f and g are respectively of exponential order α_1 and α_2 then h is of exponential order $\max\{\alpha_1, \alpha_2\} + \varepsilon$, for any $\varepsilon > 0$. Moreover, as

$$\int_0^t f(\tau)g(t - \tau)\,d\tau = \int_0^t g(\tau)f(t - \tau)\,d(-\tau) = \int_0^t g(\tau)f(t - \tau)\,d\tau,$$

it follows that h does not depend on the order in which the functions f and g are operated. Therefore, the operation $*$ is *commutative*. In fact, it is *associative* and also *distributes over sum*.

Theorem 11.18 (Convolution Theorem) *Let $f, g \in PC[0, \infty)$ be functions of exponential order, with $\mathscr{L}[f(t)] = F(s)$ and $\mathscr{L}[g(t)] = G(s)$. Then we have*

$$\mathscr{L}[f * g] = \int_0^\infty (f * g)(t)e^{-st}\,dt = F(s)G(s).$$

That is, the Laplace transform $\mathscr{L}$ is multiplicative with respect to convolution product.

Proof By using the above definition, we have

$$\mathscr{L}\big((f*g)(t)\big) = \int\limits_0^\infty (f*g)(t)e^{-st}\,dt$$

$$= \int\limits_0^\infty \left(\int\limits_0^t f(\tau)g(t-\tau)\,d\tau\right)e^{-st}\,dt$$

The region of integration is the area in the first quadrant of $t\tau$-plane bounded by the t-axis and the line $\tau = t$. The limits of the inner integral varies from $\tau = 0$ to $\tau = t$. Changing the order of integration, we see that the variable τ varies from $\tau = 0$ to $\tau = \infty$ and the limits of t are between $t = \tau$ and $t = \infty$. Hence, by using $t - \tau = u$, we obtain

$$\mathscr{L}\big((f*g)(t)\big) = \int\limits_0^\infty \left(\int\limits_\tau^\infty e^{-st}g(t-\tau)\,dt\right)f(\tau)\,d\tau$$

$$= \int\limits_0^\infty \left(\int\limits_0^\infty e^{-su}g(u)\,du\right)f(\tau)e^{-s\tau}\,d\tau$$

$$= \left(\int\limits_0^\infty e^{-su}g(u)\,du\right)\left(\int\limits_0^\infty e^{-s\tau}f(\tau)\,d\tau\right)$$

$$= F(s)G(s). \quad \square$$

In particular, taking $g : [0, \infty) \to \mathbb{R}$ as a constant function given by $g \equiv 1$, we have $G(s) = 1/s$. Therefore, Theorem 11.10 is a special case of the convolution theorem. That is, we have

$$\mathscr{L}\left[\int\limits_0^t f(t)dt\right] = \frac{\mathscr{L}[f(t)]}{s}.$$

Further, replacing f by f' in the above relation, we obtain

$$\mathscr{L}[f'(t)] = s\mathscr{L}\left[\int\limits_0^t f(t)dt\right]$$

$$= s\mathscr{L}[f(t) - f(0)]$$

$$= s\mathscr{L}[f(t)] - f(0),$$

as shown earlier by direct method. As said above, Theorem 11.18 is a powerful tool to find the inverse Laplace transform of various important functions. For, notice that

$$\mathcal{L}^{-1}[F(s)G(s)] = f(t) * g(t) = \int_0^t f(\tau)g(t - \tau)\,d\tau.$$

Example 11.30 To find the inverse Laplace transform of the function $H(s) = 1/(s(s + 1)^2)$, we may write it as a product of the functions $F(s) = 1/s$ and $G(s) = 1/(s + 1)^2$. We already know that the inverse Laplace transforms of F and G are respectively given by the functions $f(t) = 1$ and $g(t) = te^{-t}$. Hence, by Theorem 11.18, we find that

$$h(t) = \mathcal{L}^{-1}\left[\frac{1}{s(s + 1)^2}\right]$$

$$= \int_0^t f(t - \tau)g(\tau)\,d\tau$$

$$= \int_0^t \tau e^{-\tau}\,d\tau$$

$$= -(1 + \tau)e^{-\tau}\big|_0^t$$

$$= 1 - (1 + t)e^{-t}.$$

Example 11.31 To find the inverse Laplace transform of the function $H(s) = 1/(s^2 + a^2)^2$, we may write it as a product of the functions $F(s) = 1/(s^2 + a^2)$ and $G(s) = 1/(s^2 + a^2)$. We already know that the inverse Laplace transform of both F and G is given by the functions $f(t) = g(t) = \sin(at)/a$. Hence, by Theorem 11.18, we have

$$h(t) = \mathcal{L}^{-1}\left[\frac{1}{(s^2 + a^2)^2}\right]$$

$$= \frac{1}{a^2}\int_0^t \sin(a\tau)\sin(a(t - \tau))\,d\tau.$$

The identity $2 \sin x \sin y = \cos(x - y) - \cos(x + y)$ yields

$$2 \sin(a\tau)\sin(a(t - \tau)) = \cos(a(2\tau - t)) - \cos(at).$$

It thus follows that the integral in the last equation is given by

$$\int_0^t 2\sin(a\tau)\sin(a(t-\tau))d\tau = \frac{1}{2a}(\sin(at) - \sin(-at)) - t\cos(at).$$

Finally, by using the above relation, we obtain

$$h(t) = \mathcal{L}^{-1}\left[\frac{1}{(s^2+a^2)^2}\right] = \frac{1}{2a^3}\left[\sin(at) - at\cos(at)\right].$$

11.4 Applications to Differential Equations

We first apply Laplace transform to solve initial value problems for ordinary differential equations that are typically used to model the law of natural growth or decay concerning some natural process. Consider initial value problem as given below:

$$y' + p\,y = f(t), \quad \text{for } t > 0, \text{ and with } y(0) = a,$$

where both p and a are constants, and f is a *forcing function*. Recall that

$$\mathcal{L}[y'] = s\,Y(s) - y(0), \quad \text{where } Y(s) = \mathcal{L}[y] = \int_0^\infty e^{-st}y(t)\,dt.$$

Applying Laplace transform to the above problem, we obtain

$$s\,Y(s) - y(0) + pY(s) = F(s) \quad \text{or} \quad Y(s) = \frac{a}{s+p} + \frac{F(s)}{s+p}.$$

Therefore, by applying the inverse Laplace transform and the convolution theorem, a unique solution of the given initial value problem is obtained as

$$y(t) = a\,e^{-pt} + \int_0^t f(t-\tau)\,e^{-p\tau}\,d\tau.$$

The first term on the right side corresponds to the response to the assigned initial condition, and the second to the external force represented by the function f, according to $p > 0$ or $p < 0$. We consider next initial value problem for a second order ordinary differential equation, with forcing function given by $f(t) = t$, such as

$$y'' + y = t, \quad \text{with } y(0) = 0 \text{ and } y'(0) = 2.$$

Recall that we have

$$\mathcal{L}[y''] = s^2 Y(s) - s\, y(0) - y'(0),$$

so that by applying Laplace transform to the above problem, we obtain

$$s^2 Y(s) - sy(0) - y''(0) + Y(s) = \frac{1}{s^2}$$

so that it follows by using the assigned initial conditions that

$$s^2 Y(s) - 2 + Y(s) = \frac{1}{s^2} \quad \Longrightarrow \quad Y(s) = \frac{1}{s^2(s^2+1)} + \frac{2}{s^2+1}$$

Therefore, by using the partial fractions technique, we may write

$$Y(s) = \frac{1}{s^2} + \frac{1}{s^2+1}.$$

Hence, by applying the inverse Laplace transform, a unique solution of the given initial value problem is obtained as $y(t) = t + \sin t$. The same techniques apply to initial value problems for higher order ordinary differential equations with constant coefficients. Notice that application of Laplace transform method to first order system of linear ordinary differential equations with constant coefficients leads to a problem about solving a system of linear equations.

We now apply Laplace transform to solve some initial-boundary value problems for partial differential equations involving a function of two variable, say $u = u(x, t)$, with $x \geq 0$ and $t > 0$. In this case, treating the variable x as a parameter, we may write

$$U(x, s) := \mathcal{L}[u(x, t)] = \int_0^\infty e^{-st} u(x, t) \mathrm{d}t. \tag{11.4.1}$$

Further, assuming that the function $u(x, t)$ satisfies all the requisite conditions, we have

$$\mathcal{L}[u_t(x, t)] = s\, U(x, s) - u(x, 0); \tag{11.4.2}$$

$$\mathcal{L}[u_{tt}(x, t)] = s^2\, U(x, s) - su(x, 0) - u_t(x, 0). \tag{11.4.3}$$

Similarly, treating the variable t as a parameter, we have

$$\mathscr{L}[u_x(x,t)] = \int_0^\infty e^{-st} u_x(x,t)dt = \frac{dU(x,s)}{dx}; \qquad (11.4.4)$$

$$\mathscr{L}[u_{xx}(x,t)] = \int_0^\infty e^{-st} u_{xx}(x,t)dt = \frac{d^2U(x,s)}{dx^2}. \qquad (11.4.5)$$

As said earlier, the basic idea is to apply the Laplace transform with respect to variable t, and then solve the resulting initial value problem for ordinary differential equation by using variation of parameter technique. Finally, the inverse Laplace transform is applied to obtain the solution. The simple initial-boundary value problem given in the next example illustrates the procedure.

Example 11.32 Consider the initial-boundary value problem:

$$u_x(x,t) + u_t(x,t) = x, \quad \text{for } 0 \le x < \infty \text{ and } t > 0;$$
$$u(x,0) = 0 \quad \text{and} \quad u(0,t) = 0.$$

Applying the Laplace transform with respect to variable t, we have

$$\frac{dU(x,s)}{dx} + s\,U(x,s) - u(x,0) = \frac{x}{s}$$

so that, by using the assigned initial condition, we obtain

$$\frac{dU(x,s)}{dx} + s\,U(x,s) = \frac{x}{s}$$

$$\implies \quad \frac{d}{dx}\left[e^{sx}U(x,s)\right] = \frac{x\,e^{sx}}{s}.$$

Therefore, the method of *integrating factor* implies

$$U(x,s) = \frac{e^{-sx}}{s}\left(\int_0^x e^{su}u\,du\right) + A\,e^{-sx}$$

$$= \frac{e^{-sx}}{s}\left(\frac{x e^{sx}}{s} - \frac{e^{sx}}{s^2}\right) + A(s)\,e^{-sx}$$

$$= \frac{x}{s^2} - \frac{1}{s^3} + A(s)\,e^{-sx}.$$

The assigned boundary condition implies $A = s^{-3}$. Hence, we obtain

$$U(x,s) = \frac{x}{s^2} - \frac{1}{s^3} + \frac{e^{-sx}}{s^3}.$$

Finally, by applying the inverse Laplace transform, a unique solution of the given initial-boundary value problem is obtained as

$$u(x, t) = xt - \frac{t^2}{2} + \frac{(t-x)^2}{2} = \frac{x^2}{2}.$$

The same technique implies that a unique solution of the initial-boundary value problem:

$$u_x(x, t) + u_t(x, t) = x, \quad \text{for } 0 \le x < \infty \text{ and } t > 0;$$
$$u(x, 0) = 0 \quad \text{and} \quad u(0, t) = 0.$$

is given by the function
$$u(x, t) = x(1 - e^{-t}).$$

Wave Equations

We first solve initial value problem for wave equation as given below:

$$u_{tt} = c^2 u_{xx}, \quad \text{for } -\infty < x < \infty \text{ and } t > 0$$
$$u(x, 0) = f(x) \quad \text{and} \quad u_t(x, 0) = g(x). \tag{11.4.6}$$

Recall that we have

$$\mathcal{L}[u_{tt}] = s^2 U(x, s) - s\, u(x, 0) - u_t(x, 0),$$

where $U(x, s)$ is as defined in (11.4.1). Therefore, by applying the Laplace transform to (11.4.8), we have

$$s^2 U(x, s) - s\, u(x, 0) - u_t(x, 0) = c^2 U_{xx}(x, s)$$
$$\implies \quad s^2 U(x, s) - s\, f(x) - g(x) = c^2 U_{xx}(x, s)$$
$$\implies \quad -s\, f(x) - g(x) = c^2 U_{xx}(x, s) - s^2 U(x, s)$$
$$\implies \quad U_{xx}(x, s) - \frac{s^2}{c^2} U(x, s) = -\frac{s}{c^2} f(x) - \frac{1}{c^2} g(x), \tag{11.4.7}$$

which is an ordinary differential equation, with parameter s. The two linearly independent solutions of the associated homogeneous equation are given by

$$y_1(x, s) = e^{-sx/c} \quad \text{and} \quad y_2(x, s) = e^{sx/c}.$$

Therefore, by variation of parameters, a particular solution is given by

$$y_p(x, s) = \int_0^x \left[\frac{y_1(\xi, s)y_2(x, s) - y_1(x, s)y_2(\xi, s)}{y_1(\xi, s)y_2'(\xi, s) - y_1'(\xi, s)y_2(\xi, s)} \right] h(\xi, s)\, d\xi,$$

where the function $h(\xi, s)$ is given by

$$h(\xi, s) = -\frac{s}{c^2} f(\xi) - \frac{1}{c^2} g(\xi).$$

As $y_1(\xi, s)y_2'(\xi, s) - y_1'(\xi, s)y_2(\xi, s) = 2s/c$, and

$$y_1(\xi, s)y_2(x, s) - y_1(x, s)y_2(\xi, s) = e^{-(s/c)[\xi - x]} - e^{-(s/c)[x - \xi]},$$

it follows that the general solution of Eq. (11.4.7) is given by

$$
\begin{aligned}
U(x, s) &= y(x, s) \\
&= A(s)y_1(x, s) + B(s)y_2(x, s) + y_p(x, s) \\
&= A(s)e^{-sx/c} + B(s)e^{sx/c} + \frac{c}{2s} \int_0^x \left[e^{-(s/c)[\xi - x]} - e^{-(s/c)[x - \xi]} \right] h(\xi, s) \, d\xi \\
&= e^{-sx/c} \left[A(s) - \frac{c}{2s} \int_0^x e^{s\xi/c} h(\xi, s) \, d\xi \right] + e^{sx/c} \left[B(s) - \frac{c}{2s} \int_0^x e^{-s\xi/c} h(\xi, s) \, d\xi \right].
\end{aligned}
$$

Therefore, if $\lim_{x \to -\infty} U(x, s)$ is finite, we must have

$$A(s) - \lim_{x \to -\infty} \frac{c}{2s} \int_0^x e^{s\xi/c} h(\xi, s) \, d\xi = 0.$$

That is,

$$A(s) = \frac{c}{2s} \int_0^{-\infty} e^{s\xi/c} h(\xi, s) \, d\xi = -\frac{c}{2s} \int_{-\infty}^0 e^{s\xi/c} h(\xi, s) \, d\xi.$$

On the other hand, by similar argument, we have

$$B(s) = -\frac{c}{2s} \int_0^{\infty} e^{-s\xi/c} h(\xi, s) \, d\xi.$$

Substituting the above expressions of $A(s)$ and $B(s)$ into the general solution, it follows by simple algebraic manipulations that

$$U(x, s) = -\frac{c}{2s} \int_{-\infty}^{\infty} e^{-(s/c)|x-\xi|} h(\xi, s)\, d\xi$$

$$= -\frac{c}{2s} \int_{-\infty}^{\infty} e^{-(s/c)|x-\xi|} \left[-\frac{s}{c^2} f(\xi) - \frac{1}{c^2} g(\xi) \right] d\xi$$

$$= \frac{1}{2c} \int_{-\infty}^{\infty} e^{-(s/c)|x-\xi|} f(\xi)\, d\xi + \frac{1}{2c} \int_{-\infty}^{\infty} \frac{e^{-(s/c)|x-\xi|}}{s} g(\xi)\, d\xi$$

As we have

$$\mathscr{L}\left[\int_{-\infty}^{\infty} f(\xi)\delta\left(t - \frac{|x-\xi|}{c}\right) d\xi \right] = \int_{0}^{\infty} \left[\int_{-\infty}^{\infty} f(\xi)\delta\left(t - \frac{|x-\xi|}{c}\right) d\xi \right] e^{-st}\, dt$$

$$= \int_{-\infty}^{\infty} \left[\int_{0}^{\infty} e^{-st}\delta\left(t - \frac{|x-\xi|}{c}\right) dt \right] f(\xi)\, d\xi,$$

by changing the order of integration. Therefore, by using the relation

$$\int_{-\infty}^{\infty} e^{-st}\delta\left(t - \frac{|x-\xi|}{c}\right) dt = e^{-(s/c)|x-\xi|},$$

it follows that

$$\mathscr{L}\left[\int_{-\infty}^{\infty} f(\xi)\delta\left(t - \frac{|x-\xi|}{c}\right) d\xi \right] = \int_{-\infty}^{\infty} e^{-(s/c)|x-\xi|} f(\xi)\, d\xi,$$

or, equivalently, we may write

$$\mathscr{L}^{-1}\left[\int_{-\infty}^{\infty} e^{-(s/c)|x-\xi|} f(\xi)\, d\xi \right] = \int_{-\infty}^{\infty} f(\xi)\delta\left(t - \frac{|x-\xi|}{c}\right) d\xi.$$

Taking H for the Heaviside function, we can write

$$\mathscr{L}\left[\int_{-\infty}^{\infty} g(\xi) H\left(t - \frac{|x - \xi|}{c}\right) d\xi\right] = \int_{0}^{\infty}\left[\int_{-\infty}^{\infty} g(\xi) H\left(t - \frac{|x - \xi|}{c}\right) d\xi\right] e^{-st} dt$$

$$= \int_{-\infty}^{\infty}\left[\int_{0}^{\infty} e^{-st} H\left(t - \frac{|x - \xi|}{c}\right) dt\right] f(\xi) d\xi,$$

by changing the order of integration. Now, as we have

$$\int_{0}^{\infty} e^{-st} H\left(t - \frac{|x - \xi|}{c}\right) dt = \int_{|x-\xi|/c}^{\infty} e^{-st} dt = \frac{e^{-(s/c)|x-\xi|}}{s},$$

it thus follows that

$$\int_{-\infty}^{\infty}\left[\int_{0}^{\infty} e^{-st} H\left(t - \frac{|x - \xi|}{c}\right) dt\right] f(\xi) d\xi = \int_{-\infty}^{\infty} \frac{e^{-(s/c)|x-\xi|}}{s} g(\xi) d\xi$$

Therefore, we obtain

$$\mathscr{L}^{-1}\left[\int_{-\infty}^{\infty} e^{-(s/c)|x-\xi|} g(\xi) d\xi\right] = \int_{-\infty}^{\infty} g(\xi) H\left(t - \frac{|x - \xi|}{c}\right) d\xi$$

Finally, we obtain

$$u(x, t) = \mathscr{L}^{-1}\left[U(x, s)\right]$$

$$= \mathscr{L}^{-1}\left[\frac{1}{2c}\int_{-\infty}^{\infty} e^{-(s/c)|x-\xi|} f(\xi) d\xi + \frac{1}{2c}\int_{-\infty}^{\infty} \frac{e^{-(s/c)|x-\xi|}}{s} g(\xi) d\xi\right]$$

$$= \frac{1}{2c}\left\{\mathscr{L}^{-1}\left[\int_{-\infty}^{\infty} e^{-(s/c)|x-\xi|} f(\xi) d\xi\right] + \mathscr{L}^{-1}\left[\int_{-\infty}^{\infty} \frac{e^{-(s/c)|x-\xi|}}{s} g(\xi) d\xi\right]\right\}$$

$$= \frac{1}{2c}\left[\int_{-\infty}^{\infty} f(\xi)\delta\left(t - \frac{|x - \xi|}{c}\right) d\xi + \int_{-\infty}^{\infty} g(\xi) H\left(t - \frac{|x - \xi|}{c}\right) d\xi\right].$$

Taking $u = \xi/c$ so that $du = d\xi/c$, the first integral in the last equation can be written as

$$\frac{1}{2}\int_{-\infty}^{\infty} f(\xi)\delta\left(t - \frac{|x - \xi|}{c}\right)\frac{d\xi}{c} = \frac{1}{2}\int_{-\infty}^{\infty} f(cu)\delta\left(t - \frac{|x - cu|}{c}\right) du.$$

Further, as

$$t - \frac{|x - \xi|}{c} = 0, \quad \text{for } u = \frac{x \pm ct}{c},$$

it follows that

$$\frac{1}{2} \int_{-\infty}^{\infty} f(\xi) \delta\left(t - \frac{|x - \xi|}{c}\right) \frac{d\xi}{c} = \frac{1}{2} \int_{-\infty}^{\infty} f(cu) \delta\left(t - \frac{|x - cu|}{c}\right) du$$

$$= \frac{1}{2}\left[f\left(c\left(\frac{x - ct}{c}\right)\right) + f\left(c\left(\frac{x + ct}{c}\right)\right)\right]$$

$$= \frac{1}{2}\left[f(x - ct) + f(x + ct)\right].$$

On the other hand, as

$$H\left(t - \frac{|x - \xi|}{c}\right) = \begin{cases} 0, & \text{for } |x - \xi| > ct \\ 1, & \text{for } |x - \xi| < ct \end{cases},$$

we have

$$\frac{1}{2c} \int_{-\infty}^{\infty} g(\xi) H\left(t - \frac{|x - \xi|}{c}\right) d\xi = \frac{1}{2c} \int_{x-ct}^{x+ct} g(\xi) \, d\xi.$$

Hence,

$$u(x, t) = \frac{1}{2}\left[f(x - ct) + f(x + ct)\right] + \frac{1}{2c} \int_{x-ct}^{x+ct} g(\xi) \, d\xi,$$

which is d'Alembert's formula, as obtained earlier in Chap. 6.

Example 11.33 Let $y(x, t)$ be the transverse displacement of a vibrating semi-infinite string that is initially at rest in the equilibrium position. Suppose the end at $x = 0$ is constrained to move at $t = 0$ so that we have $y(0, t) = \alpha f(t)$, for $t \geq 0$, with α being a constant. Consider the initial-boundary value problem for the wave equation as given below:

$$y_{tt} = c^2 y_{xx}, \quad \text{for } 0 \leq x < \infty \text{ and } t > 0;$$
$$y(x, 0) = y_t(x, 0) = 0 \text{ and } y(0, t) = \alpha f(t),$$

further assuming that $y(x, t) \to 0$ as $x \to \infty$. Applying the Laplace transform with respect to t, we obtain the BVP given by

$$\frac{d^2 Y(x, s)}{dx * 2} - \frac{s^2}{c^2} Y(x, s) = 0, \quad \text{for } 0 \le x < \infty;$$

$$Y(x, s) = \alpha F(s) \text{ and } Y(x, s) \to 0, \text{ as } x \to \infty.$$

It follows from the standard method as discussed in Chap. 3 that

$$Y(x, s) = \alpha F(s) e^{-(xs)/c}.$$

Therefore, an application of the inverse Laplace transform and convolution theorem gives

$$y(x, t) = A f(t - (x/c)) H(t - (x/c)) = \begin{cases} A f(t - (x/c)), & \text{for } t > (x/c) \\ 0 f(t - (x/c)), & \text{for } t < (x/c) \end{cases}$$

As before, the above solution reflects a wave propagation at speed c along the characteristic $x = ct$.

Heat Equations

Let $u = u(x, t)$ represents the temperature distribution of a properly insulated rod of length $\ell = 2$, with ends at $x = 0$ and $x = 2$ maintained at the temperature zero. We may take the *thermal diffusivity* of the rod to be unity. Suppose the initial temperature distribution of the rod is given by $f(x) = \sin(2\pi x)$. Therefore, we are led to consider the initial-boundary value problem as given below:

$$u_t = u_{xx}, \quad \text{for } 0 < x < 2 \text{ and } t > 0;$$
$$u(0, t) = u(2, t) = 0 \text{ and } u(x, 0) = \sin(2\pi x),$$

Applying the Laplace transform with respect to t, we obtain the BVP

$$\frac{d^2 U(x, s)}{dx^2} - s U(x, s) = -\sin(2\pi x),$$

with boundary conditions given by

$$U(0, s) = 0 \quad \text{and} \quad U(2, s) = 0.$$

Using the standard solution method, as discussed in Chap. 3, the *fundamental solution* of the ordinary differential equation is given by

$$U_c(x, s) = A \cosh(\sqrt{s}\, x) + B \sinh(\sqrt{s}\, x),$$

where the constants A, B need to be determined by applying the assigned boundary conditions. Further, by Theorem 3.6, the *particular integral* is given by

$$U_p(x, s) = \frac{1}{\sqrt{s}} \int \sinh(\sqrt{s}(u - x)) \sin(2\pi u)du.$$

Hence, the complete solution obtained from the boundary conditions is given by

$$U(x, s)$$
$$= \left[\frac{1}{\sqrt{s}} + \frac{1}{\sqrt{s} \sinh(2\sqrt{s})} - \frac{\cosh(\sqrt{s}\, x)}{\sqrt{s}} - \frac{\sinh(\sqrt{s}\, x)}{\sqrt{s}}\right] \int \sinh(\sqrt{s}(u - x)) \sin(2\pi u)du.$$

We next consider a simple initial value problem as given below:

$$u_t = u_{xx}, \quad \text{for } -\infty < x < \infty \text{ and } t > 0;$$
$$u(x, 0) = f(x), \quad \text{for } x \in \mathbb{R},$$

where the function $u = u(x, t)$ is assumed bounded. Notice that

$$|u(x, t)| \leq M \quad \Longrightarrow \quad |U(x, s)| \leq \frac{M}{s}.$$

That is, $U(x, s)$ is also bounded, for all $x \in \mathbb{R}$ and $s > 0$. By the method of *variation of parameters* (Theorem 3.6), the general solution $U(x, s)$ is then given by

$$\left(c_1 - \frac{1}{2\sqrt{s}} \int_0^x e^{-\sqrt{s}\, u} f(u)du\right)e^{\sqrt{s}\, x} + \left(c_2 + \frac{1}{2\sqrt{s}} \int_0^x e^{\sqrt{s}\, u} f(u)du\right)e^{-\sqrt{s}\, x}.$$

For the first term in the above solution to stay bounded, as $x \to \infty$, we must have

$$\lim_{x \to \infty} \left(c_1 - \frac{1}{2\sqrt{s}} \int_0^x e^{-\sqrt{s}\, u} f(u)du\right) = 0,$$

which gives

$$c_1 = \frac{1}{2\sqrt{s}} \int_0^\infty e^{-\sqrt{s}\, u} f(u)du.$$

Similarly, for the second term to stay bounded as $x \to -\infty$, we must have

$$\lim_{x \to -\infty} \left(c_2 + \frac{1}{2\sqrt{s}} \int_0^x e^{\sqrt{s}\, u} f(u)du\right) = 0,$$

which gives

$$c_2 = -\frac{1}{2\sqrt{s}} \int\limits_0^{-\infty} e^{\sqrt{s}\,u} f(u)\mathrm{d}u = \frac{1}{2\sqrt{s}} \int\limits_{-\infty}^0 e^{\sqrt{s}\,u} f(u)\mathrm{d}u.$$

Hence, by simple algebraic manipulation, we obtain

$$U(x, s) = \frac{1}{2\sqrt{s}} \int\limits_{-\infty}^{\infty} e^{-\sqrt{s}|x-u|} f(u)\mathrm{d}u.$$

Taking $a = |x - u|$, we are led to use the relation

$$\mathscr{L}^{-1}\left(\frac{e^{-a\sqrt{s}}}{2\sqrt{s}}\right) = \frac{e^{-a^2/(4t)}}{\sqrt{4\pi t}} := K(a, t).$$

It thus follows that the final solution is given by

$$u(x, t) = \mathscr{L}^{-1}(U(x, s))$$

$$= \mathscr{L}^{-1}\left(\frac{1}{2\sqrt{s}} \int\limits_{-\infty}^{\infty} e^{-\sqrt{s}|x-u|} f(u)\mathrm{d}u\right)$$

$$= \int\limits_{-\infty}^{\infty} \mathscr{L}^{-1}\left(\frac{e^{-\sqrt{s}|x-u|}}{2\sqrt{s}}\right) f(u)\mathrm{d}u$$

$$= \frac{1}{\sqrt{4\pi t}} \int\limits_{-\infty}^{\infty} e^{-|x-u|^2/(4t)} f(u)\mathrm{d}u$$

$$= \int\limits_{-\infty}^{\infty} K(|x - u|, t) f(u)\mathrm{d}u,$$

which is the *fundamental solution* of the heat equation over $\mathbb{R}^+$, where the *Green function*

$$K(x, t) := \frac{e^{-x^2/(4t)}}{2\sqrt{\pi t}}$$

is called the *fundamental heat kernel*.

In general, we may apply the Laplace transform technique to solve the following two types of initial-boundary value problems respectively over *infinite* or *semi-infinite domain*, and possible with some additional *limit conditions*:

$$u_t = u_{xx}, \quad \text{for} \ -\infty < x < \infty \ \text{and} \ t > 0$$
$$u(x, 0) = f(x) \ \text{and} \ u(0, t) = g(t); \tag{11.4.8}$$

$$u_t = u_{xx}, \quad \text{for} \ 0 < x < \infty \ \text{and} \ t > 0$$
$$u(x, 0) = f(x) \ \text{and} \ u(0, t) = g(t), \tag{11.4.9}$$

One may find it tempting to apply the Laplace transform with respect to both the variables in either situation. However, for the initial-boundary value problems of the type (11.4.8), it applies only with respect to variable t. For initial-boundary value problems of the type (11.4.9), it is possible to apply the Laplace transform with respect to both variables. As shown in the previous chapter, application of the Fourier transform with respect to the spatial variable x is valid for all types of problems. But, for the Laplace transform, the function itself must be zero when $x < 0$.

Example 11.34 We solve the initial-boundary value problems as given below:

$$u_t = k u_{xx}, \quad \text{for} \ 0 < x < \infty \ \text{and} \ t > 0;$$
$$u(x, 0) = 0 \ \text{and} \ u(0, t) = g(t), \tag{11.4.10}$$

further assuming that $u(x, t) \to 0$ as $x \to \infty$. As before, the general solution of the transformed second order ordinary differential equation is given by

$$U(x, s) = A\, e^{-x\sqrt{s/k}} + B\, e^{x\sqrt{s/k}}, \tag{11.4.11}$$

where A and B are constants of integration. Now, for solution to be bounded, we must have $B \equiv 0$, and so using $U(0, s) = G(s)$ it follows that

$$U(x, s) = G(s)\, e^{-x\sqrt{s/k}}. \tag{11.4.12}$$

Therefore, by applying the inverse Laplace transform and convolution theorem, we obtain

$$u(x, t) = \frac{x}{2\sqrt{k\pi}} \int_0^t g(t - \tau)\tau^{-3/2} e^{-x^2/4k\tau} d\tau, \tag{11.4.13}$$

which can be simplified by using the substitution $\ell = x/(2\sqrt{k\tau})$ so that we have $d\ell = -(x/4\sqrt{k})\tau^{-3/2} d\tau$. Therefore, we can write

$$u(x, t) = \frac{2}{\sqrt{\pi}} \int_{x/(2\sqrt{kt})}^{\infty} g\left(t - \frac{x^2}{4k\ell^2}\right) e^{-\ell^2} d\ell, \tag{11.4.14}$$

which is a unique solution of the given problem. Notice that, when $g(t) = T_0 =$ constant, the above solution gives $u(x, t) = T_0 \, \text{erfc} \left(x/(2\sqrt{kt}) \right)$. Clearly, u approaches T_0 asymptotically as $t \to \infty$.

Exercises 11

11.1. Suppose the derivative of a function $f : [0, \infty) \to \mathbb{R}$ is of finite exponential order $\alpha > 0$. Show that the function f itself is of finite exponential order $\alpha > 0$.

11.2. Suppose functions f and g are respectively of finite exponential orders $\alpha_1 > 0$ and $\alpha_2 > 0$. Show that the product function $f \cdot g$ is of finite exponential order $\max\{\alpha_1, \alpha_2\} + \varepsilon$, for any $\varepsilon > 0$.

11.3. Compute the Laplace transform of the function $f(t) = \sin^3(at)$.

11.4. Compute the Laplace transform of the function given by

$$
f(t) = \begin{cases} \alpha \sin(\omega t), & 0 \leq t \leq \pi/\omega, \\ 0 \sin(\omega t), & t \geq \pi/\omega \end{cases}, \quad \text{with } f(t + 2) = f(t)
$$

11.5. Suppose $f : [0, \infty) \to \mathbb{R}$ is the periodic extension of the function

$$
f(t) = \begin{cases} t, & 0 \leq t < 1, \\ 2 - t, & 1 \leq t < 2 \end{cases}, \quad \text{with } f(t + 2) = f(t)
$$

Show that $\mathscr{L}[f] = (1/s^2) \tanh(s/2)$.

11.6. Suppose the function $g : [0, \infty) \to \mathbb{R}$ is given by

$$
g(t) = \begin{cases} 0, & \text{for } 0 \leq t \leq 1 \\ t, & \text{for } t > 1 \end{cases}
$$

Show that $\mathscr{L}[g(t)] = [(s + 1)/s^2]e^{-s}$.

11.7. Suppose the function $g : [0, \infty) \to \mathbb{R}$ is given by

$$
g(t) = \begin{cases} e^{-t}, & \text{for } 0 \leq t \leq 4 \\ 0, & \text{for } t > 4 \end{cases}.
$$

Show that $\mathscr{L}[g(t)] = (1 - e^{4(s-1)})/(s + 1)$.

11.8. Let f be the *error function* defined by $f(t) = \text{erf} \left(a/(2\sqrt{t}) \right)$. Show that

$$
\mathscr{L}[f(t)] = \frac{1}{s} \left(1 - e^{-a\sqrt{s}} \right).
$$

11.9. Let erf and erfc(x) be respectively the *error* and *complementary error functions*. Prove the related formulas as given in (11.2.4).

11.10. Let $f(t) = J_0(\alpha t)$ and $g(t) = J_1(\alpha t)$ be respectively the Bessel's functions of order zero and one. Show that

$$\mathcal{L}[f] = -\frac{1}{\sqrt{s^2 + \alpha^2}} \quad \text{and} \quad \mathcal{L}[g] = \frac{1}{s}\left[\frac{s}{\sqrt{s^2 + \alpha^2}} - 1\right].$$

11.11. Give a proof of Theorem 11.14 by using the Fourier Inversion Formula.

11.12. Compute the integral $f * g$ when $f(t) = g(t) = \sin t$.

11.13. Compute the integral $f * g$ when $f(t) = t^2$ and $g(t) = e^t$.

11.14. By applying Theorem 11.18, find the inverse Laplace transforms of the functions

(a) $\dfrac{1}{s(s^2 + a^2)}$; (b) $\dfrac{s}{(s^2 + 1)^2}$; (c) $\dfrac{e^{-as}}{s^3}$; (d) $\dfrac{1}{(s+1)^2}$.

11.15. Show that

$$(u_0 * f)(t) = \int_0^t f(\tau)d\tau \quad \text{and} \quad (u_0 * u_0 * f)(t) = \int_0^t (t - \tau)f(\tau)d\tau$$

11.16. Give an example to show that $1 * f \neq f$, in general.

11.17. Apply the Laplace transform to solve the initial value problem

$$y'' + 2y' + 5y = 0, \quad \text{where } y(0) = 2 \text{ and } y'(0) = -4.$$

11.18. Apply the Laplace transform to solve the integral equation given by

$$y' + \int_0^t y(t - \tau)e^{-2\tau}d\tau = 1, \quad \text{with } y(0) = 1.$$

11.19. With notations as in Example 3.14, consider the initial value problem

$$m\,y''(t) + k\,y(t) = -mg, \quad \text{where } y(0) = y_0 \text{ and } y'(0) = 0.$$

Apply the Laplace transform to show that

$$y(t) = \left(y_0 + \frac{mg}{k}\right)\cos\sqrt{k/mt} - \frac{mg}{k}.$$

11.20. Apply the Laplace transform to solve the initial value problem

$$y'' - 2y' - 3y = 0, \quad \text{where } y(0) = 1 \text{ and } y'(0) = 7.$$

11.21. Apply the Laplace transform to solve the initial-boundary value problem as given below:

$$y'' + 2y' + 2y = r(t); \quad \text{with } y(0) = 0 \text{ and } y(0) = 0,$$

where r is the *tooth function* given by

$$r(t) = \begin{cases} t, & \text{for } 0 \le t < 1 \\ 0, & \text{for } t \ge 1 \end{cases}, \quad \text{for } t \in [0, \infty).$$

11.22. Apply the Laplace transform to solve the initial-boundary value problem as given below:

$$y'' + 2y' + 2y = f(t); \quad \text{with } y(0) = 0 \text{ and } y'(0) = 0,$$

where the function f is given by

$$f(t) = \begin{cases} 1, & \text{for } 0 \le t < 1 \\ 0, & \text{for } t \ge 1 \end{cases}, \quad \text{for } t \in [0, \infty).$$

11.23. Apply the Laplace transform to solve the initial-boundary value problem as given below:

$$u_x + u_t + u(x, t) = 0; \quad \text{with } u(x, 0) = \sin x \text{ and } u(0, t) = 0.$$

11.24. Apply the Laplace transform to solve the initial-boundary value problem as given below:

$$u_x + x u_t = 0; \quad \text{with } u(x, 0) = 0 \text{ and } u(0, t) = t.$$

Notice that the method of separation of variables fails in this case.

11.25. Apply the Laplace transform to solve the initial-boundary value problem as given below:

$$u_{tt} = c^2 u_{xx}, \quad \text{for } 0 < x < \infty \text{ and } t > 0$$

$$u(x, 0) = 0, \quad u_t(x, 0) = 0, \quad \text{and} \quad u(0, t) = f(t), \quad \text{for } t \in [0, \infty).$$

assuming further $\lim_{x \to \infty} u(x, t) = 0$.

11.26. Apply the Laplace transform to solve the initial-boundary value problem as given below:

$$u_{tt} = c^2 u_{xx}, \quad \text{for } 0 < x < \pi \text{ and } t > 0$$

$$u(x, 0) = \sin x, \quad u_t(x, 0) = 0; \quad \text{and} \quad u(0, t) = u(\pi, t) = 0.$$

11.27. The problem is about waggling of a semi-infinite perfectly elastic string. Let $y(x, t)$ be the transverse displacement of an string such that the following three conditions hold:

   a. The string is initially at rest along the x-axis;
   b. For some $t > 0$, the end at $x = 0$ is moved according to the function given by

$$y(0, t) = \begin{cases} \sin t & \text{if } 0 \le t \le 2\pi \\ 0 & \text{otherwise} \end{cases} ;$$

   c. $\lim\limits_{x \to \infty} y(x, t) = 0$, for $t \ge 0$.

   Find an explicit expression for $y(x, t)$.

11.28. Apply the Laplace transform to solve the initial-boundary value problem as given below:

$$u_t = u_{xx}, \quad \text{for } 0 < x < \infty \text{ and } t > 0$$
$$u(x, 0) = 0 \text{ and } u(0, t) = \sin t,$$

   assuming further $\lim\limits_{x \to \infty} u(x, t) = 0$.

11.29. Apply the Laplace transform to solve the initial-boundary value problem as given below:

$$u_t = u_{xx}, \quad \text{for } 0 < x < \infty \text{ and } t > 0$$
$$u(x, 0) = 0 \text{ and } u_x(0, t) = f(t),$$

   assuming further $\lim\limits_{x \to \infty} u(x, t) = 0$.

11.30. Apply the Laplace transform to solve the initial-boundary value problem as given below:

$$u_t = u_{xx}, \quad \text{for } 0 < x < \infty \text{ and } t > 0$$
$$u(x, 0) = 1 \text{ and } u(0, t) = \begin{cases} 1, & \text{for } 0 < t < 2 \\ 0, & \text{otherwise} \end{cases}$$

Reference

1. Brown JW, Churchill RV (2001) Fourier series and boundary value problems, 6th ed. McGraw-Hill, New York

Appendix A
Supplements

> Mathematics directs the flow of the universe, lurks behind its
> shapes and curves, holds the reins of everything from tiny atoms
> to the biggest stars.
>
> Edward Frenkel, In: Love and Math—The Heart of Hidden
> Reality

A.1 Banach Fixed Point Theorem

An important property of complete metric spaces X is that every contraction of X has a unique fixed point. This is known as the *Banach contraction principle*. Among many other applications, it is used to prove the existence of a local solution of a nonlinear system of first order initial value problems of the form (3.5.6). The reader is referred to [1] or [2], for further details.

Recall that a (nonempty) set X is a *Metric space* if there is a function $d : X \times X \to \mathbb{R}$, called *distance function* (or a *Metric*, such that, for all $x, y, z \in X$,

1. Positivity $d(x, y) = 0$ if and only if $x = y$,
2. Symmetric $d(x, y) = d(y, x)$,
3. Triangle Inequality $d(x, z) \leq d(x, y) + d(y, z)$.

A metric space is usually denoted by the pair (X, d). For example, $X = \mathbb{R}^n$ is a metric space with respect to Euclidean distance $d_2 : \mathbb{R}^n \times \mathbb{R}^n \to \mathbb{R}$ defined by

$$d_2(x, y) := \sqrt{(x_1 - y_1)^2 + \cdots + (x_n - y_n)^2}.$$

A. K. Razdan and V. Ravichandran, *Fundamentals of Partial Differential Equations*, https://doi.org/10.1007/978-981-16-9865-1

It is easy to see that d_2 satisfies *positivity* and *symmetric* properties. And, *triangle inequality* follows from the following Cauchy–Schwartz inequality:

$$|\langle x, y \rangle| \leq \|x\| \|y\|, \quad \text{for all } x, y \in \mathbb{R}^n.$$

Also, if $B(I)$ is the space of real-valued bounded functions defined on a set I then the function $d_\infty : B(I) \times B(I) \to \mathbb{R}$ given by

$$d_\infty(f, g) := \sup\{|f(t) - g(t)| : t \in I\}$$

defines a metric on $B(I)$. Again, *positivity* and *symmetric* properties follow easily, and *triangle inequality* follows because, for any $h \in B(I)$ and $t \in I$,

$$\begin{aligned}
|f(t) - g(t)| &= |f(t) - h(t) + h(t) - g(t)| \\
&\leq |f(t) - h(t)| + |h(t) - g(t)| \\
&\leq \sup_{s \in I} |f(s) - h(s)| + \sup_{s \in I} |h(s) - g(s)| \\
&= d_\infty(f, h) + d_\infty(h, g).
\end{aligned}$$

In particular, if $I \subset \mathbb{R}$ is a closed and bounded set, then the space $C(I)$ of real-valued continuous functions defined on a set I is a metric space with respect to the *sup metric* d_∞.

Let (X, d) be a metric space, and $x \in X$. Then, for any $r > 0$,

$$B(x; r) := \{y \in X : d(y, x) < r\}$$

is called an *open ball* of radius r centered at x. We may also write $B_d(a; r)$ if some argument involves more than one metrics. Replacing $<$ by $\leq$, we have the concept of a *closed ball* of radius r centered at x, denoted by $\overline{B}(x; r)$. The set $S(x; r) = \overline{B}(x; r) \setminus B(x; r)$ is called a *circle* of radius r centered at $x \in X$. Since distinct points in a metric space can be separated by open balls, *every metric spaces is Hausdorff*. Also, every metric d on a set X defines a *topology*, where a set $U \subseteq X$ is *open* if for any point $u \in U$ there is some $r > 0$ such that $B(u; r) \subset U$. In fact, collection of all ball at $x \in X$ forms a *neighbourhood base* for the metric topology determined by the metric d. In particular, a set $U \subseteq X$ is open if and only if U is (countable) union of open balls.

Let $S \subseteq X$. An element $x \in X$ is called a *limit point* of the set S if

$$S \cap (B(a, r) \setminus \{a\}) \neq \emptyset, \quad \text{for every } r > 0.$$

The set $B(a, r) \setminus \{a\}$ is called a *punctured ball* at a with radius r. The set of all limit points of a set S is called the *derived set* of S, denoted by S', and the *closure* $\overline{S}$ of S is the union $S \cup S'$. Notice that a limit point of S may or may not belong to S. Notice that every ball around of a limit point of S contains infinitely many points of S. So, a set $S \subseteq X$ has no limit point if it is finite set. A set $F \subseteq X$ is said to be *closed* if $U = X \setminus F$ is open in X. Clearly then, F is closed if and only if $F' \subset F$. In particular, a set $F \subseteq X$ is closed if and only if $F = \overline{F}$. It follows easily that if a set $F \subset \mathbb{R}$ is bounded above, with $a = \sup F$, then $s \in \overline{F}$.

The concept of convergence of real sequences extends naturally to metric spaces: A sequence $\langle x_n \rangle$ in a metric space (X, d) is *convergent* if there exists $x \in X$ such that

$$a_n = d(x_n, x) \rightarrow 0, \quad \text{as } n \rightarrow \infty,$$

with respect to the usual absolute metric on $\mathbb{R}$. In topological terms, a sequence $\langle x_n \rangle$ converges to $x \in X$ if for any $\epsilon > 0$, $x_n \in B(x; \epsilon)$ *eventually*. Thus, if limit exists, it is unique by Hausdorff property. We also say that $x \in X$ is the *limit* of the sequence $\langle x_n \rangle$, and write $\lim_{n \to \infty} x_n = x$ or $x_n \rightarrow x$ as $n \rightarrow \infty$. Clearly, every convergent sequence is *bounded* i.e., for some $x_0 \in X$ and $M > 0$, we have $x_n \in B(x_0; M)$ for all n. Notice that a sequence $\langle x_n \rangle$ converges to $x \in X$ if and only if every subsequence $\langle x_{n_k} \rangle$ converges to x. It is important to know that, for a set S in a metric space (X, d), $x \in S'$ if and only if there is a sequence $\langle s_n \rangle$ in the set S such that $x_n \rightarrow x$.

For example, using *projections*, it follows easily that convergence in $(\mathbb{R}^n, d_2)$, and in some other similar metric spaces, is a straightforward extension of idea of convergence in $\mathbb{R}$. A sequence $\langle f_n \rangle$ in the metric space $(C[I], d_\infty)$ converges to some $f \in C(I)$ if is a *uniform limit* of the sequence $\langle f_n \rangle$:

$$\sup\{|f_n(t) - f(t)| : t \in I\} \rightarrow 0 \text{ as } n \rightarrow \infty.$$

We write $f_n \rightrightarrows f$ over the interval I. Also, a sequence $f_n \in C[I]$ is said to *converge pointwise* to a function $f \in C[I]$ if the sequence of real numbers $\langle f_n(t) \rangle$ converges to the real number $f(t)$, for each $t \in I$. We then say f is a *pointwise limit* of the sequence $\langle f_n \rangle$, and write

$$f(t) := \lim_{n \to \infty} f_n(t), \quad \text{for } t \in I.$$

Notice that there doesn't exists any metric on the set $C(I)$ that induces the *pointwise convergence*. So, *not every type of convergence in mathematics is metric based*.[1] In any situation, terminology chosen for a type of convergence always reminds us about the purpose it is expected to serve.

There are many situations wherein it is important to test a sequence for convergence, without the knowledge about the limit, if one exists. A sequence $\langle x_n \rangle$ in a metric space (X, d) is called a *Cauchy sequence* if, for a given $\epsilon > 0$, there is a natural number $N = N(\epsilon)$ such that

$$d(x_n, x_m) < \epsilon, \quad \text{for all } n, m \geq N.$$

---

[1] This aspect of convergence could be explained using the idea of **path homotopy** in a general topological setting.

Clearly, every Cauchy sequence is bounded. Notice that, if $Y \subset X$ and a sequence $\langle y_n \rangle$ in Y is Cauchy in X, then it would remain so in Y; but, the converse need not be true in general. Since $x_n \to x$ implies that most terms of the sequence $\langle x_n \rangle$ are eventually very close to the limit x, so the terms of the sequence are eventually very close to each other as $n \to \infty$. Thus every convergent sequence is a Cauchy sequence. There exists metric spaces wherein not every Cauchy sequence is convergent. In general, convergence of a Cauchy sequence depends on the *topology* of the space X. So, Cauchy condition separates topological aspects of a metric space (X, d) from the geometric properties induced by the metric d.

A metric space (X, d) is said to be *complete* if every Cauchy sequence of X converges in X. It follows easily that $(\mathbb{R}^n, d_2)$ and $(C(I), d_\infty)$ are complete metric spaces. In some situations, it helps to use the fact that a Cauchy sequence $\langle x_n \rangle$ converges to $x \in X$ if it has a subsequence $\langle x_{n_k} \rangle$ that converges to x.

For a metric space (X, d), a function $T : X \to X$ is called a *contraction* if for some real number $c \in (0, 1)$, known as the *contraction factor* of the function T, we have

$$d(Tx, Ty) \leq c\, d(x, y), \quad \text{for all } x, y \in X. \tag{A.1.1}$$

Clearly, every contraction is a uniformly continuous function.

Theorem A.1 (Banach Contraction Principle) *Let* (X, d) *be a complete metric space, and* $T : X \to X$ *be a contraction, with contraction factor* $c \in (0, 1)$. *Then, there is a unique* $x_0 \in X$ *such that* $T(x_0) = x_0$. *In fact, for any* $x \in X$, $T^n(x) \to T(x_0)$ *as* $n \to \infty$.

Proof Uniqueness of x_0 follows trivially; for, if x_0' is such that $Tx_0' = x_0'$ then $d(x_0, x_0') \leq \lambda\, d(Tx_0, Tx_0') = d(x_0, x_0') \Rightarrow x_0 = x_0'$, because $\lambda < 1$. To prove the existence, for an arbitrary $x \in X$ (but otherwise fixed), consider the sequence $\langle T^n x \rangle$. Consider the iterates $T^n = T^{n-1} \circ T$, $n \geq 1$ so that

$$d(T^n x, T^n y) \leq \lambda\, d(T^{n-1} x, T^{n-1} y)$$

$$\leq \cdots$$

$$\leq \lambda^n d(x, y), \quad \text{for all } x, y \in X,$$

using (A.1.1) n-times. That is, for each $n \geq 1$, the nth iterate of T is a contraction, with *contraction factor* λ^n. So, for any $n \in \mathbb{N}$,

$$d(T^n x, T^{n+1} x) \leq \lambda^n\, d(x, Tx).$$

Then, triangle inequality and repeated use of the last inequality gives, for any $n, k \in \mathbb{N}$,

$$d(T^n x, T^{n+k} x) \leq d(T^n x, T^{n+1} x) + \cdots + d(T^{n+k-1} x, T^{n+k} x)$$
$$\leq \lambda^n d(x, Tx) + \lambda^{n+1} d(x, Tx) + \cdots + \lambda^{n+k-1} d(x, Tx)$$
$$\leq (\lambda^n + \lambda^{n+1} + \cdots + \lambda^{n+k-1}) d(x, Tx)$$
$$= \lambda^n \frac{1 - \lambda^k}{1 - \lambda} d(x, Tx)$$
$$\leq \frac{\lambda^n}{1 - \lambda} d(x, Tx) \longrightarrow 0 \quad \text{as} \quad n \to \infty,$$

because $\lambda < 1$. We thus conclude that $\langle T^n x \rangle$ is a Cauchy sequence. Since (X, d) is complete, take $x_0 = \lim_{n \to \infty} T^n x$. Finally, note that the continuity of T implies that

$$T x_0 = T \left(\lim_{n \to \infty} T^n x \right) = \lim_{n \to \infty} T^{n+1} x = x_0.$$

This completes the proof. $\qquad\square$

This theorem fails for $\lambda = 1$. For example, consider the closed set $Y = [1, \infty) \subset \mathbb{R}$ viewed as a metric space under the induced *absolute value* metric. Since closed subsets of a complete metric space are always complete, so is the space Y. Notice that the function $T : Y \to Y$ defined by $T(y) = y + 1/y$ satisfies

$$|T(y_1) - T(y_2)| < |y_1 - y_2|, \quad \text{for all } y_1, y_2 \in Y,$$

but T has *no fixed point*.

A.2 Maxwell and Helmholtz Equations

Faraday is, and must always remain, the father of that enlarged science of electromagnetism.
James R. Maxwell (1831–1879)

The purpose here is to briefly discuss the *Maxwell* and *Helmholtz equations* that help explain many different kinds of electromagnetic phenomena. Such types of equations are used to describe the propagation of electromagnetic wave such as light waves, X-rays, etc., through a region $\Omega \subset \mathbb{R}^3$. Our main source for the exposition given here are the standard texts [3, 4]. The reader may refer Chap. 2 to brush up the preliminary concepts on vector calculus, and [5] or [6], for details on the topic at an advanced level.

As usual, *boldface* symbols denote the vector quantities defined over a region $\Omega \subseteq \mathbb{R}^3$. The main leads of the story are the five 3-dimensional *vector fields* as given below: The vector function **E** representing an *electric field*, with the field of *electric flux density* represented by a vector function **D**; the vector function **H** representing a *magnetic field*, with the field of *magnetic flux density* represented by a vector function **B**; and, the vector function **J** representing the field of *current density* of a current flow through a medium. It also involves the scalar function ρ representing the electric charge density of the medium.

In practical terms, $\mathbf{J}$ and ρ are respectively the macroscopic mean values of current and charge densities of elements within the region Ω. In the conducting medium, the *Ohm's Law* given by

$$\mathbf{J} = \sigma \, \mathbf{E} + \mathbf{J}_e, \qquad\qquad (A.2.1)$$

is a linear approximation of the induced current due to the electric field $\mathbf{E}$, where $\mathbf{J}_e$ denote the external current density, and $\sigma : \Omega \to \mathbb{R}$ is the *conductivity* for an isotropic medium. Recall that we say a material is **dielectric** if $\sigma = 0$. All physical quantities are evaluated at a point $\mathbf{r} \in \Omega$ and at a time $t \geq 0$. The involved boundary surface $S = \partial\Omega$ is assumed to be sufficiently smooth and oriented. The prerequisites for the concepts presented in the subsequent discussion, and the notations used, are provided in Chap. 2.

First, we discuss the following *constitutive relations*:

$$\mathbf{D} = \epsilon \, \mathbf{E} \quad \text{and} \quad \mathbf{B} = \mu \, \mathbf{H}, \qquad\qquad (A.2.2)$$

where $\epsilon : \Omega \to \mathbb{R}^{3\times3}$ and $\mu : \Omega \to \mathbb{R}^{3\times3}$ are respectively called the *dielectric tensor* (measuring *permittivity*) and the *permeability tensor* of inhomogeneous and anisotropic media. We know that electric properties of the material depend on molecular structure as well as on macroscopic quantities such as the density and the temperature. In addition, we also need to consider time-dependent *hysteresis effect*, i.e., the field values at any time depend on values it had in immediate past. Taking into consideration the mean values of macroscopic effects in the material, we can write the approximate relations as

$$\mathbf{D} = \mathbf{E} + 4\pi \, \mathbf{P} \quad \text{and} \quad \mathbf{B} = \mathbf{H} - 4\pi \, \mathbf{M},$$

where $\mathbf{P}$ and $\mathbf{M}$ are respectively the fields of *electric polarization* and *magnetization of material*. So, ignoring ferro-electric and ferro-magnetic effects, and also assuming that the fields are *small*, we can work with dependencies relation as in (A.2.2). For an *isotropic medium*, the quantities $\mathbf{P}$ and $\mathbf{M}$ don't depend on the direction so that ϵ and μ are real-valued, which are usually assumed least continuous. In simplest case, we take these to be constants.

Recall that the three component functions of $\mathbf{E}$ specify the direction and strength of electric field, and $\mathbf{D}$ gives the density of electric field lines through the area of the surface S. A similar physical interpretation stands for the pair $(\mathbf{H}, \mathbf{B})$, but for a magnetic field.

1. The *Ampére's Law* due to André Marie Ampére (1775–1836) states that an electric current (external and induced) or a changing electric flux through a surface S produces a circulating magnetic field $\mathbf{B}$ around any loop on the boundary ∂S. That is, in integral form, we have

$$\oint_{\partial S} \mathbf{H} \cdot d\mathbf{r} = \frac{d}{dt} \int_S \mathbf{D} \cdot \mathbf{n} da + \int_S \mathbf{J} \cdot \mathbf{n} da, \qquad (A.2.3)$$

By Theorem 2.27, we can write

$$\int_S \nabla \times \mathbf{H} \cdot \mathbf{n} da = \frac{d}{dt} \int_S \mathbf{D} \cdot \mathbf{n} da + \int_S \mathbf{J} \cdot \mathbf{n} da. \qquad (A.2.4)$$

So, if the law holds for arbitrary surface S enclosed by ∂S, we obtain

$$\nabla \times \mathbf{H} = \frac{\partial \mathbf{D}}{\partial t} + \mathbf{J}. \qquad (A.2.5)$$

This is the first Maxwell equation in differential form. It says electrical current and time-varying electric fields create magnetic fields curling around them.

2. The *Faraday's Law* (of *induction*) due to Michael Faraday (1791–1867) states that changing magnetic flux through a surface S induces an inner voltage in any loop on the boundary ∂S, and a changing magnetic field through a surface S induces a circulating electric field. That is,

$$u_i = -\int_S \frac{\partial \mathbf{B}}{\partial t} \cdot \mathbf{n} da = \int_S \nabla \times \mathbf{E} \cdot \mathbf{n} da.$$

So, by argument as above, it follows that

$$\nabla \times \mathbf{E} = -\frac{\partial \mathbf{B}}{\partial t}. \qquad (A.2.6)$$

This is second Maxwell equation in differential form, which says that a time-varying magnetic fields create electric fields $\mathbf{E}$ curling around them.

3. Considering the fact that electric charge produces an electric field, the *Gauss's Law* due to Carl Friedrich Gauss (1777–1855) states that the flux of the electric field passing through any closed surface S is proportional to the total charge contained within the surface. Thus, the integral formulation of Gauss's (electric) law is given by

$$\int_{\partial \Omega} \mathbf{D} \cdot \mathbf{n} da = \int_\Omega \rho dv,$$

so that by divergence theorem, we have

$$\int_\Omega \nabla \cdot \mathbf{D} dv = \int_\Omega \rho dv.$$

So, by usual argument, it follows that

$$\nabla \cdot \mathbf{D} = \rho, \tag{A.2.7}$$

proving that there are no magnetic currents. This is third Maxwell equation in differential form, which says that charge creates electric fields diverging from it. So, it describes the source of the electric displacement.

4. The fourth equation follows from a crucial modification of Gauss's (magnetic) law by *James Maxwell*. It states that the total magnetic flux passing through any closed surface is zero, provided we have magnetic north and south poles:

$$\int_{\partial \Omega} \mathbf{B} \cdot \mathbf{n} \mathrm{d}a = 0.$$

Therefore, once again, divergence theorem implies

$$\nabla \cdot \mathbf{B} = 0. \tag{A.2.8}$$

So, the magnetic fields do not diverge from anything, they only curl around.

Notice that the first and third Maxwell equations put together imply the following *equation of continuity*:

$$\frac{\partial \rho}{\partial t} = \nabla \cdot \frac{\partial \mathbf{D}}{\partial t} = \nabla \cdot \left[\nabla \times \mathbf{H} - \mathbf{J} \right] = -\nabla \cdot \mathbf{J}, \tag{A.2.9}$$

which explain why electromagnetic waves exist and can carry energy really very far away from their source. Also, if the electric field $\mathbf{E}$ is independent of time in some *simply connected* region Ω so that by Faraday's induction law $\nabla \times \mathbf{E} = 0$. Then, there exists a scalar (potential) function $u : \Omega \to \mathbb{R}$ such that $\mathbf{E} = -\nabla u$ holds over Ω. But then, by Gauss's (electric) law, it follows that in the homogeneous medium the function u satisfies the *Poisson equation*

$$\rho = \nabla \cdot \mathbf{D} = -\nabla \cdot [\epsilon \mathbf{E}] = -\epsilon \nabla^2 u. \tag{A.2.10}$$

This is a basic partial differential equation that describes the *electrostatic potential*. Further, Eq. (A.2.8) is satisfied for the magnetic field

$$\mathbf{B} = \nabla \times \mathbf{A}, \tag{A.2.11}$$

where $\mathbf{A}$ is a *vector potential* of the field $\mathbf{B}$. Also, using the fact that ∇ can be taken inside the time-derivative by a change in the order of differentiation, Eq. (A.2.6) is satisfied for the electric field

$$\mathbf{E} = -\nabla \phi - \frac{\partial \mathbf{A}}{\partial t}, \tag{A.2.12}$$

where ϕ is called a *scalar potential*. Notice that, for an arbitrary scalar field ψ, magnetic and electric fields in (A.2.11) and (A.2.12) remain invariant under the transformations

$$\mathbf{A} \mapsto \mathbf{A} - \nabla\psi \quad \text{and} \quad \phi \mapsto \phi + \frac{\partial\psi}{\partial t}.$$

The prescriptions in these two equations can be assumed unique by adopting a convention usually known as the *gauge condition*. Also, since Maxwell equations are *Lorentz invariant*, it makes sense to use a gauge condition that is also Lorentz invariant. So, we use

$$\epsilon\mu \frac{\partial\phi}{\partial t} + \nabla \cdot \mathbf{A} = 0, \tag{A.2.13}$$

called *Lorenz gauge condition*.

Next, we use these four fundamental Maxwell equations to derive the Helmholtz equation. The conducting material is assumed to be uniform and that we are in a bounded region Ω so that the permittivity (ϵ) and the permeability (μ) are constant, both with respect to the space variable and the time. To get our equation, we solve Maxwell's equations in the region Ω, under the assumption that the region has *zero conductivity* so that the electrical current density $\mathbf{J}$ is zero by (A.2.1). To start with, we rewrite Eq. (A.2.5) as

$$\frac{1}{\mu} \nabla \times \mathbf{B} = \epsilon \frac{\partial\mathbf{E}}{\partial t} \quad \text{i.e., } \nabla \times \mathbf{B} = \mu\epsilon \frac{\partial\mathbf{E}}{\partial t}, \tag{A.2.14}$$

using the constitutive relations (A.2.2) and the assumption $\mathbf{J} = 0$. Next, taking curl of both sides of Eq. (A.2.6), we have

$$\nabla \times \nabla \times \mathbf{E} = -\frac{\partial \nabla \times \mathbf{B}}{\partial t},$$

where ∇ is taken inside the time-derivative by a change in the order of differentiation. It thus follows from the previous equation that

$$\nabla \times \nabla \times \mathbf{E} = -\mu\epsilon \frac{\partial^2\mathbf{E}}{\partial t^2}.$$

Using the vector identity

$$\nabla \times \nabla \times \mathbf{E} = \nabla(\nabla \cdot \mathbf{E}) - \nabla^2\mathbf{E},$$

we obtain

$$\nabla(\nabla \cdot \mathbf{E}) - \nabla^2\mathbf{E} = -\mu\epsilon \frac{\partial^2\mathbf{E}}{\partial t^2}.$$

Now, assuming that the region Ω has no *free* (unbounded) electric charge, we have that the electric charge density $\rho = 0$, and so Eq. (A.2.7) becomes

$$\nabla \cdot \mathbf{D} = 0 = \epsilon \nabla \cdot \mathbf{E} \quad \Rightarrow \quad \nabla \cdot \mathbf{E} = 0. \tag{A.2.15}$$

Using this in the last equation, it follows that

$$\nabla^2 \mathbf{E} - \mu \epsilon \frac{\partial^2 \mathbf{E}}{\partial t^2} = 0, \tag{A.2.16}$$

which is the *Helmholtz equation* for the electric field **E**. Similarly, it follows that the components of the field **H** satisfy the linear wave equation

$$\nabla^2 \mathbf{H} - \mu \epsilon \frac{\partial^2 \mathbf{H}}{\partial t^2} = 0, \tag{A.2.17}$$

which is the *Helmholtz equation* for the magnetic field **H**. Further, combining (A.2.13) and (A.2.11), with Eqs. (A.2.14) and (A.2.15), it follows that

$$\frac{1}{c^2} \frac{\partial^2 \phi}{\partial t^2} - \nabla^2 \phi = \frac{\rho}{\epsilon}; \quad \text{and,} \tag{A.2.18}$$

$$\frac{1}{c^2} \frac{\partial^2 \mathbf{A}}{\partial t^2} - \nabla^2 \mathbf{A} = \mu \mathbf{J}, \quad \text{with } c^2 = \frac{1}{\epsilon \mu}. \tag{A.2.19}$$

The constant c has the dimension of velocity, and it is called the *speed of light*. Therefore, Maxwell's equations essentially boil down to these two *Poisson equations* that model time-harmonic wave propagation in free space due to a localized source. A nonhomogeneous *Helmholtz wave equation* for a function $u = u(x, y)$ is a differential equation of the form

$$\nabla^2 u + k^2 u = -f(x, y),$$

where ∇^2 is the Laplacian in two variables, and the constant $k > 0$ is called the *wave number*. The function $f = f(x, y)$ represents the source wave. See Sect. 10.5 for a derivation of this equation from the 2-dimensional wave equation, using Fourier transform.

A.3   Generalised Functions

All the mathematical sciences are founded on the relations between physical laws and laws of numbers.   James R. Maxwell (1831–1879)

We start with a simple illustration that explains why we need to study *function like*[2] mathematical objects that vanish everywhere except at a point where it is infinite,

---

[2] As we define below, such types of *"functions"* are called *generalised functions*.

and yet they have a *nonzero finite integral* about every neighbourhood of the point. For, let $\mathbf{r} = (x, y, z)$, with $r = \|\mathbf{r}\| = \sqrt{x^2 + y^2 + z^2}$. It follows easily that

$$\mathbf{v} = \frac{1}{r^2}\hat{\mathbf{r}} \quad \Rightarrow \quad \nabla \cdot \mathbf{v} = 0, \quad \text{at every point } \mathbf{r} \neq 0.$$

On the other hand, over the sphere $S_\epsilon = \mathbb{S}(0; \epsilon)$, we have

$$\oint_{S_\epsilon} \mathbf{v} \cdot d\mathbf{a} = \iint \frac{1}{\epsilon^2}\hat{\mathbf{r}} \cdot (\epsilon^2 \sin\theta d\theta d\phi \hat{\mathbf{r}}) = 4\pi.$$

The reason why the vector $\mathbf{v}$ does not contradict the *divergence theorem* is the fact that the function given by

$$\varphi(\mathbf{r}) := \nabla \cdot \mathbf{v}, \quad \text{for } \mathbf{r} = (x, y, z),$$

has a spike at the origin $\mathbf{r} = 0$, and so $\nabla \cdot \mathbf{v} = 0$ at the origin is not a valid statement. For a point $(a, b, c) \in \mathbb{R}^3$, we write

$$\varphi(\mathbf{r}) = \delta^3(\mathbf{r}) = \delta(x - a)\delta(y - b)\delta(z - c),$$

which is called a 3-dimensional *Dirac delta function*. A 1-dimensional (Dirac) delta function about the origin may be defined very roughly as the symbol $\delta(x)$ given by the conditions

$$\delta(x) = 0, \quad \text{for } x \neq 0; \tag{A.3.1a}$$

$$\int_{-\infty}^{\infty} \delta(x)\,dx = 1. \tag{A.3.1b}$$

Clearly, it is not possible to treat the symbol $\delta(x)$ as a function of x in usual mathematical sense because the integral in the second condition must be zero, by using the first condition. For a simple physical interpretation, suppose $\delta_\epsilon(x)$ denote the *rectangular pulse* about the origin given by the function

$$\delta_\epsilon(x) = \begin{cases} \epsilon/2, & \text{for } |x| \leq 1/\epsilon \\ 0, & x \notin [-1/\epsilon, 1/\epsilon] \end{cases} \tag{A.3.2}$$

Clearly, as $\epsilon \to \infty$, the duration of the pulse $\delta_\epsilon(x)$ tends to zero, the *amplitude* increases without bounds, i.e., $\delta_\epsilon(x_0)$ is *infinite*, but all through the limit process the integral of $\delta_\epsilon(x)$ remains unity. This explains why

$$\delta(x) = \lim_{\epsilon \to \infty} \delta_\epsilon(x) \tag{A.3.3}$$

is also known as the *impulse function* at time $x = 0$. In electrical engineering, $\delta(x - x_0)$ is known as an *impulse delayed* in time by x_0, for $x_0 > 0$; and, as an *impulse advanced* in time by x_0, for $x_0 < 0$. We may also be view the quantity $q\delta(x - x_0)$ as the *charge q* accumulated at the point $x = x_0$.

Remark A.1 In some cases, it is more useful to define the symbol $\delta(x)$ by using the functions such as Fejér kernels, Gaussian pulses, sinc functions, or top-hat functions in place of the function $\delta_\epsilon(x)$ as in (A.3.2). For example, we may take $\delta(x) = \lim_{\epsilon \to \infty} \gamma_\epsilon(x)$, where γ_ϵ is the smooth *Gaussian pulse* given by

$$\gamma_\epsilon(x) = \frac{\epsilon}{\sqrt{\pi}} e^{-\epsilon^2 x^2}, \quad \text{for } \epsilon > 0.$$

Notice that we may start with a definition of the symbol $\delta(x)$ about the origin, and then obtain $\delta(x - x_0)$ from $\delta(x)$ by shifting the *spike* to the point $x = x_0$. That is, we may define the symbol $\delta(x - x_0)$ as given by the conditions

$$\delta(x - x_0) = 0, \quad \text{for } x \neq x_0; \tag{A.3.4a}$$

$$\int_{-\infty}^{\infty} \delta(x - x_0)\, dx = 1. \tag{A.3.4b}$$

Theorem A.2 *If $g : \mathbb{R} \to \mathbb{R}$ is a sufficiently well-behave continuous, and $x_0 \in \mathbb{R}$, then*

$$\int_{-\infty}^{\infty} g(x)\delta(x - x_0)\, dx = g(x_0). \tag{A.3.5}$$

Proof By using the continuity of g at x_0, it follows from the previous discussion that

$$\int_{-\infty}^{\infty} g(x)\delta(x - x_0)\, dx = \lim_{\epsilon \to \infty} \int_{-\infty}^{\infty} g(x)\delta_\epsilon(x - x_0)\, dx$$

$$= g(x_0) \lim_{\epsilon \to \infty} \int_{-\infty}^{\infty} \delta_\epsilon(x - x_0)\, dx$$

$$= g(x_0).$$

Notice that, for ϵ sufficiently large, the value of the integral on the right side of the first equality depends on how g behaves in a neighbourhood of the point the point x_0, and the error $|g(x) - g(x_0)|$ approaches zero. $\square$

The argument used in the proof of the above theorem does not require us to integrate over whole line $\mathbb{R}$. We only need the domain of integration to contain the point of continuity x_0 of the function g. Therefore, if a function $g : [x_0 - a, x_0 + b] \to \mathbb{R}$ is continuous at x_0 for some $a > 0, b > 0$, then we have

$$\int_{x_0-a}^{x_0+b} g(x)\delta(x - x_0) \, dx = g(x_0). \tag{A.3.6}$$

Theorem A.2 is called the *shifting property* for the symbol $\delta(x)$. In practical terms, the product function $g(x)\delta(x - x_0)$ represents a *time-continuous signal* obtained from the impulse $\delta(x - x_0)$ at x_0, by *scaling* it using the value $g(x_0)$. That is, for any function g continuous at x_0, the following *symbolic identity* holds in the sense that the both sides give the same output when integrated over any interval containing the point x_0:

$$g(x)\,\delta(x - x_0) = g(x_0)\,\delta(x - x_0). \tag{A.3.7}$$

Therefore, when $x_0 = 0$, we have

$$\int_{x_0-a}^{x_0+b} g(x)\delta(x) \, dx = g(0).$$

The above observation is significant mainly due to the fact that two expressions $D_1(x)$ and $D_2(x)$ involving the symbol $\delta(x)$ are equal if, for every *test function* g in some suitable function space, we have

$$\int_{-\infty}^{\infty} g(x)D_1(x) \, dx = \int_{-\infty}^{\infty} g(x)D_2(x) \, dx. \tag{A.3.8}$$

For example, given any $0 \neq \alpha \in \mathbb{R}$, since

$$\int_{-\infty}^{\infty} g(x)\delta(\alpha x) \, dx = \pm\frac{1}{\alpha} \int_{-\infty}^{\infty} g(u/\alpha)\delta(u) \, du$$

$$= \pm\frac{1}{\alpha} g(0) = \frac{1}{|\alpha|} g(0),$$

holds for every sufficiently well-behaved function g that is continuous at $x_0 = 0$, it follows from (A.3.8) that

$$\int_{-\infty}^{\infty} g(x)\delta(\alpha x) \, dx = \frac{1}{|\alpha|} g(0), \quad \text{for all } 0 \neq \alpha \in \mathbb{R}. \tag{A.3.9}$$

This is known as the *scaling property* of the delta function. In particular, we have $\delta(-x) = -\delta(x)$. That is, delta function is an *even function*.

For some applications, we also need to use the derivatives of the delta function $\delta(x)$, which are defined as described below. Once again, we take $\delta(x) = \lim_{\epsilon \to \infty} \delta_\epsilon(x)$, where $\langle \delta_\epsilon \rangle$ is some function sequence as mentioned above. For an arbitrary $g \in C^1(\mathbb{R})$, taking $y = g(x)$ and $dy = \delta'_\epsilon(x)\,dx$, it follows by using *integration by parts* that

$$\lim_{\epsilon \to \infty} \int_{-\infty}^{\infty} g(x)\delta'_\epsilon(x)\,dx = -\lim_{\epsilon \to \infty} \int_{-\infty}^{\infty} g'(x)\delta_\epsilon(x)\,dx,$$

because $\delta_\epsilon(\pm\infty) = 0$, as $\epsilon \to \infty$. Hence, we obtain

$$\int_{-\infty}^{\infty} g(x)\delta'(x)\,dx = -g'(0). \tag{A.3.10}$$

Similarly, by repeating the same argument n-times, we obtain

$$\int_{-\infty}^{\infty} g(x)\delta^{(n)}(x)\,dx = (-1)^n g^{(n)}(0). \tag{A.3.11}$$

More generally, we have

$$g(x)\delta'(x - x_0) = g(x)\delta'(x - x_0) - g'(x)\delta(x - x_0). \tag{A.3.12}$$

In particular, for $g(x) = x$, it follows that

$$x \frac{d}{dx}\big(\delta(x)\big) = -\delta(x). \tag{A.3.13}$$

Notice that, since $g(x) = x$ is an odd function and $\delta(x)$ is an even function, we have $\delta'(x)$ is an *odd function*, i.e., $\delta'(-x) = -\delta'(x)$. Also, taking $g(x) = x^n$, (A.3.13) implies a general formula

$$x^n \frac{d^n}{dx^n}\big(\delta(x)\big) = (-1)^n n!\,\delta(x), \quad \text{for } n \geq 1. \tag{A.3.14}$$

Remark A.2 Geometrically, we may also define the symbol $\delta(x)$ by using the Heaviside *unit step function* $u(x)$ given by

$$u(x) = \begin{cases} 1, & \text{for } x > 0 \\ 0, & \text{for } x < 0 \end{cases}. \tag{A.3.15}$$

Notice that the limit of the continuous-time *signal* given by

$$\sigma_\epsilon(x) = \begin{cases} 0, & \text{for } x \leq 0 \\ \epsilon x, & \text{for } x \in (0, 1/\epsilon), \\ 1, & \text{for } x > 1/\epsilon \end{cases}$$

as $\epsilon \to \infty$, is the function $u(x)$ as in (A.3.15). And, since the square pulse function $\delta_\epsilon(x)$ is the derivative of the function $\sigma_\epsilon(x)$, it follows that

$$\delta(x) = \lim_{\epsilon \to \infty} \delta_\epsilon(x) = \frac{d}{dx}\left[\lim_{\epsilon \to 0} \sigma_\epsilon(x)\right] = u'(x). \qquad \text{(A.3.16)}$$

Therefore, the unit step function $u(x)$ can also be viewed as the integral

$$u(x) = \int_{-\infty}^{x} \delta(t)\, dt, \quad x \neq 0, \qquad \text{(A.3.17)}$$

which follows by using (A.3.6) and (A.3.16).

For $\mathbf{r} = (x, y, z)$, the 3-dimensional delta function $\delta^3(\mathbf{r})$ is given by

$$\delta^3(\mathbf{r}) = \delta(x)\delta(y)\delta(z), \qquad \text{(A.3.18)}$$

which is zero everywhere except at the origin, where it blows up, and has the volume integral 1, As before, for any suitable function $f = f(\mathbf{r})$ we have

$$\int_{-\infty}^{\infty} f(\mathbf{r})\delta^3(\mathbf{r} - \mathbf{r}_0)\, dx = f(\mathbf{r}_0). \qquad \text{(A.3.19)}$$

Notice that, in view of above formulation, we can write

$$\nabla\left(\frac{\hat{\mathbf{r}}}{r^2}\right) = 4\pi\delta^3(\mathbf{r}).$$

More generally, treating $\mathbf{r}_1$ as a fixed reference point, we have

$$\nabla\left(\frac{\hat{\mathbf{r}}_{12}}{r^2}\right) = 4\pi\delta^3(\mathbf{r}_{12}), \quad \text{with } \mathbf{r}_{12} \equiv \mathbf{r}_2 - \mathbf{r}_1.$$

Also, using $\nabla(1/r) = -(\hat{\mathbf{r}}_{12}/r^2)$, it follows that $\nabla^2(1/r) = -4\pi\delta^3(\mathbf{r}_{12})$.

A.4   Signals and (LTI) Systems

There is no philosophy which is not founded upon knowledge of the phenomena, but to get any profit from this knowledge it is absolutely necessary to be a mathematician.
                                                                    Daniel Bernoulli (1700–1782)

The purpose here is to describe briefly some basic *mathematical tools* that are commonly used to study a progressive elastic wave or an electromagnetic wave, both viewed as a *signal* over an interval $I \subseteq \mathbb{R}$. Elastic waves of high amplitude are known as the *shock-waves*, and the one with low amplitude is a (longitudinal) *sound wave*. A standing wave is created when waves interfere with each other so that addition of waves of different frequencies results into an *intensity modulation* as a function of time. In this way, a mathematical description of wave phenomena is fundamental to a wide range of practical applications.

Understanding about sound as disturbance in air pressure goes back to antiquity. For example, a musical tone is a periodic succession of pulses transmitted through the air that are received by our eardrum. Such type of *disturbances* transmitted through space over a time period are studied mathematically using some measurable complex-valued function $f(t)$ with finite energy given by

$$\int_{-\infty}^{\infty} |f(t)|^2 dt,$$

so $f \in L^2(\mathbb{R})$. In general, a complex-valued function in the Hilbert space $L^2(\Omega)$ is called a *continuous-time* (or analog) *signal*, where $\Omega \subseteq \mathbb{R}^n$ is an open set. For example, a speech signal is acoustic pressure as a function of time $t \in I \subseteq \mathbb{R}$; an image is a signal given by a function of brightness over two spatial variables $(x, y) \in \Omega \subseteq \mathbb{R}^2$; and, a video signal is a function defined over a set of the form $\Omega \times \mathbb{R}$. Signals used in geophysics are functions of density, porosity, and electric registivity. In metrological studies, signals are functions of air pressure, temperature, and wind speed (at an altitude).

A *linear time-invariant* (in short LTI) system is a *transformation* that takes *signals* (periodic or aperiodic) as inputs and produces a simpler type of a *signal*. For example, each dynamical system modeled by a *system of linear ordinary differential equations* is a LTI system. Also, the integral transforms such as described in the last two chapters of the book are LTI systems. Notice that the concept of Fourier series is an example of a *discrete* LTI system.

Signals that we come across in practical situations are mostly *periodic functions* $f : \Omega \to \mathbb{C}$ representing wave propagations with a *measurable speed*. A signal is called *time-harmonic* if it has a single frequency. Such type of signals constitutes building blocks for the Fourier and Laplace transforms. Notice that while dealing with a periodic signal $f : (-\ell, \ell) \to \mathbb{C}$ of *finite power*, i.e. when

$$\frac{1}{\ell} \int_0^{\ell} |f(t)|^2 \mathrm{d}t < \infty,$$

we still have $f \in L^2((-\ell, \ell))$. In technical terms, a signal $f \in L^2(\Omega)$ represents an information about a physical quantity related to some phenomenon. It is called a *discrete-time signal* if its domain is a subset of the set $\mathbb{Z} \times \cdots \times \mathbb{Z}$ (n-copies). A very important class of discrete-time signals arise while taking samples of continuous-time signals over uniformly spaces nodes. For example, if $f = f(t)$ is a signal defined over $\mathbb{R}$ then a discrete-time signal is given by

$$s_n = f(n\,T), \quad n \in \mathbb{Z},$$

where T is a *sampling period*. It is often possible to normalize the sampling period so that $T = 1$. We may take s_n to represent a finite number of bits as in a finite state machine; indeed, all signals processed in binary machines are truly digital signals.

Among other things, waves are described in terms of *sinusoidal signals* represented by a linear combination of sine and cosine functions. In addition, other commonly used periodic signals are as follows:

1. a *triangular wave* given by

$$T(t) := \begin{cases} t, & \text{for } 0 \le t \le 1 \\ 2 - t, & \text{for } 1 \le t \le 2 \end{cases}, \quad T(t + 2) = T(t), \quad \text{for all } t \in \mathbb{R};$$

2. a *unit pulse* of period 1 given by

$$r(t) := \begin{cases} 0, & \text{for } 0 < t < 1/3 \\ 1, & \text{for } 1/3 \le t \le 2/3 , \\ 0, & \text{for } 2/3 < t < 1 \end{cases} \quad T(t + 1) = T(t), \quad \text{for all } t \in \mathbb{R}$$

3. an *damped exponential* given by $f(t) = A e^{\alpha + i\beta t}$, with $e^{i\beta T} = 1$, or its close relative the T_0-periodic *sinusoidal signal*[3]

$$y(t) = A \cos(\beta t + \phi), \quad \text{with } |\beta| = 2\pi / T_0,$$

of *amplitude* A, where T_0 is the fundamental period and $\nu_0 = 2\pi / T_0$ is the *fundamental frequency*.

Recall that the number of wave cycles formed per unit time is called the *frequency*, and the maximum vertical distance a wave is displaced from the equilibrium state is

---

[3] The last equation represents a typical simple *harmonic oscillator*, which first appeared in Newton's derivation of the velocity of sound.

the *amplitude*. Also, the length over which a single peak and trough are formed is the *wavelength* (λ).

In context of present discussion, a *linear system* is a linear transformation taking a signal to a signal.

In an actual application, if a periodic signal $f \in L^2((-\ell, \ell))$ is represented by the Fourier series

$$\frac{a_0}{2} + \sum_{n=1}^{\infty} [a_n \cos(\omega_n t) + b_n \sin(\omega_n t)],$$

then $\omega_n = n\pi/\ell$ is the frequency of the nth mode of the signal $f(t)$ approximated by a superposition of *harmonics*

$$h_n(t) = a_n \cos(\omega_n t) + b_n \sin(\omega_n t);$$

the following coefficients represent *amplitudes* of moving waves:

$$a_n = \frac{1}{\ell} \int_{-\ell}^{\ell} f(x) \cos(\omega_n x) dx,$$

$$b_n = \frac{1}{\ell} \int_{-\ell}^{\ell} f(x) \sin(\omega_n x) dx,$$

Further, the sequence of *partial sums*

$$S_k(f) := \frac{a_0}{2} + \sum_{n=1}^{k} [a_n \cos(\omega_n t) + b_n \sin(\omega_n t)],$$

of the Fourier series helps to find *appropriate frequencies to represent the signal* $f(t)$ *accurately*. In general, *Fourier analysis* is about extracting *frequencies* present in a signal. Fourier series works only for cases of periodic signals and Fourier transforms are applied to *aperiodic* signals. According to Hamming, written in his 1977 book on *Digital Filters*, if a signal $f(t)$ is viewed as a light ray then the Fourier transform (like a prism) breaks it into frequencies p (colors), of intensity $\widehat{f}(p)$. That is, the Fourier transform $\widehat{f}(p)$ gives the (color) spectrum of the waveform $f(t)$.

For a *linear time-invariant* (LTI) system, signals and spectra are actually Fourier transforms of each other. More precisely, for a continuous-time signal $f(t)$, the absolute value $|\widehat{f}(p)|$ represents the *amplitude spectrum*, $\mathrm{Arg}(\widehat{f}(p))$ is the *phase spectrum*, and $|\widehat{f}(p)|^2$ gives the *spectral energy density* of the signal $f(t)$. This process is called the *spectrum analysis*, and the reverse process described in terms of Fourier inverse transforms (Definition 10.2) is called the *spectrum synthesis*.

In most applications related to optimisation of the performance of a digital system, the most important question is to know the number of bits required to represent the digital signals. Many modern systems consist of both analog and digital subsystems, with appropriate analog-to-digital (A/D) and digital-to-analog (D/A) devices at the interfaces. For example, it is common to use a digital computer in the control loop of an analog plant. Analytical difficulties often occur at the boundaries between the analog and digital portions of the system because the mathematics used on the two sides of the interface must be different. It is often useful to assume that a sequence $\langle s_n \rangle$ is derived from an analog signal $s_a(t)$ by ideal sampling, i.e. $s_n = s_a(t)\big|_{t=nT}$. An alternative model for the sampled signal is denoted by $s * (t)$ and defined by

$$s * (t) = \sum_{-\infty}^{\infty} s_a(t)\delta_a(t - nT),$$

where $\delta_a(t)$ is an analog impulse function. Both s_n and $s * (t)$ are used throughout the literature to represent an ideal sampled signal. Notice that even though s_n and $s * (t)$ represent the same essential information, s_n is a digital-tine signal and $s * (t)$ is a continuous-time signal. Hence, they are not mathematically identical.

Example A.1 Like in Example 10.7, we see that the *damped harmonic wave* $f(t) = ke^{(ip_0-\alpha)t}$, $t > 0$, has the Fourier transform

$$\widehat{f}(p) = \frac{k}{\alpha + i(p - p_0)},$$

so that the modulus of the *Lorentzian* $\widehat{f}(p)$ gives

$$|\widehat{f}(p)| \approx \frac{1}{\sqrt{\alpha^2 + (p - p_0)^2}}.$$

Clearly, as $\alpha \to 0$, only the fundamental frequency p_0 remains, and $\widehat{f}(p)$ gives a delta function (see (10.3.1)). Hence, the *width* of Lorentzian vanishes with the damping $\alpha \to 0$.

The main result in this context, due to *Shannon-Whittaker*, says that if ℓ is the maximum frequency of a continuous-time signal $f(t)$ then to capture all the information the sampling must be done at a rate $\geq (\ell/\pi)$, i.e. sampling interval must be π/ℓ-seconds. We start with the following definition.

Definition A.1 An admissible function $f : \mathbb{R} \to \mathbb{R}$ **band-limited**, with a *band width* $\ell > 0$, if the Fourier transform $\widehat{f}(p) = 0$ vanishes outside the interval $[-\ell, \ell]$.

For example, the *box function* in Example 10.5 is *band-limited*. Equation (10.2.7) implies that the *damped oscillations* modeled by a normalized sinc function is the spectrum of the box function. The conclusion is very significant in the context of LTI systems. In fact, it applies directly while reconstructing a continuous-time *band-limited* signal from (uniformly spaced) sampled values of the signal.

Theorem A.3 (Sampling Theorem) *Let $f : \mathbb{R} \to \mathbb{R}$ be an admissible band-limited function of band length ℓ, and $p_0 = \pi/\ell$. Then*

$$f(t) = \frac{1}{2\ell} \sum_{n=-\infty}^{\infty} f\left(\frac{n\pi}{\ell}\right) \int_{-\infty}^{\infty} e^{ip[t-np_0]} \, dp = \sum_{n=-\infty}^{\infty} f\left(\frac{n\pi}{\ell}\right) \frac{\sin(\ell t - n\pi)}{(\ell t - n\pi)}.$$

Proof Since the Fourier transform

$$\widehat{f}(p) = \int_{-\infty}^{\infty} e^{-ipt} f(t) \, dt$$

of $f(t)$ is a continuous function defined in the interval $[-\ell, \ell]$, we may regard it as a 2ℓ-periodic function defined over $\mathbb{R}$, with the *fundamental frequency* $p_0 = \pi/\ell$. Then, in view of Fourier *inversion formula*, $\widehat{f}(p)$ corresponds to a discrete aperiodic function in the time domain. That is, $f(t) \leftrightarrow \widehat{f}(p)$, as discrete-time Fourier transform pair. Thus, by Fourier theorem, we may write

$$\widehat{f}(p) = \sum_{n=-\infty}^{\infty} c_n \, e^{-inp_0 p}, \quad \text{where} \tag{A.4.1}$$

$$c_n = \frac{1}{2\ell} \int_{-\ell}^{\ell} e^{inp_0 p} \widehat{f}(p) \, dp. \tag{A.4.2}$$

On the other hand, by Fourier *inversion formula*, we have

$$f(t) = \frac{1}{2\pi} \int_{-\infty}^{\infty} e^{ipt} \widehat{f}(p) \, dp = \frac{1}{2\pi} \int_{-\ell}^{\ell} e^{ipt} \widehat{f}(p) \, dp. \tag{A.4.3}$$

By comparing this with previous equations, we get

$$c_n = \frac{\pi}{\ell} f\left(\frac{n\pi}{\ell}\right).$$

Putting the value of c_n in Eq. (A.4.1), we obtain

$$\widehat{f}(p) = \frac{\pi}{\ell} \sum_{n=-\infty}^{\infty} f\left(\frac{n\pi}{\ell}\right) e^{-inp_0 p}.$$

In turn, substituting $\widehat{f}(p)$ in Eq. (A.4.3), we have

$$f(t) = \frac{1}{2\ell} \int_{-\ell}^{\ell} e^{i\,pt} \left[\sum_{n=-\infty}^{\infty} f\left(\frac{n\pi}{\ell}\right) e^{-inp_0 p} \right] dp$$

$$= \frac{1}{2\ell} \sum_{n=-\infty}^{\infty} f\left(\frac{n\pi}{\ell}\right) \int_{-\ell}^{\ell} e^{i\,p(t-np_0)} dp \qquad\qquad (A.4.4)$$

$$= \sum_{n=-\infty}^{\infty} f\left(\frac{n\pi}{\ell}\right) \frac{\sin(\ell t - n\pi)}{(\ell t - n\pi)}. \quad \square$$

References

1. Braun, M. (1993). *Differential equations and their applications* (4th ed.). Springer.
2. Xie, W.-C. (2010). *Differential equations for engineers*. Cambridge University Press.
3. Griffiths, D. J. (2013). *Introduction to electrodynamics* (5th ed.). Pearson.
4. Purcell, E. M., & Morin, D. A. (2013). *Electricity and magnetism* (3th ed.). Cambridge University Press.
5. Jänich, K. (2013). *Vector analysis*. Springer.
6. Perwass, D. C. (2009). *Geometric algebra with applications to engineering*. Springer.

Index

Printed in the United States
by Baker & Taylor Publisher Services

Printed in the United States
by Baker & Taylor Publisher Services